Latin Squares and their Applications

Latin Squares and their Applications

by

J. DÉNES
A.D. KEEDWELL

1974

ACADEMIC PRESS
NEW YORK AND LONDON

QA
165
.D42

ISBN 0—12—209350—X
LCCN—73—19244

First printed 1974

Copyright © 1974, Akadémiai Kiadó, Budapest

All rights reserved. No part of this publication may be reproduced or transmitted in any form or by any means, electronic or mechanical, including photocopy, recording, or any information storage and retrieval system, without permission in writing from the publisher.

Academic Press Inc. New York
A subsidiary of Harcourt Brace Jovanovich, Publishers

Joint edition with Akadémiai Kiadó, Budapest

Printed in Hungary by

Akadémiai Nyomda, Budapest

Foreword

The subject of latin squares is an old one and it abounds with unsolved problems, many of them up to 200 years old. In the recent past one of the classical problems, the famous conjecture of Euler, has been disproved by Bose, Parker, and Shrikhande.

It has hitherto been very difficult to collect all the literature on any given problem since, of course, the papers are widely scattered. This book is the first attempt at an exhaustive study of the subject. It contains some new material due to the authors (in particular, in chapters 3 and 7) and a very large number of the results appear in book form for the first time. Both the combinatorial and the algebraic features of the subject are stressed and also the applications to Statistics and Information Theory are emphasized. Thus, I hope that the book will have an appeal to a very wide audience.

Many unsolved problems are stated, some classical, some due to the authors, and even some proposed by the writer of this foreword. I hope that, as a result of the publication of this book, some of the problems will become theorems of Mr. So and So.

PAUL ERDŐS

Contents

Preface

The concept of the latin square probably originated with problems concerning the movement and disposition of pieces on a chess board. However, the earliest written reference to the use of such squares known to the authors concerned the problem of placing the sixteen court cards of a pack of ordinary playing cards in the form of a square so that no row, column, or diagonal should contain more than one card of each suit and one card of each rank. An enumeration by type of the solutions to this problem was published in 1723. The famous problem of similar type concerning the arrangement of thirty-six officers of six different ranks and regiments in a square phalanx was proposed by Euler in 1779, but not until the beginning of the present century was it shown that no solution is possible.

It is only comparatively recently that the subject of latin squares has attracted the serious attention of mathematicians. The cause of the awakening of this more serious interest was the realization of the relevance of the subject to the algebra of generalized binary systems, and to the study of combinatorics, in particular to that of the finite geometries. An additional stimulus has come from practical applications relating to the formation of statistical designs and the construction of error correcting codes. Over the past thirty years a great number of papers concerned with the latin square have appeared in the mathematical journals and the authors felt that the time was ripe for the publication in book form of an account of the results which have been obtained and the problems yet to be solved.

Let us analyse our subject a little further. We may regard the study of latin squares as having two main emphases. On the one hand is the study of the properties of single latin squares which has very close connections with the theory of quasigroups and loops and, to a lesser extent, with the theory of graphs. On the other is the study of sets of mutually orthogonal latin squares. It is the latter which is most closely connected with the theory of finite projective planes and with the construction of statistical designs. We have organized our book in accordance with this general scheme. However, each of these two branches of the subject has many links with the other, as we hope that the following pages will clearly show.

We have tried to make the book reasonably self-contained. No prior knowledge of finite geometries, loop theory, or experimental designs has been assumed on the part of the reader, but an acquaintance with elementary group theory and with the basic properties of finite fields has been taken for granted where such knowledge is needed. Full proofs of several major results in the subject have been included for the first time outside the original research papers. These include the Hall–Paige theorems (chapter 1), two major results due to R. H. Bruck (chapter 9) and the proof of the falsity of Euler's conjecture (chapter 11).

We hope that the text will be found intelligible to any reader whose standard of mathematical attainment is equivalent to that of a third year mathematics undergraduate of a British or Hungarian University. Probably the "deepest" theorem given in the book is theorem 1.4.8 which unavoidably appears in the first chapter. However, the reader's understanding of the remainder of the book will not be impaired if he skips the details of the proof of this theorem.

Part of the manuscript is based on lectures given by one of the authors at the Loránd Eötvös University, Budapest.

The bibliography of publications on latin squares has been made as comprehensive as possible but bibliographical references on related subjects have been confined to those works actually referred to in the text.

Decimal notation has been used for the numbering of sections, theorems, diagrams, and so on. Thus, theorem 10.1.2 is the second theorem of section 10.1 and occurs in chapter 10. The diagram referred to as Fig. 1.2.3 is the third diagram to be found in section 1.2. The use of lemmas has been deliberately avoided as it seemed to at least one of the authors better for the purposes of cross-reference to present a single numbering system for the results (theorems) of each section.

As regards references to the Bibliography, these are in the form of an author reference followed by a number. For example, the statements "In [2], R. H. Bruck has shown . . ." or "R. H. Bruck (see [2]) has shown . . ." or a reference to "R. H. Bruck [2]" are all references to the second publication listed under that author at the back of the book.

A list of unsolved problems precedes the bibliography and each is followed by a page reference to the relevant part of the text.

Acknowledgements

The authors wish to express grateful thanks for their helpful comments and constructive criticism both to their official referee, Prof. N. S. Mendelsohn, and also to Profs V.D. Belousov, D.E. Knuth, C.C. Lindner, H. B. Mann, A. Sade and J. Schönheim all of whom read part or all of the draft manuscript and sent useful comments. They owe a particular debt of gratitude to Prof. A. Sade for his very detailed commentary on a substantial part of the manuscript and for a number of valuable suggestions.[1]

The Hungarian author wishes also to express thanks to his former Secretary, Mrs. E. Szentes, to Prof. S. Csibi of the Research Institute of Telecommunications[2], to Mr. E. Gergely and to several of his former students who took part in the work of his seminars given at the Loránd Eötvös University since 1964 (some of whom read parts of the manuscript and made suggestions for improvement) and to the staff of the Libraries of the Hungarian Academy of Sciences and its Mathematical Institute.

The British author wishes to express his very sincere thanks for their helpfulness at all times to the Librarian of the University of Surrey and to the many members of his library staff whose advice and assistance were called upon over a period of nearly five years and who sometimes spent many hours locating copies of difficult-to-obtain books and journals. He wishes to thank the Librarian of the London Mathematical Society for granting him permission to keep on loan at one time more than the normal number of Mathematical Journals. He also thanks the Secretarial Staff of the University of Surrey Mathematics Department for many kindnesses and for being always willing to assist with the retyping of short passages of text, etc., often at short notice.

(1) It is with deep regret that the authors have to announce the sudden death of Prof. Albert Sade on 10th February 1973 after a short illness.

(2) Now appointed to a Chair at the Technical University of Budapest.

Finally, both authors wish to thank Mr. G. Bernát, General Manager of Akadémiai Kiadó, Publishing House of the Hungarian Academy of Sciences, and Messrs B. Stevens and A. Scott of the Editorial Staff of English Universities Press for their advice and encouragement throughout the period from the inception of the book to its final publication. They are also grateful to Mrs. E. Róth, Chief Editor, and to Mrs. E. K. Kállay and Miss D. Bodoky of the Editorial Staff of Akadémiai Kiadó for their care and attention during the period in which the manuscript was being prepared for printing.

CHAPTER 1

Elementary properties

In this preliminary chapter, we introduce a number of important concepts which will be used repeatedly throughout the book. In the first section, we describe briefly the history of the latin square concept and its equivalence to that of a quasigroup. Next, we explain how those latin squares which represent group multiplication tables may be characterized. We mention briefly the work of A. Ginzburg, D. Tamari and others on the reduced multiplication tables of finite groups. In the third and fourth sections respectively, we introduce the important concepts of isotopy and complete mapping and develop their basic properties in some detail. In the final section of the chapter (which may be omitted on a first reading), we discuss the interrelated notions of subquasigroup and latin subsquare.

1.1. The multiplication table of a quasigroup

As we remarked in our preface, the concept of the latin square is of very long standing but, so far as we are aware, was first systematically developed by Euler. A *latin square* was regarded by Euler as a square matrix with n^2 entries of n different elements, none of them occurring twice within any row or column of the matrix. The integer n is called the *order* of the latin square. (We shall, when convenient, assume the elements of the latin square to be the integers $0, 1, \ldots, n-1$ or, alternatively, $1, 2, \ldots, n$, and this will entail no loss of generality.)

Much later, it was shown by Cayley, who investigated the multiplication tables of groups, that a multiplication table of a group is in fact an appropriately bordered special latin square. (See A. Cayley [1] and [2].) The multiplication table of a group is called its *Cayley table*.

Later still, in the 1930's, latin squares arose once again in the guise of multiplication tables when the theory of quasigroups and loops began to be developed as a generalization of the group concept. A set S is called a *quasigroup* if there is a binary operation $(\cdot)$ defined in S and if, when any two elements a, b of S are given, the equations $ax = b$ and $ya = b$ each have exactly one solution. A *loop* L is a quasigroup with an identity element: that is, a quasigroup in which there exists an element e of L with the property that $ex = xe = x$ for every x of L.

However, the concept of quasigroup had actually been considered in some detail much earlier than the 1930's by E. Schroeder who, between 1873 and 1890, wrote a number of papers on "formal arithmetics": that is, on algebraic systems with a binary operation such that both the left and right inverse operations could be uniquely defined. Such a system is evidently a quasigroup.

A list of E. Schroeder's papers and a discussion of their significance(1) will be found in S. G. Ibragimov [1].

In 1935, Ruth Moufang published a paper (see R. Moufang [1]) in which she pointed out the close connection between non-desarguesian projective planes and non-associative quasigroups.

The results of Euler, Cayley and Moufang made it possible to characterize latin squares both from the algebraic and the combinatorial points of view.

Until recently, only a very few authors have attempted to point out the close relationship that exists between the algebraic and combinatorial results when dealing with latin squares. However, discussion of such relationships may be found in J. R. Barra and R. Guérin [1], J. Dénes [2], J. Dénes and K. Pásztor [1], D. Fog [1], E. Schönhardt [1] and H. Wielandt [1].

Particularly in practical applications it is important to be able to exhibit results in the theory of quasigroups and groups as properties of the Cayley tables of these systems and of the corresponding latin squares. This becomes clear when we prove:

THEOREM 1.1.1. *The multiplication table of a quasigroup is a latin square.*

PROOF. Let $a_1, a_2, \ldots, a_n$ be the elements of the quasigroup and let its multiplication table be as shown in Fig. 1.1.1, where the entry a_{rs} which occurs in the rth row of the sth column is the product $a_r a_s$ of the elements a_r and a_s. If the same entry occurred twice in the rth row, say in the sth and tth columns so that $a_{rs} = a_{rt} = b$ say, we would have two solutions to the equation $a_r x = b$ in contradiction to the quasigroup

(1) It is interesting to note that this author was also the first to consider generalized identities. (These are defined and discussed in section 2.2.)

axioms. Similarly, if the same entry occurred twice in the sth column, we would have two solutions to the equation $ya_s = c$ for some c. We conclude that each element of the quasigroup occurs exactly once in each row and once in each column, and so the unbordered multiplication table (which is a square array of n^2 rows and n^2 columns) is a latin square.

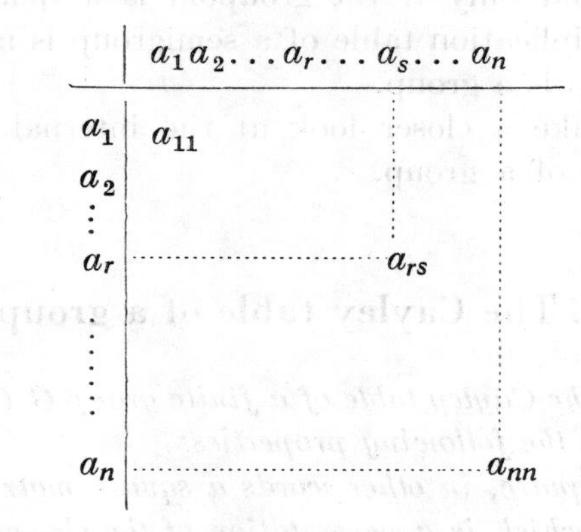

Fig. 1.1.1

As a simple example of a finite quasigroup, consider the set of integers modulo 3 with respect to the operation defined by $a * b = 2a + b + 1$. The multiplication of this quasigroup is shown in Fig. 1.1.2 and we see at once that it is a latin square.

(*)	0	1	2
0	1	2	0
1	0	1	2
2	2	0	1

Fig. 1.1.2

More generally, the operation $a * b = ha + kb + l$, where addition is modulo n and h, k and l are fixed integers with h and k prime to n, defines a quasigroup on the set $Q = \{0, 1, \cdot \cdot \cdot, n - 1\}$.

We end this preliminary section by drawing the reader's attention to the fact that quasigroups, loops and groups are all examples of the primitive mathematical structure called a groupoid.

Definition. A set S forms a *groupoid* $(S, \cdot)$ with respect to a binary operation $(\cdot)$ if, with each ordered pair of elements a, b of S is associated a uniquely determined element $a.b$ of S called their *product*. (If a product is defined for only a subset of the pairs a, b of elements of S, the system is sometimes called a *half-groupoid*. See, for example, R. H. Bruck [5].)

A groupoid whose binary operation is associative is called a *semigroup*.

Theorem 1.1.1 shows that the multiplication table of a groupoid is a latin square if and only if the groupoid is a quasigroup. Thus, in particular, the multiplication table of a semigroup is not a latin square unless the semigroup is a group.

Next, we shall take a closer look at the internal structure of the multiplication table of a group.

1.2. The Cayley table of a group

THEOREM 1.2.1. *The Cayley table of a finite group G (with its bordering elements deleted) has the following properties:*

(1) *It is a latin square, in other words a square matrix $||a_{ik}||$ each row and each column of which is a permutation of the elements of G.*

(2) *The quadrangle criterion holds; which means that, for any indices $i, j, k, l, \ldots$, it follows from the equations $a_{jk} = a_{j_1k_1}$, $a_{ik} = a_{i_1k_1}$, $a_{il} = a_{i_1l_1}$ that $a_{jl} = a_{j_1l_1}$.*

PROOF. Property (1) is an immediate consequence of Theorem 1.1.1.

Property (2) is implied by the group axioms, since by definition $a_{ik} = a_i a_k$ and hence, using the conditions given, we have

$$a_{jl} = a_j a_l = a_j(a_k a_k^{-1})(a_i^{-1} a_i)a_l = (a_j a_k)(a_i a_k)^{-1}(a_i a_l) =$$
$$= a_{jk}\, a_{ik}^{-1}\, a_{il} = a_{j_1k_1}\, a_{i_1k_1}^{-1}\, a_{i_1l_1} = (a_{j_1}a_{k_1})(a_{i_1}a_{k_1})^{-1}(a_{i_1}a_{l_1}) = a_{j_1}a_{l_1} = a_{j_1l_1}.$$

Conversely, we shall show that any matrix $||a_{ik}||$ with the properties (1) and (2) is an unbordered Cayley table of a group G.

To prove this, a bordering procedure has to be found which will show that the Cayley table thus obtained is, in fact, the multiplication table of a group. If we use as borders the first row and the first column of the latin square, the invertibility of the multiplication defined by the Cayley table thus obtained is easy to show and is indeed a consequence merely of property (1). For, in the first place, when the border is so chosen, the leading element of the matrix acts as an identity element, e. In the second place, since this element occurs exactly once in each row and column of the matrix, the equations $a_r x = e$ and $y a_s = e$ are soluble for every choice of a_r and a_s.

Now, only the associativity has to be proved. Let us consider arbitrary elements a, b and c. If one of them is identical with e, it follows directly that $(ab)c = a(bc)$. If, on the other hand, each of the elements a, b and c differs from e, then the subsquare determined by the rows e and a and by the columns b and bc of the multiplication table is

$$\begin{array}{|cc} \hline b & bc \\ ab & a(bc) \end{array}$$

while the subsquare determined by the rows b and ab and by the columns e and c is

$$\begin{array}{|cc} \hline b & bc \\ ab & (ab)c \end{array}$$

Hence, because of property (2), $a(bc) = (ab)c$.

COROLLARY. *If $a_1, a_2, \ldots, a_n$ are distinct elements of a group of order n, and if b is any fixed element of the group, then each of the sets of products, $ba_1, ba_2, \ldots, ba_n$ and $a_1b, a_2b, \ldots, a_nb$ comprises all the n group elements, possibly rearranged.* (In fact, they will be rearranged whenever b is not the identity element.)

REMARK. The property (2) was first observed by M. Frolov in [1] in 1890. Later H. Brandt (in [1]) showed that it was sufficient to postulate the quadrangle criterion to hold only for quadruples in which one of the four elements is the identity element. Text-books on the theory of finite groups (see for example A. Speiser [1]) adopted the criterion established by H. Brandt. Recently, J. Aczél and A. Bondesen have published papers in which they have rediscovered the quadrangle criterion (see J. Aczél [3] and A. Bondesen [1]). Also, in [1], A. Hammel has suggested some ways in which testing the validity of the quadrangle criterion may in practice be simplified when it is required to test the multiplication tables of finite quasigroups of small orders for associativity.

In F. D. Parker [1], an algorithm for deciding whether a loop is a group was proposed but the author later found an error in his paper and his method turned out to give only a necessary condition, not a sufficient one.

A. Wagner proved in [1] that to test whether a finite quasigroup Q of order n is a group it is sufficient to test only about $3n^3/8$ appropriately chosen ordered triples of elements for associativity. However, if a minimal set of generators of Q is known, then it is sufficient to test the validity of at most $n^2[\log_2 2n]$ associative statements, provided that these are appropriately selected.

Wagner also showed in the same paper that every triassociative quasigroup Q (that is, every quasigroup whose elements satisfy $xy.z = x.yz$ whenever x, y, z are *distinct*) is a group, and the same result has been proved independently by D. A. Norton in [3].

These results lead us to ask the question "What is the maximum number of associative triples which a quasigroup may have and yet not be a group?"

T. Faragó has proved in [1] that the validity of any of the following identities in a loop guarantees both its associativity and commutativity:

$$\begin{array}{rlrl} (i) & (ab)c = a(cb), & (ii) & (ab)c = b(ac) \\ (iii) & (ab)c = b(ca), & (iv) & a(bc) = b(ca) \\ (v) & a(bc) = c(ab), & (vi) & a(bc) = c(ba) \\ (vii) & (ab)c = (ac)b, & (viii) & (ab)c = (bc)a \end{array}$$

$$(ix)\ (ab)c = (ca)b$$

In fact, the identities (*iv*) and (*v*) are isomorphic and so also are the identities (*viii*) and (*ix*), as A. Sade has pointed out in [18].

Theorem 1.2.2. *A finite group is commutative if, and only if, its multiplication table (with row and column borders taken in the same order) has the property that products located symmetrically with respect to the main diagonal represent the same group element.*

Proof. By the commutative law, $ab = ba = c$ for any arbitrary pair of elements a, b and so the cells in the ath row and bth column and in the bth row and ath column are both occupied by c. If this were not the case for some choice of a and b, we would have $ab \neq ba$ and the commutativity would be contradicted.

We note that theorem 1.2.2 remains valid if the word "group" is replaced by the word "quasigroup".

A Cayley table of a group is called *normal* if every element of its main left-to-right diagonal (from the top left-hand corner to the bottom right-hand corner) is the identity element of the group (see H. J. Zassenhaus [1], page 4).

If the notation of theorem 1.1.1 is used, it follows as a consequence of the definition that a normal multiplication table $||a_{ij}||$ of a group has to be bordered in such a way that $a_{ij} = a_i a_j^{-1}$ holds. Thus, if the element bordering the ith row is a_i, the element bordering the jth column must be a_j^{-1}.

Obviously, the following further conditions are satisfied:

(*i*) $a_{ij}a_{jk} = a_{ik}$ (since $a_ia_j^{-1}a_ja_k^{-1} = a_ia_k^{-1}$); and

(*ii*) $a_{ji}^{-1} = a_{ij}$ (since $(a_ja_i^{-1})^{-1} = a_ia_j^{-1}$).

For example, the normal multiplication table of the cyclic group of order 6, written in additive notation, is shown in Fig. 1.2.1.

(+)	0	5	4	3	2	1
0	0	5	4	3	2	1
1	1	0	5	4	3	2
2	2	1	0	5	4	3
3	3	2	1	0	5	4
4	4	3	2	1	0	5
5	5	4	3	2	1	0

Fig. 1.2.1

As was first suggested by an example which appeared in H. J. Zassenhaus' book on Group Theory (H. J. Zassenhaus [1], page 78, example 1), the normal multiplication table of a finite group has a certain amount of redundancy since every product $a_ia_j^{-1}$ can be found n times in the table, where n is the order of the group. In fact, $a_ia_j^{-1} = a_{ij} = a_{ik}a_{kj}$ for $k = 0, 1, \ldots, n-1$. Consequently, it is relevant to seek smaller tables that give the same information. A multiplication table having this property is called a *generalized normal multiplication table* if it has been obtained from a normal multiplication table by the deletion of a number of columns and corresponding rows.

The idea of such generalized normal multiplication tables was first mentioned by D. Tamari in [1] and this author subsequently gave some illustrative examples in [2] but without proof. Thus, as one of his examples, he stated that the table given in Fig. 1.2.2 is a generalized normal multiplication table of the cyclic group of order 6, obtained from the complete table displayed in Fig. 1.2.1 by deleting the rows bordered by 3 and 5 and the columns bordered by $3^{-1} = 3$ and $5^{-1} = 1$.

(+)	0	5	4	2
0	0	5	4	2
1	1	0	5	3
2	2	1	0	4
4	4	3	2	0

Fig. 1.2.2

The same idea was mentioned again by A. Ginzburg in [2] who there gave a reduced multiplication table for the quaternion group of order 8. Later, in [3], the latter author developed the concept in much more detail and gave full proofs of his results. This paper contains, among other things, a complete list of the minimal generalized normal multiplication tables for all groups of orders up to 15 inclusive.

It will be clear to the reader that of especial importance to the theory is the determination of the minimal number of rows and columns of a generalized normal multiplication table. If r denotes the minimal number of rows (or columns), then P. Erdős and A. Ginzburg have proved in [1] that $r < C\,(n^2 \log n)^{1/3}$ (where C is a sufficiently large absolute constant) while, in A. Ginzburg [3], it has been shown that, in general, $r > n^{2/3}$ and that, for the cyclic group C_n of order n, $r < (6n^2)^{1/3}$.

For further generalizations of the concept of a generalized normal multiplication table and for discussion of some of the mathematical ideas relevant to it, the reader should consult A. Ginzburg [1], D. Tamari and A. Ginzburg [1], A. Ginzburg and D. Tamari [2], and D. Tamari [3].

THEOREM 1.2.3. *A subsquare of a Cayley table of a group or quasigroup of order n contains every group element at least once if its size is $k \times k$ and $k \geq [1 + n/2]$.*

PROOF. The result can be deduced from a theorem due to H. B. Mann and W.A. McWorter which we shall prove in section 1.5 (theorem 1.5.4) and which can be stated as follows:

Let G be a group or quasigroup and let A and B be two subsets of G such that not every element of G can be written in the form ab with a in A and b in B. Then $|G| \geq |A| + |B|$. Here $|S|$ denotes the cardinal of the set S.

Mann and McWorter's theorem implies that, when $|G| < |A| + |B|$ holds, then for every c in G there exists an element a in A and an element b in B such that $ab = c$.

Since $n < [1 + n/2] + [1 + n/2]$, the validity of theorem 1.2.3 is immediate.

1.3. Isotopy

Let $(G, \cdot)$ and $(H, *)$ be two quasigroups. An ordered triple (θ, φ, ψ) of one-to-one mappings θ, φ, ψ of the set G onto the set H is called an *isotopism* of $(G, \cdot)$ upon $(H, *)$ if $(x\theta) * (y\varphi) = (x . y)\psi$ for all x, y in G. The quasigroups $(G, \cdot)$ and $(H, *)$ are then said to be *isotopic*. (It is worth remarking that the same definition holds for any two groupoids.)

The concept of isotopy seems to be very old. In the study of latin squares the concept is so natural as to creep in unnoticed; and latin squares are simply the multiplication tables of finite quasigroups. It was consciously applied by E. Schönhardt in [1], by R. Baer in [1] and [2], and independently by A. A. Albert in [1] and [2]. A. A. Albert had earlier borrowed the concept from topology for application to linear algebras; and it had subsequently been virtually forgotten except for applications to the theory of projective planes.

A latin square becomes a multiplication table as soon as it has been suitably bordered. For example the latin square

1	2	3	4
2	3	4	1
3	4	1	2
4	1	2	3

becomes the Cayley table of the cyclic group of order 4 if the bordering elements are written in, and takes the form

	1	2	3	4
1	1	2	3	4
2	2	3	4	1
3	3	4	1	2
4	4	1	2	3

Of the permutations θ, φ, ψ introduced in the definition of isotopy, ψ operates on the elements of the latin square, while θ and φ operate on the borders.

Let us suppose, for example, that the elements, the row border, and the column border of the Cayley table exhibited above are transformed in the manner prescribed by the following permutations ψ, θ, φ, respectively

$$\psi = \begin{pmatrix} 1 & 2 & 3 & 4 \\ 2 & 1 & 4 & 3 \end{pmatrix}, \quad \theta = \begin{pmatrix} 1 & 2 & 3 & 4 \\ 3 & 2 & 4 & 1 \end{pmatrix}, \quad \varphi = \begin{pmatrix} 1 & 2 & 3 & 4 \\ 2 & 4 & 3 & 1 \end{pmatrix}.$$

Then the Cayley table of the cyclic group of order 4 is transformed into that of an isotopic quasigroup, namely:

	2	4	3	1
3	2	1	4	3
2	1	4	3	2
4	4	3	2	1
1	3	2	1	4

If $\theta = \varphi = \psi$, such a transformation is an *isomorphism*. It is easy to see that isotopy is an equivalence relation between quasigroups (or between groupoids). Geometrically speaking, as we shall show, isotopic quasigroups are quasigroups which co-ordinatize the same 3-net. We shall discuss this aspect of isotopy in detail in chapter 8. Let us notice at this point that there may also exist isotopisms of a quasigroup $(G, \cdot)$ onto itself.

The foregoing remarks should make it clear that the concept of isotopy is fundamental to our subject and so we shall need to develop some of its basic properties for future application.

THEOREM 1.3.1. *Every groupoid that is isotopic to a quasigroup is itself a quasigroup.*

PROOF. Let $(G, \cdot)$ be a quasigroup and let $(H, *)$ be a groupoid isotopic to $(G, \cdot)$ with $(x\theta) * (y\varphi) = (x \,.\, y)\psi$ for all x, y in G.

Let a, b be arbitrary elements in H. We require to show that there exists a unique x in H such that $a * x = b$. Since $a\theta^{-1}$ and $b\psi^{-1}$ belong to G and $(G, \cdot)$ is a quasigroup, the equation $a\theta^{-1}.y = b\psi^{-1}$ has a unique solution c in G. Write $x = c\varphi$. Then, $a * x = a * c\varphi = (a\theta^{-1})\,\theta * (c\varphi) = (a\theta^{-1}.c)\,\psi = (b\psi^{-1})\,\psi = b$, so the equation $a * x = b$ is soluble. Further, if $a * x' = b$ we have $(a\theta^{-1}.x'\varphi^{-1})\psi = b$ or equivalently, $a\theta^{-1}.x'\varphi^{-1} = b\psi^{-1}$. Since the equation $a\theta^{-1}.y = b\psi^{-1}$ has a unique solution, $x'\varphi^{-1} = c$ whence $x' = c\varphi$. Thus, the equation $a * x = b$ is uniquely soluble in H. By a similar argument, we may show that the equation $z * a = b$ is uniquely soluble for z. This proves the theorem.

REMARK. As an alternative to the formal proof above one can easily see that theorem 1.3.1 is an immediate consequence of the fact that isotopy rearranges the rows and columns and permutes the elements of a latin square. The result of applying such operations to the latin square is a latin square again; that is, another quasigroup.

Definition. If $(G, \cdot)$ is a given quasigroup (or groupoid) and σ, τ are one-to-one mappings of G onto G, then the isotope $(G, \otimes)$ defined by $x \otimes y = (x\sigma) \cdot (y\tau)$ is called a *principal isotope* of $(G, \cdot)$.

The mappings θ, φ, ψ of the general definition are here replaced by σ^{-1}, τ^{-1} and the identity mapping respectively.

Theorem 1.3.2. *Every isotope $(H, *)$ of a quasigroup $(G, \cdot)$ is isomorphic to a principal isotope of the quasigroup.*

Proof. Let θ, φ, ψ be one-to-one mappings of G onto H which define the isotopism between $(G, \cdot)$ and $(H, *)$ so that $(x\theta) * (y\varphi) = (xy)\psi$ for all x, y in G. Then $\psi\theta^{-1}$ and $\psi\varphi^{-1}$ are one-to-one mappings of G onto G, so the operation $\otimes$ given by $x \otimes y = (x\psi\theta^{-1}).(y\psi\varphi^{-1})$ defines a principal isotope $(G, \otimes)$ of G.

Also $(x\psi) * (y\psi) = (x\psi\theta^{-1})\theta * (y\psi\varphi^{-1})\varphi = (x\psi\theta^{-1}).(y\psi\varphi^{-1})\,\psi = (x\otimes y)\psi$ so $(H, *)$ and $(G, \otimes)$ are isomorphic under the mapping $G \xrightarrow{\psi} H$.

Theorem 1.3.3. *Among the principal isotopes of a quasigroup $(G, \cdot)$ there always exist loops.* [Such loops are called *LP-isotopes* (*loop-principal isotopes*) of $(G, \cdot)$.]

Proof. Define mappings σ, τ of G onto G by $x\sigma^{-1} = x.v$, $x\tau^{-1} = u.x$, where u and v are fixed elements of G, and write $e = uv$. Then $(G, \otimes)$, where $x \otimes y = (x\sigma).(y\tau)$, is a loop with e as identity element. For, let a be in G. Since $(G, \cdot)$ is a quasigroup, $a = ua'$ and $a = a''v$ for some elements a', a'', in G. Then

$$e \otimes a = uv \otimes ua' = u\sigma^{-1} \otimes a'\tau^{-1} = u.a' = a,$$

and

$$a \otimes e = a''v \otimes uv = a''\sigma^{-1} \otimes v\tau^{-1} = a''.v = a.$$

Conversely, every LP-isotope of $(G, \cdot)$ is obtained by mappings σ, τ of the type defined above.

For, if $(G, \otimes)$ is an LP-isotope of $(G, \cdot)$ and has identity element e then $x \otimes y = x\sigma . y\tau$, so $x\sigma^{-1} = x\sigma^{-1} \otimes e = x . e\tau = x . v$ say, where $v = e\tau$. Also $x\tau^{-1} = e \otimes x\tau^{-1} = e\sigma . x = u . x$ say, where $u = e\sigma$.

Remark. The proof of the above theorem can be formulated in terms of latin squares. It is equivalent to the statement that any latin square can be bordered in such a way that the borders are identical to one of the rows and one of the columns of the latin square.

An unsolved problem is that of finding necessary and sufficient conditions on a loop G in order that every loop isotopic to G be isomorphic to G. (See R. H. Bruck [5] page 57.) Associativity is sufficient, as our next theorem will show, but is not necessary.

LP-isotopes of a quasigroup have been further investigated by B. F. Bryant and H. Schneider in [1].

The preceding three theorems will be found in A. A. Albert's paper [1] on "Quasigroups". The following theorem is due to R. H. Bruck (see [2]) and, independently, N. J. S. Hughes (see [1]).

THEOREM 1.3.4. *If a groupoid* $(G, \cdot)$ *with an identity element* e *is isotopic to a semigroup, then the groupoid and semigroup are isomorphic and so both are associative and both have an identity element.*

PROOF. Let $(H, *)$ be the semigroup and let the isotopism be defined by mappings θ, φ, ψ from G onto H such that $(x\theta)*(y\varphi) = (x.y)\psi$. Since $(H, *)$ is a semigroup, we have

$$(a'*b')*c' = a'*(b'*c')$$

for all a', b', c' in H, which implies

$$[(a'\theta^{-1}.b'\varphi^{-1})\psi\theta^{-1}.c'\varphi^{-1}]\psi = [a'\theta^{-1}.(b'\theta^{-1}.c'\varphi^{-1})\psi\varphi^{-1}]\psi.$$

Thus,

$$(a'\theta^{-1}.b'\varphi^{-1})\psi\theta^{-1}.c'\varphi^{-1} = a'\theta^{-1}.(b'\theta^{-1}.c'\varphi^{-1})\psi\varphi^{-1} \quad \text{......(1)}$$

for all a', b', c' in H. In particular, when $a'\theta^{-1} = e$ and $c'\varphi^{-1} = e$, we get

$$b'\varphi^{-1}\psi\theta^{-1} = b'\theta^{-1}\psi\varphi^{-1} \quad \text{...............(2)}$$

and this must hold for all b' in H.

Now, put $a'\theta^{-1} = e$ in (1). We get

$$b'\varphi^{-1}\psi\theta^{-1}.c'\varphi^{-1} = (b'\theta^{-1}.c'\varphi^{-1})\psi\varphi^{-1}.$$

Using (2),

$$b'\theta^{-1}\psi\varphi^{-1}.c'\varphi^{-1} = (b'\theta^{-1}.c'\varphi^{-1})\psi\varphi^{-1}.$$

Therefore,

$$b\psi\varphi^{-1}.c = (b.c)\psi\varphi^{-1}. \quad \text{..................(3)}$$

for all b, c in G.

Next, put $c'\varphi^{-1} = e$ in (1). We get

$$(a'\theta^{-1}.b'\varphi^{-1})\psi\theta^{-1} = a'\theta^{-1}.b'\theta^{-1}\psi\varphi^{-1}.$$

Using (2),

$$(a'\theta^{-1}.b'\varphi^{-1})\psi\theta^{-1} = a'\theta^{-1}.b'\varphi^{-1}\psi\theta^{-1}.$$

Therefore,

$$(a\,.\,b)\,\psi\theta^{-1} = a\,.\,b\,\psi\theta^{-1} \quad \ldots\ldots\ldots\ldots\ldots\ldots (4)$$

for all a, b in G.

Thence, $(a' * b')\,\varphi^{-1}\psi\theta^{-1} = (a'\,\theta^{-1}.\,b'\,\varphi^{-1})\,\psi\varphi^{-1}\psi\theta^{-1}$

$$= (a'\,\theta^{-1}\psi\varphi^{-1}.\,b'\,\varphi^{-1})\,\psi\theta^{-1} \quad \text{using (3)},$$

$$= (a'\,\theta^{-1}\psi\varphi^{-1}.\,b'\,\varphi^{-1}\psi\theta^{-1}) \quad \text{using (4)},$$

$$= (a'\,\varphi^{-1}\psi\theta^{-1}.\,b'\,\varphi^{-1}\psi\theta^{-1}) \quad \text{using (2)}.$$

Thus, $(a' * b')\,\sigma^{-1} = (a'\,\sigma^{-1}.\,b'\,\sigma^{-1})$ where $\sigma^{-1} = \varphi^{-1}\psi\theta^{-1}$, so σ is a one-to-one mapping of G onto H such that $(a\,.\,b) = (a\sigma * b\,\sigma)\sigma^{-1}$ for all a, b in G. That is, σ maps G isomorphically onto H. In particular,

$$(a\,.\,b)\,.\,c = [(a\sigma * b\sigma)\,\sigma^{-1}\sigma * c\sigma]\,\sigma^{-1} = [(a\sigma * b\sigma) * c\sigma]\,\sigma^{-1}.$$

Similarly, $a\,.\,(b\,.\,c) = [a\sigma * (b\sigma * c\sigma)]\,\sigma^{-1}$, whence $(ab)c = a(bc)$.

Corollary 1. *If a loop is isotopic to a group then the loop is a group isomorphic to the given group.*

Corollary 2. *If groups are isotopic, they are isomorphic as well.*

The first corollary is a consequence of the facts that a quasigroup with identity is a loop and that any isotope of a quasigroup is also a quasigroup as shown in theorem 1.3.1. It was first proved by A. A. Albert in [1]. The second corollary follows immediately from the first.

Certain non invariants of principal isotopy may be illustrated in terms of the two loops $(G, \cdot)$ and $(G, *)$ shown in Fig. 1.3.1, where G is a set of six elements; as has been pointed out by R. H. Bruck in [5].

(·)	1	2	3	4	5	6
1	1	2	3	4	5	6
2	2	1	6	3	4	5
3	3	4	5	2	6	1
4	4	5	1	6	2	3
5	5	6	4	1	3	2
6	6	3	2	5	1	4

(∗)	2	3	4	5	6	1
2	2	3	4	5	6	1
3	3	2	6	4	1	5
4	4	6	2	1	5	3
5	5	4	1	2	3	6
6	6	1	5	3	2	4
1	1	5	3	6	4	2

Fig. 1.3.1

(1) Commutativity: $(G, *)$ is commutative but $(G, \cdot)$ is not.

(2) Number of generating elements: $(G, \cdot)$ can be generated by any one of the elements 3, 4, 5, or 6. On the other hand, no single element generates $(G, *)$ but any two of 3, 4, 5, 6, 1 will generate it.

(3) Automorphism group: The automorphism group of $(G, *)$ has order 4 and is generated by the permutation $(3\ 4\ 5\ 6)$. In contrast, that of $(G, \cdot)$ has order 20 and is generated by the two permutations $(3\ 4\ 5\ 6\ 1)$ and $(3\ 4\ 6\ 5)$.

1.4. Transversals and complete mappings

Definitions. A *transversal* of a latin square of order n is a set of n cells, one in each row, one in each column, and such that no two of the cells contain the same symbol.

This concept has very close connections with the theory and construction of orthogonal latin squares and will be referred to in that connection in Chapter 5.

A *complete mapping* of a group, loop, or quasigroup $(G, \otimes)$ is a biunique mapping $x \to \theta(x)$ of G upon G such that the mapping $x \to \eta(x)$ defined by $\eta(x) = x \otimes \theta(x)$ is again a biunique mapping of G upon G.

Theorem 1.4.1. *If Q is a quasigroup which possesses a complete mapping, then its multiplication table is a latin square with a transversal. Conversely, if L is a latin square having a transversal, then at least one of the quasigroups which have L as multiplication table has a complete mapping.*

Proof. Let us suppose that Q has a complete mapping, say

$$\theta = \begin{pmatrix} 1 & 2 & \dots & n \\ a_1 & a_2 & \dots & a_n \end{pmatrix}, \eta = \begin{pmatrix} 1 & 2 & \dots & n \\ b_1 & b_2 & \dots & b_n \end{pmatrix}$$

then its multiplication table has at least one transversal since

$$\begin{aligned} 1 \otimes a_1 &= b_1 \\ 2 \otimes a_2 &= b_2 \\ \dots &= . \\ \dots &= . \\ \dots &= . \\ n \otimes a_n &= b_n \end{aligned}$$

implying that the cell of the ith row and a_ith column has b_i as entry, for $i = 1, 2, \ldots n$, and these entries are all distinct.

Conversely, if L is a latin square having a transversal, comprising the elements $b_1, b_2, \ldots, b_n$ occupying the cells $(1, a_1), (2, a_2), \ldots, (n, a_n)$ then there exists a quasigroup $(Q, \otimes)$ having L as its multiplication table for which

$$\begin{aligned} 1 \otimes a_1 &= b_1 \\ 2 \otimes a_2 &= b_2 \\ \cdot \quad \cdot &= \cdot \\ \cdot \quad \cdot &= \cdot \\ \cdot \quad \cdot &= \cdot \\ n \otimes a_n &= b_n. \end{aligned}$$

This quasigroup Q has a complete mapping, characterized by mappings θ and η defined as follows:

$$\theta = \begin{pmatrix} 1 & 2 & \ldots & n \\ a_1 & a_2 & \ldots & a_n \end{pmatrix}, \; \eta = \begin{pmatrix} 1 & 2 & \ldots & n \\ b_1 & b_2 & \ldots & b_n \end{pmatrix}.$$

The notion of *transversal* was first introduced by Euler under the title *formule directrix* (see L. Euler [2]). The concept was used extensively by H. W. Norton in [1] under the name of *directrix*. It has been called a *1-permutation* by J. Singer in [2] and, in J. Dénes and K. Pásztor [1], it was given the name *diagonal*.

The concept of *complete mapping* was introduced by H. B. Mann in [1].

In the rest of this section we shall give some of the results concerning these concepts which are contained in these and other papers.

THEOREM 1.4.2. *If L is a latin square of order n which satisfies the quadrangle criterion and possesses at least one transversal, then L has a decomposition into n disjoint transversals.*

PROOF. By theorem 1.2.1, there exists a group G whose multiplication table is L.

If L has a transversal formed by taking the symbol c_1 from the first row, c_2 from the second row, $\ldots$, c_n from the nth row, then it follows easily from the group axioms that another transversal can be obtained by taking c_1g from the first row, c_2g from the second row, $\ldots$, c_ng from the nth row, where g is any fixed element of the group. As g varies through the n elements of the group, we shall thus obtain n disjoint transversals.

To see this, suppose that $c_i = g_i g_{j(i)}$ where the sequences $c_1, c_2, \ldots, c_n$ and $g_1, g_2, \ldots, g_n$ both represent orderings of the elements of G, the latter corresponding to the ordering of the rows and columns of L in the multiplication table of G. In other words, c_i is the element to be found in the cell which occurs in the ith row and jth column of L. Also, since the c_i form a transversal, the integer j is a function of i such that $j(i_1) \neq j(i_2)$ if $i_1 \neq i_2$. Then, because G is a group,

$$c_i\, g = (g_i\, g_{j(i)})\, g = g_i\, (g_{j(i)}\, g) = g_i\, g_{k(i)}$$

where, as g_j varies through the elements of G, so does g_k. Consequently, $c_{i_1} g$ and $c_{i_2} g$ are always in distinct columns and so the $c_i g$ form a transversal. Moreover, $g_{j(i)} \neq g_{k(i)}$ for any value of i and so the transversal formed by the c_i's is disjoint from that formed by the $c_i g$'s. Similarly, the transversals corresponding to two different choices of the element g are disjoint.

Note that the validity of the associative law is an essential requirement for the proof.

Remark. The converse statement is not true. We shall exhibit a latin square of order 10 as a counter example. For the labelled elements in Fig. 1.4.1 the quadrangle criterion does not hold, but the square has a decomposition into n disjoint transversals.

0	4	1	7	2	9	8	3	6	5
8	1	5	2	7	3	9	4	0	6
9	8	2	6	3	7	4	5	1	0
5	9	8	3	0	4	7	6	2	1
7	6	9	8	4	1	5	0	3	2
6	7	0	9	8	5	2	1	4	3
3	0	7	1	9	8	6	2	5	4
1	2	3	4	5	6	0	7	8	9
2	3	4	5	6	0	1	8	9	7
4	5	6	0	1	2	3	9	7	8

Fig. 1.4.1

It is also worthwhile to point out that there are many examples of latin squares which have no transversals and which do not satisfy the quadrangle criterion. One of order 6 is shown in Fig. 1.4.2.

THEOREM 1.4.3. *If G is a group of odd order, then G has a complete mapping.*

PROOF. If G is a group of odd order, it is well known that every element of G has a unique square root in G. To prove this, let $g \in G$ be an element of order $2n - 1$. Then $h = g^n$ satisfies $h^2 = g^{2n} = g$ and so h is a square root of g. Further, if $k \in G$ satisfies $k^2 = g$, we have $k^{4n-2} = e$, the identity element of G. Since k necessarily has odd order, $k^{2n-1} = e$. Then $k = k^{2n} = g^n = h$, so h is the unique square root of g.

It follows that, in a group G of odd order, $g_i^2 = g_j^2$ only if $i = j$. Consequently, the mapping $\eta(g_i) = g_i^2$ $(i = 1, 2, \ldots, n)$ is a biunique mapping of G upon G. Thus, the identity mapping $\theta(g_i) = g_i$ satisfies the definition of a complete mapping of G.

In the notation of theorem 1.4.2, the entries $g_i^2 = g_i g_i$ $(i = 1, 2, \ldots, n)$ of the leading diagonal of the multiplication table L of G form a transversal of L.

COROLLARY. *Every finite group of odd order is isotopic to an idempotent quasigroup.*

PROOF. The proof is due to R. H. Bruck (see [1]), whose argument is as follows. Define a new operation $(*)$ on G by the equation $g_i * g_j = \sigma(g_i g_j)$ where σ is the permutation of G which maps g^2 onto g for every g in G. Then $(G, *)$ is an idempotent quasigroup isotopic to the group $(G, \cdot)$.

We observe that the proof of theorem 1.4.3 is solely dependent on the fact that every element of a group of odd order has a unique square root.

A quasigroup with the property that every element has an exact square root has been called *diagonal* by A. Sade (see [15]). He has shown in [19] that *the necessary and sufficient condition for a commutative quasigroup to be diagonal is that it be of odd order* by the following simple argument.

Let Q be the quasigroup of order n and let a be a fixed element of it. If $xy = a$, so also $yx = a$. But, since a occurs exactly n times in the latin square representing the Cayley table of Q, there are $[n/2]$ pairs x, y with $x \neq y$ such that $xy = a = yx$. If n is odd, this leaves one element z such that $z^2 = a$. Thus, theorem 1.4.3 holds for all commutative quasigroups of odd order. Certain types of non-commutative quasigroups are also known to be diagonal.

A loop is called a *Bruck loop* if it satisfies the identities $[(xy)z]y = x[(yz)y]$ and $(xy)^{-1} = x^{-1}y^{-1}$. It is called *Moufang* if it satisfies the identity $[(xy)z]y = x[y(zy)]$ or either of the identities $(xy)(zx) = [x(yz)]x$, $x[y(xz)] = [(xy)x]z$. Such a loop satisfies the identity $(xy)^{-1} = y^{-1}x^{-1}$. Every commutative Moufang loop is a Bruck loop but a non-commutative Moufang loop is not a Bruck loop.

For the justification of the latter statements, see R. H. Bruck [5] and [7].

It follows from the results of D. A. Robinson (see [1]) and G. Glaubermann (see [1] and [2]) that every element of a Moufang loop of odd order or of a Bruck loop of odd order has a unique square root. Consequently, theorem 1.4.3 remains true for such loops.

In fact, D. A. Robinson's results imply that, even in Bol loops (that is, loops satisfying the identity $[(xy)z]y = x[(yz)y]$ alone) of odd order, every element has a unique square root and so theorem 1.4.3 holds. This follows from the facts that such loops are power associative and satisfy a weak form of Lagrange's theorem (namely, that the order of every element divides the order of the loop). The class of Bol loops includes both all Moufang loops and all Bruck loops.

In [3], H. J. Ryser posed the question whether there exist any quasigroups of odd order which do not possess a complete mapping.

Certainly there exist quasigroups of even order which have no complete mappings.

In particular, H. B. Mann has proved that, if a quasigroup Q of order $4k + 2$ has a subquasigroup of order $2k + 1$, then the multiplication table of Q is without transversals. (For the proof, see theorem 5.1.4.) An example of such a quasigroup of order 6 is given in Fig. 1.4.2.

In this connection it is relevant to point out that

THEOREM 1.4.4. *If the latin square representing the Cayley table of a quasigroup Q has a transversal, so does that of any isotope of Q.*

PROOF. The isotopy replaces the latin square L by one obtained from L by rearranging rows, rearranging columns, and renaming elements. None of these processes affects the existence of a transversal.

COROLLARY. *If Q is a quasigroup which has a complete mapping, then any isotopic quasigroup has one also.*

In fact, V. D. Belousov (see [8]) has proved the stronger result that *if a quasigroup Q has a complete mapping then so do all the parastrophes of Q.* (For the definition of parastrophic quasigroups, see section 2.1.)

	1	2	3	4	5	6
1	1	2	[3]	[4]	5	6
2	2	3	[1]	[5]	6	4
3	3	1	2	6	4	5
4	[4]	5	6	1	2	[3]
5	[6]	4	5	2	3	[1]
6	5	6	4	3	1	2

Fig. 1.4.2

THEOREM 1.4.5. (due to L. J. Paige, see [3]). *If $(G, \cdot)$ is a finite group of order n which has a complete mapping, there exists an ordering of its elements, say $a_1, a_2, \ldots, a_n$, such that $a_1a_2 \ldots a_n = e$, where e denotes the identity element of G.*

PROOF. Let $x \to \theta'(x)$ be a complete mapping for G. Then the mapping θ such that $\theta(x) = \theta'(x)\,\theta'(e)^{-1}$ is also a complete mapping of G and $\theta(e) = e$, where e denotes the identity element of G. To see this, note that as $\theta'(x)$ ranges over G so does $\theta(x)$. Moreover, as $\eta'(x) = x\theta'(x)$ ranges over G, so does $x\theta(x) = \eta'(x)\,\theta'(e)^{-1}$.

For the remainder of the proof, we suppose that $x \to \theta(x)$ is a complete mapping such that $\eta(e) = e\theta(e) = e$, and we assume that $a_1 = e$.

Let us consider the product $a_2\theta(a_2)$. Since $\eta(a_2) \neq e$, we have $a_2{}^{-1} \neq$ $\neq \theta(a_2)$ and so $\theta(a_2)^{-1}$ occurs among the elements of G distinct from a_2. Let $\theta(a_2)^{-1} = a_3$ and form the product $a_2\theta(a_2)\,a_3\theta(a_3)$ where we may suppose that $\theta(a_3)^{-1} = a_4 \neq a_3$, since $a_3\theta(a_3) = \eta(a_3) \neq e$. We continue in this manner and ultimately reach a product (or "cycle")

$$a_2\theta(a_2)\,a_3\theta(a_3) \ldots a_s\theta(a_s) = e,$$

where $\theta(a_{i-1}) = a_i^{-1}$ $(i = 1, \ldots, s)$ and $\theta(a_s) = a_2^{-1}$.

If $s = n$, the theorem is proved, since

$$\eta(a_1)\,\eta(a_2) \ldots \eta(a_n) = e.$$

If not, we repeat the process beginning with $a_{s+1}\,\theta(a_{s+1})$, where a_{s+1} is an element of G distinct from $a_1, a_2, \ldots, a_s$ and finally we shall arrive at a series of cycles like that above whose product is the identity element. Namely, we shall have $\eta(a_1)\,\eta(a_2), \ldots, \eta(a_n) = e$.

COROLLARY. *If G possesses a complete mapping, the product of the elements of G in any order is an element of the commutator subgroup of G.*

It has been proved by L. J. Paige (see [1]) that, for abelian groups, the necessary condition given in theorem 1.4.5 for the existence of a complete mapping is also sufficient. In the same paper, he has also proved that *the product of the elements of an abelian group G is the identity element e unless G contains exactly one element of order 2. In this latter case the product is equal to the unique element of order 2.* (The same result has been obtained in M. Hall [5] as a deduction from a more general theorem about permutations.)

It follows from these results of Paige that the only finite abelian groups which have no complete mapping are those with a unique element of order 2. (More recently this result has been re-proved by L. Carlitz in [2].)

The following theorem is also due to L. J. Paige (see [3]):

THEOREM 1.4.6. *A sufficient condition for a finite non-abelian group G of order n to have a complete mapping is that there exist an ordering $a_2, a_3, \ldots, a_n$ of the elements of $G \setminus \{e\}$ such that the partial products $a_2, a_2a_3, a_2a_3a_4, \ldots, a_2a_3 \ldots a_n$ are all distinct and such that the complete product $a_2a_3 \ldots a_n$ is equal to the identity element e of G.*

PROOF. We are given that the $n - 1$ elements $a_2, a_2a_3, \ldots, a_2a_3 \ldots, a_n$ are all distinct and that the last of them is the identity element e of G. It follows that there is just one non-identity element of G which is not equal to any one of the above partial products. Let us denote this element by b_2 and construct a mapping $g \to \theta(g)$ of G as follows: $\theta(e) = e$, $\theta(b_2)$ is the solution of the equation $b_2x = a_2$ and, for each j, successively define $b_{j+1} = \theta(b_j)^{-1}$ and then let $\theta(b_{j+1})$ be the solution of the equation $b_{j+1}x = a_{j+1}$, $j = 2, 3, \ldots, n-1$.

All of $\theta(b_2), \theta(b_3), \ldots, \theta(b_n)$ are distinct and different from the identity: for if $\theta(b_r) = \theta(b_s)$, $r \neq s$, we would have

$$\begin{aligned} a_2a_3 \ldots a_r &= b_2\theta(b_2)\, b_3\theta(b_3) \ldots b_r\theta(b_r) = b_2\theta(b_r) = b_2\theta(b_s) = \\ &= b_2\theta(b_2)\, b_3\theta(b_3) \ldots b_s\theta(b_s) = a_2a_3 \ldots a_s \end{aligned}$$

which contradicts the distinctness of the partial products $a_2a_3 \ldots a_r$ and $a_2a_3 \ldots a_s$. Moreover, if $\theta(b_r) = e$, we would have $b_2 = a_2a_3 \ldots a_r$ which would contradict our choice of b_2.

From our mode of construction of $\theta(b_j)$ it follows that $b_j\,\theta(b_j) = a_j$ for each j and so θ is a complete mapping.

The following problem is related to theorem 1.4.5 and its corollary.

Let G be a finite group of order n and G' its commutator subgroup. Since the elements of G commute mod G', every product of n distinct elements belongs to the same coset mod G'. L. Fuchs suggested the study of groups G for which it is true, conversely, that every element of this coset can be represented as a product of n distinct elements?

Groups which have the latter property have been called *P-groups* by one of the authors in a lecture given at the International Mathematical Congress in Moscow (see also J. Dénes and É. Török [1]).

Let us show, as an example, that the dihedral group D_3 of order 6 is a P-group. When regarded as a permutation group, D_3 consists of the following six permutations:

$$\alpha_1 = (1)(2)(3)(4)(5)(6), \qquad \alpha_2 = (1\ 2\ 3)(4\ 5\ 6),$$
$$\alpha_3 = (1\ 3\ 2)(4\ 6\ 5), \qquad \alpha_4 = (1\ 4)(2\ 6)(3\ 5),$$
$$\alpha_5 = (1\ 5)(2\ 4)(3\ 6), \qquad \alpha_6 = (1\ 6)(2\ 5)(3\ 4).$$

The commutator subgroup of D_3 comprises the three permutations $\alpha_1, \alpha_2, \alpha_3$ and so, in order to show that D_3 is a P-group, it is sufficient to show that each of the elements of the coset $\alpha_4, \alpha_5, \alpha_6$ can be represented as a product of n distinct elements. We have

$$\alpha_4 = \alpha_1\alpha_4\alpha_6\alpha_2\alpha_5\alpha_3$$
$$\alpha_5 = \alpha_1\alpha_4\alpha_2\alpha_6\alpha_3\alpha_5$$
$$\alpha_6 = \alpha_1\alpha_2\alpha_4\alpha_6\alpha_3\alpha_5$$

Up to the present time, no example of a finite group which is not a P-group is known.

A. R. Rhemtulla has proved in [1] that every soluble group is a P-group, and the reader can find examples of non-soluble P-groups in J. Dénes [3], [5].

In section 9.1, we shall explain an application of P-groups to a graph theoretical interpretation of the Feit–Thompson theorem concerning the solubility of groups of odd order.

Theorem 1.4.5 gave a necessary condition for a finite group to possess a complete mapping and this, as we remarked above, is also a sufficient condition if the group is abelian. Theorem 1.4.6, on the other hand, gave a sufficient condition which is not always necessary. The problem of finding conditions under which a finite group has a complete mapping has been further investigated in M. Hall and L. J. Paige [1]. In that paper, these authors proved that for soluble groups the necessary and sufficient, and for groups of even order the necessary, condition for the existence of a transversal in the Cayley table of the group is that its Sylow 2-subgroups should be non-cyclic. They conjectured further that this condition is also sufficient for non-soluble groups. Their main theorems are as follows:

Theorem 1.4.7. *A finite group G of order n which has a cyclic Sylow 2-subgroup does not possess a complete mapping.*

Proof. Let S be a cyclic Sylow 2-subgroup of G. Then we shall show first that S is in the centre of its normalizer $N(S)$.

If S is cyclic of order 2^m, its automorphism group is a 2-group of order 2^{m-1}. (See page 146 of H. J. Zassenhaus [1].) Let Z be the centre of $N(S)$. If $Z = N(S)$ then S lies in the centre of its normalizer and so the theorem is true in this case. We may suppose that $Z \neq N(S)$, and if this is the case, we may choose an arbitrary element $x \in N(S)$ and write $x = yz = zy$ where y and z are some powers of x and so belong to $N(S)$ and the order of z is a power of 2 (and so $z \in S$) while the order of y is prime to 2. To prove our result, it is sufficient to show that if $y \in N(S)$ where the order of y is r say and r is prime to 2 then $yay^{-1} = a$ for every $a \in S$. Since $ySy^{-1} = S$, the mapping $\varphi : a \to yay^{-1}$ is an automorphism of S. The order of the automorphism group of S is 2^{m-1} and hence $y^{2^{m-1}} a y^{-2^{m-1}} = a$ for every $a \in S$. Since r is prime to 2, there exists an integer t such that $(y^{2^{m-1}})^t = y$ and this implies that $yay^{-1} = a$ for every $a \in S$.

By a theorem of Burnside (see page 203 of M. Hall [9] for a proof), if a Sylow p-subgroup P of a finite group G is in the centre of its normalizer, then G has a normal subgroup H which has the elements of P as its coset representatives. Thus, in the present case, our group G has a normal subgroup H which has the elements of S as its coset representatives. If $G = H + Hs_2 + \ldots + Hs_u$, each $g \in G$ can be written in the form $g = hs_i$ for some $h \in H$ and $s_i \in S$, and we have $u = \text{ord}\, S$ and $\text{ord}\, H$ is odd. Hence, $\prod_{g \in G} g = h^* \prod_{i=1}^{u} (s_i)^{\text{ord}\, H}$ where h^* is some element of H. Here, we have used the fact that $G/H \cong S$ is abelian and that exactly $\text{ord}\, H$ elements of G occur in each coset. Since S is abelian and has a unique element of order 2, $\prod_{i=1}^{u} s_i$ is this element of order 2 and so $\prod_{i=1}^{u} (s_i)^{\text{ord}\, H} \notin H$. But because G/H is abelian, the derived group (commutator subgroup) G' of G is contained in H. Thus $\prod_{g \in G} g \notin G'$ and so, by the corollary to theorem 1.4.5, G does not possess a complete mapping.

Corollary. *If G is an arbitrary group of order $n = 4k + 2$, then G has no complete mapping. In particular, the symmetric group on three elements has no complete mapping.*

Remark. The result of the corollary can be deduced from an earlier theorem of H. B. Mann, as we shall explain in section 5.1. Also, it can be obtained as a special case of a more general result of R. H. Bruck which we shall prove in chapter 9.

Theorem 1.4.8. *A finite soluble group G whose Sylow 2-subgroups are non-cyclic has a complete mapping.*

Proof. The proof is obtained by a series of steps. In the first place, Hall and Paige have shown that every non-cyclic 2-group has a complete mapping. They have also shown that if a group G can be written in the form $G = AB$ where A and B are subgroups such that $A \cap B = \{e\}$ and if A and B both have complete mappings, so does G. But, by a theorem of P. Hall (see [1]), if A is a Sylow 2-subgroup of a soluble group, then A has a 2-complement B and we then have $G = AB$ and $A \cap B = \{e\}$. Also, B has odd order and so it has a complete mapping as shown in theorem 1.4.3 and A, being a non-cyclic 2-group, has a complete mapping also. Hence, the result of the theorem follows.

Hall and Paige have proved further that all finite alternating groups have complete mappings and that for $n > 3$ the symmetric group S_n has a complete mapping. As we noted above, the symmetric group S_3 has no complete mapping and so we may deduce from the last result that the property of having a complete mapping is not invariant under homomorphic mappings.

It has been conjectured by Hall and Paige that every finite group whose Sylow 2-subgroups are non-cyclic has a complete mapping, but this question remains an open one.

Using the above two theorems and the fact that, in a finite group G of order n, every product of the n distinct elements of G is in the same coset mod G', it follows that the necessary condition given in theorem 1.4.5 for G to have a complete mapping is also sufficient if G is soluble. For, if such a group G has no complete mapping, it has a non-trivial cyclic Sylow

2-subgroup by theorem 1.4.8 and then, as in the proof of theorem 1.4.7, $\prod_{g \in G} g \notin G'$ and so there does not exist an ordering of the n elements of G which is equal to the identity element e.

Paige himself has conjectured that the condition of theorem 1.4.5 may be sufficient for arbitrary groups but his best result in this connection is that of theorem 1.4.6.

He has proved two further useful results as follows. (See L. J. Paige [3].)

THEOREM 1.4.9. *Let H be a normal subgroup of a finite group G. If G/H admits a complete mapping θ_1, H a complete mapping θ_2, then G has a complete mapping too.*

PROOF. Let $G/H \cong K$ and let the elements of K be $e, k_1, k_2, k_3, \ldots$. Let $u_e, u_{k_1}, u_{k_2}, u_{k_3}, \ldots$, be a set of coset representatives of H in G so that, in the isomorphism $G/H \to K$, $u_k H \to k$, and so on. Then, every element of G is uniquely expressible in the forms $u_k h$ or $h' u_k$ for $k \in K$ and $h, h' \in H$. Equality of $u_{k_1} h_1$ and $u_{k_2} h_2$ implies that $u_{k_2} \in u_{k_1} H$ and so $k_2 = k_1$ and also $h_2 = h_1$.

Let us define a mapping θ of G by $\theta(u_k h) = \theta_2(h)\, u_{\theta_1(k)}$. It is easy to check that this is one-to-one and onto G. Let $\eta(u_k h) = u_k h \theta(u_k h)$. We wish to show that η is also a one-to-one mapping of G onto itself. Then θ will be a complete mapping of G, as required. Suppose, on the contrary that $\eta(u_{k_1} h_1) = \eta(u_{k_2} h_2)$. Then, $u_{k_1} h_1 \theta_2(h_1)\, u_{\theta_1(k_1)} = u_{k_2} h_2 \theta_2(h_2)\, u_{\theta_1(k_2)}$, or we may write $u_{k_1} h_3 u_{\theta_1(k_1)} = u_{k_2} h_4 u_{\theta_1(k_2)}$ where $h_3 \neq h_4$ unless $h_1 = h_2$ since θ_2 is a complete mapping. This can be rewritten as $u_{k_1} u_{\theta_1(k_1)} h_3' = u_{k_2} u_{\theta_1(k_2)} h_4'$ for some h_3' and $h_4' \in H$. Since the mapping $u_k H \to k$ is an isomorphic mapping, $u_{k_1} u_{\theta(k_1)} H = u_{k_1 \theta(k_1)} H$, and, since the u_{k_i} are a complete set of coset representatives, we deduce that $k_1 \theta_1(k_1) = k_2 \theta_1(k_2)$. Because θ_1 is a complete mapping of K, we then have $k_1 = k_2$ whence also $h_1 = h_2$. The result of the theorem follows.

THEOREM 1.4.10. *Let G be a finite group which contains a normal subgroup H of odd order such that the factor group G/H is abelian and has no complete mapping. Then G has no complete mapping.*

PROOF. Since G/H is abelian and has no complete mapping it possesses exactly one element of order two, say $g_2 H$, and the product of all the elements of G/H is equal to this element $g_2 H$. That is, $H(g_2 H)(g_3 H) \ldots (g_n H) = g_2 H$ or we may write $e g_2 g_3 \ldots g_n H = g_2 H$. Thus, the product $\prod_{g \in G} g$ is not in H. But since G/H is abelian, H contains the commutator subgroup of G. That G has no complete mapping now follows from the corollary to theorem 1.4.5.

It is proved in P. T. Bateman [1] that all infinite groups possess complete mappings.

THEOREM 1.4.11. *Let Q_n be a quasigroup of order n and let $Q_{n_1}, Q_{n_2}, \ldots Q_{n_r}$ be subquasigroups of Q_n of orders $n_1, n_2, \ldots, n_r$ respectively such that the sets Q_{n_i} are disjoint and Q_n is their set theoretical union. Then if, for each k $(k = 1, 2, \ldots, r)$ Q_{n_k} has a complete mapping, Q_n has one too.*

PROOF. Without loss of generality one can suppose that the multiplication table of Q_n has the following form

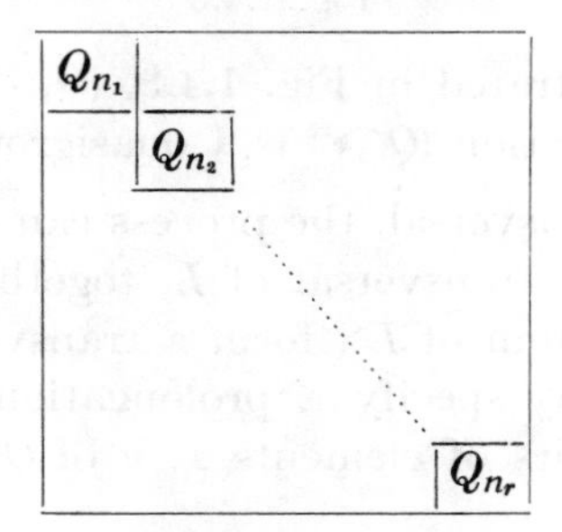

Since $Q_{n_1}, Q_{n_2}, \ldots, Q_{n_r}$ have transversals, Q_n has at least one too.

REMARK. For further results on quasigroups similar to that just proved see J. Dénes and K. Pásztor [1], S. Haber and A. Rosenfeld [1], J. Csima and A. Rosa [1] and also section 1.5.

Let $(Q, \cdot)$ be a given quasigroup of order n which possesses a complete mapping θ. By a process which V. D. Belousov has called *prolongation* (see [8]) we shall show how, starting from $(Q, \cdot)$, a quasigroup $(Q', *)$ of order $n + 1$ can be constructed, where the set Q' is obtained from Q by the adjunction of one additional element.

Before presenting a formal algebraic definition of a prolongation, we shall explain how the construction may be carried out in practice. We suppose that the elements of Q are $1, 2, \ldots, n$ as usual and let L be the latin square formed by the multiplication table of the quasigroup $(Q, \cdot)$. Since $(Q, \cdot)$ has a complete mapping, L possesses at least one transversal (theorem 1.4.1). We replace the elements in all the cells of this transversal by the additional element $n + 1$ and then, without changing their order, adjoin the elements of the transversal to the resulting square as its $(n + 1)$th row and $(n + 1)$th column. Finally, to complete the enlarged square L', we adjoin the element $n + 1$ as the entry of the cell which lies at the intersection of the $(n + 1)$th row and $(n + 1)$th column. The square L' is then latin (see Fig. 1.4.3 for an example) and defines the multiplication table of a quasigroup $(Q', *)$ of order one greater than that of $(Q, \cdot)$.

$(\cdot)$	1	2	3
1	1	2	$\boxed{3}$
2	$\boxed{2}$	3	1
3	3	$\boxed{1}$	2

$(*)$	1	2	3	4
1	1	2	4	3
2	4	3	1	2
3	3	4	2	1
4	2	1	3	4

Fig. 1.4.3

In the example illustrated in Fig. 1.4.3, $(Q, \cdot)$ is the cyclic group of order 3 and its prolongation $(Q', *)$ is a quasigroup of order 4.

If L has a second transversal, the process can be repeated; since then the cells of this second transversal of L, together with the cell of the $(n + 1)$th row and column of L', form a transversal of L'.

Algebraically, we may specify a prolongation by defining the products $x * y$ of all the pairs of elements x, y of Q'. If $x . \theta(x) = \eta(x)$ we get the following:

$$\begin{aligned} x * y &= x . y && \text{if } x, y \in Q, y \neq \theta(x) \\ &= n + 1 && \text{if } x, y \in Q, y = \theta(x) \\ &= \eta(x) && \text{if } x \in Q, y = n + 1 \\ &= \eta[\theta^{-1}(y)] && \text{if } y \in Q, x = n + 1 \\ &= n + 1 && \text{if } x = y = n + 1. \end{aligned}$$

In our example (Fig. 1.4.3) we have

$$\begin{aligned} \theta(1) &= 3, \qquad & \eta(1) &= 3 \\ \theta(2) &= 1, & \eta(2) &= 2 \\ \theta(3) &= 2, & \eta(3) &= 1. \end{aligned}$$

The construction of prolongation was first studied by R. H. Bruck (see [1]) who discussed only the case in which $(Q, \cdot)$ is an idempotent quasigroup. The construction for arbitrary quasigroups has been defined in J. Dénes and K. Pásztor [1] and in J. M. Osborn [1]. We shall make use of the construction in section 1.5 and again in section 3.2. K. Yamamoto has used the same concept, under the name *1-extension,* in connection with the construction of pairs of mutually orthogonal latin squares. He has also defined the inverse construction and called it a *1-contraction*. We shall give an account of his work in section 12.2.

We mention two further properties of prolongation without proof:

(1) The necessary and sufficient condition that a prolongation of a quasigroup $(Q, \cdot)$ be isotopic to a group G is that G be a generalized Klein group; that is, a group with every element of order two, direct sum of cyclic groups each of order two.

(2) Let G be a group which has at least one element of order greater than two. If G possesses two complete mappings θ_1 and θ_2 then the prolongations of G constructed by means of θ_1 and θ_2 respectively are isotopic if and only if $\theta_1 = \theta_2$.

The reader will find the proofs lacking here and further results of similar type in V. D. Belousov [7] and [8], V. D. Belousov and G. B. Belyavskaya [1], and G. B. Belyavskaya [1].

A generalization of the notion of prolongation for latin hypercubes (the latter concept will be defined later in section 5.4) has been introduced by B. Elspas, R. C. Minnick and R. A. Short in [1].

Further generalizations of the concept of prolongation have been discussed in G. B. Belyavskaya [3], [4] and [5] and also referred to in [2]. In particular, in the second of these papers, the inverse of a prolongation (which had earlier been defined by K. Yamamoto under the name *1-contraction*, as we have already mentioned above) has been re-introduced under the title of *compression*. Thus, by means of a compression one can obtain a latin square of order $n - 1$ from a given latin square of order n.

1.5. Latin subsquares and subquasigroups

The concepts of subquasigroup and latin subsquare are very closely connected.

Definition. Let the square matrix A shown in Fig. 1.5.1 be a latin square. Then, if the square submatrix B shown in Fig. 1.5.2 (where $1 \leq i, j, \ldots, l, p, q, \ldots, s \leq n$) is again a latin square, B is called a *latin subsquare* of A.

$$A = \begin{pmatrix} a_{11} & a_{12} & \cdots & a_{1n} \\ a_{21} & a_{22} & \cdots & a_{2n} \\ \cdot & \cdot & \cdots & \cdot \\ \cdot & \cdot & \cdots & \cdot \\ a_{n1} & a_{n2} & \cdots & a_{nn} \end{pmatrix} \qquad B = \begin{pmatrix} a_{ip} & a_{iq} & \cdots & a_{is} \\ a_{jp} & a_{jq} & \cdots & a_{js} \\ \cdot & \cdot & \cdots & \cdot \\ \cdot & \cdot & \cdots & \cdot \\ a_{lp} & a_{lq} & \cdots & a_{ls} \end{pmatrix}$$

Fig. 1.5.1 Fig. 1.5.2

Thus, the latin square corresponding to the Cayley table of a subquasigroup Q' of any quasigroup Q is a latin subsquare of the latin square defined by the Cayley table of Q. Conversely, any latin subsquare of the latin square given by the Cayley table of a quasigroup Q becomes, when bordered appropriately, the Cayley table of a subquasigroup of a quasigroup isotopic to Q. (The reason for its being not the same as Q but only isotopic to it is that the bordering elements contained in the rows and columns defining the latin subsquare may be different from those of the latin subsquare.)

In [1], H. W. Norton called a latin subsquare of order 2 an *intercalate*. He has made use of this concept in connection with the enumeration of latin squares of order 7. (See section 4.3.)

In Fig. 1.5.3, the Cayley table of a quasigroup of order 10 is shown which has a subquasigroup of order 4 (consisting of the elements 1, 2, 3, 4) and also one of order 5 (with elements 3, 4, 5, 6, 7) the intersection of which is a subquasigroup of order 2 (with elements 3, 4).

	0	8	9	1	2	3	4	5	6	7
5	1	9	2	8	0	6	7	4	5	3
6	8	2	1	0	9	7	5	3	4	6
7	2	1	0	9	8	5	6	7	3	4
3	0	8	9	1	2	3	4	6	7	5
4	9	0	8	2	1	4	3	5	6	7
1	5	6	7	3	4	1	2	0	8	9
2	6	7	5	4	3	2	1	8	9	0
0	7	4	3	5	6	0	9	1	2	8
8	3	5	4	6	7	8	0	9	1	2
9	4	3	6	7	5	9	8	2	0	1

Fig. 1.5.3

We come now to our first theorem:

Theorem 1.5.1. *For given integers n and k, n arbitrary and $k \leq n/2$, there exists a quasigroup of order n which contains at least one subquasigroup of order k.*

Proof. We shall prove our theorem (which was first published in T. Evans [2]) by using the procedure of prolongation which we explained in section 1.4 above. (See also J. Dénes [6], J. Dénes and K. Pásztor [1].)

Let us consider a quasigroup G of order $n - k$ which has $n - k$ disjoint transversals (we show below that such a quasigroup always exists when $n - k \neq 2$ or 6).

Let us prolongate the multiplication table. The procedure of prolongation can be iterated at least k times, since $k \leq n/2$; that is, $k \leq n - k$. The result of the procedure is an $n \times n$ latin square in which the k^2 cells of the last k rows and columns form a latin subsquare of order k. When this latin square is bordered by its own first row and first column, it becomes the multiplication table of a quasigroup of order n with a subquasigroup of order k as required.

Since the necessary condition that there exists a latin square orthogonal to a given square L of order n is that L possesses n disjoint transversals (see section 5.1) and since orthogonal pairs of latin squares exist for all values of n except 2 and 6, as we shall prove in chapter 11 (which is devoted mainly to this subject), our construction works for all integers n and k except those for which $n - k = 2$ or 6.

Since $k \leq n/2$, $n - k = 2$ and $k > 1$ hold if and only if $n = 4$ and $k = 2$. However, the cyclic group of order 4 is an example of a quasigroup of order 4 which has a subquasigroup of order 2, so the theorem is true for this case.

Next let us suppose that $n - k = 6$. Since $k \leq n/2$ if and only if $n \leq 12$, we have to check the following further cases: $n = 8$, $k = 2$; $n = 9$, $k = 3$; $n = 10$, $k = 4$; $n = 11$, $k = 5$; $n = 12$, $k = 6$.

By Cauchy's theorem, any group of order 8 has a subgroup of order 2 and any group of order 9 has a subgroup of order 3 so the theorem is true for these cases. The cyclic group of order 12 has a subgroup of

1	2	3	4	5	6	7	8	9	10
2	3	4	1	6	7	8	9	10	5
3	4	1	2	7	8	9	10	5	6
4	1	2	3	8	9	10	5	6	7
5	6	7	8	4	10	2	1	3	9
6	7	8	9	10	5	3	4	2	1
7	8	9	10	1	3	5	6	4	2
8	9	10	5	2	4	6	7	1	3
9	10	5	6	3	1	4	2	7	8
10	5	6	7	9	2	1	3	8	4

Fig. 1.5.4

1	2	3	4	5	6	7	8	9	10	11
2	3	4	5	1	7	8	9	10	11	6
3	4	5	1	2	8	9	10	11	6	7
4	5	1	2	3	9	10	11	6	7	8
5	1	2	3	4	10	11	6	7	8	9
6	7	8	9	10	11	1	2	3	4	5
7	8	9	10	11	2	6	3	4	5	1
8	9	10	11	6	4	3	7	5	1	2
9	10	11	6	7	1	5	4	8	2	3
10	11	6	7	8	3	2	5	1	9	4
11	6	7	8	9	5	4	1	2	3	10

Fig. 1.5.5

order 6 generated by its unique element of order 6, so the theorem is valid for this case also. The cases $n = 10$, $k = 4$ and $n = 11$, $k = 5$ are settled by means of the examples given in Figs 1.5.4 and 1.5.5 respectively. This completes the proof.

Corollary. *Let L_k be an arbitrary latin square of order k. Then, if $k \leq n/2$, there exists at least one latin square L_n of order n such that L_k is a latin subsquare of L_n.*

Proof. We may rearrange the symbols in a $k \times k$ latin subsquare of a latin square of order n without affecting its property of being latin.

In contrast to the above theorem, A. J. W. Hilton has conjectured recently that, for all sufficiently large n, there exist quasigroups of order n which contain no proper subquasigroups.

Latin subsquares can be used for the characterization of non-simple groups. We remind the reader that a group G is simple if and only if it has no non-trivial normal subgroup.

Let A_0 be the latin square corresponding to the Cayley table of a normal subgroup N of a group G and let $A_1, A_2, \ldots, A_{k-1}$ be the latin squares corresponding to the cosets $a_1N, a_2N, \ldots, a_{k-1}N$ of N in G. (Here $a_0 = e, a_1, a_2, \ldots, a_{n-1}$ denote the elements of G.) Then, we have the following result:

Theorem 1.5.2. *The latin square corresponding to the Cayley table of a group G of order n having a proper normal subgroup N of order m comprises the union of k^2 latin squares ($k = n/m$) consisting of k subsquares* *whose elements are those of N, k subsquares whose elements are those of a_1N, k subsquares whose elements are those of $a_2N, \ldots$, and k subsquares whose elements are those of $a_{k-1}N$, where $N, a_1N, a_2N, \ldots, a_{k-1}N$ are the cosets of N in G.*

PROOF. The theorem follows directly from the properties of a normal subgroup.

As an example, consider the dihedral group $D_3 = \{\alpha, \beta \mid \alpha^3 = \beta^2 = e, \alpha\beta = \beta\alpha^{-1}\}$ of order 6 and its normal subgroup $N = \{e, \alpha, \alpha^2\}$. The result of theorem 1.5.2 for this case is exhibited in Fig. 1.5.6.

	e	α	α^2	β	$\beta\alpha$	$\beta\alpha^2$
e	e	α	α^2	β	$\beta\alpha$	$\beta\alpha^2$
α	α	α^2	e	$\beta\alpha^2$	β	$\beta\alpha$
α^2	α^2	e	α	$\beta\alpha$	$\beta\alpha^2$	β
β	β	$\beta\alpha$	$\beta\alpha^2$	e	α	α^2
$\beta\alpha$	$\beta\alpha$	$\beta\alpha^2$	β	α^2	e	α
$\beta\alpha^2$	$\beta\alpha^2$	β	$\beta\alpha$	α	α^2	e

Fig. 1.5.6

If the latin subsquares $A_0, A_1, \ldots, A_{k-1}$ in the Cayley table are replaced successively by the coset representatives $a_0, a_1, \ldots, a_{k-1}$, the Cayley table of a group which is isomorphic to the factor group of G mod N will be obtained (for full details see L. Baumgartner [1]).

We note also that the same construction has been used by F. Šik in [1] in connection with the generalization of the Jordan–Hölder theorem and by J. Dénes and K. Pásztor in [1] for the purpose of disproving D. W. Wall's conjecture. (We shall discuss this below.)

The construction mentioned above was also utilized by H. J. Zassenhaus for the purpose of characterizing the normal Cayley tables of non-simple groups (see H. J. Zassenhaus [1]) and by R. H. Bruck in the construction of an example of a loop L which has a characteristic subloop which is not normal in L. (See R. H. Bruck [2], page 355.)

In his paper [1], D. W. Wall suggested a full investigation of the following problem: If a quasigroup Q of order n contains m subquasigroups each of order s and defined on disjoint subsets of Q, what conditions must Q satisfy if the inequality $n \geq (m+1)s$ is to hold?

He showed that if $m = 2$, the inequality holds for any quasigroup Q, but also pointed out some cases in which it is false with $m > 2$. In [1], J. Dénes and K. Pásztor have subsequently proved the following result:

THEOREM 1.5.3. *For $n > 2$ and m a divisor of n there exists a quasigroup $(Q, \cdot)$ of order n which contains m subquasigroups which are all of the same order $s = n/m$ and are defined on disjoint subsets of Q.*

PROOF. It follows from theorems 1.4.3, 1.4.4, and 1.4.2 that, for every odd n, there exists an idempotent quasigroup of order n whose multiplication table is composed of disjoint transversals. By a prolongation one can obtain from this an idempotent quasigroup of order $n+1$, where n is any odd integer greater than one. Hence, idempotent quasigroups exist of all orders $m > 2$.

Let Q be an idempotent quasigroup of order $m > 2$ and let each element a_i of its multiplication table be replaced by a latin square L_{a_i} of order s, where, if $a_i \neq a_j$, the latin squares L_{a_i} and L_{a_j} are defined over disjoint sets of elements. One thus obtains a latin square which, when bordered by a row consisting of the first rows of the latin squares $L_{a_1}, L_{a_2}, \ldots, L_{a_m}$ which appear along its main left-to-right diagonal taken in order and by a column consisting of the entries in the first columns of these squares taken in order, becomes the multiplication table of a quasigroup of order ms which is a union of m disjoint subquasigroups.

COROLLARY. *No latin square which is the union of two disjoint latin subsquares exists.*

PROOF. The corollary becomes obvious when we take into consideration the construction used in the proof of theorem 1.5.3 and the fact that no idempotent quasigroup of order 2 exists.

Another proof of the result stated in the corollary has been published in D. W. Wall [1].

The result of the corollary for the special case of latin squares which satisfy the rectangle criterion was obtained by S. Haber and A. Rosenfeld in [1]. These authors have also obtained some results on groups which are unions of proper subgroups. We shall now discuss these.

Let us observe first that it is evident that any group G which is not cyclic is expressible as a set-theoretical union of (not necessarily disjoint) proper subgroups because, for example, such a group G is the union of its cyclic subgroups which, by hypothesis, are all proper. In S. Haber and A. Rosenfeld [1] the problem of determining the minimum number of proper subgroups into which a group can be decomposed is discussed. Two of the main results obtained are as follows:

(1) Let G be a finite group of order n, and let p be the smallest prime which divides n. Then G is not the union of p or fewer of its proper subgroups.

(2) A group G is the union of three proper subgroups if and only if Klein's four-group is a homomorphic image of G.

The latter result is not new. It was in fact first obtained by G. Scorza (see [1]) for the case of finite groups and was later discovered again by M. Bruckenheimer, A. C. Bryan and A. Muir in [1].

In D. Greco [1] and [2], necessary and sufficient conditions that a finite group G be expressible as the union of precisely four and precisely five proper subgroups are given and, in D. Greco [3], some extensions of these results to the case of infinite groups are discussed.

Next, we shall prove an interesting result due to W. A. McWorter (see [1]) which we have already made use of in theorem 1.2.3.

Theorem 1.5.4. *Let Q be a quasigroup and let A and B be two subsets of Q such that not every element of Q has the form ab, with a in A and b in B. Then*

$$|Q| \geq |A| + |B|$$

where $|X|$ denotes the cardinal of the set X.

Proof. If Q is infinite, the result is obvious. Suppose therefore that Q is finite and has multiplication table M. We may rearrange the rows of M in such a way that the first $|A|$ rows of M have entries consisting of all right multiples of the elements of A and then rearrange the columns of M in such a way that the first $|B|$ columns consist of all left multiples of elements of B. Then M takes the form shown in Fig. 1.5.7

C	Y
*	Z

Fig. 1.5.7

where C is an $|A|$ by $|B|$ submatrix of M. Hence the entries of C consist precisely of the elements of the form $a.b$, with a in A and b in B.

Now let $N(x)$ be the number of times that the element x in Q appears in C. Since x appears $|A|$ times in C and Y combined, x appears $|A| - N(x)$ times in Y. But x appears $|Q| - |B|$ times in Y and Z together. Hence x appears

$$|Q| - |B| - [|A| - N(x)]$$

times in Z.

Since the number of times that x appears in Z is not negative, we must have

$$|Q| - |A| - |B| + N(x) \geq 0,$$

for all x in Q. Since by our assumption $N(x) = 0$ for some x in Q, we get

$$|Q| \geq |A| + |B|$$

and the proof is complete.

The same result for the case when Q is a group had earlier been proved by H. B. Mann (see [5]).

Our next topic is that of (n, d)-complete rectangles.

DEFINITION. A latin rectangle (that is, a rectangular array which can be completed to a latin square) is called *(n, d)-complete* if it contains n different elements each of which appears exactly d times in the array.

As an example, we note that the 6×6 latin rectangle shown in Fig. 1.5.8 is (9,4)-complete.

1	2	3	4	5	6
2	6	7	5	9	1
3	7	5	6	1	8
4	5	9	7	8	3
8	9	1	2	3	4
6	4	8	9	7	2

Fig. 1.5.8

Up to the present interest seems to have centred on the case $d = 1$. In particular, one of the authors (see J. Dénes [6]) has proved the following theorem:

THEOREM 1.5.5. *If L is the latin square representing the multiplication table of a group G of order n, where n is a composite number, then L can be split into a set of n $(n, 1)$-complete non-trivial latin rectangles.* (A latin rectangle is trivial if it consists of a single row or a single column.)

PROOF. If n is a composite number, G has at least one proper subgroup A_0, of order h say. By Lagrange's theorem, G splits into disjoint cosets of A_0, say $A_0, A_1, \ldots, A_{k-1}$, where $n = hk$.

By selecting one element from each of these k cosets, we get a set of coset representatives of G relative to A_0. It is evident that one can select (in many ways) h such sets of coset representatives which are pairwise disjoint and which together cover G. Call these sets of coset representatives $R_1, R_2, \ldots, R_h$.

Now let a multiplication table for G be formed by taking as row border the elements of the cosets $A_0, A_1, \ldots, A_{k-1}$ in order and as column border the elements of the sets of coset representatives $R_1, R_2, \ldots, R_h$ in order. By the properties of coset representatives, the rectangle whose row border comprises the elements of the coset A_i and whose column border comprises the elements of the set R_j contains each element of G exactly once and is an $(n, 1)$ rectangle. Hence the multiplication table of G is the union of n such $(n, 1)$-rectangles.

As an example, we show in Fig. 1.5.9 the result of carrying out the construction for the cyclic group of order 8 when regarded as the additive group of integers modulo 8, with the subgroup $\{0, 2, 4, 6\}$ as A_0. We have $A_1 = \{1, 3, 5, 7\}$ and we may take (for example) $R_1 = \{0, 1\}$, $R_2 = \{2, 5\}$, $R_3 = \{4, 3)$, $R_4 = \{6, 7\}$.

	0	2	4	6	1	3	5	7
0	0	2	4	6	1	3	5	7
1	1	3	5	7	2	4	6	0
2	2	4	6	0	3	5	7	1
5	5	7	1	3	6	0	2	4
4	4	6	0	2	5	7	1	3
3	3	5	7	1	4	6	0	2
6	6	0	2	4	7	1	3	5
7	7	1	3	5	0	2	4	6

Fig. 1.5.9

The converse of theorem 1.5.5 is false as is shown by the following latin square of order 27 which arose in an investigation concerning projective planes made by one of the authors (see A. D. Keedwell [3]). This latin square (Fig. 1.5.10) represents the multiplication table of a quasigroup

1	2	3	4	5	6	7	8	9	1′	2′	3′	4′	5′	6′	7′	8′	9′	1″	2″	3″	4″	5″	6″	7″	8″	9″
1′	2′	3′	6′	4′	5′	8″	9″	7″	1″	2″	3″	6″	4″	5″	8	9	7	1	2	3	6	4	5	8′	9′	7′
1″	2″	3″	5″	6″	4″	9′	7′	8′	1	2	3	5	6	4	9″	7″	8″	1′	2′	3′	5′	6′	4′	9	7	8
2	3	1	5	6	4	8	9	7	2′	3′	1′	5′	6′	4′	8′	9′	7′	2″	3″	1″	5″	6″	4″	8″	9″	7″
2′	3′	1′	4′	5′	6′	9″	7″	8″	2″	3″	1″	4″	5″	6″	9	7	8	2	3	1	4	5	6	9′	7′	8′
2″	3″	1″	6″	4″	5″	7′	8′	9′	2	3	1	6	4	5	7″	8″	9″	2′	3′	1′	6′	4′	5′	7	8	9
3	1	2	6	4	5	9	7	8	3′	1′	2′	6′	4′	5′	9′	7′	8′	3″	1″	2″	6″	4″	5″	9″	7″	8″
3′	1′	2′	5′	6′	4′	7″	8″	9″	3″	1″	2″	5″	6″	4″	7	8	9	3	1	2	5	6	4	7′	8′	9′
3″	1″	2″	4″	5″	6″	8′	9′	7′	3	1	2	4	5	6	8″	9″	7″	3′	1′	2′	4′	5′	6′	8	9	7
4	5′	6″	7	8′	9″	1	2′	3″	4′	5″	6	7′	8″	9	1′	2″	3	4″	5	6′	7″	8	9′	1″	2	3′
5	6′	4″	9	7′	8″	1″	2	3′	5′	6″	4	9′	7″	8	1	2′	3″	5″	6	4′	9″	7	8′	1′	2″	3
6	4′	5″	8	9′	7″	1′	2″	3	6′	4″	5	8′	9″	7	1″	2	3′	6″	4	5′	8″	9	7′	1	2′	3″
7	8″	9′	1	2″	3′	4	5″	6′	7′	8	9″	1′	2	3″	4′	5	6″	7″	8′	9	1″	2′	3	4″	5′	6
8	9″	7′	3	1″	2′	4″	5′	6	8′	9	7″	3′	1	2″	4	5″	6′	8″	9′	7	3″	1′	2	4′	5	6″
9	7″	8′	2	3″	1′	4′	5	6″	9′	7	8″	2′	3	1″	4″	5′	6	9″	7′	8	2″	3′	1	4	5″	6′
4′	5″	6	8′	9″	7	3	1′	2″	4″	5	6′	8″	9	7′	3′	1″	2	4	5′	6″	8	9′	7″	3″	1	2′
5′	6″	4	7′	8″	9	3″	1	2′	5″	6	4′	7″	8	9′	3	1′	2″	5	6′	4″	7	8′	9″	3′	1″	2
6′	4″	5	9′	7″	8	3′	1″	2	6″	4	5′	9″	7	8′	3″	1	2′	6	4′	5″	9	7′	8″	3	1′	2″
7′	8	9″	3′	1	2″	5″	6′	4	7″	8′	9	3″	1′	2	5	6″	4′	7	8″	9′	3	1″	2′	5′	6	4″
8′	9	7″	2′	3	1″	5′	6	4″	8″	9′	7	2″	3′	1	5″	6′	4	8	9″	7′	2	3″	1′	5	6″	4′
9′	7	8″	1′	2	3″	5	6″	4′	9″	7′	8	1″	2′	3	5′	6	4″	9	7″	8′	1	2″	3′	5″	6′	4
4″	5	6′	9″	7	8′	2	3′	1″	4	5′	6″	9	7′	8″	2′	3″	1	4′	5″	6	9′	7″	8	2″	3	1′
5″	6	4′	8″	9	7′	2″	3	1′	5	6′	4″	8	9′	7″	2	3′	1″	5′	6″	4	8′	9″	7	2′	3″	1
6″	4	5′	7″	8	9′	2′	3″	1	6	4′	5″	7	8′	9″	2″	3	1′	6′	4″	5	7′	8″	9	2	3′	1″
7″	8′	9	2″	3′	1	6′	4	5″	7	8″	9′	2	3″	1′	6″	4′	5	7′	8	9″	2′	3	1″	6	4″	5′
8″	9′	7	1″	2′	3	6	4″	5′	8	9″	7′	1	2″	3′	6′	4	5″	8′	9	7″	1′	2	3″	6″	4′	5
9″	7′	8	3″	1′	2	6″	4′	5	9	7″	8′	3	1″	2′	6	4″	5′	9′	7	8″	3′	1	2″	6′	4	5″

Fig. 1.5.10

which is not a group but, despite this, it can be split into 27 (27,1)-rectangles.

Closely related to the subject of $(n, 1)$-complete latin rectangles is that of subsquare complete latin squares. The latter concept is due to R. B. Killgrove (see R. B. Killgrove [2] and F. P. Hiner and R. B. Killgrove [1]) and we have the following definitions.

Definitions. A latin square L is *subsquare complete* if and only if, for each two distinct cells of L which contain equal elements, there is a proper latin subsquare of L which includes these two cells. We say that a latin square is an $\alpha, \beta, \gamma, \ldots$ *subsquare complete* latin square if it is subsquare complete and the order of each proper subsquare is included among the integers $\alpha, \beta, \gamma, \ldots$ (It is important to observe that this definition implies that an α-subsquare complete square is also $\alpha, \beta, \gamma, \ldots$, subsquare complete for arbitrary choice of the integers $\beta, \gamma, \ldots$)

It is easy to check, for example, that the latin square given in Fig. 1.5.10 is a 3-subsquare complete latin square.

In F. P. Hiner and R. B. Killgrove [2], the authors have proved that *the multiplication table of any finite non-cyclic group defines a subsquare complete latin square and that, if the elements of the group have orders $\alpha, \beta, \gamma, \ldots$, then the latin square is $\alpha, \beta, \gamma, \ldots$ subsquare complete.* Thus, for example, the square exhibited in Fig. 1.5.11, which represents the multiplication table of the dihedral group D_3 of six elements, is 2,3-subsquare complete. The converse of the above result is false because, for example, there exists a 2,3-subsquare complete latin square of order 7. This is shown in Fig. 1.5.12.

1	2	3	4	5	6
2	3	1	6	4	5
3	1	2	5	6	4
4	5	6	1	2	3
5	6	4	3	1	2
6	4	5	2	3	1

Fig. 1.5.11

Hiner and Killgrove have also proved that *every 2-subsquare complete latin square is the multiplication table of an elementary abelian 2-group.*

However, the corresponding result for 3-subsquare complete squares is false as already shown by the non-group example given in Fig. 1.5.10.

Hiner has made use of the principle of 1-extension and 1-contraction introduced by K. Yamamoto (see sections 1.4 and 12.2, and also K. Yamamoto [7] and [8]) to give algorithms for obtaining 2,3-subsquare complete latin squares of orders $2^\alpha - 1$ $(\alpha > 2)$ and $3^\beta + 1$ $(\beta > 1)$. These are as follows:

Algorithm A. Take the latin square representing the multiplication table of the elementary abelian 2-group of order 2^α $(\alpha > 2)$ and replace the element a_{ii} by the element a_{in} for $i = 1, 2, \ldots, n-1$, where $n = 2^\alpha$. Then delete the last row and column of the amended square.

Algorithm B. Take the latin square representing the multiplication table of the elementary abelian 3-group of order 3^β ($\beta > 1$) and adjoin to it a new row and column such that $a_{m+1,i} = a_{i,m+1} = a_{ii}$ for $i = 1, 2, \ldots, m$ where $m = 3^\beta$. In the enlarged square replace each element a_{ii} ($i = 1, 2, \ldots, m + 1$) of the leading diagonal by a new element $m + 1$.

Algorithm A amounts to effecting a 1-contraction of the latin square which represents the multiplication table of the elementary abelian group of order 2^α, while algorithm B amounts to effecting a 1-extension (prolongation) of the latin square which represents the multiplication table of the elementary abelian group of order 3^β. For proofs that these algorithms both yield 2,3-subsquare complete squares, the reader is referred to F. P. Hiner and R. B. Killgrove [2]. Examples of latin squares of orders 7 and 10 constructed by means of these algorithms are given in Figs 1.5.12 and 1.5.13.

1	2	3	4	5	6	7	8
2	1	4	3	6	5	8	7
3	4	1	2	7	8	5	6
4	3	2	1	8	7	6	5
5	6	7	8	1	2	3	4
6	5	8	7	2	1	4	3
7	8	5	6	3	4	1	2
8	7	6	5	4	3	2	1

8	2	3	4	5	6	7
2	7	4	3	6	5	8
3	4	6	2	7	8	5
4	3	2	5	8	7	6
5	6	7	8	4	2	3
6	5	8	7	2	3	4
7	8	5	6	3	4	2

Fig. 1.5.12

1	2	3	4	5	6	7	8	9
2	3	1	5	6	4	8	9	7
3	1	2	6	4	5	9	7	8
4	5	6	7	8	9	1	2	3
5	6	4	8	9	7	2	3	1
6	4	5	9	7	8	3	1	2
7	8	9	1	2	3	4	5	6
8	9	7	2	3	1	5	6	4
9	7	8	3	1	2	6	4	5

0	2	3	4	5	6	7	8	9	1
2	0	1	5	6	4	8	9	7	3
3	1	0	6	4	5	9	7	8	2
4	5	6	0	8	9	1	2	3	7
5	6	4	8	0	7	2	3	1	9
6	4	5	9	7	0	3	1	2	8
7	8	9	1	2	3	0	5	6	4
8	9	7	2	3	1	5	0	4	6
9	7	8	3	1	2	6	4	0	5
1	3	2	7	9	8	4	6	5	0

Fig. 1.5.13

The algorithms can also be applied in the cases $\alpha = 2$ and $\beta = 1$ but in these cases the 2,3-subsquare complete latin squares which they yield are also 3-subsquare complete or 2-subsquare complete respectively.

In the same joint paper, Hiner and Killgrove have shown that the 2,3-subsquare complete latin squares constructed by Hiner's algorithm A (for $\alpha > 2$) are characterized by the fact that these and only these among the universe of 2,3-subsquare complete squares have the property that, in each two rows, all but three pairs of cells which contain equal elements belong to proper subsquares of order 2. In Fig. 1.5.12 for example, taking the second and third rows, we find that the only pairs of cells which contain equal elements but do not belong to proper subsquares of order 2 are the pairs (2,2) and (3,5), (2,3) and (3,2), (2,5) and (3,3).

The particular interest in 2,3-subsquare complete latin squares first arose as a consequence of their significance in connection with a hypothesis concerning non-Desarguesian and singly-generated projective planes. For the details, see R. B. Killgrove [2]. The results given in Hiner and Killgrove's joint paper referred to above show that, up to isomorphism, there is a unique 2,3-subsquare complete latin square for each of the orders 4, 5, 7, 8, 9, and 10 (those of orders 4 and 8 being also 2-subsquare complete and that of order 9 3-subsquare complete) but that there are (for example) at least two 2,3-subsquare complete latin squares of order 12 and at least three of order 27 (each of the latter being also 3-subsquare complete).

Another topic which has connections with (n, d)-complete latin rectangles is that of factorization of a group or quasigroup.

Definition. A quasigroup Q of order n is said to be *m-factorizable* if its multiplication table contains a non-trivial (n, m)-complete rectangle. The sets A and B formed by the row and column borders respectively of this (n, m)-complete rectangle are called *factors* of Q. We write $Q = AB$. Each element q of Q has m distinct representations as a product $q = a_i b_j$ of an element of A and an element of B. (See Fig. 1.5.14.)

If $m = 1$, each element of Q has a unique representation as a product ab. If, further, each of the subsets A, B is a subquasigroup of Q, we call these subquasigroups *1-factors* with respect to Q. If they are also disjoint, the multiplication table of Q takes the form shown in Fig. 1.5.15.

If $m = 1$, Q is a group G, and each of A and B is a subgroup of G then, necessarily, $A \cap B = \{e\}$ where e denotes the identity element of G. The group is said to *factorize* and to have A and B as its *subgroup factors*. If, moreover, one of A or B is normal in G we have $G = \{A, B\} = AB$ and, in that case, the two subgroups are called *complementary subgroups* in G.

The latter concepts are of long-standing. The extension to the case of quasigroups was made by S. K. Stein in [4]. We shall state two of his results without proof.

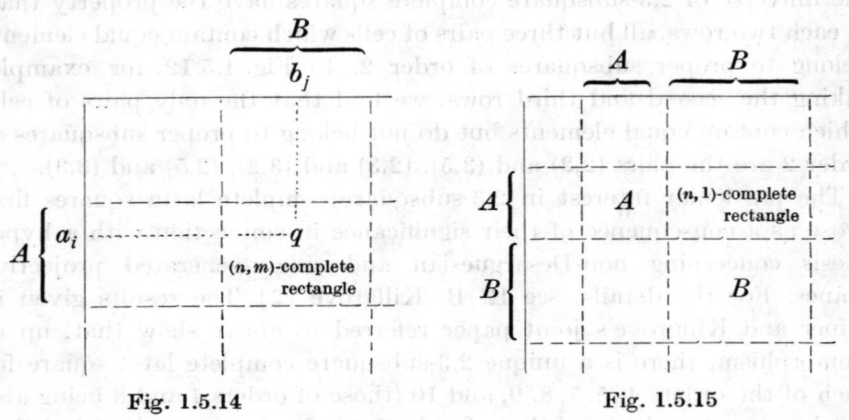

Fig. 1.5.14 Fig. 1.5.15

Let Q be a finite quasigroup and let A and B be subsets of Q such that each element of Q is uniquely expressible in the form ab, with $a \in A$ and $b \in B$, as before.

For brevity, we write $Q = (A, B)$, and we denote the number of elements in B by $|B|$.

(1) If T is a subset of $Q = (A, B)$ which has more than $(k - 1)|B|$ elements, where k is a positive integer, then there is a set of the form Tg, $g \in Q$ such that $|Tg \cap A| \geq k$.

(2) Let G be a group and A and B subsets of G. Then $G = (A, B)$ if and only if, for each choice of the element g in G, $A \cap gB^{-1}$ has exactly one element.

It seems fairly certain that the interpretation of factorization in terms of latin squares given above will allow some simplification of the proof of these two results.

In 1942, G. Hajós proved the following important result concerning factorizations of abelian groups in [1]:

Every finite abelian group G is 1-factorizable into a finite number ($\leq$ ord G) of sets $A, B_1, B_2, \ldots, B_{k-1}$ such that A is a subgroup of G and each of $B_1, B_2, \ldots, B_{k-1}$ consists of powers of an element of G.

By means of this result, he was able to solve a well-known geometrical problem concerning the partitioning of n-dimensional Euclidean space which had earlier been proposed by Minkowski. It will be evident to the reader that this result of Hajós can readily be interpreted in terms of latin squares.

We end this section by mentioning a number of miscellaneous problems and results.

First we prove a result which we shall need in later chapters.

THEOREM 1.5.6. *A latin square of odd order which satisfies the quadrangle criterion cannot contain any intercalates.*

PROOF. We suppose the contrary. Since the latin square satisfies the quadrangle criterion, it represents the multiplication of a group when bordered by its own first row and column (as proved in theorem 1.2.1). Hence, see Fig. 1.5.16, we get relations of the following form: $g_3g_1 = h = g_4g_2$ and $g_3g_2 = k = g_4g_1$.

From these relations it follows that $(g_3g_1)^{-1}(g_4g_2) = (g_4g_1)^{-1}(g_3g_2)$ and so $g_3^{-1}g_4 = g_4^{-1}g_3$ implying that $(g_3^{-1}g_4)^2 = e$. However, this is impossible since a group of odd order cannot contain an element of order two.

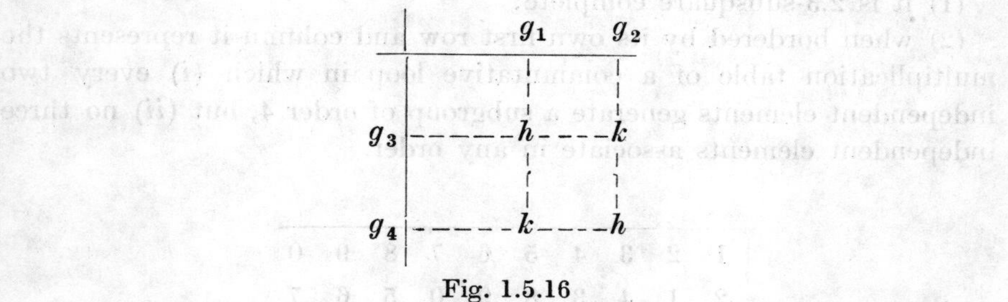

Fig. 1.5.16

Somewhat related to the above result is the question raised by L. Fuchs as to whether, given an arbitrary positive integer n, there always exists a latin square of order n which contains latin subsquares of every order m such that $m \leq n/2$. The question has not so far been answered. However, a partial answer for the case of latin squares which satisfy the quadrangle criterion has been given by C. Hobby, H. Rumsey and P. M. Weichsel in [1] and is as follows:

A finite group G of order n contains elements of every order m which divides n if and only if one of the following conditions is satisfied;

(*i*) G is cyclic;

(*ii*) G is a p-group and contains a cyclic subgroup of index p;

(*iii*) G has order $n = p^\alpha q$ for distinct primes p and q and it contains precisely one Sylow q-subgroup which is the commutator subgroup G' of G. Also if S is any Sylow p-subgroup of G, then S is cyclic, say $S = \{g\}$, and g^p is in the centre of G.

L. Fuchs has also proposed the following two problems (see J. Dénes and K. Pásztor [1]):

(1) If n is any positive integer and $n = n_1 + n_2 + \ldots + n_k$ any fixed partition of n, is it possible to find a quasigroup Q_n of order n which contains subquasigroups $Q_{n_1}, Q_{n_2}, \ldots, Q_{n_k}$ of orders $n_1, n_2, \ldots, n_k$ respectively whose set theoretical union is Q_n?

(2) The same question as in (1) but this time Q_n is required to satisfy the condition for every partition of n.

In J. Dénes and K. Pásztor [1], the authors showed that neither problem is soluble by examining the special case $n_1 = [n/2]$, $n_2 = [n/2] - 1$.

Let us complete our discussion of latin subsquares and subquasigroups by drawing the reader's attention to a particularly interesting latin square (Fig. 1.5.17) discovered by D. A. Norton (see [2]). It has the following properties:

(1) it is 2,3-subsquare complete;

(2) when bordered by its own first row and column it represents the multiplication table of a commutative loop in which (*i*) every two independent elements generate a subgroup of order 4, but (*ii*) no three independent elements associate in any order.

1	2	3	4	5	6	7	8	9	0
2	1	4	3	8	9	0	5	6	7
3	4	1	2	9	0	8	7	5	6
4	3	2	1	0	8	9	6	7	5
5	8	9	0	1	7	6	2	3	4
6	9	0	8	7	1	5	4	2	3
7	0	8	9	6	5	1	3	4	2
8	5	7	6	2	4	3	1	0	9
9	6	5	7	3	2	4	0	1	8
0	7	6	5	4	3	2	9	8	1

Fig. 1.5.17

CHAPTER 2

Special types of latin square

In view of the intimate connection between latin squares and quasigroups which the previous chapter has already made clear, a discussion of special types of latin square automatically entails a discussion of special types of quasigroup.

We begin by observing how the type of identity satisfied by a quasigroup is reflected in the structure of the corresponding latin square. We go on to discuss the concept of parastrophy and, in particular, in the second section of the chapter we consider in some detail the parastrophy invariant properties of mediality, idempotency, two-sided self-distributivity, and total symmetry. We prove that a medial quasigroup is always isotopic to an abelian group and that a finite idempotent totally symmetric quasigroup is equivalent to a Steiner triple system. Finally, we introduce the concept of generalized identities.

As an addendum to this section, we give a brief history of Steiner triple systems, pointing out that in fact they predate Steiner himself.

The last section of the chapter is devoted to the subject of complete latin squares. Squares of this type first became of interest to statisticians in the late 1940's and a number of papers on the subject were subsequently published in journals of Chemistry and Psychology as well as in more Mathematical journals. Consequently, as we point out, several of the constructions have been re-discovered two or three times. We give a detailed account of the work of B. Gordon and of later developments.

2.1. Quasigroup identities and latin squares

Two papers are known to the authors which are devoted to a detailed account of quasigroup identities: namely V. D. Belousov [4] and A. Sade [8]. The aim of the first author was to obtain a description, as complete as possible, of systems of quasigroups which satisfy certain fundamental identities: that is, to determine the structure of quasigroups satisfying a given system of identities by reducing the system to a simpler form. Sade's intention, on the other hand, was to give as complete a list as possible of those identities which had been investigated up to the time of publication of his paper. In compiling our own list (given below), we have made extensive use of that of Sade. However, we have renamed some of the identities in accordance with current usage and have added a few additional ones.

Two identities are said to be *dual* if one is obtained from the other by a reversal of the order both of the symbols which occur in it and also of the bracketing. Thus, for example, the identities (36) and (37) in our list below are dual to each other. Certain identities, such as (8) and (43) below, are *self-dual*. For completeness, we have usually included both members of each pair of dual identities.

(A) *Identities which involve one element only on each side of the equality sign:*

(1) for all $a \in Q$, $aa = a$ — the idempotent law

(2) for all $a, b \in Q$, $aa = bb$ — the unipotent law

(B) *Identities which involve two elements on one or both sides of the equality sign* (each identity is assumed to be valid for all pairs of elements $a, b \in Q$):

(3)	$ab = ba$	the commutative law
(4)	$(ab)b = a$	Sade's right "keys" law
(5)	$b(ba) = a$	Sade's left "keys" law
(6)	$(ab)b = a(bb)$	the right alternative law[1]
(7)	$b(ba) = (bb)a$	the left alternative law[1]
(8)	$a(ba) = (ab)a$	the medial alternative law (or law of elasticity)
(9)	$a(ba) = b$	the law of right semisymmetry[1]
(10)	$(ab)a = b$	the law of left semisymmetry[1]
(11)	$a(ab) = ba$	Stein's first law (or the Stein identity)[2]
(12)	$a(ba) = (ba)a$	Stein's second law[2]
(13)	$a(ab) = (ab)b$	Schröder's first law
(14)	$(ab)(ba) = a$	Schröder's second law
(15)	$(ab)(ba) = b$	Stein's third law

[1] By some authors, the adjectives "left" and "right" have been applied to these identities in reverse order.

[2] The duals of these laws are not listed.

The following two identities, although not quasigroup identities (since they contradict the axioms for a quasigroup) are listed here because they will play a role in the next chapter.

(16) $ab = a$	Sade's right translation law
(17) $ab = b$	Sade's left translation law

(C) *Identities which involve three distinct elements on one or both sides of the equality sign* (each identity is assumed to be valid for all triads of elements $a, b, c \in Q$):

(18) $(ab)c = a(bc)$	the associative law
(19) $a(bc) = c(ab)$	the law of cyclic associativity[1]
(20) $(ab)c = (ac)b$	the law of right permutability
(21) $a(bc) = b(ac)$	the law of left permutability
(22) $a(bc) = c(ba)$	Abel–Grassmann's law
(23) $(ab)c = a(cb)$	the commuting product[2]
(24) $c(ba) = (bc)a$	dual of (23)
(25) $(ab)(bc) = ac$	Stein's fourth law
(26) $(ba)(ca) = bc$	the law of right transitivity
(27) $(ab)(ac) = bc$	the law of left transitivity (called "Stein's fifth law" by Sade)
(28) $(ab)(ac) = cb$	Schweitzer's law
(29) $(ba)(ca) = cb$	dual of (28)
(30) $(ab)c = (ac)(bc)$	the law of right self-distributivity
(31) $c(ba) = (cb)(ca)$	the law of left self-distributivity
(32) $(ab)c = (ca)(bc)$	the law of right abelian distributivity
(33) $c(ba) = (cb)(ac)$	the law of left abelian distributivity
(34) $(ab)(ca) = [a(bc)]a$	the Bruck–Moufang identity
(35) $(ab)(ca) = a[(bc)a]$	dual of (34)
(36) $[(ab)c]b = a[b(cb)]$ (37) $[(bc)b]a = b[c(ba)]$	the Moufang identities
(38) $[(ab)c]b = a[(bc)b]$	the Bol identity
(39) $[b(cb)]a = b[c(ba)]$	dual of (38)
(40) $[(ab)c]a = a[b(ca)]$	the extra loop law
(41) $a[b(ca)] = cb$	Tarski's law[1]
(42) $a[(bc)(ba)] = c$	Neumann's law[1]
(43) $(ab)(ca) = (ac)(ba)$	the specialized medial law

(1) The duals of these laws are not listed.
(2) Called the "Eingewandtes Produkt" in A. Sade [8]

(D) *Identities involving four elements* $a, b, c, d \in Q$:

(44) $(ab)(cd) = (ad)(cb)$	the first rectangle rule[1]
(45) $(ab)(ac) = (db)(dc)$	the second rectangle rule[1]
(46) $(ab)(cd) = (ac)(bd)$	the medial law[2]

By means of a succession of remarks, we shall point out the structural implications of a number of the above identities in the study of the latin squares which represent multiplication tables of the appropriate types of quasigroups.

REMARKS. Any latin square which is the multiplication table of a quasigroup satisfying the identity (1) necessarily has a transversal since the cells which contain the products $aa, bb, cc, \ldots$, form such a transversal.

If the row and column borders of the Cayley table of a quasigroup satisfying the identity (2) are both ordered in the same way, then all the elements of the leading diagonal of the resulting latin square are the same. Since, by a reordering of its rows, any latin square can be transformed to one in which all the elements of the leading diagonal are the same, it follows that, given any arbitrary quasigroup, there exists a quasigroup isotopic to it which is unipotent.

The Cayley multiplication table of any quasigroup for which the identity (3) holds is symmetric.

If $xy = z$ in the multiplication table of a quasigroup Q which satisfies the identity (4) then the configuration shown in Fig. 2.1.1 exists in that table.

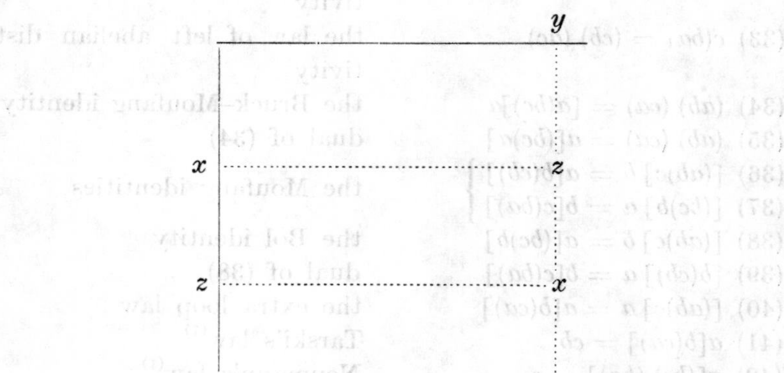

Fig. 2.1.1

[1] The duals of these laws are not listed.

[2] Many other names for this identity have been used, see page 68.

If the identities (4) and (5) both hold in a quasigroup Q, then Fig. 2.1.1 can be completed to the configuration shown in Fig. 2.1.2(a), where $xy = z$. But then, because $xz = y$, we get Fig. 2.1.2(b) by interchanging the symbols y and z; and combining this with 2.1.2(a), we have that $yz = x$ as well as $zy = x$. Consequently, we can deduce immediately a result due to A. Sade (see [7]) that the identities (4) and (5) together imply (3). An alternative algebraic proof is given later in this section.

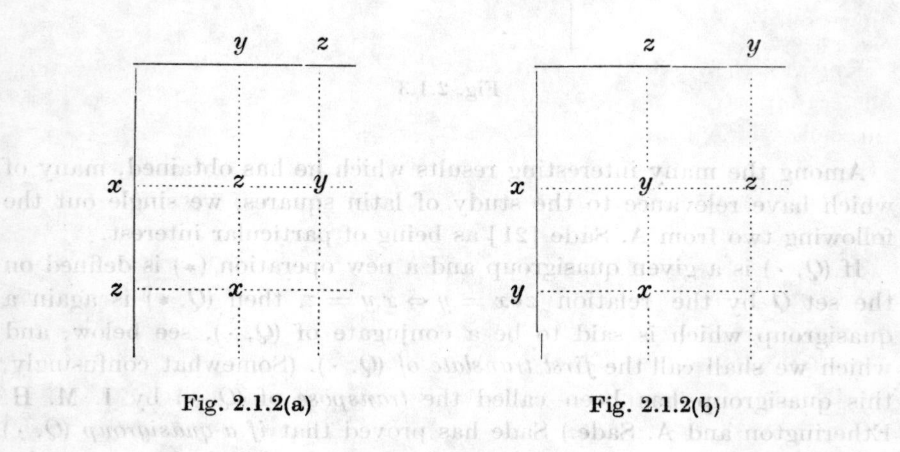

Fig. 2.1.2(a) Fig. 2.1.2(b)

Among quasigroups which satisfy the identities (6), (7), and (8) are the important class of loops known as *Moufang loops* which we mention again below.

A groupoid or quasigroup which satisfies the identity (9) has been called *demi-symétrique* by A. Sade. We have translated this as "semi-symmetric" although Sade himself considers "halfsymmetric" to be a better translation. In fact, every semisymmetric groupoid is necessarily a semisymmetric quasigroup, as has been shown in I.M.H. Etherington [1] and A. Sade [22].

In Fig. 2.1.3, c denotes the element ba whence, by virtue of the identity (9), $ac = b$. It then follows that $(ac)a = c$ so the validity of the identity (9) implies also the validity of the identity (10) and justifies the name semisymmetric quasigroup. There is a connection between semisymmetric quasigroups and balanced incomplete block designs (to be defined in section 10.2), as has been pointed out in A. Sade [22].

Sade has made a very extensive study of the properties of and constructions for such quasigroups in his papers [21], [22], [23], [24], [25], [26] and [28].

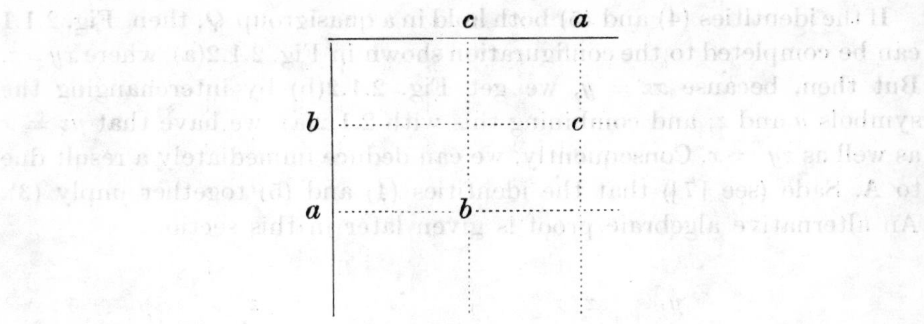

Fig. 2.1.3

Among the many interesting results which he has obtained, many of which have relevance to the study of latin squares, we single out the following two from A. Sade [21] as being of particular interest.

If $(Q, \cdot)$ is a given quasigroup and a new operation $(*)$ is defined on the set Q by the relation $z*x = y \Leftrightarrow x.y = z$, then $(Q, *)$ is again a quasigroup which is said to be a conjugate of $(Q, \cdot)$, see below, and which we shall call the *first translate of* $(Q, \cdot)$. (Somewhat confusingly, this quasigroup has been called the *transpose* of $(Q, \cdot)$ by I. M. H. Etherington and A. Sade.) Sade has proved that *if a quasigroup* $(Q, \cdot)$ *can be mapped isomorphically onto its first translate by a permutation mapping* α *of its elements* (so that $ab = c$ implies $(c\alpha)(a\alpha) = b\alpha$) *and if* α *has order* $3k$ *with* k *prime to* 3, *then* Q *is isotopic to a semisymmetric quasigroup*. He has also obtained an example of a quasigroup of lowest possible order having the properties of being isotopic to its first translate but not isotopic to any semisymmetric quasigroup. The multiplication table of this quasigroup is exhibited in Fig. 2.1.4. It is mapped onto its first translate by the autotopism ((0), (0), (1 2 3)(4 5 6)(7 8 9)). Its autotopism group reduces to a cyclic automorphism group which is generated by the permutation (1 2 3)(4 5 6)(7 8 9).

If we define the *second translate* of $(Q, \cdot)$ as being the first translate of $(Q, *)$: that is, the quasigroup $(Q, \otimes)$ such that $y \otimes z = x \Leftrightarrow x.y = z$ then the first translate of this quasigroup $(Q, \otimes)$ is again $(Q, \cdot)$. We can if we wish, give an alternative definition of a semisymmetric quasigroup by means of these concepts as being one which coincides with both its translates (cf. A. Sade [22]).

The significance of the identity (11) is that a quasigroup which satisfies it has an orthogonal complement. For the proof and for the definition of this concept see section 5.3.

	0	1	2	3	4	5	6	7	8	9
0	0	3	1	2	5	6	4	8	9	7
1	1	0	4	6	3	2	7	9	8	5
2	2	4	0	5	8	1	3	6	7	9
3	3	6	5	0	1	9	2	7	4	8
4	5	1	3	7	9	8	0	4	6	2
5	6	8	2	1	0	7	9	3	5	4
6	4	2	9	3	7	0	8	5	1	6
7	8	5	7	9	2	4	6	1	3	0
8	9	7	6	8	4	3	5	0	2	1
9	7	9	8	4	6	5	1	2	0	3

Fig. 2.1.4

If a quasigroup Q which satisfies the identity (11) has an identity element we find, on putting b equal to this identity element, that $a^2 = a$ for every element a of Q. Thus, Q cannot be a group. We may similarly show that a quasigroup which satisfies any one of the identities (12), (13), (14) or (15) cannot be a group.

An element x of a quasigroup $(Q, \cdot)$ such that, for all a and $c \in Q$, a, x and c satisfy both the identity (19) and its dual (that is, $a(xc) = c(ax)$ and $(cx)a = (xa)c$) is called a *centre-associative element*. Quasigroups which contain such elements have been studied by U. C. Guha and T. K. Hoo in [1]. These authors have proved, in particular, that a quasigroup which contains CA-elements must be a loop, but not a group unless all its elements are CA-elements. In the latter case, it is an abelian group. In any event, the number of CA-elements always divides the number of elements in Q.

The significance of many of the identities given in Sade's (and the present authors') list is best seen in the context of parastrophy or conjugacy of quasigroups. However, before introducing this concept, we shall draw attention to the special implications of the identities (34) to (40).

If a quasigroup has a two-sided identity element (that is, is a loop) and also satisfies any one of the identities (34), (35), (36), or (37), then it satisfies all four of the identities and is called a *Moufang loop*. A Moufang loop of odd order satisfies both Lagrange's theorem and Cauchy's theorem (see G. Glaubermann [2]). Consequently, in particular, the orders of all its elements and subloops divide the order of the loop and this has obvious implications in regard to the structure of the latin square which represents the Cayley table of such a loop. A loop which satisfies the identity (38) is called a *Bol loop* and such a loop satisfies the weak form of Lagrange's theorem (see D. A. Robinson [1] and also section 1.4) so again the orders of its elements divide the order of the loop. The same is true for a loop which satisfies the identity (39). There exist Bol loops which are not Moufang, as we remarked in section 1.4. However, every Moufang loop, whether of odd order or not, is also a Bol loop and so satisfies the weak form of Lagrange's theorem. To see this, notice that if a is put equal to the identity element in the identity (37), we get the identity (8) and this, together with the identity (36) implies the Bol identity (38).

A loop which satisfies the identity (40) is called an *extra loop*. The concept of an extra loop was introduced by F. Fenyves in [1]. This author has shown that every extra loop is a Moufang loop, while every commutative extra loop is an abelian group. Since there exist commutative Moufang loops which are not groups, these two results when taken together show that the class of extra loops lies properly between the class of Moufang loops and the class of groups. Fenyves has also shown that isotopic extra loops are isomorphic. (Compare corollary 2 of theorem 1.3.4.)

The reader will have noted that each of the identities (34) to (40) has the following form: both sides of the identity contain the same three symbols taken in the same order but one of them occurs twice on each side. Such an identity is said to be of *Bol-Moufang type*. In F. Fenyves [2], all possible identities of Bol-Moufang type have been listed and their inter-connections studied. The following definitions have been made.

Definitions. A loop is called a *C-loop* (central loop) if its elements all satisfy the identity $(yx.x)z = y(x.xz)$. It is called an *LC-loop* if its elements satisfy any one of the three equivalent identities $xx.yz = (x.xy)z$, $(x.xy)z = x(x.yz)$, $(xx.y)z = x(x.yz)$. It is called an *RC-loop* if its elements satisfy any one of the duals of these three equivalent identities.

Fenyves has shown that *LC-loops and RC-loops are both power associative* and that *a loop is a C-loop if and only if it is both an LC-loop and an RC-loop*.

He has also shown the validity of the implications shown in Fig. 2.1.5, each of which is irreversible. For example, we display in Fig. 2.1.6 the multiplication table of an *LC*-loop of six elements which is neither a *C*-loop nor a left Bol loop. In Fig. 2.1.7, we give the multiplication table of a C-loop of ten elements which is not a Moufang loop. Both these examples are taken from F. Fenyves [2].

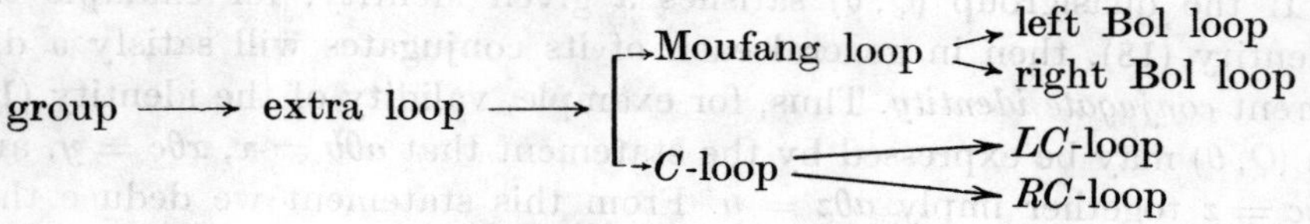

group ⟶ extra loop ⟶ Moufang loop ⟶ left Bol loop, right Bol loop

extra loop ⟶ C-loop ⟶ LC-loop, RC-loop

Fig. 2.1.5

	0	1	2	3	4	5
0	0	1	2	3	4	5
1	1	0	5	4	3	2
2	2	4	0	5	1	3
3	3	2	1	0	5	4
4	4	5	3	2	0	1
5	5	3	4	1	2	0

Fig. 2.1.6

	0	1	2	3	4	5	6	7	8	9
0	0	1	2	3	4	5	6	7	8	9
1	1	0	4	9	2	7	8	5	6	3
2	2	4	0	6	1	9	3	8	7	5
3	3	9	6	0	7	8	2	4	5	1
4	4	2	1	7	0	6	5	3	9	8
5	5	7	9	8	6	0	4	1	3	2
6	6	8	3	2	5	4	0	9	1	7
7	7	5	8	4	3	1	9	0	2	6
8	8	6	7	5	9	3	1	2	0	4
9	9	3	5	1	8	2	7	6	4	0

Fig. 2.1.7

With any given quasigroup (Q, θ), there are associated, generally speaking, five conjugate or parastrophic quasigroups defined on the same set G by the five operations θ^*, ${}^{-1}\theta$, θ^{-1}, ${}^{-1}(\theta^{-1}) \equiv ({}^{-1}\theta)^*$, $({}^{-1}\theta)^{-1} \equiv$ $\equiv (\theta^{-1})^*$ where, if $a\theta b = c$ for $a, b, c, \in Q$, we write $b(\theta^*)a = c$, $c({}^{-1}\theta)b =$ $= a$, $a(\theta^{-1})c = b$, $b\{{}^{-1}(\theta^{-1})\}\, c = a$, $c\{({}^{-1}\theta)^{-1}\}\, a = b$. It is immediately clear from the definitions that the operations $({}^{-1}\theta)^*$ and $(\theta^{-1})^*$ are respectively the same as the operations ${}^{-1}(\theta^{-1})$ and $({}^{-1}\theta)^{-1}$, as we have indicated. In S. K. Stein [1] and [2] and V. D. Belousov [4], these five operations associated with a given operation θ have been called *conjugates* of θ, while in A. Sade [12] they have been called *parastrophes* of θ. The significance of these operations from the point of view of latin squares is that, if the multiplication table of the quasigroup (Q, θ) is given by the bordered latin square L, then (Q, θ^*) has multiplication table given by the transpose of L, while in $(Q, {}^{-1}\theta)$, (Q, θ^{-1}) the roles of row and element number, column and element number, respectively, are interchanged. In the remaining two parastrophic quasigroups, the roles of row, column, and element number are all three permuted.

If the quasigroup (Q, θ) satisfies a given identity, for example the identity (18), then in general each of its conjugates will satisfy a different *conjugate identity*. Thus, for example, validity of the identity (18) in (Q, θ) may be expressed by the statement that $a\theta b = x$, $x\theta c = y$, and $b\theta c = z$ together imply $a\theta z = y$. From this statement we deduce that the relations $a\theta^{-1} x = b$, $x\theta^{-1} y = c$, and $b\theta^{-1} z = c$ together imply $a\theta^{-1} y = z$. Hence, substituting for b, z, and c in the equation $b\theta^{-1} z = c$, we get $(a\theta^{-1} x)\, \theta^{-1} (a\theta^{-1} y) = x\theta^{-1} y$. In other words, the parastrophe (Q, θ^{-1}) satisfies the identity (27) when (Q, θ) satisfies the identity (18).

Since a quasigroup which satisfies the identity (18) is a group, it is in some sense true that the theory of groups is equivalent to the theory of quasigroups which satisfy the identity (27), as has been remarked by S. K. Stein (see [2]).

The concept of conjugate or parastrophic identities seems to go back well into the last century and many other names have been used by early writers. For further details of these, see page 76 of A. Sade [12].

In the paper just cited, Sade has given some general rules for determining the identities satisfied by the parastrophes of a quasigroup (Q, θ) when (Q, θ) satisfies a given identity involving several of the operations θ, θ^*, ${}^{-1}\theta$, θ^{-1}, $({}^{-1}\theta)^{-1}$ and ${}^{-1}(\theta^{-1})$. In particular, he has proved among others the following two results:

(*i*) If a quasigroup (Q, θ) satisfies an identity I, then the quasigroup (Q, θ^*) satisfies the identity obtained from I by interchanging the pairs of operations (θ, θ^*), $({}^{-1}\theta, ({}^{-1}\theta)^{-1})$ and $(\theta^{-1}, {}^{-1}(\theta^{-1}))$ without altering the letters;

(*ii*) If a quasigroup (Q, θ) satisfies an identity I, then the quasigroup $(Q, {}^{-1}\theta)$ satisfies the identity deduced from I by interchanging the pairs of operations $(\theta, {}^{-1}\theta)$, $(\theta^*, {}^{-1}(\theta^{-1}))$ and $(\theta^{-1}, ({}^{-1}\theta)^{-1})$ without touching the letters.

He has also given a number of rules of calculation for simplifying an identity which involves more than one of the operations θ, θ^*, ${}^{-1}\theta$, θ^{-1}, $({}^{-1}\theta)^{-1}$ and ${}^{-1}(\theta^{-1})$.

In [2], S. K. Stein has listed the conjugate identities for a number of well-known identities. A somewhat similar, but slightly more extensive list is given in V. D. Belousov [4] and we reproduce it here as Fig. 2.1.8. (Note that the symbol A has been used for the quasigroup operation in place of θ.) The occurrence of the symbol A in the body of the table

	A	A^*	^{-1}A	A^{-1}	$^{-1}(A^{-1}) = (^{-1}A)^*$	$(^{-1}A)^{-1} = (A^{-1})^*$
Associativity	$xy.z = x.yz$	$xy.z = x.yz$	$yx.zx = yz$	$xy.xz = yz$	$xy.xz = yz$	$yx.zx = yz$
Mediality	$xy.uv = xu.yv$	A	A	A	A	A
Idempotence	$xx = x$	A	A	A	A	A
Commutativity	$xy = yx$	$xy = yx$	$y.yx = x$	$yx.x = y$	$xy.y = x$	$x.xy = y$
Left self-distributivity	$x.yz = xy.xz$	$yz.x = yx.zx$		$x.yz = xy.xz$		$xy.x = zx.yx$
Stein identity	$x.xy = yx$	$yx.x = xy$	$x(y.yx) = yx$	$(x.xy)y = x$	$(xy.y)x = xy$	$y(yx.x) = x$
Two-sided self-distributivity	$x.yz = xy.xz$ and $yz.x = yx.zx$	A	A	A	A	A

Fig. 2.1.8

indicates that the identity in question is the same with respect to the parastrophic operation.

It will be noted that the identities (46), (1), and (30), (31) taken jointly, appear to have a special significance in that they are parastrophy-invariant. In this connection it is worthwhile to point out, following S. K. Stein [2], that if the identities (46) and (1) both hold so also do the identities (30) and (31); that is, an idempotent medial quasigroup is both left and right self-distributive. We have $cb.ca = cc.ba$, by (46), $= c.ba$, by (1). Thus, (31) holds. Similarly, we may show that (30) holds.

Medial(1) and idempotent quasigroups do indeed play a special role in the theory of latin squares. In the first place, every principal loop isotope(2) of a medial quasigroup is an abelian group, as was first shown by D. C. Murdoch in [2]. Consequently, every medial quasigroup is isotopic to some abelian group. That is to say, medial quasigroups arise quite naturally as a consequence of relabelling the elements and rearranging the rows and columns of the latin square defined by the Cayley table of any abelian group. Moreover, R. H. Bruck has given in [1] a general method by which any medial quasigroup may be constructed from the abelian group to which it is isotopic.

Idempotent quasigroups give rise to latin squares which always possess at least one transversal, as we pointed out earlier. The class of finite totally symmetric idempotent quasigroups (which is a parastrophy invariant class) is coextensive with the class of designs known as Steiner Triple Systems.

In the next section, we give proofs of the foregoing assertions.

We end this discussion of identities by drawing the reader's attention to a remarkable theorem due to V. D. Belousov.

We say that an identity $W_1 = W_2$ defined on a quasigroup Q is *balanced* if each variable x which occurs on one side W_1 of the identity occurs on the other side W_2 too and if no variable occurs in W_1 or W_2 more than once. This definition is due to A. Sade (see [14]).

The reader can easily check that in our list given at the beginning of the chapter the following identities are balanced: (3), (18), (19), (20), (21), (22), (23), (24), (44), (46).

(1) By other authors, these quasigroups have been called *abelian quasigroups* (D. C. Murdoch [1] and [2]), *alternation quasigroups* (M. Sholander [1]) and *entropic quasigroups* (I. M. H. Etherington [2]). The medial law has been called the *symmetric law* (O. Frink [1]) or the *bisymmetric law* (J. Aczél [1] and [2]).

(2) For the definition of this concept, see section 1.3.

An identity $W_1 = W_2$ is called *reducible* (see V. D. Belousov [5]) if either (*i*) each of W_1 and W_2 contains a "free element" x so that W_1 is of the form $U_1.x$ or $x.V_1$ and W_2 likewise is of the form $U_2.x$ or $x.V_2$ (where U_i or V_i represents a subword of the word W_i for $i = 1,2$); or (*ii*) W_1 has the product xy of two free elements x and y as a subword and W_2 has one of the products xy or yx as a subword, or the dual of this statement.

An identity which is not reducible is called *irreducible*.

For example, the identity $w(x.yz) = (xy.z)w$ is reducible because each of the two words composing it has w as a free element, the identity $xy.uv = (u.yx)v$ is reducible because the left-hand side has xy as a subword and the right-hand side has yx as a subword. Of the balanced identities listed above, only the identity (3) is reducible. The remaining identities are irreducible.

Belousov has proved the following very significant theorem:

Theorem 2.1.1. *A quasigroup Q which satisfies any irreducible balanced identity is isotopic to a group.*[1]

A proof is given in V. D. Belousov [5] but uses many ideas which are outside the scope of the present book.

2.2. Quasigroups of some special types and the concept of generalized associativity

Theorem 2.2.1. *Every LP-isotope of a medial quasigroup $(Q, \cdot)$ is an abelian group.*

Proof. We begin by noting that if a medial quasigroup $(Q, \cdot)$ possesses a two-sided identity element e, then it is an abelian group. For, on putting $a = d = e$ in the medial law $ab.cd = ac.bd$ we get $bc = cb$ and so $(Q, \cdot)$ is commutative. Then putting $c = e$, we get $ab.d = a.bd$ and so $(Q, \cdot)$ is associative. The result follows.

Next, let $(Q, \cdot)$ be a given medial quasigroup, and let a new operation $\otimes$ be defined on the elements of Q by $a \otimes b = a\sigma.b$ where $a\sigma^{-1} = av$. Then we may show that $(Q, \otimes)$ is a medial quasigroup with v as unique right identity element. It is convenient to denote by R_v the unique one-to-one mapping of the set Q onto itself which is defined by $aR_v = av$ for all a in Q. Then $a \otimes b = aR_v^{-1}.b$ and it is evident that either of the

[1] For some other results concerning the conditions under which a quasigroup is isotopic to a group, see T. Evans [1] and E. Falconer [1].

relations $a\otimes b = a\otimes c$ or $b\otimes a = c\otimes a$ implies $b = c$, so that $(Q, \otimes)$ is a quasigroup. Also

$$a\otimes v = aR_v^{-1}.v = aR_v^{-1}R_v = a$$

so v is the unique right identity element of $(Q, \otimes)$. To show that $(Q, \otimes)$ is medial, let s be the unique solution of the equation $sv = v$. By the medial law in $(Q, \cdot)$, $ab.v = ab.sv = as.bv$ for all a, b in Q. That is, $(ab)R_v = aR_s.bR_v$. Replacing a by aR_s^{-1} and b by bR_v^{-1}, we get $(aR_s^{-1}.bR_v^{-1})R_v = ab$ and so $(ab)R_v^{-1} = aR_s^{-1}.bR_v^{-1}$ for all a, b in Q. Using this relation, we have

$$(a\otimes b)\otimes(c\otimes d) = (aR_v^{-1}.b)\otimes(cR_v^{-1}.d) =$$

$$= (aR_v^{-1}.b)R_v^{-1}.(cR_v^{-1}.d) = (aR_v^{-1}R_s^{-1}.bR_v^{-1}).(cR_v^{-1}.d)$$

and similarly

$$(a\otimes c)\otimes(b\otimes d) = (aR_v^{-1}R_s^{-1}.cR_v^{-1}).(bR_v^{-1}.d).$$

The medial law of $(Q, \cdot)$ shows that these are equal, and so the medial law holds in $(Q, \otimes)$. Thus, $(Q, \otimes)$ is medial and has a unique right identity element, as stated.

By a similar argument, we may show that if $(Q, \cdot)$ is a given medial quasigroup and a new operation $\oplus$ is defined on the elements of Q by $a\oplus b = a.b\tau$, where $b\tau^{-1} = ub$ then $(Q, \oplus)$ is a medial quasigroup with u as unique left identity.

Now, suppose that $(Q, \cdot)$ is a given medial quasigroup. Any *LP*-isotope $(Q, *)$ of $(Q, \cdot)$ is obtainable from it by a relation of the form $a * b = a\sigma.b\tau$, where σ, τ are one-to-one mappings of Q onto itself such that $a\sigma^{-1} = av$ and $b\tau^{-1} = ub$ for suitable fixed elements v, u of Q (see theorem 1.3.3). Also, again by theorem 1.3.3, the loop $(Q, *)$ then has $e = uv$ as identity element. Now let $a\otimes b = a\sigma.b$ as above. Then $a * b = a\otimes b\tau = a\sigma.b\tau$. Since $(Q, \cdot)$ is a medial quasigroup, $(Q, \otimes)$ is medial, as already proved. Also because $(Q, \otimes)$ is medial, so is the quasigroup $(Q, *)$ defined by $a * b = a\otimes b\tau$ and it has uv as a two-sided identity element, as already stated. [By way of confirmation of this, let us note that $b\tau^{-1} = ub = (uv)\otimes b$ whence regarding $(Q, *)$ as derived from $(Q, \otimes)$ by the definition $a * b = a\otimes b\tau$, it follows from our earlier analysis that uv is the unique left identity of $(Q, *)$. Symmetry considerations show that it is also the unique right identity.] But a medial quasigroup with a two-sided identity is an abelian group, so every *LP*-isotope of a medial quasigroup $(Q, \cdot)$ is an abelian group, as required.

We next give Bruck's method for constructing a medial quasigroup from the abelian group to which it is isotopic. This is embodied in the following theorem:

THEOREM 2.2.2. *Every medial quasigroup which is isotopic to a given abelian group* $(G, \cdot)$ *is isomorphic to some quasigroup* $(G, *)$ *obtained by a relation of the form* $a * b = w.a\sigma.b\tau$, *where* σ, τ *are commuting automorphisms of* $(G, \cdot)$ *and* w *is a fixed element of* G.

PROOF. We observe first that if the relation $(*)$ is defined as in the statement of the theorem, then

$$(a * b) * (c * d) = (w.a\sigma.b\tau) * (w.c\sigma.d\tau) =$$

$$= w.(w.a\sigma.b\tau)\sigma.(w.c\sigma.d\tau)\tau = w.w\sigma.w\tau.a\sigma^2.d\tau^2.b\tau\sigma.c\sigma\tau$$

and similarly

$$(a * c) * (b * d) = w.w\sigma.w\tau.a\sigma^2.d\tau^2.c\tau\sigma.b\sigma\tau.$$

Since $\tau\sigma = \sigma\tau$, these two expressions are equal and so every relation of the form given defines a medial quasigroup.

Conversely, let $(G,*)$ be any principal isotope of $(G, \cdot)$ defined by $a * b = a\mu.b\nu$ where μ, ν are one-to-one mappings of G onto itself. Since every isotope of $(G, \cdot)$ is isomorphic to a principal isotope (theorem 1.3.2), it is sufficient to confine our attention to the latter. If $(G, *)$ is a medial quasigroup, then

$$(a\mu^{-1} * b) * (c * d\nu^{-1}) = (a\mu^{-1} * c) * (b * a\nu^{-1}).$$

This is equivalent to

$$(a.b\nu)\mu.(c\mu.d)\nu = (a.c\nu)\mu.(b\mu.d)\nu$$

for all a, b, c, d in G. Let e be the identity element of $(G, \cdot)$ and define the (fixed) elements u, v, w by $u = e\mu$, $v = e\nu$, $w = uv$. Also let λ be the one-to-one mapping of G onto itself which maps each element a of G onto its inverse $a^{-1} = a\lambda$ in $(G, \cdot)$. Putting $c = e$ in the relation just obtained, we have

$$(a.b\nu)\mu.(ud)\nu = (av)\mu.(b\mu.d)\nu.$$

On multiplying both sides of this by $(ud)\nu\lambda.(av)\mu\lambda$, we get

$$(a.b\nu)\mu.(av)\mu\lambda = (b\mu.d)\nu.(ud)\nu\lambda.$$

Since this is to be true for all a, b, d and, since the right-hand side is independent of a, so must the left-hand side be. Similarly, the right-hand side is independent of d. Therefore, the left-hand side is equal to its expression with $a = e$, the right-hand side to its expression with $d = e$, whence

$$(a.bv)\mu.(av)\mu\lambda = bv\mu.v\mu\lambda \text{ and } (b\mu.d)v.(ud)v\lambda = b\mu v.uv\lambda.$$

These expressions can be re-written as

$$(ab)\mu = v\mu\lambda.(av)\mu.b\mu \quad \text{and} \quad (ba)v = uv\lambda.(ua)v.bv,$$

where we have postmultiplied the expressions by $(av)\mu$ and $(ud)v$ respectively and then replaced bv by b in the first expression, $b\mu$ and d by b and a respectively in the second expression.

Since $(G, \cdot)$ is commutative, $(ab)\mu = (ba)\mu$ or $v\mu\lambda.(av)\mu.b\mu = v\mu\lambda.(bv)\mu.a\mu$, whence $(av)\mu.b\mu = (bv)\mu.a\mu$. Also, $(ab)v = (ba)v$ whence $(ua)v.bv = (ub)v.av$. We can re-write these expressions in the forms $(av)\mu.a\mu\lambda = (bv)\mu.b\mu\lambda$ and $(ua)v.av\lambda = (ub)v.bv\lambda$. Hence, observing that their left-hand sides are independent of a and can be equated to their expressions with $a = e$, we get $(av)\mu.a\mu\lambda = v\mu.u\lambda$ and $(ua)v.av\lambda = uv.v\lambda$. Therefore, $(av)\mu = a\mu.v\mu.u\lambda$ and $(ua)v = uv.av.v\lambda$. On substituting these formulae for $(av)\mu$ and $(ua)v$ into our earlier expressions for $(ab)\mu$ and $(ba)v$ we get

$$(ab)\mu = v\mu\lambda.a\mu.v\mu.u\lambda.b\mu = a\mu.b\mu.u\lambda$$

and similarly $\quad (ab)v = (ba)v = av.bv.v\lambda$. In other words,

$$(ab)\mu.u\lambda = a\mu.u\lambda.b\mu.u\lambda \text{ and } (ab)v.v\lambda = av.v\lambda.bv.v\lambda.$$

So if we define $a\sigma = a\mu.u\lambda$ and $a\tau = av.v\lambda$ we have that

$$(ab)\sigma = a\sigma.b\sigma \text{ and } (ab)\tau = a\tau.b\tau$$

showing that σ and τ are both automorphisms of $(G, \cdot)$. Thus, we have shown that there exist automorphisms σ and τ of $(G, \cdot)$ such that

$$a * b = a\mu.bv = a\sigma.u.b\tau.v = w.a\sigma.b\tau.$$

Because $(G, *)$ is required to satisfy the medial law, we must have

$$w.w\sigma.w\tau.a\sigma^2.d\tau^2.b\tau\sigma.c\sigma\tau = w.w\sigma.w\tau.a\sigma^2.d\tau^2.c\tau\sigma.b\sigma\tau$$

for all a, b, c, d in G as in the first part of the theorem, and so

$$b\tau\sigma.b\sigma\tau\lambda = c\tau\sigma.c\sigma\tau\lambda.$$

Thus, the left-hand side is independent of b and is equal to its expression with $b = e$. That is,

$$b\tau\sigma.b\sigma\tau\lambda = (v.v\lambda)\sigma.(u.u\lambda)\tau\lambda = e\sigma.e\tau\lambda = e.e\lambda = e,$$

or $b\tau\sigma = b\sigma\tau$, so $\tau\sigma = \sigma\tau$ is necessary. This completes the proof of the theorem.

DEFINITION. A quasigroup $(Q, \cdot)$ is said to be *totally symmetric* if it is commutative and also semi-symmetric.

This implies that all its parastrophes coincide and so each of the equalities $bc = a$, $ca = b$, $ab = c$, $cb = a$, $ac = b$, $ba = c$ implies the other five.

The multiplication table of such a quasigroup defines a latin square which is unaffected when the roles of row, column, and element number are permuted in any way. An example of such a multiplication table is given in Fig. 2.2.1.

	1	2	3	4	5	6	7
1	1	3	2	5	4	7	6
2	3	2	1	6	7	4	5
3	2	1	3	7	6	5	4
4	5	6	7	4	1	2	3
5	4	7	6	1	5	3	2
6	7	4	5	2	3	6	1
7	6	5	4	3	2	1	7

Fig. 2.2.1

Medial quasigroups which are also totally symmetric play a role in the geometry of plane cubic curves. For the details of this application, see I. M. H. Etherington [2]. It is shown in the same paper that the construction of a medial quasigroup from the abelian group $(G, +)$ to which it is isotopic which was given in theorem 2.2.2 can be simplified considerably in the case that the quasigroup is totally symmetric as well as medial.

In any totally symmetric quasigroup, the identities (4) and (5) given at the beginning of this chapter are valid. Their validity follows from the facts that $ab = c$ implies $cb = a$ and $bc = a$ as well as $ba = c$, whence $(ab)b = a$ and $b(ba) = a$, as required. Moreover, it is easy to show that a quasigroup in which the two relations $ab.b = a$ and $b.ba = a$ hold for all a and b is commutative and hence totally symmetric (cf. page 61 above). We have

$$ab = \{(ba).(ba)a\}\, b = (ba.b)b = ba,$$

so if $ab = c$, we have $ba = c$ and hence $cb = a$, $bc = a$. Also, from $ba.a = b$ and $a.ab = b$, we get $ca = b$ and $ac = b$. In fact, a totally symmetric quasigroup may be defined as a groupoid obeying any two of the identities $ab = ba$, $ab.b = a$, and $b.ba = a$.

The finite idempotent totally symmetric quasigroups are co-extensive with the class of design known as *Steiner triple systems* (as already mentioned) and so we shall give a brief account of these designs before demonstrating the equivalence of the two concepts. A more detailed account of geometrical and statistical designs will be given in a later chapter in connection with the solution of Euler's latin square problem.

A *Steiner triple system* consists of a set of n distinct elements, each of which may occur more than once, arranged into subsets each containing exactly three distinct elements and with the property that, if a and b are any two distinct elements of the set, there is one and only one element c such that a, b, c are all different and form a triple of the system. Since the last property implies that each element occurs with each other element exactly once in a triple, every element occurs the same number r say of times, each occurrence being in a different triple. If t is the total number of distinct triples, we have $nr = 3t$ since each side of this equality represents the total number of elements occurring in all the triples. Moreover, since exactly $n(n - 1)/2$ pairs can be chosen from n elements and since each triple of the system contains three such pairs, we have $\frac{1}{2}n(n - 1) = 3t = nr$. From this relation we can deduce that $n - 1 = = 2r$, whence n is odd. Also, 3 divides r or n. If 3 divides r then $n - 1$ is a multiple of 6, say $n = 6m + 1$. If 3 divides n then, since n is odd, n is three times an odd number, say $n = 3(2m + 1) = 6m + 3$. Thus, if n is the number of distinct elements occurring in a Steiner triple system, we necessarily have $n \equiv 1$ or 3, modulo 6.

In [1], E. Netto has shown that, for every such value of n, Steiner triple systems actually exist and that for $n = 3$, 7, and 9 there is, up to isomorphism, just one such system, while, for all larger n, there are at least two non-isomorphic systems.

Two systems are *isomorphic*, if by a permutation of the n symbols, one can be transformed into the other.

THEOREM 2.2.3. *There exists a one-to-one correspondence between finite idempotent totally symmetric quasigroups*[1] *and Steiner triple systems.*

[1] Such quasigroups are sometimes called *Steiner quasigroups*. See, for example, C. C. Lindner [11].

PROOF. Let $Q = \{a, b, c, \ldots\}$ be the set of elements of a Steiner triple system. For a and b distinct, define $c = a.b$ to be the third element of the unique triple of the system which contains a and b. Also, define $a.a = a$. Then $(Q, \cdot)$ is an idempotent totally symmetric quasigroup. (The idempotent totally symmetric quasigroup whose multiplication table is given in Fig. 2.2.1 was obtained in this way.)

Conversely, let $(Q, \cdot)$ be a finite idempotent totally symmetric quasigroup. Since $a.b = c$ implies $b.c = a$, $c.a = b$, $b.a = c$, $c.b = a$, $a.c = b$, the operation $(\cdot)$ separates Q into triples, and these form a Steiner triple system because each pair a, b of distinct elements of Q are associated with a unique third element $c = a.b$.

Notice that the above correspondence implies that the number of elements in a finite idempotent totally symmetric quasigroup is necessarily an integer congruent to one or three, modulo six. The relation between these quasigroups and the Steiner triple systems was first pointed out by A. Sade in [3], page 4, and [8], page 159. It was demonstrated again by R. H. Bruck in [7].

Bruck has also pointed out that a similar relationship exists between totally symmetric loops and Steiner triple systems (see R. H. Bruck [5] page 58, and [7]).

THEOREM 2.2.4. *There exists a one-to-one correspondence between finite totally symmetric loops and Steiner triple systems.*

PROOF. If $(G, \cdot)$ is a totally symmetric loop with identity element e then, because of the totally symmetric property, $ae = a = ea$ for each element a in G implies $a^2 = e$: that is, each element of $(G, \cdot)$ has order two. If G contains $n + 1$ elements, the set $G \setminus \{e\}$ forms a Steiner triple system of n elements whose triples are given by the statement that the unique triple which contains the pair of elements a, b has $c = ab$ as its third element.

Conversely, let Q be a set of n elements forming a Steiner triple system. We can make Q into a totally symmetric loop by adjoining an additional element e and defining the loop operation $(\cdot)$ by the three rules (*i*) if $a \neq e$, $b \neq e$, then ab is the third element of the unique triple which contains a and b, (*ii*) $ae = a = ea$ for all a in Q, and (*iii*) $a^2 = e^2 = e$ for all a in Q.

The identities so far referred to have all been characterized by the fact that only one operation is involved in each. We say that such identities are of *rank 1*. More generally, by the *rank* of an identity $W_1 = W_2$, we understand the number of different binary operations that occur in the expressions W_1 and W_2. By the *length* of an expression W_i we understand the number of elements which occur in W_i. Thus, for example, the expression $a[b(cb)]$ has length 4. Evidently, the rank of the identity $W_1 = W_2$ cannot exceed the number $l = l_1 + l_2 - 2$, where l_1 and l_2 are the lengths of W_1 and W_2 respectively. For given values of l_1 and l_2 we call an identity of highest possible rank a *general identity*. Thus, in a general identity, the number of operations is fixed. In place of the term "general identity", A. Sade often uses the term *identité démosienne* (generalized identity) as for example, in A. Sade [8] and [16].

The following are examples of general identities:

(1) $(a \textcircled{1} b) \textcircled{2} c = a \textcircled{3} (b \textcircled{4} c)$	the general associative law
(2) $(a \textcircled{1} b) \textcircled{3} (c \textcircled{2} d) = (a \textcircled{4} c) \textcircled{6} (b \textcircled{5} d)$	the general medial law
(3) $(b \textcircled{1} a) \textcircled{3} (c \textcircled{2} a) = b \textcircled{4} c$	the generalized law of right transitivity
(4) $a \textcircled{1} (b \textcircled{2} c) = (a \textcircled{3} b) \textcircled{4} (a \textcircled{5} c)$	the general left distributive law
(5) $a \textcircled{1} (a \textcircled{2} b) = b$	the general (left) keys law
(6) $a \textcircled{1} b = b \textcircled{2} a$	the general commutative law

The idea of a general identity is due to R. Schauffler who introduced the concept in connection with problems of coding theory. (For the details, see R. Schauffler [1] and [2], and also section 10.1.)

Some very comprehensive survey papers on the subject of generalized identities have been written. We refer the reader particularly to V. D. Belousov [4] and A. Sade [16].

As an example of the kind of result that has been obtained, we mention that if the general medial law (identity (2) above) holds with respect to six quasigroups $(G, \textcircled{1})$, $(G, \textcircled{2})$, $(G, \textcircled{3})$, $(G, \textcircled{4})$, $(G, \textcircled{5})$, and $(G, \textcircled{6})$ all defined on the same set G, then all six of them are isotopic to one and the same abelian group. (A more general form of this result applicable to the case of multigroupoids is given in A. Sade [14].)

We shall not give the proof here. Instead, we should like to end the present section by giving two theorems which will be needed in section 10.1.

Theorem 2.2.5. *If four quasigroups $(Q, \textcircled{1})$, $(Q, \textcircled{2})$, $(Q, \textcircled{3})$, $(Q, \textcircled{4})$, defined on the same set Q, are connected by the general associative law, then they are all isotopic to one and the same group.*

Proof. Let k be a fixed element of $(Q, \mathbf{i})$.[1] Let us define mappings L_i and R_i of the set Q onto itself by the statements $xL_i = k\,\mathbf{i}\,x$ and $xR_i = x\,\mathbf{i}\,k$ where k is a fixed element of Q. (These mappings are called *translations*; xL_i is a *left translation* and xR_i a *right translation* of $(Q, \mathbf{i})$.) In the equality $(a \text{①} b) \text{②} c = a \text{③} (b \text{④} c)$ we put, in turn

$$a = c = k, \;\; a = k, \;\; b = k, \;\; c = k,$$

we get

$$(k \text{①} b) \text{②} k = k \text{③} (b \text{④} k)$$

$$(k \text{①} b) \text{②} c = k \text{③} (b \text{④} c)$$

$$(a \text{①} k) \text{②} c = a \text{③} (k \text{④} c)$$

$$(a \text{①} b) \text{②} k = a \text{③} (b \text{④} k)$$

The above equations are equivalent to

$$bL_1R_2 \;\; = bR_4L_3 \;\;\dots\dots\dots\;\; (1)$$

$$bL_1 \text{②} c = (b \text{④} c)L_3 \;\dots\dots\dots\; (2)$$

$$aR_1 \text{②} c = a \text{③} cL_4 \;\dots\dots\dots\; (3)$$

$$(a \text{①} b)R_2 = a \text{③} bR_4 \;\dots\dots\dots\; (4)$$

and the last three of these relations show that the quasigroups $(Q, \mathbf{i})$[1] are isotopic to each other; in particular, they are all isotopic to $(Q, \text{②})$.

We define a further isotope $(Q, \cdot\,)$ of $(Q, \text{②})$ by $a.b = aR_2^{-1} \text{②} bL_3^{-1}L_4^{-1}$. It then follows that all four of the quasigroups $(Q, \mathbf{i})$ are isotopic to $(Q, \cdot\,)$. Indeed, we have

$$a \text{①} b = (aR_1R_2.bR_4L_3)\,R_2^{-1} \;\dots\dots\dots\; (5)$$

$$a \text{②} b = aR_2.bL_4L_3 \;\dots\dots\dots\dots\dots\dots\; (6)$$

$$a \text{③} b = aR_1R_2.bL_3 \;\dots\dots\dots\dots\dots\dots\; (7)$$

$$a \text{④} b = (aL_1R_2.bL_4L_3)L_3^{-1} \;\dots\dots\dots\; (8)$$

Equation (6) follows from the definition of the operation $(\cdot)$. Then, from equation (3) we get $a \text{③} b = aR_1 \text{②} bL_4^{-1}$ and, using (6), this gives equation (7). From equation (2) we get $a \text{④} b = (aL_1 \text{②} b)L_3^{-1}$ and, again using (6), this gives equation (8). Finally, from equation (4) we get $a \text{①} b = (a \text{③} bR_4)R_2^{-1}$ and, using (7), this gives equation (5).

[1] The operation $\mathbf{i}$ here stands for one of the four quasigroup operations ①, ②, ③, ④.

Substituting the expressions for $a \,\mathbf{i}\, b$ so obtained in the general identity of associativity, we get

$$[(aR_1R_2.bR_4L_3)R_2^{-1}]\, R_2.cL_4L_3 = aR_1R_2.(bL_1\, R_2.cL_4\, L_3)\, L_3^{-1}\, L_3.$$

That is,

$$(aR_1R_2.bR_4L_3).cL_4L_3 = aR_1R_2.(bL_1R_2.cL_4L_3).$$

But, $bR_4L_3 = bL_1R_2$ by equation (1), so $(uv)w = u(vw)$, where $u = aR_1R_2$, $v = bR_4L_3 = bL_1R_2$ and $w = cL_4L_3$.

Since u, v, w may be arbitrary elements in Q, we conclude that $(Q, \cdot)$ is a group, and the equations (5), (6), (7), (8) show that all the $(Q, \mathbf{i})$ are isotopic to this group. This proves the theorem.

The above theorem was first formulated by V. D. Belousov in a lecture given at an Algebra Conference (Moscow University, 3rd to 6th February, 1958). It was subsequently published in V. D. Belousov [1] but without proof. Later, in 1959, M. Hosszú re-proved the theorem (see M. Hosszú [1]) and pointed out some of its applications. Then, in 1961, V. D. Belousov's paper [2] appeared and this also contains a proof of the theorem. A generalized form of the theorem has been proved by A. Sade (see page 334 of A. Sade [17]) and he has also extended the theorem to cover the case of multigroupoids in A. Sade [13]. See also S. Milić [1].

Many authors have considered special cases of the generalized associative law, see, for example, V. Devidé [1], T. Evans [1], G. G. Ford [1], and A. Suschkevitsch [1].

Definition. A set Ω_n of quasigroups of order n defined on the same set Q of elements, is called an *associative system* (see R. Schauffler [2]) if, corresponding to every two arbitrarily chosen quasigroups $(Q, ①)$, $(Q, ②)$ of Ω_n, there exist further quasigroups $(Q, ③)$, $(Q, ④)$ in Ω_n of such a kind that, for any three elements $a, b, c, \in Q$, the general associative law is satisfied.

It follows from theorem 2.2.5 that all the quasigroups contained in such a set Ω_n are isotopic to the same group.

In [2] R. Schauffler raised the following question with regard to such a system. "Under what circumstances does an associative system Ω_n comprise the set of all quasigroups defined on a given set of cardinal n?"

He proved the following theorem:

THEOREM 2.2.6. *The set Ω_n of all quasigroups of order n is an associative system if and only if $n \leq 3$.*

PROOF. It easy to show by direct enumeration that all quasigroups oi order 2 and of order 3 are isotopic to the cyclic groups of order 2 and 3 respectively. In other words Ω_2 and Ω_3 are associative systems.

Also by theorem 2.2.5 we know that all quasigroups belonging to Ω_n are isotopic to one and the same group.

On the other hand theorem 1.5.1 taken in conjunction with Lagrange's theorem for groups implies that for any $n > 4$, $n \neq 6$, there exists at least one quasigroup which is not isotopic to any group. For $n = 4$ all quasigroups are isotopic either to the cyclic group C_4 or else to $C_2 \times C_2$ but these groups are not themselves isotopic (see section 4.2), while for $n = 6$, the fact that there exist quasigroups which are not isotopic to any group is shown in section 4.2.

ADDENDUM. *Historical remarks concerning Steiner triple systems.*

It was in the year 1853 that J. Steiner (see [1]) posed the problem as to whether the necessary condition $n \equiv 1$ or 3, modulo 6, was sufficient for the existence of a triple system having n different elements and of the type described above which now bears his name. The question was answered affirmatively by M. Reiss in [1] in 1859. However, neither of these writers seems to have been aware that the problem had been both posed and solved some twelve years earlier (1847) by T.P.Kirkman (see [1]) in an article in the Cambridge and Dublin Mathematical Journal. Indeed, three years after that, in "The Lady's and Gentleman's Diary" of 1850, Kirkman had gone on to pose a more difficult but related problem which is known to this day as Kirkman's schoolgirl problem. This problem requires the construction of Steiner triple systems on $n = 6m + 3$ elements which can be resolved into $r = (n - 1)/2$ subsystems each containing every element exactly once. The general case of the latter problem was not solved until very recently. (A complete solution will be found in D. K. Ray-Chaudhury and R. M. Wilson [1] and [2].)

As already mentioned, E. Netto ([1]) was the first to show that there is just one Steiner triple system for each of the orders $n = 3, 7, 9$ and at least two non-isomorphic systems for all admissible larger values of n. F. N. Cole, A. S. White and L. D. Cummings (see [1]) found that there are 80 isomorphically distinct Steiner triple systems on 15 elements. This number was later verified by M. Hall and J. D. Swift in [1] using a computer. Some upper and lower bounds on the number of Steiner triple systems of given order have recently been obtained by J. Doyen (see [1] and [2]). Some further results on the same subject will be found in J. Doyen and G. Valette [1] and in B. Rokovska [1] and [2].

For up-to-date information both on Steiner triple systems and on Kirkman's schoolgirl problem, the interested reader is referred to M. Hall's recent book [11] and to J. Doyen and A. Rosa [1]. Also, much additional interesting information on Kirkman's schoolgirl problem and its history will be found in chapter 10 of W. W. Rouse Ball [1].

For further information on the subject matter of this and the preceding section, the reader should consult V. D. Belousov [4] and A. Sade [8] and the bibliographies contained therein.

For the general theory of quasigroups, he should consult J. Aczél [1] and [2], A. A. Albert [1] and [2], V. D. Belousov [6] and [7], R. H. Bruck [5] and A. Kertész [1], each of which contains an extensive bibliography of further papers, and also the papers of A. Sade.

2.3. Complete latin squares

As will be explained more fully in section 10.2, latin squares are made use of in the branch of Statistics known as the Design of Experiments. In an agricultural experiment, for example, adjacent cells of a latin square may represent adjacent plots of land which are to receive different treatments, the object of the experiment being to determine the relative efficiency of the various treatments. Treatments applied to adjacent plots might interact and so the problem arose as to whether latin squares exist in which each pair of distinct entries (representing distinct treatments) occur in adjacent cells just once (and necessarily only once) in some row (column). Such a latin square is called *row complete* or *horizontally complete* (*column complete* or *vertically complete*).

Precisely, a latin square with elements $1, 2, \ldots, n$ is called *horizontally complete* if, for any ordered pair of elements α, β $(1 \leq \alpha, \beta \leq n, \alpha \neq \beta)$, there exists a row of the latin square in which α and β appear as adjacent elements: that is, if c_{st} denotes the element at the intersection of the sth row and tth column of the square, there exist integers s and t such that $c_{st} = \alpha$ and $c_{s,t+1} = \beta$. Similarly, a latin square is *vertically complete* if, for any α and β there exist integers u and v such that $c_{uv} = \alpha$ and $c_{u,v+1} = \beta$. A latin square which is both horizontally and vertically complete is called a *complete latin square*.

An example of a complete latin square of order 4 is given in Fig. 2.3.1.

Notice that in this square the digits 2, 3, 4 each occur just once in an adjacent position to the left of the digit 1, just once to its right, just once above it and just once below it. A similar property holds with respect to each other digit.

1	2	3	4
2	4	1	3
3	1	4	2
4	3	2	1

Fig. 2.3.1

Row complete latin squares may also be used advantageously in experiments in which a single subject (experimental unit) is to receive a number of treatments successively, since the effect of each treatment on the subject is likely to be affected both by that which was its immediate predecessor and usually also by the number of treatments which the subject has previously received.

In an experiment on farm animals, for example, it may be desirable to apply a number of different dietary treatments to a given animal in succession. The effect of a given treatment on the animal may be affected both by the number of treatments which that animal has already received and also by the nature of the immediately preceding treatment which it has had applied to it. (As another example, in a sequence of psychological experiments on a human being, fatigue after several preceding experiments is likely to influence the reaction of the subject to later experiments. Also, the reaction to a given experiment may be affected by the outcome of the immediately preceding experiment.) If several animals are available for treatment, the first possibility can be allowed for statistically if it can be arranged that the number n of animals to be treated is equal to the number of treatments to be applied and if the order in which the treatments are to be applied to these n animals is allowed to be determined by the order of the entries in the n rows of an $n \times n$ latin square (whose n distinct elements denote the n treatments). Then any particular experiment has a different number of predecessors for each of the n different animals, since a given element of the latin square is preceded by a different number of other elements in each of the n rows of the square. The possibility of interaction between one experiment and the immediately preceding one can also be allowed for if the latin square chosen is row complete. The resulting experiment is then said to be statistically "balanced" both with respect to the effect of the immediately preceding experiment and also with respect to the number of preceding experiments.

So far as the authors are aware[1] the first author to investigate the existence of row complete latin squares was B. R. Bugelski (see [1]) who gave a 6×6 row complete latin square. Bugelski's paper was published almost at the same time as a much more comprehensive one by E. J. Williams. Both these authors were interested in designing balanced experiments of the second of the two types described above: that is, experiments in which a number of different treatments are applied successively to the same unit of experimental material.

In [1], Williams gave a very simple method for constructing row complete latin squares of any even order $n = 2m$, which we shall write as a theorem.[2]

THEOREM 2.3.1. *Let $n = 2m$ be any even positive integer and let an $n \times n$ latin square L be formed whose first row is $0, 1, 2m - 1, 2, 2m - 2, 3, 2m - 3, \ldots, \ldots, m + 1, m$, where the integers 0 to $2m - 1$ are regarded as residues modulo n, and whose following rows are formed by the rule that the elements of row k are each one greater than the elements of row $k - 1$. Then L is a row complete latin square.*

PROOF. Since the differences between the $n - 1$ pairs of elements (of the first row) chosen from consecutive columns are all different (modulo n), it follows automatically that the same is true in each row. Every element occurs just once in each column and so no two of the successors of a given element h in the $n - 1$ rows distinct from the one in which h appears in the last column differ from it by the same amount. Consequently, these successors are all different.

It is easy to see from the method of construction just described that if the rows of the row complete latin square L are now reordered in such a way that the order of the elements in the first column becomes the same as the order of the elements in the first row, then the arguments which we have just used can be applied again with the words row and column interchanged. Consequently, the amended square will be column complete as well as row complete.

(1) The authors are indebted to Prof. Van Lint for kindly drawing their attention to the existence of the papers of E. J. Williams referred to below.

(2) The graph-theoretical equivalent of this theorem has been proved in L. W. Beineke [1].

Up to the present, no row complete latin square has been discovered which cannot be made column complete as well by a suitable permutation of its rows, and it is not known whether such squares exist.[1]

The above theorem has recently been rediscovered independently by J. Dénes and É. Török (see [1]), and these authors and others have also given an alternative ordering of the elements 0 to $2m - 1$ such that the construction above will give a $2m \times 2m$ row complete latin square. Their alternative first row is $0, 2m - 1, 1, 2m - 2, 2, 2m - 3, \ldots, \ldots, m - 1, m$. The latter alternative ordering was first described by J. V. Bradley (see [1]). A generalized form of theorem 2.3.1 (which includes both the above orderings as special cases) has been stated and proved by T. R. Houston in [1]. (It appears as theorem 3.1.3 in the next chapter and will be used there.)

Williams was unable to construct any row complete latin square of odd order n, but instead gave a balanced solution to his statistical problem with the aid of a pair of $n \times n$ latin squares, so that for n odd he required $2n$ experimental units (corresponding to the $2n$ rows) for an experiment involving the application of only n successive treatments. For the case $n = 5$, for example, he gave the solution exhibited in Fig. 2.3.2, where each treatment h of the set of n treatments is succeeded by every other exactly twice. (Much later, in his paper [1] referred to above, T. R. Houston obtained, and proved the validity of, another very similar construction using a pair of latin squares.)

0	1	3	4	2	0	2	1	4	3
1	2	4	0	3	1	3	2	0	4
2	3	0	1	4	2	4	3	1	0
3	4	1	2	0	3	0	4	2	1
4	0	2	3	1	4	1	0	3	2

Fig. 2.3.2

In the same paper [1] and in a subsequent paper [2] Williams went on to discuss the design of balanced experiments in which not only the interaction between the treatment last applied and its immediate predecessor is to be allowed for but also the interaction between the treat-

[1] One of the authors has recently shown (Autumn 1973) that any row complete latin square which is the multiplication table of a group or of an inverse property loop which satisfies the identity $(gh)(h^{-1}k) = gk$ can be made column complete as well as row complete by a suitable reordering of its rows. (See A. D. Keedwell [10].)

ment last applied and that which was applied two steps earlier. He then distinguished between cases in which any interaction between these two predecessors themselves could be ignored in assessing their effect on the treatment which they preceded, and others in which this effect too was allowed for. The former type of experiment requires that each treatment be preceded by each other treatment an equal number of times among the set of sequences applied to the experimental units (represented by the rows of the latin squares providing the solution) as above, and also that the same be true for the next to immediate predecessors. The latter type of experiment requires that every ordered triad of treatments occurs exactly once. Thus, the set of three latin squares of order 4 given in Fig. 2.3.3 provides a balanced experiment of the first kind in that, for example, the treatment 0 is preceded by every other exactly three times among the rows and also has every other treatment as a next to immediate predecessor exactly twice. However, of the 24 ordered triads of treatments, six occur twice, twelve occur once, and six occur not at all among the rows of the squares, so this experiment does not fulfil the conditions necessary for it to be of the second kind.

0	1	3	2	0	3	2	1	0	2	1	3
1	2	0	3	1	2	3	0	1	0	3	2
2	3	1	0	2	0	1	3	2	3	0	1
3	0	2	1	3	1	0	2	3	1	2	0

Fig. 2.3.3

As regards balanced experiments of the second kind, Williams first found a solution valid for all prime values of n. That is, he gave a construction for a set of $n - 1$ latin squares of order n, n prime, in which each ordered triad of elements occurred exactly once among the rows of the set. His construction made use of the complete set of mutually orthogonal(1) latin squares of the same prime order n. He also gave a construction which worked for the smallest prime powers: namely $n = 4, 8, 9$. He put forward the conjecture that the problem was soluble for every integer n for which a complete set of mutually orthogonal latin squares exists.

In a much later paper, B. S. Alimena (see [1]) discussed the more general situation of an experiment in which the interactions of any number of preceding treatments on that last applied are allowed for.

(1) For the definition of this concept, see section 5.2.

For an experiment involving $2k$ treatments, he gave a construction for a $2k \times 2k$ latin square which would counterbalance not only immediate sequential effects but also all remote ones as well, valid whenever $2k+1$ is prime; in particular, therefore, for experiments involving 2, 4, 6, 10, 12, 16, 18, 22, etc., treatments. Later the same problem for more general values of k was discussed in E. N. Gilbert [1]. Gilbert also gave a construction for complete latin squares of even order n which is effectively equivalent to that of Williams and of another author, B. Gordon, whose work we are about to describe, in that it is based on a cyclic group of order n in a way which is made clear by our next theorem.

THEOREM 2.3.2. A *sufficient condition for the existence of a complete latin square L of order n is that there exist a finite group G of order n with the property that its elements can be arranged into a sequence $a_1, a_2, \ldots, a_n$ in such a way that the partial products $a_1, a_1a_2, a_1a_2a_3, \ldots, \ldots, a_1a_2 \ldots a_n$ are all distinct.*

PROOF. Let $b_1 = a_1$, $b_2 = a_1a_2$, $b_3 = a_1a_2a_3, \ldots, \ldots, b_n = a_1a_2 \ldots a_n$. Then we shall show that the $n \times n$ matrix $\| c_{st} \|$ where $c_{st} = b_s^{-1} b_t$ is a complete latin square.

By hypothesis, the b_i' s are all distinct and so, evidently, $c_{rt} \neq c_{st}$ if $r \neq s$. Thus, each row of the matrix contains each element of G exactly once. Similarly, the same is true of the columns, so the matrix is a latin square. It remains to show that it is complete.

We have to show firstly that, for any ordered pair of distinct elements α and β, it is possible to determine s and t uniquely so that $b_s^{-1} b_t = \alpha$ and $b_s^{-1} b_{t+1} = \beta$. By definition of b_t and b_{t+1}, $\alpha a_{t+1} = \beta$. This equation determines a_{t+1}, and hence t, uniquely. (We have $t > 0$ since the identity element of G is necessarily a_1 if the b_i are all distinct, and so $t = 0$ would imply $\alpha = \beta$.) When t has been determined, s is uniquely defined by the equation $b_s^{-1} b_t = \alpha$. Thus, L is horizontally complete.

Finally, we require unique solutions for s and t of the equations $b_s^{-1} b_t = \alpha$ and $b_{s+1}^{-1} b_t = \beta$. When these hold, $\alpha^{-1} a_{s+1} = \beta^{-1}$ which determines s and then the equation $b_s^{-1} b_t = \alpha$ determines t. Thus, L is also vertically complete.

The above theorem and proof are due to B. Gordon (see [1]). Gordon has called a group *sequenceable* if it has the property stated in the above theorem. We shall say that the complete latin square L obtained by the above construction from a sequenceable group G is *based on the group G*.

In the same paper [1], Gordon has proved the following important condition for a finite abelian group G to be sequenceable.

Theorem 2.3.3. *A finite abelian group G is sequenceable if and only if it is the direct product of two groups A and B such that A is a cyclic group of order 2^k, $k > 0$, and B is of odd order.*

Proof. To see the necessity of the condition, suppose that G is sequenceable, and let $b_1 = a_1$, $b_2 = a_1a_2$, $b_3 = a_1a_2a_3, \ldots \ldots, b_n = a_1a_2 \ldots a_n$ be an ordering of the elements of G such that the b_i are all distinct. Then, as above, it is immediately seen that we necessarily have $b_1 = a_1 = e$, the identity element of G. Therefore $b_n \neq e$. But we pointed out in section 1.4 (see page 34) that an abelian group G in which the product of all the elements is not equal to the identity has a unique element of order two and so has the form given in the statement of the theorem.

To prove sufficiency of the condition, suppose that $G = A \times B$ with A and B as in the statement of the theorem. We show that G is sequenceable by constructing an ordering $a_1, a_2, \ldots, a_n$ of its elements which has distinct partial products. From the general theory of abelian groups it is known that G has a basis of the form $c_0, c_1, \ldots, c_m$, where c_0 is of order 2^k and where the orders $\delta_1, \delta_2, \ldots, \delta_m$ of $c_1, c_2, \ldots, c_m$ are odd positive integers each of which divides the next; that is $\delta_i \backslash \delta_{i+1}$ for $0 < i < m$. If j is any positive integer, we shall show that there exist unique positive integers $j_0, j_1, \ldots, j_m$ such that

$$\left.\begin{array}{ll} & j \equiv j_0 \pmod{\delta_1 \delta_2 \ldots \delta_m} \\ \text{and} & j_0 = j_1 + j_2\,\delta_1 + j_3\,\delta_1\delta_2 + \ldots + j_m\,\delta_1\delta_2 \ldots \delta_{m-1} \\ \text{where} & 0 \leq j_1 < \delta_1, 0 \leq j_2 < \delta_2, \ldots, 0 \leq j_m < \delta_m \end{array}\right\} \text{formulae (1).}$$

Firstly, the remainder on division of j by $\delta_1\delta_2 \ldots \delta_m$ is j_0 and we have $0 \leq j_0 < \delta_1\,\delta_2 \ldots \delta_m$.

Next, by division of j_0 by δ_1 we get

$$j_0 = j_1 + j_1'\delta_1, \text{ where } 0 \leq j_1 < \delta_1.$$

Then, by division of j_1' by δ_2 we get

$$j_1' = j_2 + j_2'\delta_2, \text{ where } 0 \leq j_2 < \delta_2;$$

and hence

$$j_0 = j_1 + j_2\,\delta_1 + j_2'\,\delta_1\delta_2.$$

By division of j_2' by δ_3 we get $j_2' = j_3 + j_3'\,\delta_3$ and so

$$j_0 = j_1 + j_2\delta_1 + j_3\delta_1\delta_2 + j_3'\delta_1\delta_2\delta_3,$$

where $0 \leq j_3 < \delta_3$. It is clear that, by successive divisions, we shall eventually get

$$j_0 = j_1 + j_2\delta_1 + j_3\delta_1\delta_2 + \ldots + j'_{m-1}\,\delta_1\,\delta_2 \ldots \delta_{m-1},$$

where $j_1, j_2, \ldots, j_{m-1}$ satisfy the inequalities given in formulae (1) and, because $0 \leq j_0 < \delta_1\delta_2 \ldots \delta_m$, we shall also have $0 \leq j'_m < \delta_m$, as required. For consistency of notation, we replace j'_{m-1} by j_m as in formulae (1).

We are now in a position to define the desired sequencing of G. It is convenient to define the products $b_1, b_2, \ldots, b_n$ directly, to prove that they are all distinct, and then to verify that the corresponding a_i, as calculated from the formula $a_1 = e$, $a_i = b_{i-1}^{-1}\, b_i$, are all distinct. If i is of the form $2j + 1$ $(0 \leq j < n/2)$, let

$$b_{2j+1} = c_0^{-j}\, c_1^{-j_1}\, c_2^{-j_2} \ldots c_m^{-j_m},$$

where $j_1, j_2, \ldots, j_m$ are the integers defined in (1). On the other hand, if i is of the form $2j + 2$ $(0 \leq j < n/2)$, let

$$b_{2j+2} = c_0^{j+1}\, c_1^{j_1+1}\, c_2^{j_2+1} \ldots c_m^{j_m+1}.$$

The elements $b_1, b_2, \ldots, b_n$ thus defined are all distinct. For if $b_s = b_t$ with $s = 2u + 1$, $t = 2v + 1$, then

$$\left.\begin{array}{l} u \equiv v (\text{mod } 2^k) \\ u_1 \equiv v_1 \,(\text{mod } \delta_1) \\ \cdot \\ \cdot \\ \cdot \\ u_m \equiv v_m \,(\text{mod } \delta_m) \end{array}\right\} \text{equations (2).}$$

From the inequalities in (1) we conclude that $u_1 = v_1, \ldots, u_m = v_m$. Hence $u_0 = v_0$, so that $u \equiv v$ (mod $\delta_1 \ldots \delta_m$); coupled with the first of equations (2), this gives $u \equiv v$ (mod n), which implies $u = v$. Similarly $b_{2u+2} = b_{2v+2}$ implies $u = v$, so that the "even" b's are distinct.

Next suppose that $b_{2u+1} = b_{2v+2}$. Then

$$\begin{array}{l} -u \equiv v + 1 \;(\text{mod } 2^k) \\ -u_1 \equiv v_1 + 1 \;(\text{mod } \delta_1) \\ \cdot \\ \cdot \\ -u_m \equiv v_m + 1 \;(\text{mod } \delta_m) \end{array}$$

or equivalently,

$$\left.\begin{array}{l} u + v + 1 \equiv 0 \pmod{2^k} \\ u_1 + v_1 + 1 \equiv 0 \pmod{\delta_1} \\ \cdot \\ \cdot \\ \cdot \\ u_m + v_m + 1 \equiv 0 \pmod{\delta_m} \end{array}\right\} \text{equations (3).}$$

Since $0 < u_1 + v_1 + 1 \leq 2(\delta_1 - 1) + 1 < 2\delta_1$, we must have $u_1 + v_1 + 1 = \delta_1$. Reasoning similarly for $i = 2, \ldots, m$ we obtain

$$\begin{array}{l} u_1 + v_1 + 1 = \delta_1 \\ u_2 + v_2 + 1 = \delta_2 \\ \cdot \\ \cdot \\ u_m + v_m + 1 = \delta_m . \end{array}$$

Multiplying the $(i + 1)$th equation of this system by $\delta_1\delta_2 \ldots \delta_i$ $(1 \leq i < m)$ and adding, we get $u_0 + v_0 + 1 = \delta_1 \ldots \delta_m$, which implies $u+v+1\equiv 0 \pmod{\delta_1 \ldots \delta_m}$. Combining this with the first of equations (3), we find that $u + v + 1 \equiv 0 \pmod{n}$, which, on account of the inequality $0 < u + v + 1 < n$, is impossible. Hence, $b_1, b_2, \ldots, b_n$ are all distinct.

Next, we calculate $a_1, a_2, \ldots, a_n$. If $i = 2j + 2$ $(0 \leq j < n/2)$, then

$$a_i = b_{i-1}^{-1}\ b_i = c_0^{2j+1}\ c_1^{2j_1+1} \ldots c_m^{2j_m+1}.$$

These are all different by the same argument as above. If $i = 2j + 1$, and $j_1 \neq 0$ then

$$a_i = c_0^{-2j}\ c_1^{-2j_1} c_2^{-2j_2-1} \ldots c_m^{-2j_m-1}.$$

If $i = 2j + 1$ and $j_1 = 0$, but $j_2 \neq 0$, then

$$a_i = c_0^{-2j}\ c_2^{-2j_2} c_3^{-2j_3-1} \ldots c_m^{-2j_m-1},$$

while if $j_1 = j_2 = 0$ but $j_3 \neq 0$, then

$$a_i = c_0^{-2j}\ c_3^{-2j_3} c_4^{-2j_4-1} \ldots c_m^{-2j_m-1},$$

and so on. These a_i's are obviously distinct from each other by the same reasoning as before. Because of the exponent of c_0 they are also distinct from the a_i with i even. This completes the proof of the theorem.

Let us note here that B. Gordon's paper [1] predates that of E. N. Gilbert by four years and that B. Gordon's construction of complete latin squares includes that of Gilbert as a special case.

As the foregoing makes clear, the study of sequenceable groups started quite recently and consequently presents many unsolved problems.

Some of them were pointed out by B. Gordon himself in [1]: in particular, he asked

(i) Does any complete latin square of odd order exist?

(ii) What is the necessary and sufficient condition that a non-commutative group be sequenceable?

Also, in A. Rényi [1] the question is raised as to whether all complete latin squares satisfy the quadrangle criterion.

It follows from theorem 2.3.3 that the finite abelian groups can be classified into two mutually exclusive kinds: namely sequenceable groups and groups whose Cayley tables possess a transversal. This is a consequence of L. J. Paige's theorem, see page 34 of section 1.4, to the effect that, except when a finite abelian group contains exactly one element of order 2, its Cayley table will always possess a transversal and (see theorem 1.4.2) will consequently be separable into a set of disjoint transversals.

However, the same result is not true for non-abelian groups. For, on the one hand, the symmetric group S_3 of degree three is neither sequenceable nor contains a transversal in its Cayley table, as was pointed out by B. Gordon himself. On the other hand, the non-abelian group of order 21 and the non-abelian group of order 27 on two generators are both sequenceable and both have complete mappings. (The latter fact is shown in section 1.4.) A sequencing of the non-abelian group of order 21 was first published in N. S. Mendelsohn [1], and a sequencing for the non-abelian group of order 27 has been obtained recently by one of the present authors.

Our next theorems follow the line of investigation given in J. Dénes and É. Török [1]. The first two of them are almost direct consequences of theorem 2.3.2.

Theorem 2.3.4. *A finite group* $(G, \cdot)$ *is sequenceable if and only if there exists a permutation* $a_1, a_2, \ldots, a_n$ *of its elements such that* $a_1 = e$ (the identity of the group) *and such that the product* $a_{i+1} \ldots a_{i+j} \neq e$ *for any choices of the integers* $i, j, 1 < i < n - 1, 2 < i + j < n$.

Proof. Define $b_h = a_1 a_2 \ldots a_h$ for each integer $h, 1 \leq h \leq n$. Then $a_{i+1} \ldots a_{i+j} = b_i^{-1} b_{i+j}$. The condition given is the condition that

$b_i \neq b_{i+j}$ for each choice of i and j in the given ranges. Thus, if and only if the condition holds, the products b_h, for $h = 1, 2, \ldots, n$, are all distinct and the given ordering of the elements of the group then provides a sequencing.

Theorem 2.3.5. *A finite group $(G, \cdot)$ is sequenceable if and only if there exists a permutation $b_1, b_2, \ldots, b_n$ of its elements such that the equation $b_{i-1}^{-1} b_i = b_{j-1}^{-1} b_j$ holds only when $i = j$.*

Proof. Suppose first that the group is sequenceable and that the ordering $a_1, a_2, \ldots, a_n$ of its elements is a sequencing. Then, if the elements b_h are defined as above, the statement is that the equation $a_i = a_j$ holds only when $i = j$, and this is obviously true.

Conversely, suppose that the equation $b_{i-1}^{-1} b_i = b_{j-1}^{-1} b_j$ holds only when $i = j$. Define $a_i = b_{i-1}^{-1} b_i$ for $i = 2, \ldots, n$ and let $a_1 = e$, the identity element of the group. Then the a_i are certainly all distinct and we have $a_1 a_2 \ldots a_h = b_1^{-1} b_h$ for $h = 1, 2, \ldots, n$. Since $b_1, b_2, \ldots, b_h$ are the elements of G, they are all distinct and so also are the elements $b_1^{-1} b_1 = e, b_1^{-1} b_2, \ldots, b_1^{-1} b_n$. Thus, the ordering $a_1, a_2, \ldots, a_n$ of the elements of G provides a sequencing and so the group $(G, \cdot)$ is sequenceable as required.

Theorem 2.3.6. *Let $(G, \cdot)$ be a group of order n with identity element e and let S_k denote the set of all ordered sequences of k distinct elements of $G \setminus \{e\}$. We define a subset E_k of S_k as follows. The sequence $a_1, a_2, \ldots, a_k$ is a member of E_k if the partial product $a_h a_{h+1} \ldots a_{k-1} a_k$ is equal to e for some integer h, $1 \leq h \leq k$ and if $a_i a_{i+1} \ldots a_{j-1} a_j \neq e$ whenever $1 \leq i \leq j < k$. In this way, we define sets $E_1, E_2, \ldots, E_{n-1}$ where E_1 is empty, E_2 comprises all products of the form $a_p a_p^{-1}$ $(a_p \neq e)$, E_3 comprises all products of the form $a_p a_q a_q^{-1}$ and $a_p a_q a_r (a_q \neq a_p^{-1})$ where $a_r = (a_p a_q)^{-1}$, and so on. The cardinal of the set E_k will be denoted by ε_k. Then the group $(G, \cdot)$ is sequenceable if and only if $\varepsilon_2(n-3)! + \varepsilon_3(n-4)! + \ldots + \varepsilon_{n-2}(1!) + \varepsilon_{n-3} < (n-1)!$*

Proof. Let us consider the set $G \setminus \{e\}$ and let us call an arrangement $a_1, a_2, \ldots, a_{n-1}$ of the distinct elements of $G \setminus \{e\}$ *wrong* if there exist positive integers h and k, $1 \leq h < k \leq n-1$ such that $a_h a_{h+1} \ldots a_{k-1} a_k = e$. Let E^* denote the set of all such wrong arrangements.

The total number of distinct arrangements of the set $G \setminus \{e\}$ is $(n-1)!$ and, by theorem 2.3.4, the group is sequenceable if and only if the number of wrong arrangements is less than $(n-1)!$

Let the ordered sequence $a_1, a_2, \ldots, a_k$ belong to E_k, then this sequence can be completed to a permutation of all the elements of the set $G \setminus \{e\}$ in $(n - 1 - k)!$ different ways, and all these permutations are different wrong arrangements. Thus, there exist $\varepsilon_k(n - 1 - k)!$ permutations of the $n - 1$ distinct elements of $G \setminus \{e\}$ for which the subsequence formed by their first k elements is a member of E_k. Moreover, given any wrong arrangement $a_1, a_2, \ldots, a_{n-1}$, there exists exactly one positive integer k such that $a_1, a_2, \ldots, a_k \in E_k$: for, if $a_1, a_2, \ldots, a_{n-1}$ is a wrong arrangement, there exists at least one pair of positive integers h, k, $1 \leq h < k \leq$ $\leq n - 1$, such that $a_h a_{h+1} \ldots a_{k-1} a_k = e$. Let us choose the pair h, k for which k is minimal. Then $a_1, a_2, \ldots, a_k \in E_k$, but $a_1, a_2, \ldots, a_l \notin E_l$ for $l > k$ since the condition $a_i a_{i+1} \ldots a_{j-1} a_j \neq e$ whenever $1 \leq i \leq j < l$ is violated. Also, since k is minimal, $a_1, a_2, \ldots, a_l \notin E_l$ for $l < k$. Hence, the total number of wrong arrangements is $\varepsilon_2(n - 3)! + \varepsilon_3(n - 4)! +$ $+ \ldots + \varepsilon_{n-2}(1!) + \varepsilon_{n-1}$, and both the necessity and the sufficiency of our condition follow.

The application of this theorem to a computer search for sequenceable groups is described in section 13.1. In particular, J. Dénes and É. Török have found a number of sequencings for the dihedral groups D_5, D_6, D_7 and D_8, where D_m denotes the non-commutative group of order $2m$ generated by two elements a and b with defining relations $a^m = b^2 = e$, $ab = ba^{m-1}$; and have shown that there are no other non-abelian sequenceable groups of orders less than or equal to 14. They have also found a large number of sequencings for the non-commutative group of order 21 generated by two elements a and b with the defining relations $a^7 = b^3 = e$, $ab = ba^2$.

N. S. Mendelsohn (see [1]) was the first to publish a sequencing for this latter group and thus disproved an earlier conjecture according to which no latin square of odd order is complete. The second author of the present book found that it was not difficult to obtain such sequencings by trial and error once their existence was known. The sequencings of N. S. Mendelsohn and A. D. Keedwell are given in Figs 2.3.4 and 2.3.5, respectively.

The complete latin squares which correspond to the last two of the five sequencings obtained by Mendelsohn are exhibited in Figs 2.3.6 and 2.3.7.

Complete latin squares also have some connection with graph theory, as we shall explain in section 9.1.

(1) $e\ b\ a\ b^2\ ba^2\ b^2a^4\ b^2a\ a^2\ ba^4\ a^3\ b^2a^2\ a^6\ a^5\ b^2a^6\ ba^6\ b^2a^5\ ba^5$

$e\ b\ ba\ a^4\ ba^3\ a^2\ b^2a^2\ b^2a^4\ a^5\ a\ b^2a^6\ b^2a^5\ b^2a^3\ ba^4\ b^2\ ba^5\ b^2a$

$ba\ ba^3\ a^4\ b^2a^3$

$a^3\ ba^2\ ba^6\ a^6$

(2) $e\ a\ b\ ba^2\ b^3\ b^2a^4\ a^2\ ba^4\ a^3\ b^2a\ b^2a^5\ a^4\ b^2a^3\ a^6\ ba^5\ b^2a^6\ ba$

$e\ a\ ba^2\ b^2a^6\ ba^3\ a^2\ a^4\ ba^5\ ba\ a^5\ b^2a^4\ b^2a\ b\ ba^6\ b^2a^3\ ba^4\ b^2a^2$

$ba^6\ b^2a^2\ a^4\ ba^3$

$a^3\ b^2\ b^2a^5\ a^6$

(3) $e\ a\ b\ ba^2\ b^2\ b^2a^4\ a^2\ ba^4\ a^3\ b^2a\ ba\ ba^6\ ba^3\ b^2a^3\ a^4\ a^6\ a^5$

$e\ a\ ba^2\ b^2a^6\ ba^3\ a^2\ a^4\ ba^5\ ba\ a^5\ ba^4\ b^2\ a^3\ b^2a\ b^2a^5\ b^2a^4\ b^2a^2$

$b^2a^5\ ba^5\ b^2a^2\ b^2a^6$

$ba^6\ b^2a^3\ b\ a^6$

(4) $e\ a^3\ b\ a\ ba^2\ b^2\ b^2a^4\ b^2a\ a^2\ ba^4\ ba\ b^2a^5\ ba^6\ a^6\ ba^5\ b^2a^3\ b^2a^6$

$e\ a^3\ ba^6\ b\ b^2a^2\ ba\ a\ b^2a^5\ b^2\ a^4\ ba^2\ a^6\ ba^4\ ba^3\ b^2a^4\ ba^5\ a^5$

$b^2a^2\ a^5\ a^4\ ba^3$

$b^2a\ b^2a^6\ b^2a^3\ a^2$

(5) $e\ a^3\ b\ a\ ba^2\ b^2\ b^2a^4\ b^2a\ a^2\ ba^4\ b^2a^2\ b^2a^3\ a^5\ a^6\ ba^6\ b^2a^6\ ba^5$

$e\ a^3\ ba^6\ b\ b^2a^2\ ba\ a\ b^2a^5\ b^2\ a^4\ b^2a^4\ ba^5\ ba^3\ ba^2\ b^2a^3\ ba^4\ b^2a^6$

$ba\ b^2a^5\ ba^3\ a^4$

$a^6\ b^2a\ a^5\ a^2$

Fig. 2.3.4

$e\ b^2\ b^2a^3\ ba^2\ a^4\ b^2a\ a^2\ a^6\ a^5\ b^2a^6\ b^2a^5\ ba^3\ a^3\ b^2a^2\ ba^5\ ba^6\ a\ b$

$e\ b^2\ ba^3\ b^2a\ b^2a^5\ b\ ba^2\ ba\ ba^6\ a^2\ b^2a^6\ a\ a^4\ b^2a^4\ ba^6\ ba^4\ ba^5\ b^2a^3$

$ba^4\ b^2a^4\ ba$

$a^3\ b^2a^2\ a^5$

Fig. 2.3.5

01	02	03	04	05	06	07	08	09	10	11	12	13	14	15	16	17	18	19	20	21
04	03	05	21	11	14	02	09	13	07	18	16	19	08	20	01	06	10	17	12	15
07	09	19	03	06	12	14	21	20	17	08	05	16	01	10	18	15	13	04	02	11
18	14	09	07	19	15	17	01	21	13	06	02	20	12	11	05	04	08	16	10	03
19	04	12	08	21	07	20	02	18	01	16	13	03	11	09	14	05	15	10	06	17
05	17	14	18	09	04	13	12	01	08	19	10	21	15	03	02	16	06	20	11	07
12	10	07	16	02	19	18	06	14	11	03	15	08	17	04	20	13	05	09	21	01
17	21	16	09	15	02	12	03	10	04	01	19	05	18	13	08	11	20	07	14	06
08	12	21	17	16	11	04	18	03	20	15	14	10	02	06	19	07	01	05	13	09
16	07	02	01	03	17	10	14	08	18	05	20	09	06	21	12	19	11	13	15	04
02	13	17	05	14	16	08	15	12	06	09	11	01	04	07	10	20	19	21	03	18
21	05	11	15	18	08	03	13	19	02	10	01	17	09	12	04	14	07	06	16	20
11	06	08	10	13	21	19	16	04	09	17	07	15	20	05	03	01	14	12	18	02
10	08	13	02	17	20	06	04	15	19	14	03	12	16	18	11	21	09	01	07	05
14	20	04	19	12	05	01	11	02	15	21	06	18	07	17	13	10	16	03	09	08
03	19	06	11	08	01	09	20	16	14	13	18	04	21	02	07	12	17	15	05	10
20	18	10	12	07	13	11	17	06	05	02	21	14	19	01	15	09	03	08	04	16
06	15	01	13	20	03	16	05	07	21	04	17	11	10	19	09	18	12	02	08	14
13	01	20	14	04	10	15	07	11	16	12	09	02	05	08	06	03	21	18	17	19
09	16	15	06	01	18	21	10	05	12	20	08	07	03	14	17	02	04	11	19	13
15	11	18	20	10	09	05	19	17	03	07	04	06	13	16	21	08	02	14	01	12

Fig. 2.3.6

ADDED IN PROOF. As we remarked on page 89, the non-abelian groups of orders 21 and 27 on two generators are both sequenceable. Recently, L. L. Wang has shown that the non-abelian groups on two generators with similar structure having orders 39, 55 and 57 are also sequenceable (see A. D. Keedwell [10] for details) and this adds strength to an earlier conjecture of one of the present authors that every finite non-abelian group of odd order on two generators is sequenceable. (See A. D. Keedwell [8]).

01	02	03	04	05	06	07	08	09	10	11	12	13	14	15	16	17	18	19	20	21
04	03	05	21	11	14	02	09	13	07	18	16	19	08	20	01	06	10	17	12	15
07	09	19	03	06	12	14	21	20	17	08	05	16	01	10	18	15	13	04	02	11
18	14	09	07	19	15	17	01	21	13	06	02	20	12	11	05	04	08	16	10	03
19	04	12	08	21	07	20	02	18	01	16	13	03	11	09	14	05	15	10	06	17
05	17	14	18	09	04	13	12	01	08	19	10	21	15	03	02	16	06	20	11	07
12	10	07	16	02	19	18	06	14	11	03	15	08	17	04	20	13	05	09	21	01
17	21	16	09	15	02	12	03	10	04	01	19	05	18	13	08	11	20	07	14	06
08	12	21	17	16	11	04	18	03	20	15	14	10	02	06	19	07	01	05	13	09
16	07	02	01	03	17	10	14	08	18	05	20	09	06	21	12	19	11	13	15	04
14	20	04	19	12	05	01	11	02	15	21	06	18	07	17	13	10	16	03	09	08
03	19	06	11	08	01	09	20	16	14	13	18	04	21	02	07	12	17	15	05	10
10	08	13	02	17	20	06	04	15	19	14	03	12	16	18	11	21	09	01	07	05
02	13	17	05	14	16	08	15	12	06	09	11	01	04	07	10	20	19	21	03	18
09	16	15	06	01	18	21	10	05	12	20	08	07	03	14	17	02	04	11	19	13
11	06	08	10	13	21	19	16	04	09	17	07	15	20	05	03	01	14	12	18	02
13	01	20	14	04	10	15	07	11	16	12	09	02	05	08	06	03	21	18	17	19
21	05	11	15	18	08	03	13	19	02	10	01	17	09	12	04	14	07	06	16	20
06	15	01	13	20	03	16	05	07	21	04	17	11	10	19	09	18	12	02	08	14
20	18	10	12	07	13	11	17	06	05	02	21	14	19	01	15	09	03	08	04	16
15	11	18	20	10	09	05	19	17	03	07	04	06	13	16	21	08	02	14	01	12

Fig. 2.3.7

CHAPTER 3

Generalizations of latin squares

The generalizations of the latin square concept which we now discuss are latin rectangles, row latin squares and various kinds of incomplete latin square.

3.1. Latin rectangles and row latin squares

A *latin rectangle* of c columns and $n > c$ rows is an array of n symbols such that each column contains all the symbols, and no row contains any symbol more than once.

Any latin rectangle can be extended to a latin square, as we show in theorem 3.1.1 below. The number of ways in which this can be done is discussed in P. Erdős and I. Kaplansky [1], and also in section 4.3.

Our first two theorems both make use of P. Hall's theorem on representatives of subsets, namely: "The necessary and sufficient condition that a system of distinct representatives, one for each member of a set $\{T_1, T_2, \ldots, T_m\}$ of subsets of a given set S, can be chosen simultaneously is that, for each $k = 1, 2, \ldots, m$, any selection of k of the subsets shall contain between them at least k distinct elements of S". (For the proof, see P. Hall [2].)

THEOREM 3.1.1. *An $n \times c$ rectangular matrix containing n distinct symbols and such that each symbol occurs once in every column and at most once in every row can always be extended to a latin square.*

PROOF. Let us form n sets S_i, $i = 1, 2, \ldots, n$, where, for each i, the set S_i consists of those $n - c$ symbols of our given set of n symbols which do not occur among the c symbols of the ith row of the rectangular matrix. Since each symbol occurs once in each of the c columns of the given matrix, each symbol occurs $n - c$ times among the S_i.

Any selection of k of the S_i will contain $k(n - c)$ symbols all together and at least k of these symbols must be distinct since each symbol occurs at most $n - c$ times among all the S_i. Thus, the requirements of P. Hall's theorem are satisfied and distinct representatives $a_1, a_2, \ldots, a_n$ can be chosen for all the sets S_i simultaneously. By definition of the sets S_i, these n distinct representative symbols a_i, $i = 1, 2, \ldots, n$, may be used to form a $(c + 1)$th column of the given rectangular matrix and yield an $n \times (c + 1)$ rectangular matrix of the same type.

By repetition of the process $n - c$ times, we eventually obtain an $n \times n$ latin square.

The above theorem will be found in M. Hall [2]. The following extension of it is due to H. J. Ryser (see [1]).

THEOREM 3.1.2. *An $r \times c$ rectangular matrix can be extended to a latin square of order n if the matrix contains n symbols in such a way that no symbol appears twice in row or in any column and provided that the number $N(i)$ of times that the symbol i appears in the rectangle is at least $r + c - n$, for every symbol i.*

PROOF. The necessity of the condition is easy to see. If we want to extend the given rectangle to a latin rectangle of n rows and c columns then we shall have, in the latter, c representations of every symbol. But in the additional $n - r$ rows we cannot include any symbol more than $n - r$ times, and so it must already appear $c + r - n$ times in the original matrix.

To prove the sufficiency, let us form c sets S_j, $j = 1, 2, \ldots, c$, where, for each j, the set S_j consists of those elements which are not included in the jth column of the original matrix. Each of these sets contains $n - r$ elements, while, in the completed $n \times c$ rectangle each element i will, altogether, appear $M(i) = c - N(i)$ additional times. To write down the elements of these S_j in the order in which they are to appear in the completed rectangle, we must be able to choose a system of c distinct elements, one from each set, which we then use as a further row. Such a system exists by P. Hall's theorem given above, because any k of the sets S_j together contain $k(n - r)$ elements, none more often than $n - r$ times, in view of the fact that $M(i) = c - N(i) \leq c - (c + r - n)$ for each i. Continuing thus, we can construct $n - r$ additional rows to complete an n by c latin rectangle. And this in turn can be completed to a latin square by theorem 3.1.1.

From theorem 3.1.2, one can deduce at once the fact that an incomplete latin square of side t (≥ 4) with n different elements (that is, a square of side t such that a subset of its t^2 cells are occupied by numbers from among 1, 2, . . . , n and no element occurs twice in the same row or column) can be embedded in a latin square of side n provided that $n \geq 2t$. This result can be found in T. Evans [2] and again in J. Dénes and K. Pásztor [1] where the proof is formulated in terms of the transversals of a latin square.

This result of T. Evans has been generalized in C. Treash [1] where it is proved that any finite incomplete loop which is also totally symmetric can be embedded in a finite complete totally symmetric loop. The definition of an incomplete loop is given later in this chapter in section 3.3 and, in that section, a number of further results concerning the embedding of incomplete latin squares and quasigroups will be found.

In section 2.3 we defined a complete latin square. Analogously a latin rectangle is said to be *horizontally complete* if the unordered pairs of adjacent elements appearing in its rows are all distinct. If we replace the word "rows" by the word "columns" in the preceding definition, it becomes that of a *vertically complete* latin rectangle. A latin rectangle which is both horizontally and vertically complete is called *complete*.

The following theorem was first stated and proved by T. R. Houston (see [1]) and has already been referred to in section 2.3. We include it here because of its relevance to latin rectangles (see theorem 3.1.4. below).

THEOREM 3.1.3. *If there exists a permutation* (defined on the natural numbers 1, 2, . . . , n) *with the property that the differences* (taken modulo n) *between pairs of adjacent elements are all distinct, then there exists a horizontally complete latin square of order* n.

The proof is the same as that of theorem 2.3.1 and will be omitted.

COROLLARY. *Permutations of odd degree* n *such that the differences* (taken modulo n) *between pairs of adjacent elements are all distinct do not exist*.

PROOF. The complete latin square L constructed by the method of theorem 3.1.3 is a Cayley table for the cyclic group of order n. However, by theorem 2.3.3, complete latin squares based on abelian groups of odd order do not exist.

It follows from theorem 3.1.3 that, in particular, a latin square of order $n = 2m$ whose first row is $0,1,2m-1,2,2m-2,\ldots,k,\ 2m-k,\ldots,m+1,m$ and whose ith row is obtained from the first by adding $(i-1)$, $i = 1,2,\ldots,n$, taken modulo n, is a horizontally complete latin square; as was earlier proved in theorem 2.3.1. When we construct this latin square, we find that its last m rows are the same as the first but in reverse order. It follows that the first m rows form a horizontally complete $m \times 2m$ latin rectangle. Consequently, we have

THEOREM 3.1.4. *Horizontally complete $m \times 2m$ latin rectangles exist for every positive integer m.*

A detailed proof of this result is given in theorem 9.1.2, where it is also shown that each such latin rectangle defines a decomposition of the complete undirected graph on $2m$ vertices into m disjoint Hamiltonian paths. Fig. 3.1.1 illustrates this fact for the special case $m = 3$.

We may explain Fig. 3.1.1 as follows. Each of the three rows of the horizontally complete latin rectangle defines an ordering of the vertices of the graph and hence defines a Hamiltonian path. The Hamiltonian paths labelled by ———, – – – –, –·–·–· correspond to the first, second and third rows of the latin rectangle respectively.

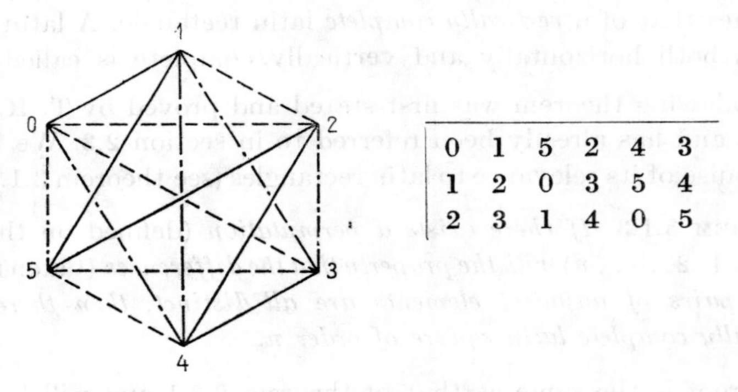

0	1	5	2	4	3
1	2	0	3	5	4
2	3	1	4	0	5

Fig. 3.1.1

We should like to pose two problems connected with complete latin rectangles:

(1) If n is a given odd integer, what is the maximum value of m_n such that a complete latin rectangle of size $m_n \times n$ exists?

(2) For which values of k_n is it true that an arbitrary complete latin rectangle of size $k_n \times n$ can be completed to a complete latin square of order n?

Using Lagrange's theorem it is easy to see that a latin square of composite order which satisfies the quadrangle criterion can be decomposed into disjoint latin rectangles. (In section 1.5 we gave a more precise statement.) Consequently, we shall find it relevant when discussing the enumeration of latin squares to be able first to enumerate latin rectangles. However, we shall defer consideration of problems of enumeration until the next chapter.

The concept of a *transversal* of a latin square is defined in section 1.4. One can similarly define a *partial transversal* or a *transversal* of a latin rectangle.

K. K. Koksma proposed the following problem in [1]:

What is the largest number of cells of an arbitrarily chosen latin square or latin rectangle of given size which can be contained in a single partial transversal?

As a contribution to the solution of the problem, Koksma proved the following:

THEOREM 3.1.5. *An arbitrary latin square A of order $n(n \geq 7)$, has at least one partial transversal of length t with $t \geq (2n+1)/3$.*

PROOF. Let A be divided into four submatrices (LU, RU, LL, RL) as shown in Fig. 3.1.2, and let us assume that the elements of the latin square A are the integers $1, 2, \ldots, n$.

Let t be an integer having the properties that

(1) there is no partial transversal of A of order greater than t, and

(2) there is at least one partial transversal of A of order t.

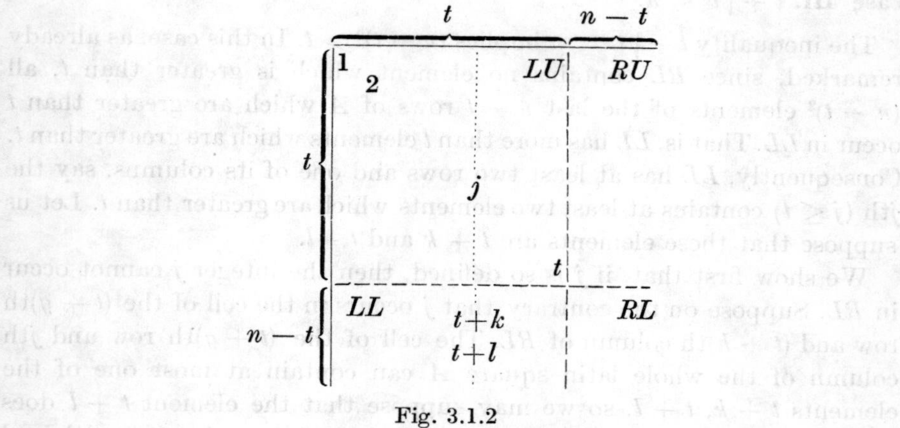

Fig. 3.1.2

Without loss of generality, we can suppose the rows and columns of A to be rearranged in such a way that the elements $1, 2, \ldots, t$ occur in the cells of the main left-to-right diagonal of LU. Then, from property (1) it is clear that RL must consist entirely of elements which are less than or equal to t. Therefore, each row of LL contains all the elements of the set $t+1, t+2, \ldots, n$.

The total number of elements in LL which are greater than t is therefore exactly $(n-t)^2$.

We now separate the argument into three cases according as $t+\sqrt{t}$ is greater than, equal to. or less than n.

Case I. $t+\sqrt{t} > n$.

We have $t+\sqrt{t} \geq (\frac{2}{3}n+\frac{1}{3})+(\frac{1}{3}n+\frac{2}{3})$. If we show that $\sqrt{t} \leq \frac{1}{3}n+\frac{2}{3}$, it will follow that $t \geq \frac{2}{3}n+\frac{1}{3}$ in this case. Since $n \geq 7$, $t+\sqrt{t} \geq n$ implies that $t \geq 5$ and, for such values of t, we have $\sqrt{t} \leq \frac{1}{3}t+\frac{2}{3}$ since the right-hand side increases with t more rapidly than the left-hand side. Also, by definition of t, we necessarily have $t \leq n$. Hence $\sqrt{t} \leq \frac{1}{3}t+\frac{2}{3}$, as required.

Case II. $t+\sqrt{t} = n$.

We have $t+\sqrt{t} = (\frac{2}{3}n+\frac{1}{3})+(\frac{1}{3}n-\frac{1}{3})$. We show that, for this case, $\sqrt{t} \leq \frac{1}{3}n-\frac{1}{3}$ and it will then follow as in case I that $t \geq \frac{2}{3}n+\frac{1}{3}$. Since $n \geq 7$ and $t+\sqrt{t} = n$, the smallest possible value of n is 12, so $n > t \geq 9$. For such values of t, $\sqrt{t} \leq \frac{1}{3}(t+1)-\frac{1}{3}$ since the right-hand side increases with t more rapidly than the left-hand side. That is, $\sqrt{t} \leq \frac{1}{3}n-\frac{1}{3}$, as required.

Case III. $t+\sqrt{t} < n$.

The inequality $t+\sqrt{t} < n$ implies $(n-t)^2 > t$. In this case, as already remarked, since RL contains no element which is greater than t, all $(n-t)^2$ elements of the last $n-t$ rows of A which are greater than t occur in LL. That is, LL has more than t elements which are greater than t. Consequently, LL has at least two rows and one of its columns, say the jth ($j \leq t$) contains at least two elements which are greater than t. Let us suppose that these elements are $t+k$ and $t+l$.

We show first that, if j is so defined, then the integer j cannot occur in RL. Suppose on the contrary that j occurs in the cell of the $(t+g)$th row and $(t+h)$th column of RL. The cell of the $(t+g)$th row and jth column of the whole latin square A can contain at most one of the elements $t+k$, $t+l$, so we may suppose that the element $t+l$ does not appear in this cell (Fig. 3.1.3). Let us interchange the $(t+g)$th and jth rows of the latin square to obtain Fig. 3.1.4. Next, let us interchange the $(t+h)$th and jth columns of the latin square to obtain Fig. 3.1.5. Neither of these interchanges will affect the length of any partial transversal of the latin square. But in Fig. 3.1.5 the elements of the main left-to-right diagonal of LU together with the element $t + l$ in the $(t + h)$th column form a partial transversal of length $t + l$, contrary to our hypothesis that there is no partial transversal of length greater than t.

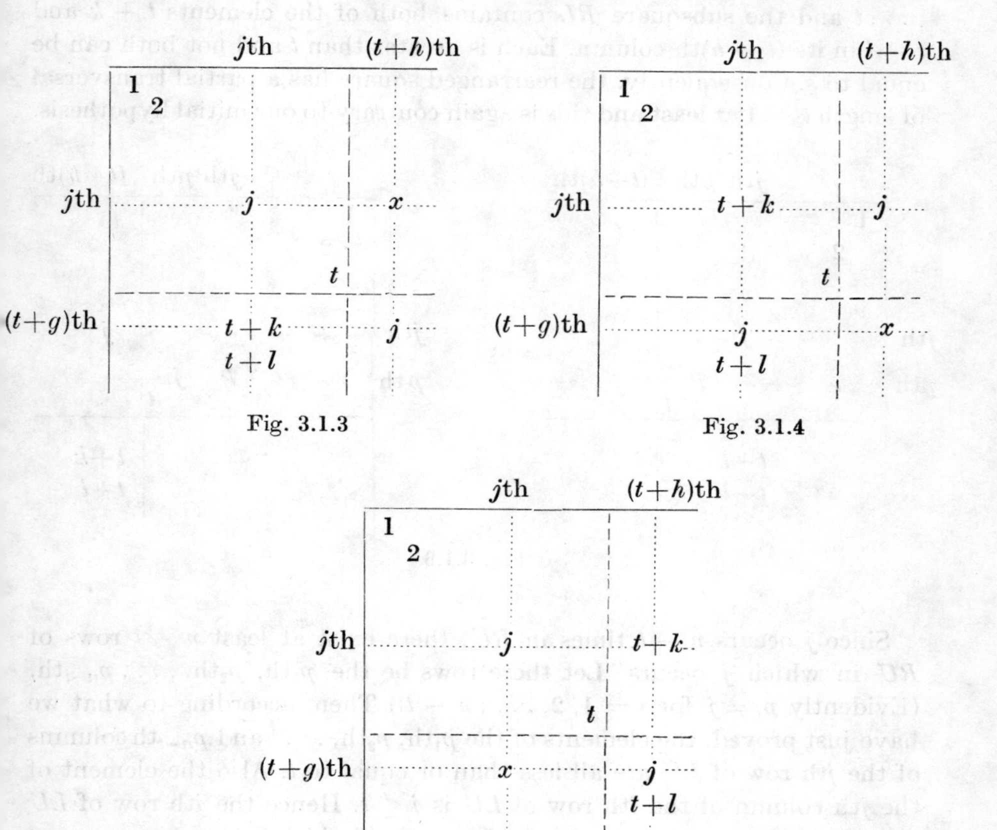

Fig. 3.1.3

Fig. 3.1.4

Fig. 3.1.5

Thus, we may suppose that the integer j does not occur in RL. If not, j occurs $n - t$ times in RU. Suppose in particular that j occurs in the pth row $(p \leq t)$ of RU in the $(t + h)$th column, say, of that row. Then we shall show that the pth element of the jth row of the whole square (which is in fact an element of LU) is less than or equal to t.

Suppose not, and let this element be $s(s > t)$. In that case, let us rearrange the jth, pth and $(t + h)$th columns of the whole square cyclically as follows. Replace the jth column by the pth, the pth by the $(t+h)$th, and the $(t + h)$th by the jth. This interchange of columns will not affect the length of any partial transversal of the latin square. But, in the rearranged square (Fig. 3.1.6) the main left-to-right diagonal of LU contains the elements $1, 2, \ldots, j-1, s, j+1, \ldots, p-1, j, p+1, \ldots, t$ and the subsquare RL contains both of the elements $t+k$ and $t+l$ in its $(t+h)$th column. Each is greater than t and not both can be equal to s. Consequently, the rearranged square has a partial transversal of length $t+1$ at least and this is again contrary to our initial hypothesis.

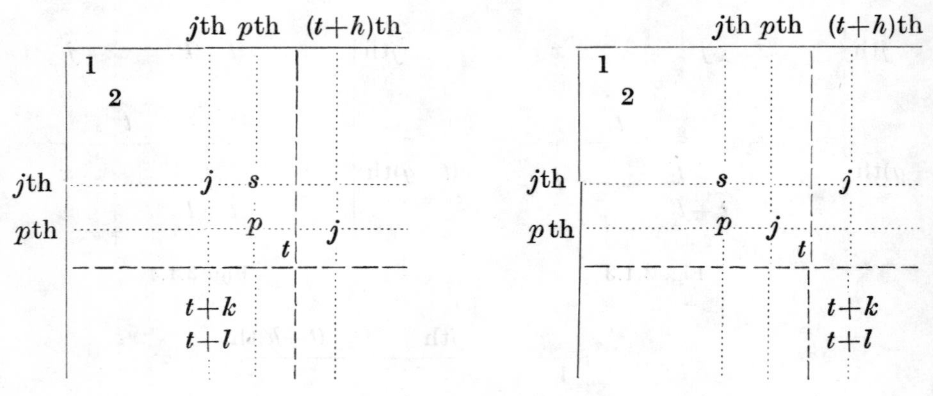

Fig. 3.1.6

Since j occurs $n-t$ times in RU, there exist at least $n-t$ rows of RU in which j occurs. Let these rows be the p_1th, p_2th, ..., p_{n-t}th. (Evidently $p_i \neq j$ for $i = 1, 2, \ldots, n-t$.) Then, according to what we have just proved, the elements of the p_1th, p_2th, ..., and p_{n-t}th columns of the jth row of LU are all less than or equal to t. Also the element of the jth column of the jth row of LU is $j \leq t$. Hence the jth row of LU contains at least $(n-t)+1$ elements each of which is less than or equal to t. But it is easy to see that all the elements in the jth row of RU also are less than or equal to t. For suppose that one such element, say s, occurring in the $(t+h)$th column were not. In that case, we could interchange the jth and $(t+h)$th columns of the whole square and obtain a new main left-to-right diagonal of LU as follows: $1, 2, \ldots, j-1, s, j+1, \ldots, t,$ and in the $(t+h)$th column of the new subsquare RL would occur the elements $t+k$ and $t+l$, not both of which are equal to s. The new main left-to-right diagonal of LU together with whichever of the elements $t+k$ or $t+l$ was distinct from s would provide a partial transversal of length $t+1$, again contrary to hypothesis.

We conclude that the only possibility is that the jth row of the whole latin square A contains at least $2(n-t)+1$ elements which are less than or equal to t. This number cannot exceed t and so $2(n-t)+1 \leq t$, whence $t \geq (2n+1)/3$ as required.

As Koksma has himself remarked, theorem 3.1.5 is still valid when $3 \leq n \leq 6$. The example exhibited in Fig. 3.1.7 shows that cases in which $t=(2n+1)/3$ can occur and so the inequality given by theorem 3.1.5 is the best possible.

$\boxed{1}$	2	4	3
2	$\boxed{3}$	1	4
3	4	$\boxed{2}$	1
4	1	3	2

Fig. 3.1.7

In fact, even in cases when a latin square has no transversals, it is very often (and R. A. Brualdi has conjectured that it may be always) the case that partial transversals of $n-1$ elements occur in it. (An example of this situation is given in Fig. 3.1.8.)

$\boxed{1}$	6	3	7	4	9	2	5	0	8
2	$\boxed{0}$	4	6	5	8	3	1	9	7
3	9	$\boxed{5}$	0	1	7	4	2	8	6
4	8	1	$\boxed{9}$	2	6	5	3	7	0
5	7	2	8	$\boxed{3}$	0	1	4	6	9
6	1	8	2	9	$\boxed{4}$	7	0	5	3
7	5	9	1	0	3	$\boxed{8}$	6	4	2
8	4	0	5	6	2	9	$\boxed{7}$	3	1
9	3	6	4	7	1	0	8	$\boxed{2}$	5
0	2	7	3	8	5	6	9	1	4

Fig. 3.1.8

The reader can find further results on the general subject of latin rectangles in H. B. Mann [4], H.J. Ryser [2], J. Riordan [5] and W.T. Federer [1].

We turn now to another generalization of the latin square concept which D. A. Norton has called a *row latin square*. (See D. A. Norton [1].)

A *row latin square* is a square matrix of order n say, each of whose rows is a permutation of the same n elements. A *column latin square* is similarly defined. These concepts are related to that of a latin rectangle by the fact that a union of latin rectangles each having n columns (and such that the total number of rows is n) is obviously a row latin square. However, the converse of this remark is *false*.

It is clear that a square matrix which is both a row latin square and a column latin square is a latin square.

Let R be a groupoid satisfying Sade's left translation law (identity 17 of section 2.1), then clearly the multiplication table of R is a row latin square. Similarly, the multiplication table of a groupoid R' satisfying Sade's right translation law is a column latin square.

Let R be a row latin square bordered by its elements taken in natural order. Then the ith row of the square determines a permutation P_i of the row border, and the square is completely determined by giving the permutations $(P_1, P_2, \ldots, P_n)$. The product of two row latin squares A and B, which are represented by the permutations $(P_1, P_2, \ldots, P_n)$ and $(Q_1, Q_2, \ldots, Q_n)$ may be defined as the square matrix $C = AB = (P_1Q_1, P_2Q_2, \ldots, P_nQ_n)$ whose ith row is given by the product permutation P_iQ_i. Then C is again a row latin square.

Such a representation seems likely to be very useful for quasigroups since, as A. Sutschkewitsch observed (see [1]), there are quasigroups whose right (or left) representation forms a group. Such a quasigroup is isomorphic to a quasigroup $(G, \cdot)$ obtained from an appropriate group $(G, \circ)$ by a relation of the form

$$x . y = (x\alpha^{-1} \circ y)\alpha$$

where α is an arbitrary fixed permutation of G. If both the right and left representations of a quasigroup form groups, the quasigroup is a group.

We shall give an example of a quasigroup Q whose right representation is a group but whose left representation is not a group.

Take Q to be the quasigroup of order 4 whose multiplication table

	1	2	3	4
1	1	4	2	3
2	2	3	1	4
3	3	2	4	1
4	4	1	3	2

Fig. 3.1.9

is given by Fig. 3.1.9. Then clearly the right representation of Q consists of the permutations (1)(2)(3)(4), (1 4)(2 3), (1 2)(3 4), (1 3)(2 4) and these form a group. However, the left representation of Q contains an element (1)(2 4 3) which is of order 3 so it is not a group. We constructed Q from the Klein group of order 4 using the rule of A. Sutschkewitsch and taking $\alpha = \begin{pmatrix} 1 & 2 & 3 & 4 \\ 2 & 3 & 1 & 4 \end{pmatrix}$.

The representation of a quasigroup by means of the row permutations defined by the corresponding latin square has been found useful by one of the authors in work on finite projective planes.

Some of the results published in D. A. Norton [1] will now be stated without proof.

The set of all row latin squares of order n forms a group of order $(n!)^n$ under the product operation defined above. This group is isomorphic to the direct product of n symmetric groups of degree n.

The number of row latin squares L_i of order n which have the property that every row of L_i^2 is represented by the identity permutation (that is,

$$L_i^2 = \left| \begin{matrix} 1 & 2 & \dots & n \\ 1 & 2 & \dots & n \\ . & . & \dots & . \\ 1 & 2 & \dots & n \end{matrix} \right. \quad)$$

is $[p\langle n\rangle]^n$, where $p\langle n\rangle$ is given inductively by $p\langle 1\rangle = 1$, $p\langle 2\rangle = 2$, and $p\langle n\rangle = p\langle n-1\rangle + (n-1)\, p\langle n-2\rangle$ for $n > 2$. The number of row latin squares of standard form(1) which have this same property is $[p\langle n-2\rangle]^{n-1}$.

In D. A. Norton [1], it has been shown that the existence of sets of mutually orthogonal latin squares (see chapter 5 for the definition of this concept) is dependent upon the parallel problem for row latin squares. Consequently, existence problems concerning sets of the former squares may be studied in terms of the latter.

Theorem 3.1.6. *A row (column) latin square which satisfies the quadrangle criterion and is such that at least one among its elements occurs in one of the cells of every column (row) is necessarily a latin square.*

(1) A latin square is said to be in *standard form* if its first row and first column contain the elements in their natural order.

PROOF. Theorem 3.1.6 is an immediate consequence of the well known theorem that an associative groupoid with a left identity element and which is such that each of its elements has a left inverse is a group.

In the present case, we take the element which occurs in every column as the left identity element (by bordering the latin square appropriately) and we easily see that then every element has a left inverse with respect to this identity.

3.2. L. Fuchs' problems

In his book [1], L. Fuchs raised the following problem: "Let k elements be deleted at random in the Cayley table of a finite Abelian group G of order n. Determine the greatest $k = k(n)$ for which

(a) the rest of the table always determines G up to isomorphism;

(b) the table can be reconstructed uniquely from its remaining part."

One of the authors has solved problem (b) without the restriction that the group is abelian (see J. Dénes [2]).

He has shown that for any given group G of order $n \neq 4$ we have

$$k(n) = 2n - 1.$$

Thus, unexpectedly, $k(n)$ does not depend on the structure of the group but depends only on its order. For the case $n \neq 4$ he obtained the value $k(n) = 3$.

For the proof, let us begin by pointing out that an abstract group is completely known when each of its elements has been represented by a symbol and the product of any two symbols in each order has been exhibited.

For a finite group, the products may be exhibited conveniently by means of the Cayley table of the group and we may represent the elements of the group by means of the natural numbers $1, 2, \ldots, n$. Moreover, as proved in section 1.1, a matrix $\| a_{ik} \|$ whose entries are the natural numbers $1, 2, \ldots, n$, will represent the Cayley table of such a group if and only if (1) it is a latin square and (2) the quadrangle criterion holds, that is, for all indices $i, j, \ldots$, the equalities

$$a_{ik} = a_{i_1k_1}, \quad a_{il} = a_{i_1l_1}, \quad a_{jk} = a_{j_1k_1},$$

imply $a_{jl} = a_{j_1l_1}$.

Further, we may choose an arbitrary row and column, say the jth row and the lth column, and consider them as being the products of the elements by the group identity from the left and from the right, respectively. Then a_{jl} will be the identity of the system G_{jl} thus arising, and necessarily $a_{il}\, a_{jk} = a_{ik}$. Now (1) ensures that G_{jl} is a loop and (2) implies associativity, thus G_{jl} is a group. Clearly, every group with the same Cayley table arises in this way.

All these G_{jl} are isomorphic, for a transition from G_{jl} to G_{rs} means simply that we take three permutations θ, φ, ψ such that $ab = c$ in G_{rs} if and only if $a\theta . b\varphi = c\psi$ in G_{jl}: that is, G_{jl} and G_{rs} are isotopic and hence, by theorem 1.3.4, isomorphic groups.

Note that different multiplication tables of a group can be transformed into one another by row and column interchanges.

If $\| a_{ik} \|$ and $\| b_{ik} \|$ are two Cayley tables, with the same number of rows and columns, then we call the ith rows *corresponding rows*, the kth columns *corresponding columns*, and a_{ik} and b_{ik} *corresponding elements*.

Any permutation may be written as a product of disjoint cycles. If all these cycles have the same length, the permutation is called *regular*. If for the permutations σ, τ we have $a\sigma \neq a\tau$ for exactly k letters a, then we say that they differ in k places.

We shall need the following result concerning permutations.

Let π and ϱ denote two distinct regular permutations of degree n and of orders l, m $(m \leq l \leq n)$, respectively. If n is even and π, ϱ are of the form

$$\pi = (i_1\, i_2 \ldots i_n), \;\; \varrho = (i_1\, i_2 \ldots i_{n/2})(i_{1+n/2} \ldots i_n),$$

then they differ in two places. In all other cases they differ at least in three places.

For the proof, let us suppose that π and ϱ differ in two places. Then there is a transposition τ such that $\varrho = \pi\tau$. Both π and ϱ are products of disjoint cycles. If the letters of τ belong to the same cycle of π, then in the product $\pi\tau$ this cycle splits into two.

On the other hand, if the symbols of τ belong to two distinct cycles of π, these cycles are replaced by one in ϱ. Since, π and ϱ are both regular, the only possibility is that one consists of one cycle and the other of two. That is, they have the form indicated.

Theorem 3.2.1. *Two different Cayley tables, A and A', of a given group G differ from each other in at least $2n$ places.*

Proof. If no two corresponding rows of the two Cayley tables are the same then every row differs from the corresponding one at least in two places, and the two Cayley tables differ at least in $2n$ places. The same argument applies if the corresponding columns are different.

For the rest of the proof we may assume that the fth rows and the gth columns of the two Cayley tables are equal. By the quadrangle criterion,

$$a_{fg} = a'_{fg}, \quad a_{fj} = a'_{fj}, \quad a_{ig} = a'_{ig}$$

imply $a_{ij} = a'_{ij}$ for all indices i, j; whence $A = A'$. Thus, this case cannot occur.

THEOREM 3.2.2. *If G and G' are different groups of the same order n ($n \neq 4$) then their Cayley tables, A and A', differ from each other in at least $2n$ places.*

PROOF. We may suppose that at least one pair of corresponding rows of A and A' are equal. Otherwise we could use the same inference as that in the proof of theorem 3.2.1 to obtain the desired conclusion.

Every finite group may be represented as a group of regular permutations (see, for example, C. Jordan [1]). Such representations of G and G' are

$$x_i \to \begin{pmatrix} x \\ xx_i \end{pmatrix} \quad (x_i \in G)$$

and

$$x'_i \to \begin{pmatrix} x \\ xx'_i \end{pmatrix} \quad (x'_i \in G')$$

where we may suppose that x varies over the elements of (either of) the pair of equal corresponding rows.

Let P, P' denote the regular permutation groups corresponding to G, G'. They are different and so their intersection $H = P \cap P'$, which is obviously a subgroup of both P and P', necessarily has order s, where $s \leq n/2$ and is a divisor of n. (Thus, if $s \neq n/2$ then $s \leq n/3$.) Consequently, $n-s$ rows of the Cayley tables of G and G' are different, where $n-s \geq n/2$.

First we consider the case when $s \leq n/3$ and the set $(P \cup P') \setminus H$ does not contain permutations of both of the forms $\pi = (i_1\, i_2 \ldots i_n)$ and $\varrho = (i_1\, i_2 \ldots i_{n/2})(i_{1+n/2} \ldots i_n)$.

In that case, each of the $n - s$ rows differ in at least three places and so the squares differ in $3(n - s) \geq 2n$ places all together.

Next, if $s = n/2$ and the set $(P \cup P') \setminus H$ does not contain permutations of both of the forms π and ϱ, then we shall show that each pair of unequal corresponding rows differ from each other at least in four places. That is, we wish to show that ψ and σ ($\psi \in P \setminus H$, $\sigma \in P' \setminus H$) differ from each other in at least four places. If they differed from each other only in three places, then there would be a cycle φ of order 3 such that $\sigma\varphi = \psi$. Taking into consideration the regularity of ψ and σ and the fact that φ is a product of two transpositions, the difference between the numbers of cycles of ψ and σ is either 2 or 0. Then ψ and σ must have the form (1) or (2) given below:

(1) $\psi = (i_1\, i_2 \ldots i_n), \quad \sigma = (i_1 \ldots i_{n/3})\, (i_{1+n/3} \ldots i_{2n/3})\, (i_{1+2n/3} \ldots i_n) =$
$= \psi(i_1\, i_{1+2n/3}\, i_{1+n/3})$

(2) $\psi = (i_1 \ldots i_j\, i_{j+1} \ldots i_{j+k}\, i_{j+k+1} \ldots i_m)\, (\ldots)\, (\ldots) \ldots$
$\sigma = (i_1 \ldots i_j\, i_{j+k+1} \ldots i_m\, i_{j+1} \ldots i_{j+k})\, (\ldots)\, (\ldots) \ldots =$
$= \psi(i_1 \quad i_{j+1} \quad i_{j+k+1})$.

As H is a subgroup of index 2 in P and P', all even powers of ψ and σ are in H. Since, up to isomorphism, there exists only one group for each of the orders $n = 2$, 3, and 5, we may now restrict ourselves to the case $n > 6$. If ψ and σ take the form (1), then $\psi^{n-2}\sigma^2 \in H$ and $\psi^{n-2}\sigma^2$ maps i_n onto itself. Since H is regular it follows that $\psi^{n-2}\sigma^2$ must be equal to the identical permutation. But because $\psi^{n-2}\sigma^2$ maps $i_{2+n/3}$ onto i_2, this is impossible. If ψ and σ take the form (2), we may suppose that they do not consist of a single cycle, for otherwise $\{\psi\} = P$ and $\{\sigma\} = P'$ would be cyclic groups and our statement would be trivial. We may assume that $j \geq 1$ since $\sigma = \psi(i_1\, i_{j+1}\, i_{j+k+1})$. As $\sigma^2 \in H, \psi^{m-2}\sigma^2$ is regular. Also $\psi^{m-2}\sigma^2$ maps i_3 onto itself, so it is the identical permutation. But because $\psi^{m-2}\sigma^2$ maps i_2 onto i_{j+2} $(= i_2$ for $j \geq 1)$, this is impossible.

This proves the theorem for the case when $s = n/2$ (under the assumption that $(P \cup P') \setminus H$ does not contain permutations of both of the forms π and ϱ) since we have shown that the squares differ in at least $4\ (n - n/2) = 2n$ places.

Finally, if $s \leq n/2$ and π and ϱ both belong to $(P \cup P') \setminus H$ then $P = \{\pi\}$ and $P' = \{\sigma, \varrho\}$ where σ is an arbitrary element of P' not in $\{\varrho\} = R$. Clearly $P \cap R = \varepsilon$ (the identity permutation) as every cycle of a power π^k $(\neq \varepsilon)$ contains symbols with suffices both larger and smaller than $n/2$, and this is impossible for the elements of R. So $H \cap R = \varepsilon$. Therefore all the products $\alpha\beta$ $(\alpha \in H, \beta \in R)$ are distinct. As the index of R is 2, the order of H is at most 2. Therefore we may have only two equal rows in arbitrary Cayley tables of G and G'.

Let us suppose that among the elements of $P \setminus H$ and $P' \setminus H$ there are pairs, other than π and ϱ and their inverses, whose letters differ from each other in two places. If π^k, ϱ' $(\pi^k \neq \pi,\ \varrho' \neq \varrho)$ were such a pair, then ϱ' would not be an element of $\{\varrho\}$ and $\varrho'\varrho$ would not be regular. We conclude that the elements of the sets $P \setminus H$ and $P' \setminus H$ differ at least in three places from each other in each row except in the rows containing π and ϱ. Consequently, arbitrary Cayley tables of G and G' differ from each other at least in $3(n-4)+2.2$ places. This number is $\geq 2n$ when $n>7$.

It remains to consider the cases when $n \leq 7$. If n is a prime number, then all groups of order n are isomorphic cyclic groups and so our statement follows from theorem 3.2.1.

The only case that remains is $n=6$. In view of theorem 3.2.1, we may suppose without loss of generality that G and G' are not isomorphic, and so the groups in question are the cyclic and the dihedral groups of order six. Now

$$P=\{(1\ 2\ 3\ 4\ 5\ 6)\}, \quad \text{and}$$

$$P'=\{(1\ 2\ 3)(4\ 5\ 6),\ (16)(25)(34)\}.$$

H cannot contain the only permutation $(14)(25)(36)$ of order 2 in P, because $\varrho(14)(25)(36)=(1\ 2\ 3)(4\ 5\ 6)(14)(25)(36)=(1\ 5\ 3\ 4\ 2\ 6)$ and $(1\ 5\ 3\ 4\ 2\ 6) \notin P'$. Thus $s=1$ and therefore arbitrary Cayley tables of G and G' differ from each other in

$$3(n-3)+2.2=3n-5=13>2.6$$

places. This completes the proof of theorem 3.2.2.

Theorem 3.2.3. *For a group G of order n we have $k(G)=2n-1$ $(n \neq 4)$.*

Proof. Let us delete $2n-1$ arbitrary elements in a Cayley table A of the group G of order n $(n \neq 4)$. Suppose that there is a Cayley table A' $(A \neq A')$ of G having the property that the rest of A may be completed to A'. Then, clearly, A and A' differ in $2n-1$ places, which is impossible because of theorems 3.2.1, 3.2.2.

We have to prove now that we can delete $2n$ elements of a Cayley table A of a group G of order n in such a way that the rest of the table may be completed to a Cayley table A' different from A. If we interchange two arbitrary symbols, a and b, throughout in A, then we shall obtain a new Cayley table differing from A in exactly $2n$ places. So the proof of our statement is completed.

Corollary. *Two different Cayley tables of arbitrary groups of order n $(n \neq 4)$ differ from each other in at least $2n$ places.*

Theorem 3.2.4. *An arbitrary Cayley table of the cyclic group of order 4 differs in at least four places from an arbitrary Cayley table of Klein's group.*

Proof. Let us first observe that the two Cayley tables given in Fig. 3.2.1 are different in four places.

e	a	b	c
a	b	c	e
b	c	e	a
c	e	a	b

(cyclic group of order 4)

e	a	b	c
a	e	c	b
b	c	e	a
c	b	a	e

(Klein's group)

Fig. 3.2.1

Then, in order to complete the proof, we have to show that any two distinct latin squares of order 4 differ from each other in at least four places. If the two latin squares represent multiplication tables of the same group, they differ in at least 8 places by theorem 3.2.1. If not, the result follows from the fact that if two corresponding rows or columns are unequal, then these rows or columns differ at least in two places. Thus, if there is a pair of unequal corresponding rows, then there are at least two pairs of unequal columns, and therefore the squares differ in at least four different places.

Remarks.

The following statements are immediate consequences of the above theorems:

(1) The result remains the same if we restrict the class of groups to any one of the following classes of finite groups: (*i*) soluble groups; (*ii*) nilpotent groups; (*iii*) abelian groups; (*iv*) cyclic groups.

(2) If we add the condition that the Cayley tables have to be normal, then $k(G) = 2n - 1$ for all n (including the case $n = 4$).

It seems to be natural to raise the following problem:

What is the maximum number of squares which a set of latin squares satisfying the quadrangle criterion and all of the same order n can contain if each pair of squares in the set are to differ from each other in at most m places?

At the beginning of this section, we stated the problem of L. Fuchs as a problem concerning groups. Let us end this section by discussing the solution to the same problem relative to quasigroups. For this purpose, we shall make use of the concept of distance between two matrices.

The *distance* between two square matrices $\| a_{ij} \|$ and $\| b_{ij} \|$, $i, j = 1, 2, \ldots, n$, is equal to the number of cells in which the corresponding elements a_{ij} and b_{ij} are not equal.

The distance so defined is denoted by $D(\| a_{ij} \|, \| b_{ij} \|)$ and is a generalization of the original notion of Hamming distance between vectors having binary components which we shall make use of in section 10.1. The above generalized distance was introduced by C. Y. Lee (see [1]).

Let us denote the multiplication table of an arbitrary group G of order n by $\| G \|$. Then the result proved in theorem 3.2.2 may be stated as follows:

If G is an arbitrary group of order n $(n \neq 4)$ and $\| G \|$ and $\| G' \|$ are two Cayley tables of G and G' which are not identical then $D(\| G \|, \| G' \|) \geq 2n$ unless G and G' are of order 4, in which case $D(\| G \|, \| G' \|) \geq 4$.

The corresponding result for quasigroups, which was first proved by J. Dénes and K. Pásztor in [1], may be deduced as a corollary to the theorem which follows. For this, we need to remember the fact that the multiplication table of an arbitrary quasigroup is a latin square (proved in theorem 1.1.1).

THEOREM 3.2.5. *For arbitrary n $(n \geq 4)$ there exists a latin square of order n having a latin subsquare of order 2.*

PROOF. If n is even then the theorem is a trivial consequence of Cauchy's theorem: namely that, if G is a finite group of order n and p a prime divisor of n, then G contains at least one element of order p. Thus for the rest of our proof we may suppose that n is odd and $n > 3$.

We shall use the notion of *transversal* which was introduced in section 1.4. It is well known that a cyclic group of odd order m, $m = n - 2 > 1$, has m disjoint transversals. Using this fact, one can obtain a latin square of order $m + 2$ $(= n)$ from the Cayley table of a cyclic group of order m using the method of prolongation introduced in section 1.4.

We choose two arbitrary disjoint transversals from the Cayley table of a cylic group of order m. We replace each element of the first transversal by $m + 1$ (without loss of generality, we may suppose that the elements of the latin square defined by the Cayley table are the natural numbers $1, 2, \ldots, m$) and we replace each element of the second transversal by $m+2$. The new configuration so obtained is not a latin square, but after adding the first and second transversals to this configuration as its $(m+1)$th and $(m+2)$th rows and also as its $(m+1)$th and $(m+2)$th columns, we can complete the resulting configuration to a latin square of order $m+2\,(=n)$ by adjoining $m+1, m+2$ to the $(m+1)$th and $(m+2)$th rows and columns respectively. This enlarged latin square clearly contains a latin subsquare of order 2: namely that formed by the last two of its rows and columns.

COROLLARY. *For arbitrary n, ($n \neq 3$, $n > 1$), there exist two latin squares of order n whose distance is exactly* 4; *that is, so that at most* 3 *arbitrary elements may be deleted from one of the squares if it is still to be possible to reconstruct that square uniquely from the remaining elements.*

PROOF. This is immediate by theorem 3.2.5. For let T be a latin square of order n having a latin subsquare of order 2, say

$$T = \begin{array}{|lllll} \hline a_1 & a_2 & a_3 & \cdots & a_n \\ a_2 & a_1 & & & \\ \cdot & & & & \\ \cdot & & & & \\ \cdot & & & & \end{array}$$

Let us denote the following latin square

$$\begin{array}{|lllll} \hline a_2 & a_1 & a_3 & \cdots & a_n \\ a_1 & a_2 & & & \\ \cdot & & & & \\ \cdot & & & & \\ \cdot & & & & \end{array}$$

by T'; then $D(T, T') = 4$.

For an analogous result concerning semigroups, see J. Dénes [4].

3.3. Incomplete latin squares and partial quasigroups

A latin rectangle of size $r \times s$ is called *incomplete* or *partial* if less than rs of its cells are occupied. An *incomplete* or *partial* latin square is analogously defined. Precisely, we have:

DEFINITION. An $n \times n$ *incomplete* or *partial latin square* is an $n \times n$ array such that in some subset of the n^2 cells of the array each of the cells is occupied by an integer from the set $1, 2, \ldots, n$ and such that no integer from the set 1, 2, . . . , n occurs in any row or column more than once.

In section 3.2 we investigated incomplete latin squares whose elements were not arbitrarily assigned but had been obtained by deleting elements from a given latin square. The distinction will be clear from Fig. 3.3.1.

In the first part of this section we shall consider incomplete latin squares and rectangles whose elements will be arbitrarily chosen and the question of whether can they be completed to a latin square or rectangle will be investigated.

$$\begin{array}{|cccc}
\hline
1 & . & . & . \\
. & 2 & 3 & 4
\end{array}$$

Fig. 3.3.1

Figure 3.3.1 shows an incomplete latin rectangle of size 2×4 which cannot be completed to a latin rectangle of the same size. By generalizing this, it is easy to see that, for any n ($n \geq 2$), there exists an incomplete latin rectangle with $2n$ cells occupied which cannot be completed to an $n \times 2n$ latin rectangle. On the other hand, C. C. Lindner proved in [1] that an incomplete $n \times 2n$ latin rectangle with $2n - 1$ cells occupied can always be completed to an $n \times 2n$ latin rectangle.

In [2], T. Evans posed the following problem:

"What conditions suffice to enable an incomplete $n \times n$ latin square to be embedded in a complete latin square of order n? In particular, can an $n \times n$ incomplete latin square which has $n - 1$ or less places occupied

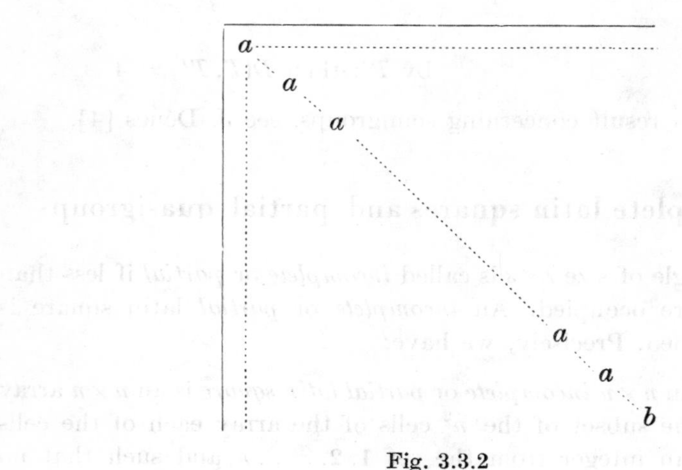

Fig. 3.3.2

be completed to a latin square of order n?" Exactly the same problem has been posed by D. A. Klarner on page 1167 of P. Erdős, A. Rényi and V. Sós [1], and independently by J. Dénes in a lecture given at the University of Surrey.

It is easy to see that there do exist incomplete latin squares with n cells occupied which cannot be so completed, as Figs 3.3.1 and 3.3.2 illustrate.

The general case of T. Evans' problem is not yet solved, but some results exist which can be regarded as partial solutions to the problem. They will be listed below.

Firstly, J. Marica and J. Schönheim have proved that an incomplete latin square containing $n - 1$ arbitrarily chosen elements can be completed to a latin square of order n provided that the chosen elements are in different rows and columns (see J. Marica and J. Schönheim [1]).

Secondly, C. C. Lindner (using the same technique as Marica and Schönheim) has proved that an incomplete latin square containing $n - 1$ *distinct* elements can be completed to a latin square of order n provided that the chosen elements are either in different rows or in different columns (see C. C. Lindner [2]).

Thirdly, Lindner (in C. C. Lindner [1]) has also proved the following theorem: Let L be an $n \times n$ incomplete latin square with $n - 1$ cells occupied. Let r and c denote respectively the number of rows and the number of columns in which occupied cells occur. Then, if $r \leq [n/2]$ or $c \leq [n/2]$, L can be completed to a latin square of order n.

In the same connection, one can ask for the solution of the following problem: "How many elements of a latin square of order n and which satisfies the quadrangle criterion can be located arbitrarily subject only to the condition that no row or column shall contain any element more than once?" Since a latin square of odd order which satisfies the quadrangle criterion cannot contain a latin subsquare of order 2 (see section 1.5), it is easy to see that, for n odd, this number is at most 3. Thus, for example, the latin square shown in Fig. 3.3.3 cannot satisfy the quadrangle criterion if it is of odd order.

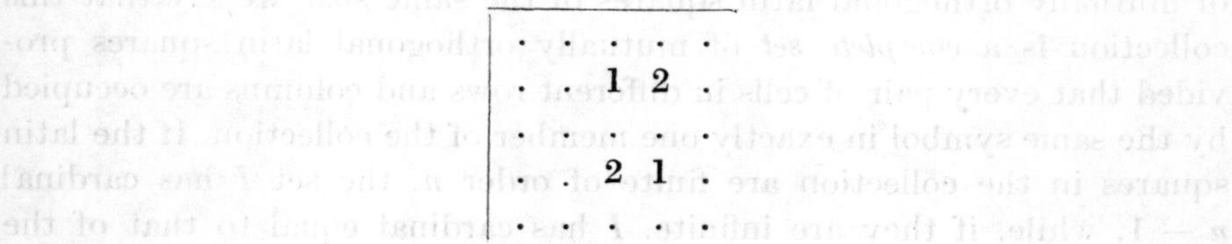

Fig. 3.3.3

Moreover, in consequence of A. Sade's theorem to the effect that every commutative quasigroup of odd order is diagonal (see section 1.4), not even *two* arbitrarily chosen elements can be arbitrarily placed in a symmetric latin square of odd order: for Sade's result implies that no element can appear more than once in the main left-to-right diagonal of such a square.

In [1], J. Csima has translated T. Evans' problem of finding under what conditions a partial latin square can be embedded in a complete latin square into a problem about the combinatorial structure which he has called a *pattern*. For details, the reader is referred to Csima's paper.

In his paper [2], already referred to, T. Evans has posed the following further question which has not so far been answered. "Can a pair of $n \times n$ incomplete latin squares which are orthogonal (insofar as the condition for orthogonality applies to the incomplete squares) be respectively embedded in a pair of $t \times t$ orthogonal latin squares; and, if so, what is the smallest value of t for each value of n?"

Somewhat related to this question is an interesting result proved by C. C. Lindner in [9]: namely, *any finite collection of mutually orthogonal $n \times n$ partial latin squares can be embedded in a complete set of mutually orthogonal infinite latin squares.*

By an *infinite latin square* is meant a countably infinite array of rows and columns such that each positive integer occurs exactly once in each row and exactly once in each column. If P is a finite (partial) latin square, we will denote by C_p the set of all the cells which are occupied in P. If P and Q are finite (partial) latin squares of the same size, we say that P and Q are *orthogonal* if the cardinals of the sets $C_p \cap C_q$ and $\{(p_{ij}, q_{ij}) : (i, j) \in C_p \cap C_q\}$ are equal. If P and Q are infinite latin squares, we say that P and Q orthogonal provided that the set $\{(p_{ij}, q_{ij}) : (i, j) \in C_p \cap C_q\}$ has the same cardinal as the Cartesian product set $Z^+ \times Z^+$ (where Z^+ denotes the set of positive integers) and that every pair of cells in different rows and columns are occupied by the same symbol in at most one of P and Q. If $\{P_i\}_{i \in I}$ is a collection of mutually orthogonal latin squares of the same size, we say that this collection is a *complete set* of mutually orthogonal latin squares provided that every pair of cells in different rows and columns are occupied by the same symbol in exactly one member of the collection. If the latin squares in the collection are finite of order n, the set I has cardinal $n - 1$, while, if they are infinite, I has cardinal equal to that of the set Z^+.

The reader new to the subject is advised that he will find the above concepts easier to understand if he first reads sections 5.1 and 5.2.

For the proof of his result stated above, Lindner uses the geometrical concepts of *projective plane* and *partial projective plane*. (The connection between orthogonal latin squares and projective planes is explained in section 5.2.)

Definitions. An $n \times n$ *partial idempotent latin square* is an $n \times n$ partial latin square with the additional property that, for each i, $i = 1, 2, \ldots, n$, the cell (i, i) is either empty or else is occupied by the integer i.

An $n \times n$ *partial symmetric latin square* is an $n \times n$ partial latin square with the additional property that, if the cell (i, j) is occupied by the symbol k so also is the cell (j, i). (In other words, it is a latin square which represents the multiplication table of a commutative quasigroup when suitably bordered.)

In a series of papers, Lindner has solved a number of embedding problems concerning squares of these kinds. We have the following results:

(1) Any finite partial idempotent latin square can be embedded in a finite idempotent latin square (C. C. Lindner [6]).

(2) Any finite partial symmetric latin square can be embedded in a finite symmetric latin square (C. C. Lindner [7]).

(3) Any finite partial idempotent and symmetric latin square can be embedded in a finite idempotent and symmetric latin square (C. C. Lindner [7]).

Subsequently, A. J. W. Hilton was able to improve the result (1) to the more precise statement that *any $n \times n$ partial idempotent latin square can be embedded in a $4n \times 4n$ idempotent latin square*. (See A. J. W. Hilton [1].) Then, making use of a technique previously employed in T. Evans [2], Lindner showed that his result (2) could also be improved to the corresponding statement that *any $n \times n$ partial symmetric latin square can be embedded in a $4n \times 4n$ symmetric latin square*. The proof will be found in C. C. Lindner [8].

In A. B. Cruse [1], necessary and sufficient conditions are obtained for the extendibility of an $r \times r$ symmetric latin rectangle to an $n \times n$ symmetric latin square. It is shown that these conditions imply that any partial $n \times n$ symmetric latin square can be embedded in a complete $2n \times 2n$ symmetric latin square. Also any partial $n \times n$ diagonal symmetric latin square can be embedded in a complete diagonal symmetric latin square of order $2n + 1$. See also A. J. W. Hilton [5].

By a *partial quasigroup* we shall mean a groupoid G such that if the equations $ax = b$ or $ya = b$ have solutions for x and y in G, then these solutions are unique. A *partial loop* is a partial quasigroup with an identity element such that the product of this element with each element a of the loop is defined and is equal to a. (By other authors, partial quasigroups have been called *incomplete quasigroups* or *half-quasigroups*, see R. H. Bruck [5] and T. Evans [2].)

Another result on partial quasigroups given in T. Evans' paper [2] is as follows: An incomplete loop (or quasigroup) containing n elements can always be embedded in a loop (quasigroup) containing t elements for any $t \geq 2n$.

Evans has also succeeded in deducing as a consequence of this result a stronger one as follows:

A properly incomplete loop (quasigroup) Q of order n (that is, one such that the product of at least one pair of its elements is undefined) can always be embedded in a loop of order t which is generated by the elements of Q if t is any assigned integer $\geq 2n$. Some other results concerning the embedding of partial quasigroups of certain particular kinds will be found in C. C. Lindner [7].

In [1], H. Brandt introduced a special kind of partial groupoid R (which has subsequently been called a *Brandt groupoid*) satisfying the following postulates:

(1) If any three elements a, b, c, satisfy an equation $ab = c$ then each of the three elements is uniquely determined by the other two.

(2) If ab and bc both exist, then $(ab)c$ and $a(bc)$ also exist; if ab and $(ab)c$ both exist then bc and $a(bc)$ also exist; if bc and $a(bc)$ both exist then ab and $(ab)c$ also exist; and in all of these cases the equation $(ab)c = = a(bc)$ is valid, and consequently the expression abc has an unambiguous meaning.

(3) To any given element a there corresponds a unique right-identity element e such that $ae = a$, a unique left-identity element e' such that $e'a = a$ and an inverse $\bar{a}$ of a such that $\bar{a}a = e$.

(4) If e and e' are any two members of the set of one-sided identity elements, there exists an element a such that e and e' are respectively right and left identities with respect to a.

Postulate (1) implies that the multiplication table of a Brandt groupoid is an incomplete latin square, and we shall devote the remainder of this section to mentioning some results and conjectures concerning such multiplication tables.

Theorem 3.3.1. *The multiplication table of a Brandt groupoid is an incomplete latin square satisfying the quadrangle criterion.*

Proof. As we have just remarked, postulate (1) implies that the multiplication table of a Brandt groupoid is an incomplete latin square, and it then follows immediately from postulate (2) that the multiplication table of a Brandt groupoid satisfies the quadrangle criterion.

We would like to point out by means of an example that incomplete latin squares exist which satisfy the quadrangle criterion but which can be completed to latin squares in which the quadrangle criterion does not hold. The incomplete latin square exhibited in Fig. 3.3.4 has these properties, as is indicated by Fig. 3.3.5.

3	4	2	1	.
1	2	.	3	4
.	.	1	2	.
.	.	3	4	.
.	.	.	.	.

Fig. 3.3.4

3	4	2	1	5
1	2	5	3	4
4	5	1	2	3
5	1	3	4	2
2	3	4	5	1

Fig. 3.3.5

On the other hand, the authors conjecture that every Brandt groupoid B can be embedded in a quasigroup Q which is group isotopic. The conjecture is certainly valid, for example, for the Brandt groupoid given as follows.

The elements of B are the set of all binary triplets, and we shall denote them as follows:

$1 = (0\ 0\ 0)$ $2 = (1\ 0\ 0)$ $3 = (0\ 1\ 0)$ $4 = (1\ 1\ 0)$

$5 = (0\ 0\ 1)$ $6 = (0\ 1\ 1)$ $7 = (1\ 0\ 1)$ $8 = (1\ 1\ 1)$

If $(H, *)$ denotes the cyclic group of order two with elements 0, 1, the product of two triplets $(a\ b\ c)$ and $(d\ e\ f)$ is defined to be the triplet $(a\ \ b{*}e\ \ f)$ if $c = d$ and is not defined otherwise.

Then Fig. 3.3.6 shows that the multiplication table of B can be embedded in the multiplication table of a quasigroup Q which is isotopic to the generalized Klein group of order 8. (In the diagram, the products which are defined in B are enclosed in squares.)

	1	2	3	4	5	6	7	8
1	[1]	7	[3]	8	[5]	[6]	4	2
2	[2]	5	[4]	6	[7]	[8]	3	1
3	[3]	8	[1]	7	[6]	[5]	2	4
4	[4]	6	[2]	5	[8]	[7]	1	3
5	7	[1]	8	[3]	4	2	[5]	[6]
6	8	[3]	7	[1]	2	4	[6]	[5]
7	5	[2]	6	[4]	3	1	[7]	[8]
8	6	[4]	5	[2]	1	3	[8]	[7]

Fig. 3.3.6

Let us remark here that it is well known that an arbitrary Brandt groupoid B can be embedded in a semigroup S with one additional element 0 such that $a0 = 0a = 00 = 0$ for all $a \in B$ and such that $ab = 0$ in S if a, b are in B and ab is not defined in B. (See, for example, R. H. Bruck [5], page 35). This fact leads the authors of this book to propose the alternative conjecture that in fact any Brandt groupoid can be embedded in a group.

Let us end the present chapter by remarking that, although latin cubes, orthogonal arrays, and Room squares may all be regarded as generalizations of the latin square concept, we have found it more convenient to discuss these constructs in later chapters. Latin cubes and orthogonal arrays are dealt with in section 5.4 and Room squares in section 6.4.

CHAPTER 4

Classification and enumeration of latin squares and latin rectangles

As is indicated by the title, this chapter is devoted to the classification and enumeration of latin squares and latin rectangles. The notion of *autotopism* (which we define below) plays an important role in obtaining results on the subject of this chapter and so we shall begin by giving an introductory account of this concept. Most of our results first appeared in E. Schönhardt [1]. However, the terminology of that paper is very antiquated and so we have reformulated the results using present day terms.

In the second section of the chapter (on classification), we have included a table which classifies the latin squares of all orders up to order six inclusive into their main, isotopism and isomorphism classes; and in the third section (on enumeration) will be found a historical account of the development of the subject of classification and enumeration over the last two centuries.

The final section of the chapter is devoted to the enumeration of latin rectangles.

4.1. The autotopism group of a quasigroup

We begin by reminding the reader that the concept of isotopism (which includes that of autotopism) was defined in section 1.3, and we first prove

THEOREM 4.1.1. *The set of all isotopisms of a groupoid of order n form a group I_n of order* $(n!)^3$.

PROOF. An isotopism of a groupoid can be characterized by three permutations of degree n exactly as for quasigroups. (See section 1.3.) Since each of the permutations may be chosen arbitrarily out of the possible $n!$, their number is $(n!)^3$.

An operation on the set of the isotopisms, to be called their *product*, can be defined by the statement that the product of the isotopisms (α, β, γ) and $(\alpha', \beta', \gamma')$ is $(\alpha\alpha', \beta\beta', \gamma\gamma')$. It is obvious that the product so defined is again an isotopism. Moreover, it is easy to check that all the group axioms are valid. Consequently, the isotopisms of a groupoid of order n form a group I_n under the operation just defined.

COROLLARY 1. $I_n \cong S_n \times S_n \times S_n$ *where* S_n *denotes the symmetric group of degree n.*

COROLLARY 2. *The set of all isotopisms of a quasigroup of order n form a group of order* $(n!)^3$.

DEFINITION. If (α, β, γ) is an isotopism of a groupoid Q onto a groupoid Q^* and if $Q^* = Q$ then the isotopism (α, β, γ) is called an *autotopism* of Q.

THEOREM 4.1.2. *The set of all autotopisms of a groupoid Q of order n form a group which is a subgroup of* I_n.

PROOF. Since autotopisms are isotopisms as well, we may define the product of two autotopisms in the same way as for isotopisms; and then, by definition of an autotopism, the product of two of them will be an autotopism again.

Hence, the set of all autotopisms of Q form a groupoid G_Q; and G_Q has an identity element because the identity $(\varepsilon,\varepsilon,\varepsilon)$ (ε denotes the identity permutation) of I_n acts as an identity element for G_Q also.

Associativity in G_Q is ensured by the fact that G_Q is a subgroupoid of I_n.

To complete the proof we need to show that any $(\alpha, \beta, \gamma) \in G_Q$ has its inverse $(\alpha^{-1}, \beta^{-1}, \gamma^{-1})$ also in G_Q.

However, this follows from the facts that $(\alpha, \beta, \gamma) \in G_Q$ implies $(\alpha^m, \beta^m, \gamma^m) = (\alpha, \beta, \gamma)^m \in G_Q$ for any integer m and that $(\alpha^{-1}, \beta^{-1}, \gamma^{-1}) = (\alpha^m, \beta^m, \gamma^m)$ for some integer m and so belongs to G_Q.

COROLLARY. *The order of* G_Q *is a divisor of* $(n!)^3$.

By analogy with the concept of principal isotopism (see section 1.3) one can define a *principal autotopism* as an autotopism (α, β, γ) such that γ is the identity permutation.

It is an immediate consequence of theorem 4.1.2 that the principal autotopisms of a groupoid of order n form a group which is a subgroup of the group formed by the principal isotopisms contained in I_n.

DEFINITION. An autotopism (α, β, γ) is called an *automorphism* if $\alpha = \beta = \gamma$.

The following results follow almost immediately from this definition:

(*i*) The group of automorphisms and the group of principal autotopisms are contained in the group of autotopisms.

(*ii*) The group of principal autotopisms is a normal subgroup of the group of autotopisms.

(*iii*) The group of principal autotopisms and the group of automorphisms have no common elements other than the autotopism whose components are each equal to the identity permutation.

THEOREM 4.1.3. *If Q_1 and Q_2 are two isotopic groupoids then $G_{Q_1} \cong G_{Q_2}$.*

PROOF. Since Q_1 and Q_2 are isotopic, there exists an isotopism (α, β, γ) mapping Q_1 onto Q_2. It follows that if (σ, η, ϱ) is an autotopism belonging to G_{Q_1}, then $(\sigma, \eta, \varrho)(\alpha, \beta, \gamma) = (\sigma\alpha, \eta\beta, \varrho\gamma)$ also maps Q_1 onto Q_2 and hence $(\alpha^{-1}, \beta^{-1}, \gamma^{-1})(\sigma, \eta, \varrho)(\alpha, \beta, \gamma)$ is an autotopism of G_{Q_2}.

It follows immediately that

$$(\alpha^{-1}, \beta^{-1}, \gamma^{-1}) G_{Q_1}(\alpha, \beta, \gamma) \subseteq G_{Q_2}.$$

Since $(\alpha^{-1}, \beta^{-1}, \gamma^{-1}) = (\alpha, \beta, \gamma)^{-1}$ is an isotopism mapping Q_2 onto Q_1, a similar argument with the roles of Q_1 and Q_2 interchanged gives

$$(\alpha, \beta, \gamma) G_{Q_2}(\alpha^{-1}, \beta^{-1}, \gamma^{-1}) \subseteq G_{Q_1}$$

and so

$$G_{Q_2} \subseteq (\alpha^{-1}, \beta^{-1}, \gamma^{-1}) G_{Q_1}(\alpha, \beta, \gamma).$$

The equality

$$G_{Q_2} = (\alpha^{-1}, \beta^{-1}, \gamma^{-1}) G_{Q_1}(\alpha, \beta, \gamma)$$

is an immediate consequence of the above statements and implies that $G_{Q_1} \cong G_{Q_2}$.

THEOREM 4.1.4. *Two components of an autotopism of a quasigroup determine the third one uniquely.*

PROOF. Let L be the latin square which represents the multiplication table of the quasigroup Q and let (α, β, γ) and (α, β, γ') be autotopisms of Q with two components α and β in common. Then $(\alpha, \beta, \gamma)(\alpha^{-1}, \beta^{-1}, \gamma'^{-1}) = (\varepsilon, \varepsilon, \gamma\gamma'^{-1})$ is also an autotopism of Q. The components of an autotopism rearrange the row border, the column border, and the elements of L respectively. It is clear that if one of these items is altered but not the other two, then the new multiplication table cannot represent Q. Thus, the only possibility is that $\gamma' = \gamma$.

COROLLARY 1. *The order of the group of autotopisms of a quasigroup of order n cannot exceed* $(n\,!)^2$.

COROLLARY 2. *Any one of the non-identity components of a principal autotopism of a quasigroup determines the autotopism uniquely.*

COROLLARY 3. *The order of the group of principal autotopisms of a quasigroup of order n cannot exceed* $n\,!$

A detailed study of the autotopism groups of quasigroups has been made by A. Sade in [29]. Sade has shown, for example, that the upper bound given in corollary 1 to theorem 4.1.4 above can be improved to $n(n\,!)$. This follows from the fact that each autotopism of a quasigroup Q can be represented as the product of a so-called *fundamental autotopism* (A. Sade [29], page 6), of which there can be at most n^2, and an automorphism of a loop L isotopic to Q. Such a loop evidently has at most $(n - 1)\,!$ distinct automorphisms and so the order of the automorphism group of Q, which is equal to that of L by theorem 4.1.3, is at most $n^2(n - 1)\,! = n(n\,!)$.

A number of further results on the autotopism groups of particular kinds of quasigroup will be found in V. D. Belousov [7]. For an investigation of the properties of the autotopism group of a group, see also A. Sade [27].

4.2. Classification of latin squares

For enumerational purposes it is desirable to separate the set Ω_n of all latin squares of order n into isotopy classes.

Two latin squares of order n are said to be *isotopic* or, sometimes, *equivalent* (see section 5.1) if one can be transformed into the other by rearranging rows, rearranging columns, and renaming elements: that is, if the quasigroups whose multiplication tables they represent are isotopic. This relation is an equivalence relation between latin squares and separates Ω_n into subsets of equivalent squares, called *isotopy classes* of squares.

From theorem 4.1.3 it is obvious that two latin squares contained in the same isotopy class have autotopism groups of the same order. In fact, a stronger result than this is true, see theorem 4.2.3 below.

In this section we shall again follow the methods of E. Schönhardt's paper [1] using up-to-date terminology.

We begin by reminding the reader that in section 2.1 we introduced the concept of conjugate (or parastrophic) identities: that is, identities satisfied in conjugate quasigroups. We also showed that, with any quasigroup (Q, θ) are associated, in general, five conjugates $(Q, {}^{-1}\theta)$, (Q, θ^{-1}), (Q, θ^*), $(Q, ({}^{-1}\theta)^*)$, and $(Q, (\theta^{-1})^*)$. This leads us to make the following definition:

Definition. If the multiplication table of the quasigroup (Q, θ) defines a latin square L, then the latin squares ${}^{-1}L, L^{-1}, L^*$, $({}^{-1}L)^*$, and $(L^{-1})^*$, defined by the multiplication tables of its conjugates, are said to be *conjugate* to L.

We may express the relationships between the squares ${}^{-1}L$, L^{-1}, and the square L in terms of column and row permutations respectively. Let $c_0, c_1, \ldots, c_{n-1}$ and $r_0, r_1, \ldots, r_{n-1}$ be respectively the row and column borders of L in the multiplication table of Q. Each column of L is a permutation of its column border. Let the jth column of L be as shown in Fig. 4.2.1. This is obtained from the column border of L by the permutation mapping $\eta_j: r_i \to t_i$ and we have $r_i \theta c_j = t_i$. It follows that $t_i({}^{-1}\theta)c_j = r_i$, and so the jth column of the square ${}^{-1}L$ is obtained from its column border by the permutation mapping $\eta_j^{-1}: t_i \to r_i$. We call η_j the jth *column permutation* of L. Thus, ${}^{-1}L$ is the square obtained when each column permutation of L is replaced by its inverse. In an exactly similar way, we can show that L^{-1} is the square obtained when each row permutation of L is replaced by its inverse.

$$
\begin{array}{c|c}
 & c_0\ c_1 \ldots c_j \ldots c_{n-1} \\
\hline
r_0 & t_0 \\
r_1 & t_1 \\
\vdots & \vdots \\
r_i & t_i \\
\vdots & \vdots \\
r_{n-1} & t_{n-1}
\end{array}
$$

Fig. 4.2.1.

From the relationship between the operations θ and θ^* given in section 2.1, it is immediately seen that the square L^* is the transpose of the square L.

Now let R_1, R_2, R_3 denote the transformations which map the square L onto ${}^{-1}L, L^{-1}$, and L^* respectively. Then R_1, R_2, R_3 generate a group K of mappings whose elements are the identity mapping I and the mappings R_1, R_2, R_3, R_1R_3 and R_2R_3. The last two of these mappings transform L into the squares $({}^{-1}L)^*$ and $(L^{-1})^*$ respectively.

THEOREM 4.2.1. *The number of distinct conjugates of a latin square L is always* 1, 2, 3, *or* 6.

PROOF. As mentioned above, the transformations R_1, R_2, R_3 generate a group K which is isomorphic to the symmetric group of degree three. Clearly, the number of distinct conjugates of any given latin square L is equal to the order of some subgroup of K. Consequently the number of distinct conjugates is 1, 2, 3, or 6.

THEOREM 4.2.2. *If* (Q, θ) *is a group or a loop with the inverse property, the five conjugates of* (Q, θ) *are all isotopic to it.*

PROOF. A loop (Q, θ) is said to have the inverse property (see R. H. Bruck [5]) if each element $a \in Q$ has a two-sided inverse a^{-1} such that $(a^{-1})\theta(a\theta b) = b$ and $(b\theta a)\,\theta(a^{-1}) = b$ for all $b \in Q$. In such a loop, the mapping J defined by $aJ = a^{-1}$ for all $a \in Q$ is a one-to-one mapping of Q onto itself.

If $a\theta b = c$ in a loop (Q, θ) with the inverse property, then $b = a^{-1}\theta c$ and $a = c\theta b^{-1}$. That is, $(aJ)\theta c = b$ and $c\theta(bJ) = a$. Thus, the isotopisms (J, I, I) and (I, J, I), where I denotes the identity permutation on the set Q, map (Q, θ) onto its conjugates (Q, θ^{-1}) and $(Q, {}^{-1}\theta)$ respectively. (We have $a\theta b = c \Leftrightarrow a(\theta^{-1})c = b$ and $c({}^{-1}\theta)b = a$, by definition of the operations θ^{-1} and ${}^{-1}\theta$.)

Further, the conjoint (Q, θ^*) is isomorphic to (Q, θ) since $a\theta b = c \Rightarrow b = a^{-1}\theta c \Rightarrow b\theta c^{-1} = a^{-1} \Rightarrow c^{-1} = b^{-1}\theta a^{-1} \Rightarrow (bJ)\theta(aJ) = cJ$, showing that the isotopism (J, J, J) maps (Q, θ) onto (Q, θ^*).

Similarly, the conjoints $(Q,(\theta^{-1})^*)$ and $(Q,({}^{-1}\theta)^*)$ of (Q, θ^{-1}) and (Q^{-1}, θ) are respectively isomorphic to these loops. Since the product of an isotopism and an isomorphism is again an isotopism, this completes the proof.

DEFINITION. A set of latin squares which comprises all the members of an isotopy class together with their conjugates is called a *main class* of latin squares.

THEOREM 4.2.3. *The autotopism groups of conjugate quasigroups are isomorphic and so have equal orders.*

PROOF. Let (Q, θ) be a given quasigroup and let us consider for example the conjugate quasigroup $(Q, {}^{-1}\theta)$. If $\sigma = (\alpha, \beta, \gamma)$ is an autotopism of (Q, θ), then $a\theta b = c \Leftrightarrow (a\alpha)\theta(b\beta) = c\gamma$. But $a\theta b = c$ if and only if $c({}^{-1}\theta)b = a$ and $(a\alpha)\theta(b\beta) = c\gamma$ if and only if $(c\gamma)({}^{-1}\theta)(b\beta) = a\alpha$. Therefore, $c({}^{-1}\theta)b = a \Leftrightarrow (c\gamma)({}^{-1}\theta)(b\beta) = a\alpha$ and this implies that ${}_{-1}\sigma = (\gamma, \beta, \alpha)$ is an autotopism of $(Q, {}^{-1}\theta)$.

In an exactly similar way it can be shown that $\sigma_{-1} = (\alpha, \gamma, \beta)$ is an autotopism of (Q, θ^{-1}), $\sigma_* = (\beta, \alpha, \gamma)$ is an autotopism of (Q, θ^*), and that the mappings obtained from σ by permuting its components α, β, γ in the remaining two ways are autotopisms of the remaining two conjugates $(Q, ({}^{-1}\theta)^*)$ and $(Q, (\theta^{-1})^*)$ of (Q, θ).

Clearly, we can establish a one-to-one correspondence between autotopisms $\sigma = (\alpha, \beta, \gamma)$, $\tau = (\alpha', \beta', \gamma'), \ldots$ of (Q, θ) and corresponding autotopisms ${}_{-1}\sigma = (\gamma, \beta, \alpha)$, ${}_{-1}\tau = (\gamma', \beta', \alpha'), \ldots$ of $(Q, {}^{-1}\theta)$. Also, in this correspondence, products $\sigma\tau = (\alpha\alpha', \beta\beta', \gamma\gamma')$ correspond to products ${}_{-1}\sigma_{-1}\tau = (\gamma\gamma', \beta\beta', \alpha\alpha')$. That is, ${}_{-1}\sigma_{-1}\tau = {}_{-1}(\sigma\tau)$ and so the correspondence defines an isomorphism between the autotopism groups of (Q, θ) and $(Q, {}^{-1}\theta)$. It is clear that similar correspondences can be established between the autotopism group of (Q, θ) and those of (Q, θ^{-1}), (Q, θ^*), $(Q, ({}^{-1}\theta)^*)$ and $(Q, (\theta^{-1})^*)$.

COROLLARY. *Latin squares contained in the same main class have autotopism groups of the same order.*

THEOREM 4.2.4. *Ω_n splits into disjoint main classes and each such main class is a union of complete isotopy classes.*

PROOF. If two latin squares L_1 and L_2 are isotopic, then the pairs of conjugates ${}^{-1}L_1$, ${}^{-1}L_2$; L_1^{-1}, L_2^{-1}; etc., are pairs of isotopes. It follows that each main class is a sum of complete isotopy classes, and so the main classes are disjoint.

REMARK. We note that it follows from theorem 4.2.2 that if a main class of latin squares contains a latin square satisfying the quadrangle criterion, then the main class comprises a single isotopy class.

DEFINITION. A subset of an isotopy class is called an *isomorphism class* if all its members are isomorphic.

Theorem 4.2.5. *Each isotopy class of Ω_n splits into disjoint isomorphism classes.*

Proof. The relation of being isomorphic is an equivalence relation on the members of each isotopy class C of Ω_n.

Corollary 1. *Each main class of Ω_n splits into disjoint isomorphism classes.*

Corollary 2. *Ω_n splits into disjoint isomorphism classes.*

Theorem 4.2.6. *If an isotopy class contains a latin square which represents the multiplication table of a group then it consists of a single isomorphism class.*

Proof. The result follows immediately from the fact that two isotopic groups are isomorphic, proved in theorem 1.3.4.

Remark. Since theorem 4.2.6. was proved by E. Schönhardt (see [1], theorem 24) independently of theorem 1.3.4, it follows that theorem 4.2.6 implies theorem 1.3.4. Thus, Bruck's theorem 1.3.4 can be regarded as having been proved by E. Schönhardt more than ten years before R. H. Bruck and without the use of any tool other than latin squares.

Definition. A latin square is said to be *reduced* or to be *in standard form* if in the first row and column its elements 1, 2, . . . , n occur in natural order.

It follows from the definition of isotopism of latin squares that each main (isotopy, isomorphism) class contains one or more reduced latin squares belonging to that main (isotopy, isomorphism) class.

In general each such class contains more than one reduced latin square, but we can always choose one such square as class representative.

We shall now give the classification of latin squares of all orders up to and including six into their main, isotopy and isomorphism classes.

We shall give a reduced form class representative of each class and shall use decimal notation to denote the various classes. A triple k, l, m of integers will denote a reduced latin square belonging to the kth main class M_k, lth isotopy class I_l of M_k, and mth isomorphism class of I_l.

The list given on the following pages is taken from E. Schönhardt [1], and is followed by a number of explanatory remarks which the reader may find helpful to look at first.

As an illustration of our comment that each main, isotopy and isomorphism class may contain more than one reduced latin square, let us note that there exist two reduced latin squares of order 4 additional to those given in our classiffication list below but that these both belong to the class labelled 2.1.1 in that list.

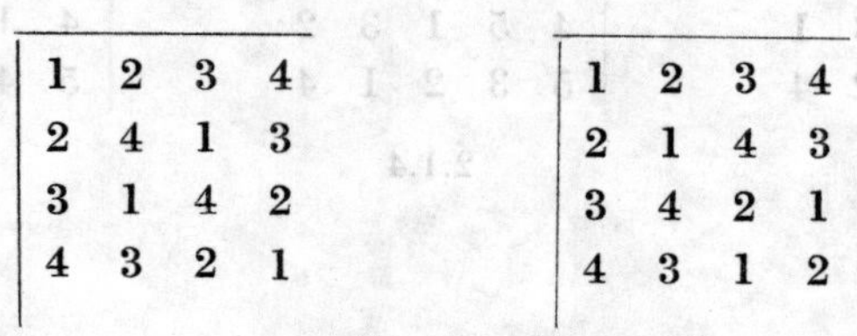

1	2	3	4
2	4	1	3
3	1	4	2
4	3	2	1

1	2	3	4
2	1	4	3
3	4	2	1
4	3	1	2

Fig. 4.2.2

The two squares in question are displayed in Fig. 4.2.2 and can be mapped onto the square exhibited as representative of the class 2.1.1 by the isomorphisms ((34), (3 4), (3 4)) and ((2 3), (2 3), (2 3)) respectively.

Classification list for latin squares of orders 2 to 6.

$n = 2$

1	2
2	1

1.1.1

$n = 3$

1	2	3
2	3	1
3	1	2

1.1.1

$n = 4$

1	2	3	4
2	1	4	3
3	4	1	2
4	3	2	1

1.1.1

1	2	3	4
2	3	4	1
3	4	1	2
4	1	2	3

2.1.1

$n = 5$

1	2	3	4	5
2	3	4	5	1
3	4	5	1	2
4	5	1	2	3
5	1	2	3	4

1.1.1

1	2	3	4	5
2	1	4	5	3
3	5	1	2	4
4	3	5	1	2
5	4	2	3	1

2.1.1

1	2	3	4	5
2	1	5	3	4
3	4	2	5	1
4	5	1	2	3
5	3	4	1	2

2.1.2

1	2	3	4	5
2	1	4	5	3
3	4	5	1	2
4	5	2	3	1
5	3	1	2	4

2.1.3

1	2	3	4	5
2	1	4	5	3
3	4	5	2	1
4	5	1	3	2
5	3	2	1	4

2.1.4

1	2	3	4	5
2	3	4	5	1
3	5	2	1	4
4	1	5	3	2
5	4	1	2	3

2.1.5

$n = 6$

1	2	3	4	5	6
2	3	4	5	6	1
3	4	5	6	1	2
4	5	6	1	2	3
5	6	1	2	3	4
6	1	2	3	4	5

1.1.1

1	2	3	4	5	6
2	1	5	6	3	4
3	6	1	5	4	2
4	5	6	1	2	3
5	4	2	3	6	1
6	3	4	2	1	5

2.1.1

1	2	3	4	5	6
2	3	1	5	6	4
3	1	2	6	4	5
4	6	5	2	1	3
5	4	6	3	2	1
6	5	4	1	3	2

3.1.1

1	2	3	4	5	6
2	1	4	3	6	5
3	4	5	6	1	2
4	3	6	5	2	1
5	6	1	2	4	3
6	5	2	1	3	4

4.1.1

1	2	3	4	5	6
2	1	4	3	6	5
3	4	5	6	1	2
4	3	6	5	2	1
5	6	2	1	4	3
6	5	1	2	3	4

5.1.1

1	2	3	4	5	6
2	1	5	6	3	4
3	6	2	5	4	1
4	5	6	2	1	3
5	4	1	3	6	2
6	3	4	1	2	5

4.1.2

1	2	3	4	5	6
2	1	4	5	6	3
3	5	1	6	2	4
4	6	5	1	3	2
5	3	6	2	4	1
6	4	2	3	1	5

5.1.2

1	2	3	4	5	6
2	1	4	3	6	5
3	5	1	6	2	4
4	6	2	5	1	3
5	3	6	2	4	1
6	4	5	1	3	2

5.1.3

1	2	3	4	5	6
2	1	5	6	3	4
3	4	1	2	6	5
4	3	6	5	1	2
5	6	4	3	2	1
6	5	2	1	4	3

5.1.4

1	2	3	4	5	6
2	1	4	3	6	5
3	5	1	6	4	2
4	6	2	5	3	1
5	3	6	1	2	4
6	4	5	2	1	3

5.1.5

1	2	3	4	5	6
2	1	4	5	6	3
3	6	2	1	4	5
4	5	6	2	3	1
5	3	1	6	2	4
6	4	5	3	1	2

6.1.1

1	2	3	4	5	6
2	1	4	5	6	3
3	4	2	6	1	5
4	6	5	3	2	1
5	3	6	1	4	2
6	5	1	2	3	4

6.1.2

1	2	3	4	5	6
2	3	4	1	6	5
3	6	2	5	1	4
4	5	6	2	3	1
5	4	1	6	2	3
6	1	5	3	4	2

6.1.3

1	2	3	4	5	6
2	1	4	5	6	3
3	4	5	6	2	1
4	6	1	2	3	5
5	3	6	1	4	2
6	5	2	3	1	4

6.1.4

1	2	3	4	5	6
2	1	4	6	3	5
3	4	5	1	6	2
4	5	6	2	1	3
5	6	2	3	4	1
6	3	1	5	2	4

6.1.5

1	2	3	4	5	6
2	5	4	6	3	1
3	1	5	2	6	4
4	6	2	3	1	5
5	4	6	1	2	3
6	3	1	5	4	2

6.1.6

1	2	3	4	5	6
2	5	1	6	4	3
3	4	5	2	6	1
4	6	2	3	1	5
5	3	6	1	2	4
6	1	4	5	3	2

6.1.7

1	2	3	4	5	6
2	3	4	5	6	1
3	6	5	2	1	4
4	1	2	6	3	5
5	4	6	1	2	3
6	5	1	3	4	2

6.1.8

1	2	3	4	5	6
2	3	6	1	4	5
3	4	5	2	5	1
4	5	2	6	1	3
5	6	1	3	2	4
6	1	4	5	3	2

6.1.9

1	2	3	4	5	6
2	1	4	3	6	5
3	5	1	6	4	2
4	6	5	1	2	3
5	3	6	2	1	4
6	4	2	5	3	1

7.1.1

1	2	3	4	5	6
2	1	4	3	6	5
3	5	2	6	4	1
4	6	5	2	1	3
5	3	6	1	2	4
6	4	1	5	3	2

7.1.2

1	2	3	4	5	6
2	1	5	6	3	4
3	6	4	5	1	2
4	5	6	3	2	1
5	4	1	2	6	3
6	3	2	1	4	5

7.1.3

1	2	3	4	5	6
2	1	6	5	4	3
3	5	4	6	1	2
4	6	5	3	2	1
5	3	1	2	6	4
6	4	2	1	3	5

7.1.4

1	2	3	4	5	6
2	1	5	6	4	3
3	4	1	5	6	2
4	3	6	2	1	5
5	6	4	3	2	1
6	5	2	1	3	4

7.1.5

1	2	3	4	5	6
2	1	4	3	6	5
3	5	1	6	4	2
4	6	5	2	3	1
5	4	6	1	2	3
6	3	2	5	1	4

7.1.6

1	2	3	4	5	6
2	1	4	5	6	3
3	5	2	6	1	4
4	6	5	3	2	1
5	3	6	1	4	2
6	4	1	2	3	5

7.1.7

1	2	3	4	5	6
2	1	5	6	3	4
3	4	2	5	6	1
4	5	6	3	1	2
5	6	1	2	4	3
6	3	4	1	2	5

7.1.8

1	2	3	4	5	6
2	5	6	1	4	3
3	4	5	2	6	1
4	6	2	3	1	5
5	3	1	6	2	4
6	1	4	5	3	2

7.1.9

1	2	3	4	5	6
2	5	4	6	3	1
3	6	5	2	1	4
4	1	2	3	6	5
5	4	6	1	2	3
6	3	1	5	4	2

7.1.10

1	2	3	4	5	6
2	1	6	5	3	4
3	6	1	2	4	5
4	3	5	6	2	1
5	4	2	1	6	3
6	5	4	3	1	2

7.1.11

1	2	3	4	5	6
2	1	5	3	6	4
3	6	2	5	4	1
4	3	6	2	1	5
5	4	1	6	3	2
6	5	4	1	2	3

7.1.12

1	2	3	4	5	6
2	1	6	5	3	4
3	6	1	2	4	5
4	5	2	1	6	3
5	3	4	6	1	2
6	4	5	3	2	1

8.1.1

1	2	3	4	5	6
2	1	5	6	4	3
3	6	4	5	1	2
4	5	6	3	2	1
5	3	2	1	6	4
6	4	1	2	3	5

8.1.2

1	2	3	4	5	6
2	1	5	6	4	3
3	4	1	5	6	2
4	3	6	1	2	5
5	6	2	3	1	4
6	5	4	2	3	1

8.2.1

1	2	3	4	5	6
2	1	6	5	3	4
3	4	5	2	6	1
4	3	2	6	1	5
5	6	1	3	4	2
6	5	4	1	2	3

8.2.2

Isomorphism classes 8.3.1 and 8.3.2 are obtained by reflecting 8.2.1 and 8.2.2 in their main left-to-right diagonals.

1	2	3	4	5	6
2	3	1	6	4	5
3	1	2	5	6	4
4	6	5	1	2	3
5	4	6	2	3	1
6	5	4	3	1	2

9.1.1

1	2	3	4	5	6
2	3	1	5	6	4
3	1	2	6	4	5
4	6	5	1	2	3
5	4	6	3	1	2
6	5	4	2	3	1

9.1.2

1	2	3	4	5	6
2	3	1	5	6	4
3	1	2	6	4	5
4	6	5	2	3	1
5	4	6	1	2	3
6	5	4	3	1	2

9.1.3

1	2	3	4	5	6
2	3	1	6	4	5
3	1	2	5	6	4
4	5	6	1	3	2
5	6	4	3	2	1
6	4	5	2	1	3

9.2.1

1	2	3	4	5	6
2	3	1	5	6	4
3	1	2	6	4	5
4	5	6	1	3	2
5	6	4	2	1	3
6	4	5	3	2	1

9.2.2

1	2	3	4	5	6
2	3	1	5	6	4
3	1	2	6	4	5
4	5	6	2	1	3
5	6	4	3	2	1
6	4	5	1	3	2

9.2.3

Isomorphism classes 9.3.1, 9.3.2, 9.3.3 are obtained by reflecting 9.2.1, 9.2.2 and 9.2.3 in their main left-to-right diagonals.

1	2	3	4	5	6
2	1	6	5	4	3
3	5	1	2	6	4
4	6	2	1	3	5
5	3	4	6	2	1
6	4	5	3	1	2

10.1.1

1	2	3	4	5	6
2	1	5	6	3	4
3	6	4	5	1	2
4	5	6	3	2	1
5	4	2	1	6	3
6	3	1	2	4	5

10.1.2

1	2	3	4	5	6
2	1	4	3	6	5
3	5	1	6	4	2
4	6	5	2	3	1
5	3	6	1	2	4
6	4	2	5	1	3

10.1.3

1	2	3	4	5	6
2	1	5	6	3	4
3	6	2	5	4	1
4	5	6	3	1	2
5	4	1	2	6	3
6	3	4	1	2	5

10.1.4

1	2	3	4	5	6
2	1	6	5	4	3
3	4	1	6	2	5
4	3	5	1	6	2
5	6	4	2	3	1
6	5	2	3	1	4

10.2.1

1	2	3	4	5	6
2	1	5	6	3	4
3	4	2	5	6	1
4	3	6	2	1	5
5	6	1	3	4	2
6	5	4	1	2	3

10.2.2

1	2	3	4	5	6
2	1	4	3	6	5
3	6	1	5	2	4
4	5	2	6	1	3
5	3	6	2	4	1
6	4	5	1	3	2

10.2.3

1	2	3	4	5	6
2	1	5	6	3	4
3	4	6	2	1	5
4	5	2	3	6	1
5	6	4	1	2	3
6	3	1	5	4	2

10.2.4

Isomorphism classes 10.3.1 to 10.3.4 are obtained by reflection of 10.2.1 to 10.2.4 in their main left-to-right diagonals.

1	2	3	4	5	6
2	1	4	5	6	3
3	4	2	6	1	5
4	5	6	2	3	1
5	6	1	3	2	4
6	3	5	1	4	2

11.1.1

1	2	3	4	5	6
2	1	4	5	6	3
3	6	5	2	4	1
4	3	1	6	2	5
5	4	6	1	3	2
6	5	2	3	1	4

11.1.2

1	2	3	4	5	6
2	1	6	5	3	4
3	6	1	2	4	5
4	5	2	1	6	3
5	3	4	6	2	1
6	4	5	3	1	2

11.1.3

1	2	3	4	5	6
2	1	5	6	4	3
3	6	4	5	1	2
4	5	6	3	2	1
5	3	1	2	6	4
6	4	2	1	3	5

11.1.4

1	2	3	4	5	6
2	1	5	6	4	3
3	6	2	5	1	4
4	5	6	3	2	1
5	3	4	1	6	2
6	4	1	2	3	5

11.1.5

1	2	3	4	5	6
2	5	6	1	4	3
3	1	5	2	6	4
4	6	2	3	1	5
5	3	4	6	2	1
6	4	1	5	3	2

11.1.6

1	2	3	4	5	6
2	5	1	6	3	4
3	6	5	2	4	1
4	1	2	3	6	5
5	4	6	1	2	3
6	3	4	5	1	2

11.1.7

1	2	3	4	5	6
2	1	6	3	4	5
3	5	4	6	1	2
4	6	2	5	3	1
5	3	1	2	6	4
6	4	5	1	2	3

11.2.1

1	2	3	4	5	6
2	1	6	3	4	5
3	5	2	1	6	4
4	6	5	2	1	3
5	3	4	6	2	1
6	4	1	5	3	2

11.2.2

1	2	3	4	5	6
2	1	5	6	4	3
3	4	1	5	6	2
4	3	6	1	2	5
5	6	4	2	3	1
6	5	2	3	1	4

11.2.3

1	2	3	4	5	6
2	1	6	5	3	4
3	4	2	6	1	5
4	3	5	2	6	1
5	6	1	3	4	2
6	5	4	1	2	3

11.2.4

1	2	3	4	5	6
2	1	6	5	3	4
3	4	5	2	6	1
4	6	2	3	1	5
5	3	1	6	4	2
6	5	4	1	2	3

11.2.5

1	2	3	4	5	6
2	4	1	6	3	5
3	6	5	2	4	1
4	1	6	5	2	3
5	3	4	1	6	2
6	5	2	3	1	4

11.2.6

1	2	3	4	5	6
2	3	4	1	6	5
3	1	5	6	2	4
4	5	6	2	1	3
5	6	2	3	4	1
6	4	1	5	3	2

11.2.7

Isomorphism classes **11.3.1** to **11.3.7** are obtained by reflection of **11.2.1** to **11.2.7** in their main left-to-right diagonals.

1	2	3	4	5	6
2	1	5	6	4	3
3	5	4	2	6	1
4	6	2	3	1	5
5	4	6	1	3	2
6	3	1	5	2	4

12.1.1

1	2	3	4	5	6
2	1	5	6	4	3
3	6	2	5	1	4
4	5	6	2	3	1
5	3	4	1	6	2
6	4	1	3	2	5

12.1.2

1	2	3	4	5	6
2	1	5	3	6	4
3	4	1	6	2	5
4	6	2	5	1	3
5	3	6	2	4	1
6	5	4	1	3	2

12.1.3

1	2	3	4	5	6
2	1	5	3	6	4
3	5	2	6	4	1
4	3	6	2	1	5
5	6	4	1	3	2
6	4	1	5	2	3

12.1.4

1	2	3	4	5	6
2	1	4	6	3	5
3	4	5	1	6	2
4	6	2	5	1	3
5	3	6	2	4	1
6	5	1	3	2	4

12.1.5

1	2	3	4	5	6
2	3	4	5	6	1
3	5	2	6	1	4
4	6	1	2	3	5
5	4	6	1	2	3
6	1	5	3	4	2

12.1.6

1	2	3	4	5	6
2	5	6	1	4	3
3	6	5	2	1	4
4	1	2	3	6	5
5	3	4	6	2	1
6	4	1	5	3	2

12.1.7

1	2	3	4	5	6
2	5	6	1	3	4
3	6	5	2	4	1
4	1	2	3	6	5
5	4	1	6	2	3
6	3	4	5	1	2

12.1.8

1	2	3	4	5	6
2	3	1	5	6	4
3	6	5	2	4	1
4	1	2	6	3	5
5	4	6	1	2	3
6	5	4	3	1	2

12.1.9

1	2	3	4	5	6
2	3	6	1	4	5
3	1	5	2	6	4
4	5	2	6	1	3
5	6	4	3	2	1
6	4	1	5	3	2

12.1.10

1	2	3	4	5	6
2	1	6	5	3	4
3	5	2	6	4	1
4	6	5	2	1	3
5	3	4	1	6	2
6	4	1	3	2	5

12.2.1

1	2	3	4	5	6
2	1	6	5	3	4
3	5	4	2	6	1
4	6	2	3	1	5
5	3	1	6	4	2
6	4	5	1	2	3

12.2.2

1	2	3	4	5	6
2	1	4	6	3	5
3	5	1	2	6	4
4	6	5	3	2	1
5	3	6	1	4	2
6	4	2	5	1	3

12.2.3

1	2	3	4	5	6
2	1	4	6	3	5
3	5	6	2	4	1
4	6	5	3	1	2
5	4	2	1	6	3
6	3	1	5	2	4

12.2.4

1	2	3	4	5	6
2	1	5	3	6	4
3	5	4	6	2	1
4	6	1	2	3	5
5	3	6	1	4	2
6	4	2	5	1	3

12.2.5

1	2	3	4	5	6
2	6	4	5	3	1
3	4	5	1	6	2
4	5	6	2	1	3
5	3	1	6	2	4
6	1	2	3	4	5

12.2.6

1	2	3	4	5	6
2	3	4	1	6	5
3	6	5	2	4	1
4	1	6	5	2	3
5	4	1	6	3	2
6	5	2	3	1	4

12.2.7

1	2	3	4	5	6
2	3	4	1	6	5
3	5	6	2	1	4
4	1	5	6	2	3
5	6	2	3	4	1
6	4	1	5	3	2

12.2.8

1	2	3	4	5	6
2	3	5	1	6	4
3	1	2	6	4	5
4	5	6	2	1	3
5	6	4	3	2	1
6	4	1	5	3	2

12.2.9

1	2	3	4	5	6
2	4	1	6	3	5
3	6	5	2	4	1
4	1	2	5	6	3
5	3	6	1	2	4
6	5	4	3	1	2

12.2.10

Isomorphism classes 12.3.1 to 12.3.10 are obtained by reflection of 12.2.1 to 12.2.10 in their main left-to-right diagonals.

Remarks. Let us observe that the latin square which corresponds to the multiplication table of the cyclic group of order 2 serves as class representative of the single main class of latin squares of order 2 and that, for $n = 3$, Ω_3 again consists of a single main class which is represented by the multiplication table of the cyclic group of order 3. In both cases, the single main class is also the sole isotopy class (by theorem 4.2.2) and consequently, by virtue of theorem 4.2.6, it is also the sole isomorphism class.

For $n = 4$, Ω_n splits into two main classes both of which are represented by latin squares which satisfy the quadrangle criterion. The main class 1.1.1 is represented by the Cayley table of the Klein group and the main class 2.1.1 by that of the cyclic group of order 4. Once again, each main class consists of a single isotopy class (by theorem 4.2.2) and, consequently, (by theorem 4.2.6) it also comprises only one isomorphism class.

For $n = 5$, there are two main classes but only the first of these is represented by a latin square satisfying the quadrangle criterion, namely the square 1.1.1. Each main class again comprises a single isotopy class.

For $n = 6$, two of the 12 main classes, namely those represented by the squares 1.1.1 and 2.1.1, contain latin squares satisfying the quadrangle criterion. The class 1.1.1 is the main class which contains the Cayley table of the cyclic group of order 6, while the main class 2.1.1 can be represented by the Cayley table of the dihedral group of order 6. We note also that this is the smallest value of n for which some of the main classes consist of more than one isotopy class.

In a recently published paper [30], A. Sade has made use of techniques previously developed by him in his papers [9], [18] and [29] to provide a systematic tabulation of all quasigroups of orders 2 to 6 inclusive by means of reduced form representatives for each main class (somewhat similar is that of E. Schönhardt given above) and has then gone on to summarize the main properties of each of these loop representatives of the main classes. Thus, for example, he has stated for each the orders of its automorphism and autotopism groups, whether or not it is isotopic to some or all of its parastrophes (conjugates), the number of distinct loops in the main class which contains it, the number of isomorphically distinct semi-symmetric quasigroups (see section 2.1) which are isotopic to it and whether it is (a) commutative, (b) isomorphic to its transpose, or (c) neither. For the orders $n = 2$, 3, and 4 he has explored the whole universe of quasigroups and has listed a number of interesting properties possessed by these quasigroups. In this way, he has in effect provided a valuable compendium of results which previously could only be found by searching among his many earlier papers on the subject of quasigroups.

In [4], A. Cayley has listed class representatives of all the main classes of Ω_n (for $7 \leq n \leq 12$) for which the quadrangle criterion holds.

A general account of the historical development of the classification problem will be found in the next section.

4.3. History and further results concerning the classification and enumeration of latin squares

In the first part of this section we give a short historical account of the many contributions which have been made to the problem of classifying and enumerating latin squares.

The problem of the enumeration of latin squares was first discussed by L. Euler in [2]. Euler showed that the number of distinct reduced latin squares of order n (that is, latin squares with first row and first column in natural order) is 1 if $n = 2$ or 3, 4 if $n = 4$, and 56 if $n = 5$. He also discussed latin squares of order 6, but did not succeed in enumerating these.

The next authors to write on the subject were A. Cayley and M. Frolov, both of whom published papers in 1890. (See A. Cayley [5] and M. Frolov [1].) Cayley stated that if the first row of a latin square is in natural order then the number of possible second rows is

$$n!\left(1-\frac{1}{1!}+\frac{1}{2!}-\ldots+(-1)^n\frac{1}{n!}\right)$$

but that to calculate the total number of possibilities for the second and third rows when considered jointly is considerably more difficult, since the number of choices for the third row varies with the particular choice of the second. He then discussed the determination of the number of distinct reduced latin squares for values of n up to 5 and obtained the same values as had been given earlier by L. Euler. M. Frolov, writing independently and undaunted by the difficulties mentioned by Cayley, attempted an enumeration of all reduced latin squares of orders 6 and 7. He obtained 9408 reduced squares of order 6 and 221,276,160 reduced squares of order 7. The former value is correct and was confirmed later by a number of authors, but the number 221,276,160 was seriously in error. However, this did not become apparent until much later on.

Frolov also stated two remarkable formulae in the second part of his paper (M. Frolov [2], page 30). The first is a recurrence relation for the number of reduced latin squares of order n. If T_n denotes the number of reduced latin squares of order n, then Frolov claimed that

$$\frac{T_n}{T_{n-1}}=\left(\frac{T_{n-1}}{T_{n-2}}\right)^2\frac{T_{n-1}}{2}.$$

This formula is valid for $n=5$ and 6 and Frolov was under the impression that it also held for $n=7$ since it gives $T_7=221{,}276{,}160$. However, it seems quite likely that even if the formula fails to give exact values for T_n it may provide an upper bound for that function.

Frolov's second formula purported to give the total number U_n of latin squares of order n as a function of T_n and the number R_n of reduced "regular" latin squares of order n. ("Regular" latin squares as defined by Frolov seem to be squares generated by the cyclic group of order n.) The formula was

$$U_n=(n!)^2(T_n-R_n)-n!(n-1)!\,R_n=(n!)^2\left(T_n-\frac{n-1}{n}R_n\right).$$

If we suppose that $R_n = T_n$, the formula is correct and agrees with that obtained later by P. A. MacMahon (see below).

Although neither of Frolov's formulae appears to be generally valid (except in the unlikely event that the presently accepted value of T_7 should prove to be in error), the paper in which they appear is now somewhat difficult to obtain and the authors consider it so remarkable that a formula for T_n should have been proposed at all that they think that by bringing these formulae to the attention of present day readers, they may stimulate a new investigations into the possibility of the existence of a valid general formula for this quantity.

About the year 1899, G. Tarry became interested in L. Euler's conjecture and, by separating the reduced latin squares of order 6 into 17 basic "families", he was able to show that no latin square of order 6 has an orthogonal mate. In other words, the problem of the thirty-six officers (see section 5.1) is insoluble. Of the 17 basic "families" obtained by Tarry, 12 were isotopy classes and the remaining 5 were unions of pairs of isotopy classes, the squares of one class of each pair being mirror images in the main left-to-right diagonal of the other. For further details of this work, see G. Tarry [1], [2], [3], [4] and [5].

We should mention here that in H. W. Norton [1] and in A. Sade [5] it has been stated that reports exist of work on the subject of enumeration of latin squares having been done much earlier in the nineteenth century, but none of the papers of these earlier authors seem to have survived. In particular, both Norton and Sade remark that, from the evidence of a letter mentioned in S. Gunther [1], it seems likely that Clausen, an assistant of the German astronomer Schumacher had correctly enumerated the 6×6 latin squares as early as 1842 and had also shown the impossibility of any having an orthogonal mate.

About the same time (1899) as Tarry was enumerating the latin squares of order 6, P. A. MacMahon published a complete algebraic solution to the problem of enumerating latin squares of finite order n. He expressed this algebraic solution in two different forms (see P. A. MacMahon [1] and [3]) both of which involve the action of differential operators on an expanded operand. If his algebraic apparatus is actually put into operation it will be found that different terms are written down corresponding to all the different ways in which each row of the square could conceivably be filled up, that those arrangements which conflict with the requirements for the formation of a latin square are ultimately eliminated and that those which conform to these requirements survive the final operation and each contribute unity to the result. The manipulation of the algebraic expression, therefore, is considerably more laborious than the direct enumeration of the possible squares by a systematic and exhaustive series of trials.

It is is probably this fact which has forced most other authors to abandon this line of investigation. (We should mention here that one attempt to simplify MacMahon's procedure was later made by P. N. Saxena and appears in P. N. Saxena [1] and [2].)

Using his own method, MacMahon again obtained the values 1, 1, 1, 4 for the numbers of reduced latin squares of orders 1, 2, 3, 4 respectively, but for the number of reduced squares of order 5 he obtained the value 52. The falsity of this value was subsequently pointed out to MacMahon by R. A. Fisher. As a result, MacMahon detected an error in his calculation and the corrected value of 56 was incorporated into later editions of P. A. MacMahon [3].

In 1930, S. M. Jacob (see [1]) carried out an enumeration of $3 \times n$ latin rectangles and in the latter part of his article on this subject he also attempted an enumeration of the 5×5 and 6×6 reduced latin squares. For this purpose, he first separated the squares into families according to the nature of the permutation which transforms the first row into the second. He obtained the correct value of 56 for the number of 5×5 reduced latin squares but found only 8192 reduced 6×6 latin squares. Later, in [1], A. Sade explained the error of Jacob which had led him to obtain the latter incorrect result.

In the same year, E. Schönhardt wrote a very comprehensive article on the subject of latin squares and loops. This included a detailed investigation of latin squares of orders 5 and 6. Schönhardt showed correctly that there exist 2 isotopy classes of 5×5 latin squares and 22 isotopy classes of 6×6 latin squares. He showed that there are 6 isomorphism classes of reduced 5×5 squares and 109 isomorphism classes of reduced 6×6 squares: further, that the total numbers of reduced squares of orders 5 and 6 are 56 and 9408 respectively. All these results are correct. For full details, the reader should see the original paper, E. Schönhardt [1].

The next two papers on the subject of enumeration were R. A. Fisher and F. Yates [1], published in 1934, and H. W. Norton [1], published in 1939, and these two papers seem to be the ones best known to, and most often quoted by, statisticians. Both papers made use of the same basic idea: namely, to make a preliminary classification of latin squares of given order according to the nature of their main left-to-right diagonal.

(See also R. A. Fisher [1] for a later development of the same idea.) They also introduced some new terminology.

An *intramutation* of a latin square L which is in standard form and has the integers $1, 2, \ldots, n$ as its elements is obtained by permuting the symbols $2, 3, \ldots, n$ and then rearranging the rows and columns so as to put the new square thus obtained back into standard form. Such a transformation does not affect the "type" of the main left-to-right diagonal.

An *intercalate* of a latin square L is a 2×2 submatrix of L which is itself a latin square. Interchange of the two elements of such a submatrix transforms L into another latin square.

The concepts of isotopy class and main class were respectively called *transformation set* and *species* by Fisher, Yates and Norton. A latin square and its five conjugates were said to form an *adjugacy* set by the latter author.

The number of intercalates possessed by a latin square L is unaffected by permutation of rows, columns, or elements separately and is also unaffected by interchanges of these three "constraints". That is, the number of intercalates is a main class invariant.

Using the above ideas, R. A. Fisher and F. Yates showed that there are 9408 reduced latin squares of order 6 which can be arranged into 22 isotopy classes or 12 main classes (species). Of the 22 isotopy classes, 10 can be arranged into pairs such that one member of each of these pairs is obtained from the other by interchange of rows and columns (that is, by transposition). If the squares of each such pair of isotopy classes are regarded as forming a single "family", the total number of families is 17, a result previously obtained both by G. Tarry and by E. Schönhardt.

H. W. Norton classified the main classes of 7×7 squares according to their numbers of intercalates and the numbers of isotopy classes and of adjugacy sets contained in each. He found 146 distinct species (main classes) of 7×7 latin squares and a total of 16,927,968 7×7 reduced latin squares all together.

Somewhat later, in 1948, A. Sade carried out an independent enumeration of reduced 7×7 latin squares which avoided the necessity of separating them into species and he obtained 16,942,080 such squares. (See A. Sade [2].) Three years later, in [4] and [5], he gave an explanation of the discrepancy between this result and that of Norton, which was due to the fact that one species had been overlooked by the latter author.

Thus, the correct number of species is 147. He also pointed out that this confirmed an earlier conjecture of S. G. Ghurye, given in [1].

Sade's method was to calculate successively for $k = 1, 2, \ldots, l$, where l is a definite integer less than or equal to 7, a complete set of reduced $k \times 7$ latin rectangles inequivalent under any combination of permutations of rows, permutations of columns, and permutations of symbols, keeping track in so doing of the number of different rectangles in each equivalence class. The $(k+1)$-rowed rectangles were formed from the k-rowed rectangles by adding a row to each k-rowed rectangle in all possible ways, eliminating equivalent rectangles as they appeared. Sade pointed out that it was not necessary, or efficient, to continue the process until $k = 7$. When k reached the value 4 ($l = 4$), Sade summed the products of the number of rectangles in an equivalence class and the number of ways a representative of that class could be completed to a square, thus obtaining the total number of reduced 7×7 squares.

Let us illustrate Sade's method by applying it to the enumeration of 4×4 reduced latin squares and taking $l = 2$. There are two equivalence classes of 2×4 latin rectangles, L_1 and L_2 below (Fig. 4.3.1) being in one class and L_3 in another. L_2 may be obtained from L_1 by first interchanging the symbols 3 and 4 and then interchanging the third and fourth columns.

$$L_1 = \begin{array}{|cccc} \hline 1 & 2 & 3 & 4 \\ 2 & 3 & 4 & 1 \end{array} \qquad L_2 = \begin{array}{|cccc} \hline 1 & 2 & 3 & 4 \\ 2 & 4 & 1 & 3 \end{array} \qquad L_3 = \begin{array}{|cccc} \hline 1 & 2 & 3 & 4 \\ 2 & 1 & 4 & 3 \end{array}$$

$$L_1' = \begin{array}{|cccc} \hline 1 & 2 & 3 & 4 \\ 2 & 3 & 4 & 1 \\ 3 & 4 & 1 & 2 \\ 4 & 1 & 2 & 3 \end{array} \qquad L_3' = \begin{array}{|cccc} \hline 1 & 2 & 3 & 4 \\ 2 & 1 & 4 & 3 \\ 3 & 4 & 1 & 2 \\ 4 & 3 & 2 & 1 \end{array} \qquad L_3'' = \begin{array}{|cccc} \hline 1 & 2 & 3 & 4 \\ 2 & 1 & 4 & 3 \\ 3 & 4 & 2 & 1 \\ 4 & 3 & 1 & 2 \end{array}$$

Fig. 4.3.1

L_1 can be completed to a reduced 4×4 latin square in only one way, since the entry in the first column of row three must be 3 and then the entry in the second column of this row must be 4. (Correspondingly, L_2 can be completed in only one way because the entry in the first column of row four must be 4 and then the entry in the second column of this row must be 3.) L_3 can be completed to a reduced 4×4 latin square in either of two ways, as shown. Thus, there exist $2.1 + 1.2 = 4$ reduced 4×4 latin squares.

Shortly after the publication of Sade's paper, K. Yamamoto showed in [3] how, using the formula given in P. Erdős and I. Kaplansky [1] for the number of ways of extending a latin rectangle by one row and by a more detailed classification of the $k \times 7$ latin rectangles, Sade's method could be made self checking. This check confirmed the accuracy of Sade's calculations (but revealed a few minor errors).

Much more recently (1967), M. B. Wells used an adaptation of A. Sade's method suitable for computer calculation and, after confirming Sade's result for the number of reduced 7×7 latin squares, used it to calculate the number of reduced 8×8 latin squares. He obtained 535,281,401,856 as the number of such squares. He also showed that the number of distinct species (main classes) of 8×8 latin squares must be more than a quarter of a million. (See M. B. Wells [1].) The value for the number of reduced 8×8 latin squares obtained by Wells confirmed A. Sade's earlier conjecture that the number lay between 45×10^{10} and 6×10^{11}. (See A. Sade [2].)

Using a method which is substantially equivalent to that of E. Schönhardt, J. W. Brown has shown in [1] that the number of isotopy classes of latin squares of order 8 is 1, 676, 257. This work was published in 1968 and completes the enumeration of isotopy classes for squares of all orders up to 8.

We may summarize the known values(1) as follows:

Order of latin square	No. of main classes	No. of isotopy classes	No. of isomorphism classes	Total number of reduced squares
1	1	1	1	1
2	1	1	1	1
3	1	1	1	1
4	2	2	2	4
5	2	2	6	56
6	12	22	109	9,408
7	147	563	—	16,942,080
8	$\geq 25 \times 10^4$	1, 676, 257	—	535,281,401,856

Fig. 4.3.2

(1) A recent enumeration by computer gives 377,597,570,964,258,816 as the total number of reduced 9×9 latin squares. See S. E. Bammel and J. Rothstein [1].

Since a reduced latin square is one which has its first row and column in natural order, any such latin square is the multiplication table of a loop and so the enumeration of loops of a given order is equivalent to enumerating the reduced latin squares of that order.

From this point of view, two further papers should be mentioned as having a place in the above history of the subject: namely A. A. Albert [2] and B. F. Bryant and H. Schneider [1].

In the first of these papers, and probably unaware of the earlier work which E. Schönhardt had done on the subject, A. A. Albert initiated a general discussion on the enumeration of loops. He pointed out first that it is easy to check that the only loops of orders 2, 3 and 4 are the groups of those orders. He went on to give a complete enumeration of the loops of order 5 and showed that every loop of order five is either a group or is isomorphic to one of the five loops whose multiplication tables are given in Fig. 4.3.3 (compare page 129). Furthermore, he showed that, with the exception of the first and fifth of these loops, which are anti-isomorphic, no other pair is either isomorphic or anti-isomorphic. That is, there exist just six isomorphism classes of loops of order 5. Also, Albert proved that these six isomorphism classes yield only two isotopism classes, one of which comprises loops isotopic to the cyclic group of order 5, and the other of which contains all other loops of order 5.

(I)	1	2	3	4	5
1	1	2	3	4	5
2	2	3	1	5	4
3	3	4	5	2	1
4	4	5	2	1	3
5	5	1	4	3	2

(II)	1	2	3	4	5
1	1	2	3	4	5
2	2	3	1	5	4
3	3	4	5	1	2
4	4	5	2	3	1
5	5	1	4	2	3

(III)	1	2	3	4	5
1	1	2	3	4	5
2	2	3	5	1	4
3	3	4	1	5	2
4	4	5	2	3	1
5	5	1	4	2	3

(IV)	1	2	3	4	5
1	1	2	3	4	5
2	2	1	5	3	4
3	3	4	1	5	2
4	4	5	2	1	3
5	5	3	4	2	1

(V)	1	2	3	4	5
1	1	2	3	4	5
2	2	1	4	5	3
3	3	4	5	1	2
4	4	5	2	3	1
5	5	3	1	2	4

Fig. 4.3.3

Albert went on to discuss the enumeration of loops of order 6 but he did not attempt to do this exhaustively. Instead, he first showed the existence of simple loops of order 6 (that is, loops with no proper subloops) and then confined his further investigations to the subclass of loops of order 6 which contain one or more subloops of order 3. He was able to establish, in particular, that every subloop of order 6 with a subloop of order 3 has only a single subloop of that order. This work was published in 1944.

Later, in 1966, B. F. Bryant and H. Schneider carried out a complete enumeration of loops of order 6 using a computer as an aid and showed that there exist altogether 109 isomorphism classes of loops of order 6. This confirms E. Schönhardt's result. Their method was first to develop theorems (of a similar nature to those given in section 4.1) which described successively the principal classes, the isotopy classes, and the isomorphism classes until they had developed the theory to a point at which a computer could be employed effectively.

To complete this historical account, we should mention A. Sade's recent paper [30], already referred to in the previous section of this chapter, in which he summarized his own enumerative work for quasigroups of all orders up to 6 inclusive and also tabulated many interesting properties of particular quasigroups of these orders.

Most of the remaining part of this section will be devoted to enumerational results concerning groups. However, before pursuing this subject, we shall make two further general remarks.

If the number of reduced latin squares of order n is denoted by T_n and the total number of latin squares of order n by U_n then $U_n = n!\,(n-1)!\,T_n$ as has been pointed out in P. A. MacMahon [3], page 248, and by a number of later authors. For the proof, it is sufficient to observe that, from any given latin square (reduced, or not), it is possible to form $n!\,(n-1)!$ others by permuting the columns in $n!$ ways and the last $n-1$ rows in $(n-1)!$ ways. In M. Hall [3], it has been shown that

$$U_n \geq n!(n-1)!\ldots(2!)\,(1!)$$

and so we have

$$T_n \geq (n-2)!\,(n-3)!\ldots(2!)\,(1!)$$

for all n.

Despite the fact that the aim of the present book is to describe properties of latin squares of finite order, we think it worthwhile to give the following result of K. Mano as a curiosity. K. Mano has proved in [1] that the number of latin squares of infinite order is equal to the cardinal number of the continuum. We remark in passing that very few authors have studied latin squares of infinite order.

The number of non-isomorphic finite abelian groups of order $n = p_1^{\alpha_1} p_2^{\alpha_2} \ldots p_r^{\alpha_r}$ is very well known and will be found, for example, in L. Fuchs [1], page 53. The number of such groups is $\prod_{i=1}^{r} P(\alpha_i)$ where the p_i are primes, the α_i are positive integers, and $P(\alpha_i)$ denotes the number of distinct partitions of the integer α_i into positive integers. For asymptotic results on the same subject, the reader is referred to P. Erdős and G. Szekeres [1], D. G. Kendall and R. A. Rankin [1] and E. Krätzel [1].

A remarkable result of L. Rédei given in [1] can be formulated as follows: If and only if $n = p_1 p_2 \ldots p_u q_1^2 q_2^2 \ldots q_v^2$ where $p_1, p_2, \ldots p_u$, $q_1, q_2, \ldots q_v$ are primes such that $(p_r, \prod_{i=1}^{u} (p_i - 1) \prod_{j=1}^{v} (q_j - 1)) = 1$ and $(q_s, \prod_{i=1}^{u} (p_i - 1) \prod_{j=1}^{v} (q_j - 1)) = 1$ for $r = 1, 2, \ldots u$ and $s = 1, 2, \ldots v$, then all groups of order n are abelian. Here, (h, k) denotes the greatest common divisor of h and k.

Although a general formula for the number of non-abelian groups of order n is not yet known, a table giving the number of such groups for each order less than 162 (except the order 128) has been published in J. K. Senior and A. C. Lunn [2]. (See also M. Hall and J. K. Senior [1].),

In Fig. 4.3.4, we give the number of non-abelian groups of each order $n < 32$. The generating relations for each of these groups can be found in H. S. M. Coxeter and W. O. J. Moser [1].

Order:	1	2	3	4	5	6	7	8	9	10	11	12	13	14	15	16	17	18	19	20	21	22	23	24	25	26	27	28	29	30	31
Number of non-abelian groups:	0	0	0	0	0	1	0	2	0	1	0	3	0	1	0	9	0	3	0	2	1	1	0	2	0	1	1	2	0	3	0

Fig. 4.3.4

In [1], G. A. Miller determined the number of groups of order 64 and he found that it is 294. In [1], T. Szele showed that a necessary and sufficient condition that the only group of a given order n is the cyclic group of that order is that $(n, \varphi(n)) = 1$ where $\varphi(n)$ denotes Euler's function.

A detailed descriptic c a.l groups up to order 64 inclusive is given in M. Hall and J. K. Senior [1]. The numbers of groups of orders 16 32 and 64 are found to be 14, 51 and 267, respectively. The number 267 conflicts with that obtained earlier by G. A. Miller.

It has been shown in P. X. Gallagher [1] that, if $M(n)$ denotes the number of non-isomorphic finite groups (abelian and non-abelian) of order n then $M(n) \leq n^{cn^{2/3}\log n}$, where $c = 2/\{1 - (\frac{1}{2})^{2/3}\}$. A result given in G. Higman [1] shows that, for the special case of soluble groups, the above estimate of P. X. Gallagher is "best possible".

An exact method of enumeration for soluble groups has been given in A. C. Lunn and J. K. Senior [1].

In [1], L. Greenberg and M. Newman considered a related question concerning the number and distribution of soluble groups generated by elements of specified odd orders $a_1, a_2, \ldots a_r$. They proved the following result:

For each positive integer n, define the value of the function $s(n)$ to be 1 if a soluble group of order n exists generated by a set of elements $x_1, x_2, \ldots, x_r$ of orders $a_1, a_2, \ldots, a_r$ respectively, and 0 otherwise. Let $S(x) = \sum_{n \leq x} s(n)$. Then $\frac{1}{x}S(x) = O((\log x)^{-1/2h})$, where $h = \varphi(a_1 a_2 \ldots a_r)$ and φ is Euler's function. Consequently, $\lim_{x \to \infty} \frac{1}{x}S(x) = 0$.

For some results concerning the enumeration of p-groups, see C. C. Sims [1] and I. J. Davies [1].

In [1], E. N. Gilbert found a formula for the number of latin squares of order n which are multiplication tables of quasigroups isotopic to the cyclic group of order n. Such latin squares were called by Gilbert *addition squares*. The number of addition squares of order n is $n![(n-1)!]^2/\varphi(n)$, where $\varphi(n)$ is Euler's function.

In the same paper, Gilbert found a formula for the number of addition squares of order n which are complete latin squares. (For the definition of a complete latin square, see section 2.3.) His result is as follows:

Let $Q(n)$ denote the number of permutations $a_1, a_2, \ldots a_n$ of the integers $1, 2, \ldots, n$ which are such that the differences $a_2 - a_1, a_3 - a_2, \ldots, a_n - a_{n-1}$ are all distinct modulo n. Then there are exactly $n![Q(n)]^2/n^2\varphi(n)$ complete addition squares of order n.

Finally, we mention an enumerative result concerning transversals which has been obtained by J. Singer. Let t_n denote the number of transversals which exist in the standard form latin square L_n of order n which represents the multiplication table of the cyclic group of order n. Then Singer proved that $t_n = 0$ if n is even and he obtained the following values for odd values of n: $t_1 = 1$, $t_3 = 3$, $t_5 = 15$, $t_7 = 133$, $t_9 = 2025$ and $t_{11} = 37{,}851$. He also showed that for all values of n, $t_n \equiv 0$ modulo n. The latter observation leads the present authors to conjecture that no two distinct sets of n transversals of L_n (n odd) have any transversal in common. Proofs of the above-mentioned results will be found in J. Singer [2].

The result $t_n \equiv 0$ modulo n is valid for a wider class of latin squares than that discussed by Singer. In particular, it is valid for all latin squares which satisfy the quadrangle criterion. This has been proved by G. B. Belyavskaya, who has recently informed the authors that she and A. F. Russu together have obtained a more general result: namely, that the number t_n of transversals of an arbitrary standard form latin square L_n of order n is congruent to zero modulo the number of elements in the left nucleus N_λ (or the right nucleus N_μ) of the loop G of which L_n represents the multiplication table. (For the definition of the left nucleus of a loop see V. D. Belousov [7] or R. H. Bruck [5].) When G is a group, N_λ coincides with G and then also the latin square L_n satisfies the quadrangle criterion. Thus, Singer's result is a special case of that of Belyavskaya.

A related result obtained by A. Hedayat in [3] is that the number of transversals of an arbitrary standard form latin square L_n of order n which have exactly one cell in common cannot exceed $n-2$ and he has given examples for which this bound is attained.

In G. B. Belyavskaya [6], a computer algorithm for determining all the transversals of a given latin square has been described and, by means of it, Belyavskaya has confirmed the result $t_7 = 133$ obtained by J. Singer for the standard form cyclic latin square of order 7. (Further details of G. B. Belyavskaya [6] will be found in section 13.2).

4.4. Enumeration of latin rectangles

Since several methods of enumerating latin squares are dependent on the enumeration of $n \times k$ latin rectangles we would like to devote the final section of the chapter to this topic.

Among the earliest writers on this subject after A. Cayley and P. A. MacMahon (see the previous section) was S. M. Jacob. This author attempted an enumeration of $3 \times n$ rectangles and tabulated the results for values of $n \leq 15$. The first five of his results are in agreement with those obtained later by S. M. Kerewala in [1] and by J. Riordan in [2] and are as follows:

$$\begin{aligned} n = 3 \quad R(3, n) &= 3 \\ n = 4 \qquad\quad &= 24 \\ n = 5 \qquad\quad &= 552 \\ n = 6 \qquad\quad &= 21,280 \\ n = 7 \qquad\quad &= 1,073,760 \end{aligned}$$

Here, $R(3, n)$ denotes the number of reduced latin rectangles of three rows. (A *reduced* latin rectangle is one whose first row and column have their elements arranged in natural order.)

A latin rectangle with only two rows defines a permutation displacing all symbols and by some authors such a latin rectangle has been called a *discordant permutation* (see J. Riordan [1]). The enumeration of permutations discordant with a given permutation is the famous *problème des rencontres*. The enumeration of permutations discordant with each of two permutations, one of which is obtained from the other by a cyclic permutation of the symbols of the form $\begin{pmatrix} 1\,2\,3\,.\,.\,.\,n \\ 2\,3\,4\,.\,.\,.\,1 \end{pmatrix}$ is known as the reduced *problème des ménages*. The latter problem was considered in I. Kaplansky [1], I. Kaplansky and J. Riordan [1], and in J. Touchard [1] and [2].

The next case in this hierarchy, the enumeration of permutations discordant with three permutations of the form

$$\begin{matrix} 1 & 2 & . & . & . & n-1 & n \\ 2 & 3 & . & . & . & n & 1 \\ 3 & 4 & . & . & . & 1 & 2 \end{matrix}$$

has been examined in J. Riordan [4].

More generally, the numbers $N_{n,k}$ of permutations which are not discordant with the above three permutations but which differ from each of them in $n - k$ places were determined.

Figure 4.4.1 contains a table which gives the values of $N_{n,k}$ for small values of n and k. The table is reproduced from J. Riordan [4]. When $k = 0$, the numbers

$$N_{n,0} = \sum_{k=1}^{n} (-1)^k (n-k)!\, r_k = R(4, n)$$

enumerate the discordant permutations, where r_k is the number of ways of putting k elements in forbidden positions subject to the compatibility conditions that no two elements may be in the same position, and no two positions have the same element.

$k = \backslash n =$	3	4	5	6	7	8	9	10
0	0	1	2	20	144	1,265	12,072	125,655
1	0	0	15	72	609	4,960	46,188	471,660
2	0	6	20	180	1,106	9,292	82,980	831,545
3	6	8	40	176	1,421	10,352	93,114	912,920
4		9	30	180	980	8,326	70,272	695,690
5			13	72	595	4,096	39,078	379,760
6				20	154	1,676	14,292	155,690
7					31	304	4,230	43,880
8						49	576	9,905
9							78	1,060
10								125

Fig. 4.4.1

A recurrence relation for the number of three-line reduced latin rectangles was obtained in J. Riordan [3]. However, his formula is rather too complicated to be given here.

Let us remark at this point that, if $L(3, n)$ denotes the total number of three-line latin rectangles and $R(3, n)$ denotes the number of three-line reduced latin rectangles then, clearly,

$$L(3, n) = n\,!\,R(3, n).$$

(See J. Riordan [1].) In J. Riordan [1], and again in J. Riordan [2] by a different method, the asymptotic result $R(3, n) \approx (n\,!)^2\, e^{-3}$ is obtained. This result has since been extended to a general asymptotic formula, as we shall point out later in this section.

In [3], J. Riordan has shown that

$$R(3, n+p) \equiv 2 . R(3, n), \bmod p,$$

where p is a prime greater than two.

In L. Carlitz [1], this result of J. Riordan has been extended to give the statement that for arbitrary m,

$$R(3, n+m) \equiv 2^m . R(3, n), \bmod m.$$

A $k \times n$ latin rectangle is said to be *very reduced* if $3 \leq k \leq n$ and the first $k-1$ rows are as shown in Fig. 4.4.2.

$$\begin{array}{ccccccccc}
1 & 2 & 3 & \dots & k-2 & k-1 & k & \dots & n \\
n & 1 & 2 & \dots & k-3 & k-2 & k-1 & \dots & n-1 \\
n-1 & n & 1 & \dots & k-4 & k-3 & k-2 & \dots & n-2 \\
. & . & . & \dots & . & . & . & . & . \\
. & . & . & \dots & . & . & . & . & . \\
. & . & . & \dots & . & . & . & . & . \\
n-k+3 & n-k+4 & n-k+5 & \dots & n & 1 & 2 & \dots & n-k+2
\end{array}$$

Fig. 4.4.2

Let $V(k, n)$ denote the number of $k \times n$ very reduced latin rectangles. Then formulae giving the values of both $V(3, n)$ and $V(4, n)$ are known. The first was obtained by I. Kaplansky in [1] and the second by W. O. J. Moser in [1]. The formulae are as follows:

$$V(3, n) = \sum_{i=0}^{n} (-1)^i \frac{2n}{2n-i} \binom{2n-i}{i} (n-i)!, \; n \geq 3,$$

$$V(4, n) = \sum_{i=0}^{n} (-1)^i g(n, 3, i)\,(n-i)!, \qquad n \geq 4,$$

where

$$g(n, 3, i) = \begin{cases} \displaystyle\sum_{\alpha=0}^{[i/2]} \sum_{\beta=0}^{m} \frac{n}{n-i} \binom{n+\alpha \;\; i-1}{\alpha} \binom{n \;\; i}{\beta} 2^{\beta} \binom{n-\alpha \;\; 1}{i-2\alpha-\beta}, \\ \qquad\qquad \text{for } 0 \leq i < n, \text{ where } m = \min\,(n-i,\; i-2\alpha) \\ \displaystyle 3 + \sum_{\alpha=1}^{[n/2]} \frac{n}{\alpha} \binom{n-\alpha-1}{\alpha-1}, \quad \text{for } i = n \end{cases}$$

Exact enumeration of $L(k, n)$, $R(k, n)$ or $V(k, n)$ for larger values of k and n seems difficult but an asymptotic formula for the number of latin rectangles has been obtained by P. Erdős and I. Kaplansky in [1] and S. M. Kerewala has also published some results on this subject in [2] and [3]. For confirmation of the results given in S. M. Kerewala [2], see K. Yamamoto [1] and [4].

The asymptotic relation obtained by P. Erdős and I. Kaplansky for the total number $L(n, k)$ of $n \times k$ latin rectangles is

$$L(n, k) \approx (n!)^k \exp\left(\frac{k(k-1)}{2}\right)$$

and they showed that this relation is valid as $n \to \infty$ not only for fixed k but also for any $k < (\log n)^{3/(2-\varepsilon)}$, where ε is an arbitrarily small positive constant.

They expressed the conjecture that the formula for $L(n, k)$ given above would be valid for values of k up to nearly $n^{1/3}$.

In [2], K. Yamamoto confirmed this conjecture, showing the validity of the result of P. Erdős and I. Kaplansky for all integers k satisfying $k < n^{(1/3)-\delta}$ where δ is an arbitrarily small positive constant, or, more generally, where δ is any positive valued function of n tending to zero in such a way that $n^{-\delta} \to 0$ (as $n \to \infty$). Some further interesting results connected with the enumeration of latin rectangles will be found in K. Yamamoto [6] and J. Riordan [5].

CHAPTER 5

The concept of orthogonality

In this chapter, we introduce the concept of orthogonality between latin squares. First we consider the case of latin squares which represent the multiplication tables of groups and then we go on to give a historical account of the famous Euler conjecture and its eventual resolution. We show that the maximum possible number of latin squares of order n in a mutually orthogonal set is $n - 1$ and that any such complete mutually orthogonal set represents a finite projective plane.

Next, we show how the concept of orthogonality between latin squares leads on to that of orthogonality between quasigroups and groupoids. We end the chapter with a short discussion of various extensions of the idea of orthogonality to other related structures: notably to latin rectangles, latin cubes and permutation cubes, and Steiner triple systems.

5.1. Existence problems for incomplete sets of orthogonal latin squares

Two latin squares $L_1 = \|a_{ij}\|$ and $L_2 = \|b_{ij}\|$ on n symbols $1, 2, \ldots, n$, are said to be *orthogonal* if every ordered pair of symbols occurs exactly once among the n^2 pairs (a_{ij}, b_{ij}), $i = 1, 2, \ldots, n$; $j = 1, 2, \ldots, n$.

As examples, the reader will find a pair of orthogonal latin squares of order 10 displayed on page 400 and another pair of the same order on page 451.

It is easy to see by trial that the smallest value of n for which two orthogonal squares exist is 3. A pair of orthogonal squares of this order is shown in Fig. 5.1.1 and the corresponding ordered pairs (a_{ij}, b_{ij}) are exhibited alongside.

$$L_1 = \begin{array}{|ccc} 2 & \boxed{3} & 1 \\ \boxed{1} & 2 & 3 \\ 3 & 1 & \boxed{2} \end{array} \qquad L_2 = \begin{array}{|ccc} 2 & 1 & 3 \\ 1 & 3 & 2 \\ 3 & 2 & 1 \end{array} \qquad \begin{array}{ccc} 2,2 & 3,1 & 1,3 \\ 1,1 & 2,3 & 3,2 \\ 3,3 & 1,2 & 2,1 \end{array}$$

Fig. 5.1.1

If we consider the (exactly) n cells of the latin square L_2 all of which contain the same fixed entry h say $(1 \leq h \leq n)$, then the entries in the corresponding cells of the latin square L_1 must all be different, otherwise the squares would not be orthogonal. Since the symbol h occurs exactly once in each row and once in each column of the latin square L_2, we see that the n entries of L_1 corresponding to the entry h in L_2 have the property that they are all different and also occur one in each row and one in each column of the square L_1. A set of n different symbols having this property in a latin square of order n is usually called a *transversal* of the latin square. See, for example, D. M. Johnson, A. L. Dulmage and N. S. Mendelsohn [2] and E. T. Parker [8], though a number of other names have been used, as explained in section 1.4.

In Fig. 5.1.1, the elements of a transversal are shown enclosed in squares.

It is immediately obvious from these remarks that

THEOREM 5.1.1. *A given latin square possesses an orthogonal mate*[1] *if and only if it has n disjoint transversals.*

If the latin square in question is the multiplication table of a group, we can say more. In that case, the existence of a single transversal is sufficient by theorem 1.4.2.

Moreover, if G is a group of odd order, the entries a_{ii} $(i = 1, 2, \ldots, n)$ of the leading diagonal of its multiplication table always form a transversal by theorem 1.4.3, so we have

THEOREM 5.1.2. *The multiplication table of any group of odd order forms a latin square which possesses an orthogonal mate.*

COROLLARY. *There exist pairs of orthogonal latin squares of every odd order.*

The latter fact was known to Euler (1782), see L. Euler [2]. In the same paper, Euler also obtained a construction for a pair of orthogonal latin squares of order equal to any multiple of four.

Let us note here that, if one of a pair of orthogonal latin squares is the Cayley table of a group, the pair is said to be *based on a group*. Since, by theorem 1.4.7, a necessary condition for the existence of a transversal in the Cayley table of a group of even order is that its Sylow 2-subgroups

[1] The descriptive term *orthogonal mate* for a latin square L_2 which is orthogonal to a given latin square L_1 was first used by E. T. Parker in [8].

be non-cyclic, a group of order $4k+2$ cannot possess such a transversal. Consequently,

Theorem 5.1.3. *No pair of orthogonal latin squares based on a group can exist when n is an odd multiple of two.*

A proof of this result was first given by H. B. Mann [1] in 1942 by a somewhat more direct argument.

Euler had conjectured that the result was true (see L. Euler [2]) as a consequence of his unsuccessful attempts to find a solution of the following problem(1) which he himself had proposed three years earlier, in 1779. "Thirty-six officers of six different ranks and taken from six different regiments, one of each rank in each regiment, are to be arranged, if possible, in a solid square formation of six by six, so that each row and each column contains one and only one officer of each rank and one and only one officer from each regiment."

As will readily be seen, the problem is soluble if and only if there exists a pair of orthogonal latin squares of order six. By systematic enumeration of cases, G. Tarry (see [2], [3], [4], [5]) proved in the year 1900 that no such pair of squares can exist. A considerably shorter proof was given by R. A. Fisher and F. Yates in 1934. (See R. A. Fisher and F. Yates [1].) Another, more recent, proof will be found in K. Yamamoto [5].

Although the evidence for it appears to us quite scanty, the Euler conjecture of 1782 was actually much bolder than that mentioned above. It asserted that no pairs of orthogonal latin squares of order n exist for any value of n that is an odd multiple of two. Certainly, one piece of evidence for this conjecture is given by the following theorem.

Theorem 5.1.4 *If a latin square L of order $4k+2$ represents the Cayley table of a quasigroup which contains a subquasigroup of order $2k+1$ then L has no orthogonal mate.*

Proof. Consider a latin square of order $4k+2$ which contains a latin subsquare of order $2k+1$ representing the multiplication table of a subquasigroup Q of order $2k+1$. To simplify the argument, we may suppose that the rows and columns of L have been rearranged so to that the latin subsquare A occupies the first $2k+1$ rows and $2k+1$ columns of L and we may further suppose that L has been partitioned into four subsquares A, B, C, D each having $2k+1$ rows and $2k+1$ columns as shown in Fig. 5.1.2. Then any transversal τ of L must involve the

(1) For a recent generalization of the problem, see C. R. Rao [4].

$2k+1$ elements x_i, $i = 1,2,\ldots, 2k+1$, of the subquasigroup Q and we may suppose that h of the cells of τ which contain members of Q occur in the subsquare A. Since no row or column of L can contain an element x_i twice, none of the x_i occur in any of the cells of the first $2k+1$ rows or columns of L which are outside the subsquare A. That is, none occur in either of the subsquares B or C. Consequently, all remaining occurences of the x_i are in the subsquare D formed by the last $2k+1$ rows and last $2k+1$ columns of L. Each x_i occurs all together $4k+2$ times in L and so it occurs $2k+1$ times outside the subsquare A. There are $2k+1$ different x_i and so the $(2k+1)\times(2k+1)$ cells of the subsquare D all contain members x_i of the subquasigroup Q. Thus, all nonmembers of Q occur in the subsquares B and C. Since no two cells of τ are in the same row of L, there are $2k+1-h$ cells of τ in the subsquare B. Likewise, since no two cells of τ are in the same column of L, there are $2k+1-h$ cells of τ in the subsquare C. The entries of these cells must include all the $2k+1$ elements of L which are non-members of Q. Thus, $2(2k+1-h) = 2k+1$. That is, $2k+1 = 2h$. But this is clearly impossible if h and k are integers. We conclude that L has no transversals, and consequently no orthogonal mate.

The result of the above theorem was obtained by H. B. Mann in [3]. In fact, Mann gave a slight refinement of it, which we shall prove later (as theorem 12.3.2). Also a generalization of the result appears as theorem 12.3.1. However, it seems certain that the basic result given in the statement of theorem 5.1.4. itself was known to Euler. (See page 156).

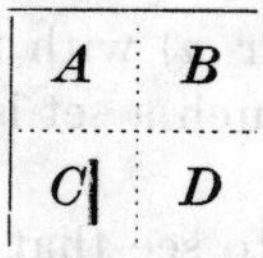

A	B
C	D

Fig. 5.1.2

Notice that the result of this theorem alone ensures that the Cayley table of a group of order $4k+2$ cannot possess a transversal, for such a group always has a subgroup of order $2k+1$ as we may easily see by considering the regular permutation representation of the group. For, certainly the group has elements of order 2. These are represented by products of $2k+1$ transpositions; that is, by odd permutations. The product of two odd permutations is even, so the regular representation contains both odd and even permutations. But, in a permutation group which contains both odd and even permutations, the even permutations form a subgroup whose order is half that of the group.

It was almost 80 years before the question raised by Euler was finally resolved, although fallacious proofs of the truth of his conjecture were published by several authors: notably by J. Petersen [1] in 1901, P. Wernicke [1] in 1910 and by H. F. MacNeish [2] in 1922. (See also E. Fleischer [1].) An explanation of the error in Wernicke's proof was given in H. F. MacNeish [1] and also in E. Witt [1]. (Further details of the early history of the problem will be found in H. W. Norton [1]).

The reader may also be interested to look at a series of notes concerning the problem which were published in the Intermédiaire des Mathématiciens in the 1890's. These are included in the bibliography of the present book and, in date order, are as follows: J. D. Loriga [1], E. Maillet [1], A. Akar [1], E. Maillet [4], L. Laugel [1], H. Brocard [1], E. Barbette [2], B. Heffter [1], H. Dellanoy and E. Barbette [1], G. Tarry [1], and E. Lemoine [1] and [2].

The first step in the resolution came in 1958 when R. C. Bose and S. S. Shrikhande [1] managed to construct an orthogonal mate for a certain latin square of order 22. Shortly afterwards, E. T. Parker (see [1] and [2]) used a different construction to obtain an orthogonal mate for a square of order 10. Then, in a combined paper [1] published in 1960, R. C. Bose, S. S. Shrikhande, and E. T. Parker proved that the Euler conjecture is false for all odd multiples of two except the values $n = 2$ and $n = 6$. All these results were obtained with the aid of statistical designs, and we shall defer their further discussion until chapter 11.

We consider next the problem of constructing sets of more than two latin squares (of the same order n) with the property that each pair of the set is an orthogonal pair. Such a set is called a set of *mutually orthogonal* latin squares.

In the first place, it is easy to see that

THEOREM 5.1.5. *Not more than $n - 1$ mutually orthogonal latin squares of order n can exist.*

PROOF. We argue as follows. If one square L_1 is given, the n symbols of any orthogonal mate can be renamed in any manner without affecting the orthogonality: for L_1 is the union of n disjoint transversals and the cells of L_2 which correspond to those of any transversal of L_1 all contain the same symbol. After the renaming this is still the case, and so every possible ordered pair of symbols still occurs exactly once among the pairs (a_{ij}, b_{ij}). We may apply this argument to any pair of the mutually orthogonal squares and so can conclude that any one of the squares may have its symbols renamed without affecting the orthogonality of the set. By such renamings, we may arrange that the symbols occurring in the first rows of all the squares are the symbols 1, 2, . . . , n in natural order. (See also H. B. Mann [1].) The symbols occurring in the first cell of the second rows of the squares must then all be different: for suppose two of them were the same, both containing the symbol r, say. Then the ordered pair (r, r) would occur in both the $(1, r)$th position and the $(2, 1)$th position in the two squares and the squares could not be orthogonal. None of the squares can have the symbol 1 as the entry in the first cell of the second row otherwise this symbol would occur twice in the first column of that square. Thus, at most $n - 1$ mutually orthogonal squares can exist corresponding to the $n - 1$ different symbols distinct from 1 which can appear in the first cells of their respective second rows.

We may remark at this point that the orthogonality of a set of mutually orthogonal latin squares is also unaffected when the same permutation of rows is made in all the squares simultaneously (but without changing the order of their entries). For, when any one square is superimposed on any other, each ordered pair of symbols occurs exactly once among the n^2 cells if the squares are orthogonal, and this property is unaffected by the simultaneous reordering of the rows. Moreover, such a reordering does not affect the "latin-ness" of any square: that is, it remains true after the reordering of rows that each symbol occurs exactly once in each row and each column of a square. By such a reordering of rows, the symbols of the first column of one of the squares can be arranged to be in natural order 1, 2, . . . , n.

A set of mutually orthogonal latin squares is said to be a *standardized set* (see R. C. Bose and K. R. Nair [2]) when, in the first rows of all the squares, the symbols are in natural order, and when, in addition, the symbols of the first column of one of the squares are in natural order. A single latin square is said to be in *standard form* or to be *reduced* when the symbols of both its first row and its first column are in natural order, as we have already mentioned in section 4.2.

As an example, we may standardize the squares of Fig. 5.1.1 in the following way: First rename the symbols 2, 3, 1 of L_1 as 1, 2, 3 respectively and also rename the symbols 2, 1, 3 of L_2 as 1, 2, 3 respectively. This gives a standardized set with the square L_2 in standard form. If it is desired to have the square L_1 in standard form, interchange the second and third rows of both the squares simultaneously.

5.2. Complete sets of mutually orthogonal latin squares

The question, for which values of n do sets of $n-1$ mutually orthogonal latin squares exist, is still an open one. However, three important facts bearing upon this question are known. In the first place, a set of $n - 1$ mutually orthogonal latin squares of order n exists if and only if there exists a finite projective plane of order n. More precisely, every finite projective plane of order n defines and is defined by such a set of squares. This result was first proved by R. C. Bose in [1]. Alternative proofs will be found in H. B. Mann [3], A. D. Keedwell [4] and G. E. Martin [1]. Secondly, it is well known that a finite projective plane of order n exists whenever n is a prime power, but, up to the present, no finite projective plane of non-prime power order has been discovered. Thirdly, R. H. Bruck and H. J. Ryser have shown in [1] that, for a certain infinite set of values of n, there cannot exist any projective plane of order n. The proofs of these three results appear in theorems 5.2.2, 5.2.3, and 5.2.5 below.

As regards the non-existence of projective planes, C. R. MacInnes proved by direct combinatorial arguments in a paper [1] published in 1907 that no finite projective plane of order 6 could exist though in fact, of course, this result was already implicit in G. R. Tarry's proof of seven years earlier that no orthogonal pair of latin squares of order 6 exists. The non-existence of a plane of order 6 was also implicit in a paper of F. H. Safford (see [1]) published in 1907, and recently a new elementary proof of the non-existence has been given by N. M. Rybnikova and A. K. Rybnikov in [1].

Safford's paper gave a proof that the following problem, proposed by O. Veblen and arising from an earlier problem of diophantine analysis, had no solutions. The problem was that of arranging, if possible, 43 distinct objects in 43 sets of seven each in such a way that every pair of objects should lie in one and only one set of seven. It would then follow that each two of the sets of seven would have in common one and only one object. The connection with the finite projective plane of order 6 is immediate from the fact that such a plane would necessarily have seven points on every line and have 43 lines altogether.

No further result concerning non-existence was obtained until 1949. In that year, R. H. Bruck and H. J. Ryser (see [1]) showed that, when n is congruent to 1 or 2 modulo 4 and the square free part of n contains at least one prime factor of the form $4k + 3$, there does not exist a finite projective plane of order n. This result excludes the possibility of the existence of a complete set of latin squares for the orders $n = 6, 14, 21, 22, 30$, and so on. (Since no larger set exists, a set of $n - 1$ mutually orthogonal latin squares of order n is called a *complete set*, see R. C. Bose and K. R. Nair [2].) We shall now give the proofs of these results.

We remind the reader that a *projective plane* comprises a set of elements called *points* and a set of elements called *lines* (which may conveniently be thought of as subsets of the points) with a relation called *incidence* connecting them such that each two points are incident with (belong to) exactly one line, and each two lines are incident with (have in common) exactly one point. The plane is *non-degenerate* if there are at least four points no three of which belong to the same line: that is, if the plane contains a proper quadrangle. It will be assumed from now on that all projective planes to be discussed are non-degenerate. If such a non-degenerate plane has a finite number $n + 1$ of points on one line, it is called a *finite* projective plane.

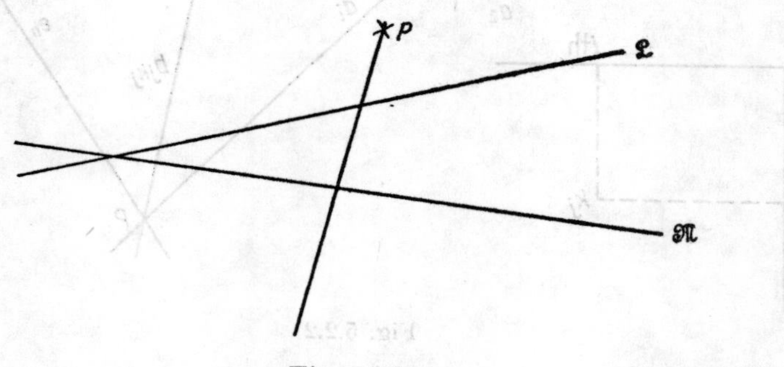

Fig. 5.2.1

THEOREM 5.2.1. *A finite projective plane π necessarily has the same number $n + 1$ of points on every line, has $n + 1$ lines through every point, and has $n^2 + n + 1$ points and $n^2 + n + 1$ lines altogether.*

PROOF. Let $\mathfrak{L}$ be the given line and $\mathfrak{M}$ any other line (Fig. 5.2.1). If P is a point not on $\mathfrak{L}$ or $\mathfrak{M}$ (such a point exists because π contains a proper quadrangle), then the joins of P to the $n + 1$ lines of $\mathfrak{L}$ intersect $\mathfrak{M}$ in $n + 1$ points. $\mathfrak{M}$ cannot contain further points otherwise there would be more than $n + 1$ lines through P and $\mathfrak{L}$ would do so also. Since every point of π is on one of the $n + 1$ lines through P (because there is a unique line joining any two points) and since each such line contains n points other than P, there are $n(n+1)+1$ points all together. Since there are $n+1$ lines through the point P, repetition of the argument with the roles of point and line interchanged shows that there are $n+1$ lines through every point and $n(n+1)+1$ lines all together.

Definition. A finite projective plane having $n+1$ points on every line is said to be of *order n*.

In O. Veblen's problem, mentioned above, the 43 sets of seven objects are required to possess the properties of lines of a projective plane having the objects as points. Safford showed in effect, therefore, that it is not possible to have a projective plane which contains $43 = 6^2 + 6 + 1$ lines containing seven points each.

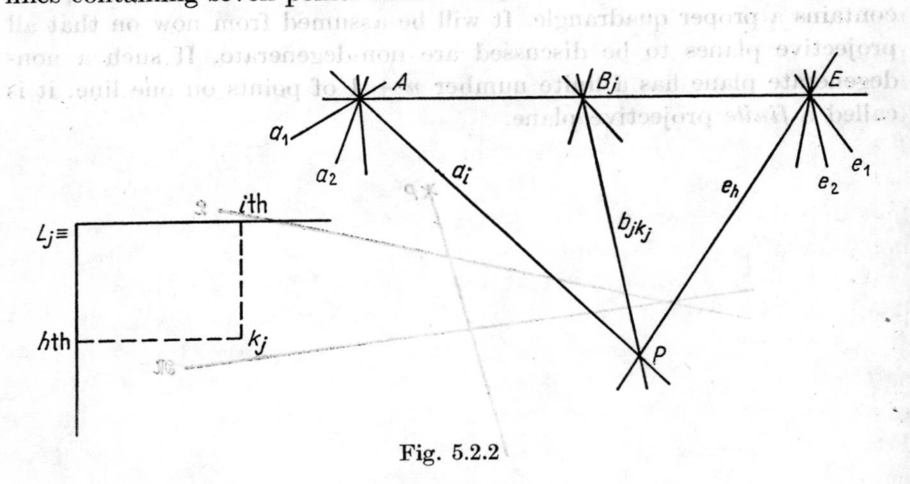

Fig. 5.2.2

We now prove:

Theorem 5.2.2. *Every finite projective plane of order n defines a set of $n-1$ mutually orthogonal latin squares; and conversely.*

Proof. Let G be a finite projective geometry with $n+1$ points on every line. We pick any line $\mathfrak{L}$ of G and call it the *line at infinity*. Let $A, E, B_1, B_2, \ldots, B_{n-1}$ be the points of $\mathfrak{L} = l_\infty$. Through each of these points there pass n lines other than l_∞. We label these lines as follows: $a_1, a_2, \ldots, a_n$ are the lines through A; $e_1, e_2, \ldots, e_n$ are the lines through E; $b_{j1}, b_{j2}, \ldots, b_{jn}$ are the lines through B_j. Every finite point $P(h, i)$ can then be identified with a set of $n+1$ numbers $(h, i, k_1, k_2, \ldots, k_{n-1})$ describing $n+1$ lines $e_h, a_i, b_{1k_1}, b_{2k_2}, \ldots, b_{n-1\,k_{n-1}}$ with which it is incident, one through each point of l_∞, and a complete set of orthogonal latin squares can be formed in the following way: in the jth square, put k_j in the (h, i)th place. Each square is latin since, as h varies with i fixed, so does k_j, and, as i varies with h fixed, so does k_j. Each two squares L_p and L_q are orthogonal: for suppose that k were to appear in the (h_1, i_1)th place of L_p and also in the (h_2, i_2)th place and that l were to appear in the corresponding places of L_q. This would imply that the lines b_{pk} and b_{ql} both passed through the two points (h_1, i_1) and (h_2, i_2) in contradiction to the axioms of G.

Conversely, we may show that a complete set of mutually orthogonal latin squares of order n defines a plane of order n.

From the given complete set of latin squares, we may define a set of n^2 "finite" points (h, i), $h = 1, 2, \ldots, n$; $i = 1, 2, \ldots, n$; where the point (h, i) is to be identified with the $(n + 1)$-ad of numbers $(h, i, k_1, k_2, \ldots, k_{n-1})$, k_j being the entry in the hth row and ith column of the jth latin square L_j. We form $n^2 + n$ lines b_{jk}, $j = -1, 0, 1, 2, \ldots, n - 1$; $k = 1, 2, \ldots, n$; where b_{jk} is the set of all points whose $(j + 2)$th entry is k and $b_{-1\,k} \equiv e_k$, $b_{0\,k} \equiv a_k$. Thus, we obtain $n + 1$ sets of n parallel[(1)] lines. From the orthogonality of the latin squares, it follows that two non-parallel lines intersect in one and only one point.

We adjoin one point to each set of parallels and let this additional point lie on every line of the set. These additional points E, A, B_1, $B_2, \ldots, B_{n-1}$ form the *line at infinity*. Two lines then intersect in one and only one point. From this, and the fact that $n + 1$ lines pass through every point, it follows that two points have at least one line in common. For, suppose on the contrary that the points Q, R have no line in common. Each of the $n + 1$ lines through Q meets each of the $n + 1$ lines through R in a point. This gives a total of $(n + 1)^2 > n^2 + n + 1$ points all together, no two of which can coincide, otherwise two lines through Q would have a second intersection. This contradiction shows that two points have at least one line in common. It follows at once that they have exactly one line in common.

Thus, we obtain a finite plane projective geometry with $n + 1$ points on every line.

As an example, a complete set of mutually orthogonal latin squares of order 4 and the corresponding projective plane of order 4 are shown

(1) We call two lines *parallel* if they have no finite point in common.

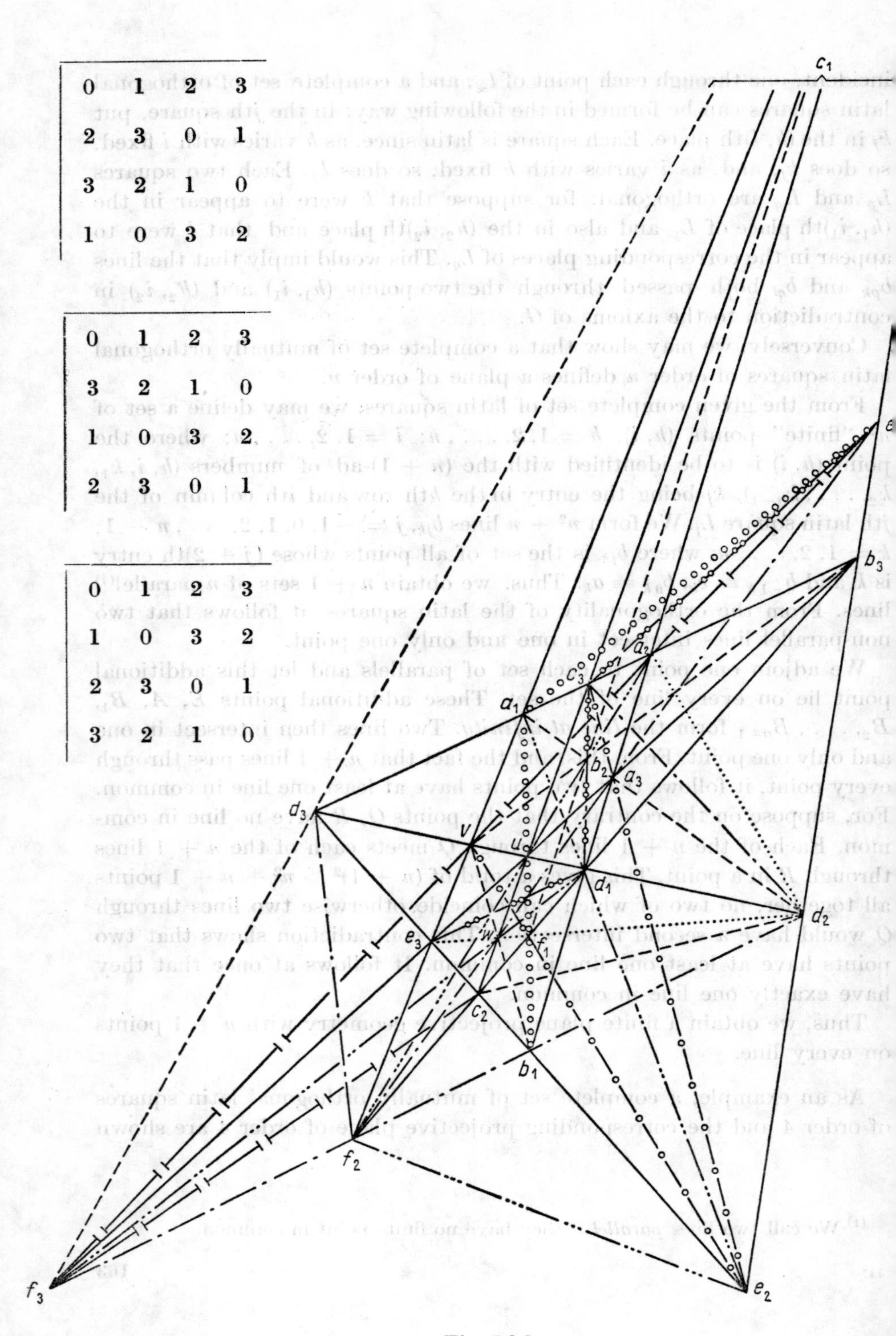

0	1	2	3
2	3	0	1
3	2	1	0
1	0	3	2

0	1	2	3
3	2	1	0
1	0	3	2
2	3	0	1

0	1	2	3
1	0	3	2
2	3	0	1
3	2	1	0

Fig. 5.2.3.

on the cover design of this book and are repeated in Fig. 5.2.3. The following 21 sets of 5 collinear points are the lines of this plane:

$a_2\ a_3\ b_1\ c_1\ d_1$	$a_1\ b_1\ e_1\ f_1\ u$	$b_1\ c_2\ d_3\ e_2\ e_3$
$a_3\ a_1\ b_2\ c_2\ d_2$	$a_2\ b_2\ e_2\ f_2\ u$	$b_2\ c_3\ d_1\ e_3\ e_1$
$a_1\ a_2\ b_3\ c_3\ d_3$	$a_3\ b_3\ e_3\ f_3\ u$	$b_3\ c_1\ d_2\ e_1\ e_2$
$b_1\ b_2\ b_3\ v\ w$	$a_1\ c_1\ e_3\ f_2\ v$	$b_1\ c_3\ d_2\ f_2\ f_3$
$c_1\ c_2\ c_3\ w\ u$	$a_2\ c_2\ e_1\ f_3\ v$	$b_2\ c_1\ d_3\ f_3\ f_1$
$d_1\ d_2\ d_3\ u\ v$	$a_3\ c_3\ e_2\ f_1\ v$	$b_3\ c_2\ d_1\ f_1\ f_2$
	$a_1\ d_1\ e_2\ f_3\ w$	
	$a_2\ d_2\ e_3\ f_1\ w$	
	$a_3\ d_3\ e_1\ f_2\ w$	

THEOREM 5.2.3. *For every integer n that is a power of a prime number, there exists at least one projective plane of order n (and consequently at least one complete set of mutually orthogonal latin squares of order n).*

PROOF. It is well known that, corresponding to each integer n which is a prime power $n = p^r$, there exists a Galois field $\mathfrak{J} = GF[p^r]$, unique up to isomorphism, which has p^r elements. For the points of our projective plane π, we take the totality of homogeneous co-ordinate vector triples $\mathbf{x} = (x_0, x_1, x_2)$ where x_0, x_1, x_2 are any three elements of $\mathfrak{J}$, not all zero, and where the triples $\mathbf{x}$ and $\lambda\mathbf{x}$ represent the same point for all $\lambda \neq 0$ in $\mathfrak{J}$. We define the line joining the points $Y(y_0, y_1, y_2)$ and $Z(z_0, z_1, z_2)$ as consisting of the set of all points whose co-ordinate vectors $\lambda\mathbf{y} + \mu\mathbf{z}$ are linear combinations of those of Y and Z, where λ and μ are in $\mathfrak{J}$. It is then easy to check that if W is a point of the line YZ, the lines WZ and YZ are the same. That is to say, a line is a well-defined concept. Since each point of the line YZ except Z itself includes among its possible co-ordinate vectors one of the form $\mathbf{y} + \nu\mathbf{z}$ and since ν takes p^r values, each line contains $p^r + 1$ points.

Since each of x_0, x_1, x_2 may take any of p^r different values, except that the vector (0, 0, 0) is to be excluded, and since each point can be represented by $p^r - 1$ different vectors, corresponding to the $p^r - 1$ possible non-zero choices of λ, there are a total of

$$(p^{3r} - 1)/(p^r - 1) = p^{2r} + p^r + 1 = n^2 + n + 1$$

points all together.

The line YZ consists of the totality of points whose co-ordinate vectors (x_0, x_1, x_2) satisfy the relation

$$\begin{vmatrix} x_0 & x_1 & x_2 \\ y_0 & y_1 & y_2 \\ z_0 & z_1 & z_2 \end{vmatrix} = 0,$$

since they are linear combinations of the co-ordinate vectors of Y and Z We say that

$$l_0 x_0 + l_1 x_1 + l_2 x_2 = 0, \text{ where } l_i = \begin{vmatrix} y_j & y_k \\ z_j & z_k \end{vmatrix}$$

and i, j, k are a cyclic rearrangement of 0, 1, 2 is the *equation* of the line and that $[l_0, l_1, l_2]$ are its *line co-ordinates*. The sets of line co-ordinates $[l_0, l_1, l_2]$ and $[\lambda l_0, \lambda l_1, \lambda l_2]$ represent the same line for all non-zero λ in $\mathfrak{J}$ and hence, by the same argument as for points, there exists a total of $n^2 + n + 1$ lines in π.

It is an immediate consequence of the definition of a line that any two points of π are incident with exactly one line and also, since just one set of ratios $x_0 : x_1 : x_2$ satisfies the equations of two distinct lines simultaneously, each two lines of π are incident with exactly one point. Moreover, it is easy to see that, even when $\mathfrak{J} = GF[2]$, there exist at least four points no three of which are on the same line. Thus, π is a projective plane of order $n = p^r$.

DEFINITION. A projective plane which is constructed in the manner just described is called a *Galois plane*.

For example, the plane of order 4 exhibited in Fig. 5.2.3 is a Galois plane and the co-ordinates of its 21 points are as follows (where α satisfies the equation $\alpha^2 = \alpha + 1$ and addition is modulo 2):

$$\begin{array}{llll} a_1 = (1, 0, 0) & c_1 = (0, 1, \alpha) & e_1 = (\alpha, 1, 1) & u = (1, 1, 1) \\ a_2 = (0, 1, 0) & c_2 = (\alpha, 0, 1) & e_2 = (1, \alpha, 1) & v = (1, \alpha, \alpha^2) \\ a_3 = (0, 0, 1) & c_3 = (1, \alpha, 0) & e_3 = (1, 1, \alpha) & w = (1, \alpha^2, \alpha) \\ b_1 = (0, 1, 1) & d_1 = (0, \alpha, 1) & f_1 = (\alpha^2, 1, 1) & \\ b_2 = (1, 0, 1) & d_2 = (1, 0, \alpha) & f_2 = (1, \alpha^2, 1) & \\ b_3 = (1, 1, 0) & d_3 = (\alpha, 1, 0) & f_3 = (1, 1, \alpha^2) & \end{array}$$

It follows from theorems 5.2.2 and 5.2.3 that, *if p is a prime number and r any integer, a set of $p^r - 1$ mutually orthogonal latin squares of order p^r can always be constructed*. R. C. Bose (in [1]) and, independently, W. L. Stevens (in [1]) have both given simple methods for carrying out the construction in a given case. Their methods are equivalent to the following:

THEOREM 5.2.4. *Let the elements of the Galois field $GF[p^r]$ be denoted by $\alpha_0 = 0$, $\alpha_1 = 1$, $\alpha_2 = x$, $\alpha_3 = x^2, \ldots, \alpha_{p^r-1} = x^{p^r-2}$, where x is a generating element of the multiplicative group of $GF[p^r]$ and $x^{p^r-1} = 1$ so that $\alpha_i \alpha_j = \alpha_t$, where $t = (i-1) + (j-1) + 1 = i + j - 1$ taken modulo $p^r - 1$ for all $i, j \neq 0$. We shall call the ordering $\alpha_0, \alpha_1, \alpha_2, \ldots, \alpha_{p^r-1}$ natural order, and the rows and columns of each latin square will be called the* 0th, 1st, 2nd, $\ldots$, $(p^r - 1)$th. *The entry in the* pth *row and* qth *column of the first square L_1 will be denoted by a_{pq}. We may then obtain a complete set of mutually orthogonal latin squares $L_1, L_2, \ldots, L_{p^r-1}$ by making the following choices:*

(1) *the elements of the* 0th *row and* 0th *column of L_1 are to be in natural order; that is, $a_{p0} = \alpha_p$ and $a_{0q} = \alpha_q$;*

(2) *the elements of the first row of L_1 are to be given by the rule $a_{1q} = \alpha_1 + \alpha_q$;*

(3) *the elements of the remaining rows of L_1 are to be given by the rules*

$$\begin{aligned}(i)\ a_{p+1,q+1} &= 0 \quad \text{if } a_{pq} = 0,\\ &= \alpha_{s+1} \text{ if } a_{pq} = \alpha_s,\ 0 < s < p^r - 1,\\ &= \alpha_1 \quad \text{if } a_{pq} = \alpha_{p^r-1},\end{aligned}$$

and (ii) the square is to be symmetrical about its main diagonal;

(4) *the columns, excluding the first, of the remaining squares are to be obtained by permuting cyclically those of L_1, and all the squares are to have the same first column.*

PROOF. Our proof that this construction does indeed give a set of $p^r - 1$ mutually orthogonal latin squares will closely parallel that in R. C. Bose [1].

We observe firstly that the above prescription is equivalent to taking $\alpha_p + \alpha_i \alpha_q$ as the entry in the pth row and qth column of the ith square for all relevant values of p, q and i. For, since $\alpha_0 = 0$, the entry $\alpha_p + \alpha_i \alpha_0$ is equal to α_p for all i; that is, the 0th column of all the squares is in natural order. Also, the entry $\alpha_0 + \alpha_i \alpha_q$ $(q \neq 0)$ is equal to α_t where $t = q + i - 1$ taken modulo $p^r - 1$. That is, the 0th row of the ith square is $\alpha_0, \alpha_i, \alpha_{i+1}, \ldots, \alpha_{i-1}$ for $i \neq p^r - 1$ and is $\alpha_0, \alpha_{p^r-1}, \alpha_1, \alpha_2, \ldots, \alpha_{p^r-2}$ for $i = p^r - 1$. The qth column of the ith square is

$$\alpha_0 + \alpha_i\alpha_q, \ \alpha_1 + \alpha_i\alpha_q, \ \alpha_2 + \alpha_i\alpha_q, \ldots, \alpha_{p^r-1} + \alpha_i\alpha_q,$$

or

$$\alpha_0 + \alpha_{q+i-1}, \ \alpha_1 + \alpha_{q+i-1}, \ \alpha_2 + \alpha_{q+i-1}, \ldots, \alpha_{p^r-1} + \alpha_{q+i-1},$$

so it is the same as the $(q + i - 1)$th column of L_1. In other words, the columns of $L_2, L_3, \ldots, L_{p^r-1}$ are obtained by cyclically permuting the columns, excluding the first (which it is easy to check is in natural order) of L_1. Next, as regards L_1 itself, notice that $a_{1q} = \alpha_1 + \alpha_1\alpha_q = \alpha_1 + \alpha_q$ as prescribed in rule (2). Also, if $a_{pq} = \alpha_p + \alpha_1\alpha_q = \alpha_s$, we have that $a_{p+1, q+1} = \alpha_{p+1} + \alpha_1\alpha_{q+1} = \alpha_p\alpha_2 + \alpha_1\alpha_q\alpha_2 = (\alpha_p + \alpha_1\alpha_q)\alpha_2 = \alpha_s\alpha_2 = 0$ if $s = 0$, $= \alpha_{s+1}$ if $0 < s < p^r - 1$, $= \alpha_1$ if $s = p^r - 1$, so rule (3) (*i*) is valid. Moreover, $a_{qp} = \alpha_q + \alpha_1\alpha_p = \alpha_q + \alpha_p = \alpha_p + \alpha_q = \alpha_p + \alpha_1\alpha_q = a_{pq}$ and so rule (3) (*ii*) is valid.

Secondly, we must prove that when $\alpha_p + \alpha_i\alpha_q$ is taken as the entry in the pth row and qth column of the ith square for all p, q, and i, we do indeed get a set of squares each of which is latin and each two members of which are orthogonal. It is immediately evident that no two elements of the qth column of L_i can be the same. If two elements of the pth row were the same, we would have $\alpha_i\alpha_{q_1} = \alpha_i\alpha_{q_2}$ for some $q_1 \neq q_2$. Since our elements belong to a field, we would then have $\alpha_{q_1} = \alpha_{q_2}$, a contradiction. Thus, each square L_i is latin. Finally, suppose that two of the squares L_i and L_j were not orthogonal. There would then be two places, say the (p, q)th and (r, s)th places, in which the same entry occurred in both the squares. We would have $\alpha_p + \alpha_i\alpha_q = \alpha_r + \alpha_i\alpha_s$ and $\alpha_p + \alpha_j\alpha_q = \alpha_r + \alpha_j\alpha_s$. By subtraction, this would lead to $(\alpha_i - \alpha_j)\alpha_q = (\alpha_i - \alpha_j)\alpha_s$ and so to $q = s$ and hence also $p = r$, since by hypothesis $i \neq j$. Thus, the supposition that some two of the squares are not orthogonal is untenable and our result is completely proved.

It is relevant to consider at this point whether other sets of $p^r - 1$ mutually orthogonal latin squares of order p^r exist which are mathematically distinct[1] from the set which we have constructed above; for

[1] Two latin squares are called *equivalent* if one can be obtained from the other by renaming the symbols and/or reordering the rows and columns: that is, if the quasigroups whose multiplication tables they represent are isotopic (see chapters 1, 2, and 4). Two sets of mutually orthogonal latin squares of the same order are *equivalent* if the numbers of squares in the two sets are the same and if the squares of the two sets can be put into one-to-one correspondence in such a way that each pair is equivalent relative to the same renaming of symbols and reordering of rows and columns.

example, a set in which the columns of $L_2, L_3, \ldots, L_{p^r-1}$ cannot be obtained by cyclically permuting the columns of L_1. In consequence of theorem 5.2.2, the answer to this question involves first asking whether there exist finite projective planes of order p^r which are isomorphically distinct from the Galois plane of that order.

It is known that for each of the prime power orders up to and including eight, no plane other than the Galois plane of that order exists. For the orders 2, 3, and 4, this was first shown by O. Veblen and J. H. M. Wedderburn (see [1]); for the order 5, by C. R. MacInnes (see [1]) by a somewhat laborious enumeration of cases. A short proof was later given in H. B. Mann [3].

R. C. Bose and K. R. Nair (see [2]) deduced the non-existence of projective planes of order 7, other than the Galois plane of that order, from an examination of H. W. Norton's (incomplete) list of 7×7 latin squares given in H. W. Norton [1]. The list was later completed by A. Sade (see [4] and [5]) who had observed the omission of one class of squares (as we explained in section 4.3), but this did not affect the result. A properly geometrical proof was not given until 1953 (corrected in 1954) and was due to the combined efforts of W. A. Pierce (see [1]) and M. Hall (see [6], [7]).

The uniqueness of the projective plane of order 8 was shown with the aid of a computer by M. Hall, J. D. Swift, and R. J. Walker (see [1]) in 1956. The method adopted by these authors again made use of the list of 7×7 latin squares compiled by Norton and completed by Sade.

As regards projective planes of order 9, a total of four isomorphically distinct planes are known(1) (at the time of writing). Also, it is known that at least two isomorphically distinct projective planes exist for all orders p^r, p odd and $r \geq 2$, and for all orders 2^r, $r \geq 4$ (see M. Hall [11], chapter 12). Certain of these planes are represented by complete sets of latin squares in which the rows (columns) of $L_2, L_3, \ldots, L_{p^r-1}$ are not a re-ordering of those of L_1. That is, there do exist complete sets of mutually orthogonal latin squares such that the quasigroups whose multiplication tables are given by these squares are not all isotopic.

We shall make further mention of these matters in later chapters of this book, but for detailed information concerning the theory of finite projective planes the reader is referred in particular to the books of

(1) In this connection, see R. A. Fisher [2] for the correction of an error made in R. A. Fisher [4].

G. Pickert (see [2]), P. Dembowski (see [1]) and F. Kárteszi (see [2]); and to two Hungarian survey papers, F. Kárteszi [1] and A. Rényi [1].

It is relevant at this point to mention two unsolved problems. Since every known finite projective plane has order p^r, p prime, and can be represented as a set of $p^r - 1$ mutually orthogonal latin squares based on the abelian group of order p^r and type $(p, p, \ldots, p)$, we may ask:

(i) Is it true that there do not exist sets of $n - 1$ mutually orthogonal latin squares of order n based on a cyclic group unless n is prime?

(ii) If m denotes the maximum number of mutually orthogonal $n \times n$ latin squares none of which satisfies the quadrangle criterion, what is the value of m?

As regards problem (ii), J. Schönheim conjectures that $m < n - 1$ for every positive integer n ($n \geq 3$).(1)

We end the present discussion of finite planes by proving:

Theorem 5.2.5. *If n is a positive integer congruent to 1 or 2 modulo 4, there cannot exist any finite projective plane of order n unless n can be expressed as a sum of two integral squares, $n = a^2 + b^2$.*

(Equivalently, if $n \equiv 1$ or 2 (modulo 4) and if the square-free part of n contains at least one prime factor p which is congruent to 3 modulo 4, there does not exist a finite projective plane of order n.)

Proof. We shall need two theorems from number theory concerning the representation of an integer as a sum of squares.

Theorem A. A positive integer n is expressible as a sum of two integer squares if and only if each prime factor of n which is of the form $4k + 3$ occurs as a factor an even number of times. In particular, every prime number of the form $4k + 1$ is representable as a sum of two squares.

Theorem B. Every natural number is representable as the sum of (at most) four integer squares.

Proofs of these two theorems will be found, for example, in chapter 5 of H. Davenport [1]. The equivalence of the two statements of theorem 5.2.5 above follows immediately from theorem A.

Theorem B may be deduced from the fact that any prime of the form $4k + 3$ is representable as the sum of four squares with the aid of the identity

$$\begin{aligned}(a^2 + b^2 + c^2 + d^2)(A^2 + B^2 + C^2 + D^2) &= \\ = (aA + bB + cC + dD)^2 &+ (aB - bA - cD + dC)^2 + \\ + (aC + bD - cA - dB)^2 &+ (aD - bC + cB - dA)^2.\end{aligned}$$

(1) Oral communication.

For the proof of theorem 5.2.5, we observe that the number of points and lines of a projective plane π of order n are each equal to $N = n^2 + n + 1$ (as was shown in theorem 5.2.1). Let variables x_i, $i = 1$ to N, be associated with the N points P_i of π, and let the lines of π be denoted by L_j, $j = 1$ to N. We define *incidence numbers* a_{ij} as follows: $a_{ij} = 1$ if P_i and L_j are incident, $a_{ij} = 0$ otherwise. The $N \times N$ matrix $A = \|a_{ij}\|$ is then called the *incidence matrix* of the plane π, and it is easy to show that

$$AA^T = A^T A + S, n = I$$

where the superscript T denotes transpose, I is the $N \times N$ identity matrix, and S is an $N \times N$ matrix consisting entirely of 1's. However, for the present proof, it is more convenient to express the content of the above mentioned identity in terms of quadratic forms.

With the line L_j we associate the linear form

$$L_j = \sum_{i=1}^{N} a_{ij} x_i \,,$$

which we may also denote by L_j without confusion. Here, the a_{ij} are defined as above. Then,

$$\begin{aligned} L_1^2 + L_2^2 + \ldots + L_N^2 &= (n+1)(x_1^2 + x_2^2 + \ldots + x_N^2) + 2 \sum_{r,s} x_r x_s = \\ &= n(x_1^2 + x_2^2 + \ldots + x_N^2) + (x_1 + x_2 + \ldots + x_N)^2. \end{aligned}$$

To see this, we observe that in the set of the L_j each x_r occurs with a coefficient 1 exactly $n + 1$ times, since each point P_r is incident with $n + 1$ lines. Also, each cross-product $2x_r x_s$ occurs in $L_1^2 + L_2^2 + \ldots + L_N^2$ exactly once, since there is exactly one line L_j containing both P_r and P_s.

The above identity may be rewritten in the form

$$\begin{aligned} L_1^2 + L_2^2 + \ldots + L_N^2 = n(x_2 + x_1/n)^2 + n(x_3 + x_1/n)^2 + \ldots \\ \ldots + n(x_N + x_1/n)^2 + (x_2 + x_3 + \ldots + x_N)^2. \end{aligned}$$

We now change the variables by writing

$y_1 = x_2 + x_3 + \ldots + x_N$, $\quad y_2 = x_2 + x_1/n$, $\quad y_3 = x_3 + x_1/n, \ldots,$
$y_N = x_N + x_1/n$, $\quad$ so that

$$L_1^2 + L_2^2 + \ldots + L_N^2 = y_1^2 + ny_2^2 + ny_3^2 + \ldots + ny_N^2,$$

where each x_i can be expressed rationally in terms of the y_i.

Next suppose that $n \equiv 1$ or $2 \pmod 4$. If $n \equiv 1 \pmod 4$, then $n^2 \equiv 1 \pmod 4$, so $N = n^2 + n + 1 \equiv 3 \pmod 4$. If $n \equiv 2 \pmod 4$, $n^2 \equiv 0 \pmod 4$, and so $N = n^2 + n + 1 \equiv 3 \pmod 4$, as before.

Thus, for $n \equiv 1$ or $2 \pmod 4$, we have $N \equiv 3 \pmod 4$. We now use the fact that every positive integer n can be written as the sum of four squares, $n = a_1^2 + a_2^2 + a_3^2 + a_4^2$.

Also,

$$\begin{aligned} n(y_i^2 + y_{i+1}^2 + y_{i+2}^2 + y_{i+3}^2) &= (a_1^2 + a_2^2 + a_3^2 + a_4^2)(y_i^2 + y_{i+1}^2 + \\ &+ y_{i+2}^2 + y_{i+3}^2) = \\ &= (a_1 y_i + a_2 y_{i+1} + a_3 y_{i+2} + a_4 y_{i+3})^2 + \\ &+ (a_1 y_{i+1} - a_2 y_i - a_3 y_{i+3} + a_4 y_{i+2})^2 + \\ &+ (a_1 y_{i+2} + a_2 y_{i+3} - a_3 y_i - a_4 y_{i+1})^2 + \\ &+ (a_1 y_{i+3} - a_2 y_{i+2} + a_3 y_{i+1} - a_4 y_i)^2 = z_i^2 + z_{i+1}^2 + z_{i+2}^2 + z_{i+3}^2. \end{aligned}$$

Taking $i = 2, 6, 10, \ldots, N-5$ successively (and remembering that $N \equiv 3 \pmod 4$), we get

$$L_1^2 + L_2^2 + \ldots + L_N^2 = y_1^2 + z_2^2 + z_3^3 + \ldots + z_{N-2}^2 + n y_{N-1}^2 + n y_N^2,$$

where each y_i can be expressed rationally in terms of the z_i and consequently each x_i can be so expressed.

We write $y_1 = z_1$, $y_{N-1} = z_{N-1}$, $y_N = z_N$ for convenience, and then

$$L_1^2 + L_2^2 + \ldots + L_N^2 = z_1^2 + z_2^2 + \ldots + z_{N-2}^2 + n(z_{N-1}^2 + z_N^2). \ldots \ldots (1)$$

Since $L_j = \sum_{i=1}^{N} a_{ij} x_i$, each L_j is a rational function of the z_i, in virtue of the fact that each x_i is such a rational function. Now, since the identity (1) is an identity in the variables z_i, it remains true if some of the z_i are specialized to be rational functions of their successors.

Suppose that $L_1 = b_1 z_1 + b_2 z_2 + \ldots + b_N z_N$ when expressed rationally in terms of the z_i. If $b_1 \neq 1$, choose z_1 so that

$$(b_1 - 1)z_1 + b_2 z_2 + \ldots + b_N z_N = 0.$$

Then $L_1^2 = z_1^2$ and z_1 is a rational function of its successors. If $b_1 = 1$, choose z_i so that

$$(b_1 + 1)z_1 + b_2 z_2 + \ldots + b_N z_N = 0.$$

Then, $L_1^2 = (-z_1)^2 = z_1^2$ and z_1 is a rational function of its successors. With this choice of z_1, the identity (1) reduces to the form

$$L_2^2 + L_3^2 + \ldots + L_N^2 = z_2^2 + z_3^2 + \ldots + z_{N-2}^2 + n(z_{N-1}^2 + z_N^2).$$

Continuing, we may choose z_2 as such a rational function of $z_1, z_3, \ldots, z_N$ (and hence of $z_3, z_4, \ldots, z_N$ alone) that $L_2^2 = z_2^2$, and so on, until, for these special choices of $z_1, z_2, \ldots, z_{N-2}$, the identity (1) reduces to $L_{N-1}^2 + L_N^2 = n(z_{N-1}^2 + z_N^2)$. This implies that the integer n is a quotient of two integers each of which is expressible as a sum of two integral squares. (Since z_{N-1} and z_N are integers when, $x_1/n, x_2, \ldots, x_n$ are taken as integers, so also must be $L_{N-1}^2 + L_N^2$.) It follows from theorem A that the integer n must also be expressible as a sum of two integral squares, say $n = a^2 + b^2$. This proves the theorem.

Our final topic in the present section concerns the function $N(n)$, which is defined as the largest possible number of $n \times n$ latin squares which can exist in a single mutually orthogonal set. For example, we have proved above that $N(p^r) = p^r - 1$ for every prime p and integer r[1]. Also, we have noted that Bose, Shrikhande and Parker have shown that $N(n) \geq 2$ for all positive integers n except 2 and 6. $N(2) = 1$ is obvious by trial and $N(6) = 1$ was proved by G. R. Tarry in [2]. Recently (1966), R. Guérin has proved that $N(n) \geq 4$ for all $n \geq 53$ and some further results have been obtained by H. Hanani and R. M. Wilson since then. (See section 11.4.) However, for most non-prime power values of n, the exact value of $N(n)$ is still not known. In particular, it is still not known whether $N(10)$ exceeds 2, despite the attempts of many authors to construct three mutually orthogonal latin squares of order 10 with the aid of computers or otherwise. Most such unsuccessful attempts remain unpublished. However, the interested reader may consult E. T. Parker [3], [4], [5] and [8] and A. D. Keedwell [4] and [6] for accounts of two of them[2] and will find an interesting discussion relating to the magnitude of the problem in a paper by L. J. Paige and C. B. Tompkins (see [1]).

In this connection, it is worth mentioning that, quite recently, in A. Hedayat, E. T. Parker and W. T. Federer [1], a pair of mutually orthogonal latin squares of order 10 which possess a common transversal has been published. The authors of that paper believe that it is the first such pair to have been obtained.

Relevant to this and to theorem 5.1.5 is the question: For a given order n which is not a prime power, what is the maximum number of

(1) A very interesting result in this connection is that given in R. H. Bruck [6] where it is shown that, if $N(n) < n - 1$ for a given positive integer n, then we can assert that $N(n) < n - 1 - (2n)^{1/4}$. For the proof, see section 9.3.

(2) Further reference to these will be found in section 13.2.

latin squares in a set with the property that all the squares of the set have a transversal in common? We note that by virtue of the discovery of Hedayat, Parker and Federer just mentioned, the number is ≥ 2 if $n = 10$. By contrast, it follows from theorem 5.1.5 that the number is $\geq p^k - 2$ if $n = p^k$ with p prime.

Several investigations of the asymptotic behaviour of the function $N(n)$ have been carried out. The earliest result of this kind is that of S. Chowla, P. Erdős, and E. G. Strauss (see [1]) who proved in a joint paper that $N(n)$ tends to infinity with n. Precisely, they showed that $N(n) > \frac{1}{3} n^{1/91}$ for all sufficiently large n. Subsequently, in [1], K. Rogers proved the sharpened result $N(n) > n^{(1/42)-\varepsilon}$ $(\varepsilon > 0)$ and later still Y. Wang (see [1] or [2]) was able to obtain the further improvement $N(n) > n^{1/25}$ for all sufficiently large n.(1) The methods used by these authors were analytic in nature and we shall not explain them in detail in this book. A number of further properties of the function $N(n)$ will be discussed in chapter 11.

5.3. Orthogonal quasigroups and groupoids

In view of the fact that any bordered latin square represents the multiplication table of a quasigroup (chapter 1), the concept of orthogonality between two latin squares leads naturally to the concept of orthogonality between two quasigroups.

DEFINITION. The two finite quasigroups $(G, \cdot)$ and $(G, *)$ defined on the same set G are said to be *orthogonal* if the pair of equations $x . y = a$ and $x * y = b$ (where a and b are any two given elements of G) are satisfied simultaneously by a unique pair of elements x and y from G.

It is clear that when $(G, \cdot)$ and $(G, *)$ are orthogonal quasigroups the latin squares defined by their multiplication tables are also orthogonal.

The above definition may be expressed in another way. We may say that $(G, \cdot)$ and $(G, *)$ are orthogonal if $x.y = z.t$ and $x * y = z * t$ together imply $x = z$ and $y = t$. The definition can then be generalized. On the one hand we can say that two binary operations $(\cdot)$ and $(*)$ defined on the same set G are *orthogonal operations* if the equations $x . y = z . t$ and $x * y = z * t$ together imply $x = z$ and $y = t$, as was done by A. Sade

(1) Very recently, R. M. Wilson has proved in his paper [1] that $N(n) \geq n^{1/17} - 2$ for all large n.

in [11] and again by V. D. Belousov in [4]. In a later paper the latter author has discussed in detail the connections between such orthogonal operations and systems of orthogonal quasigroups (see V. D. Belousov [10]). On the other hand, we can extend the above definition of orthogonality between quasigroups to the case of infinite quasigroups and also to general groupoids as was done by S. K. Stein in [2]. He made the following definition:

Definition. Two finite or infinite groupoids $(G, \cdot)$ and $(G, *)$ defined on the same set G are called *orthogonal* if the mapping σ of the cartesian product $G \times G$ onto itself defined by $(x, y) \xrightarrow{\sigma} (x.y, x*y)$ is an equivalence mapping. (If G is a finite set, this implies that σ is a one-to-one mapping of $G \times G$ onto itself.) The groupoid $(G, *)$ is said to be an *orthogonal complement* of the groupoid $(G, \cdot)$.

The following interesting results come from S. K. Stein [2].

Theorem 5.3.1. *A finite commutative groupoid $(G, \cdot)$ is a quasigroup if and only if it is orthogonal to the groupoid $(G, *)$ whose multiplication is given by $x*y = x.xy$.*

Proof. Firstly, let $(G, \cdot)$ be a quasigroup and let a and b belong to G. Then there is an element x in G such that $xa = b$ and an element y in G such that $xy = a$. Thus, the simultaneous equations $xy=a$ and $x*y = b$ have a solution for x and y, which is necessarily unique since $(G, \cdot)$ is a quasigroup. That is, the groupoids $(G, \cdot)$ and $(G, *)$ are orthogonal.

Secondly, let $(G, \cdot)$ and $(G, *)$ be orthogonal groupoids where $x*y = x.xy$. Suppose, if possible, that $ay = az$. Then $a.ay = a.az$ or $a*y = a*z$. Since, by definition of orthogonality, the equations $ay = az$ and $a*y = a*z$ have unique solutions for a and y, we get $y = z$. So, the equation $ay = b$ is uniquely soluble in $(G, \cdot)$. Since $(G, \cdot)$ is commutative, it is a quasigroup.

Corollary. *Every quasigroup possesses an orthogonal complement* (which is not necessarily a quasigroup).

Proof. If $(G, \cdot)$ is the given quasigroup, we define a groupoid $(G,*)$ by $x*y = x.xy$. Then $(G,*)$ is orthogonal to $(G, \cdot)$ by the first part of the proof of the theorem.

Theorem 5.3.2. *A quasigroup $(G, \cdot)$ which satisfies the constraint $x.xz = y.yz$ implies $x = y$ possesses an orthogonal complement which is a quasigroup.*

Proof. Define a groupoid $(G, *)$ by $x*y = x.xy$. Then the equation $x*y = x*z$ is equivalent to $x.xy = x.xz$ and implies $y = z$ since $(G, \cdot)$ is a quasigroup. The equation $y*x = z*x$ is equivalent to $y.yx = z.zx$ and implies $y = z$ by the hypothesis of the theorem. It follows that $(G, *)$ is a quasigroup and, by the first part of theorem 5.3.1, it is orthogonal to $(G, \cdot)$.

Corollary. *A quasigroup $(G, \cdot)$ which satisfies the constraint $x.xy = yx$ has an orthogonal complement which is a quasigroup.*

Proof. Such a quasigroup clearly satisfies the condition $x.xz = y.yz$ implies $x = y$.

S. K. Stein hoped that by means of the above theorems he might be enabled to construct counter examples to the Euler conjecture about orthogonal latin squares. However, in this connection he was only able to prove the following:

Theorem 5.3.3. *There exist quasigroups $(G, \cdot)$ satisfying the constraint $x.xz = y.yz$ implies $x = y$ for all orders $n \equiv 0, 1,$ or 3 (mod 4).*

Proof. Let us denote by $A(GF[2^k], \alpha, \beta)$ the groupoid constructed from the Galois field $GF[2^k]$ by the multiplication $x \otimes y = \alpha x + \beta y$, where α and β are fixed elements of $GF[2^k]$; and by $A(C_n, p, q)$ the groupoid constructed from the cyclic group C_n of order n by the multiplication $x \otimes y = x^p y^q$, where p and q are fixed integers. (See next page.)

If $\alpha\beta(1 + \beta) \neq 0$ and $k \geq 2$, the groupoid $A(GF[2^k], \alpha, \beta)$ satisfies the constraint and is a quasigroup because $x \otimes y = x \otimes z$ implies $\alpha x + \beta y = \alpha x + \beta z$, whence $y = z$ if $\beta \neq 0$, and $y \otimes x = z \otimes x$ implies $\alpha y + \beta x = \alpha z + \beta x$, whence $y = z$ if $\alpha \neq 0$. As regards the constraint, we have $x \otimes (x \otimes z) = y \otimes (y \otimes z)$ implies $\alpha x + \beta(\alpha x + \beta z) = \alpha y + \beta(\alpha y + \beta z)$. That is, $\alpha(1 + \beta)(x - y) = 0$, giving $x = y$ if $\alpha(1 + \beta) \neq 0$. The order of this quasigroup is 2^k, which is congruent to zero modulo 4 if $k \geq 2$.

By forming the direct product of a system of this type with an abelian group of odd order, it is possible to construct systems satisfying the constraint whose orders n are congruent to 1 or 3 modulo 4. (In a group of odd order, $xx = yy$ implies $x = y$, as was proved in section 1.4, so $x.xz = y.yz$ implies $x = y$.) The elements of the direct product will be ordered pairs (b_i, c_i), $b_i \in GF[2^k]$, $c_i \in (H, \cdot)$ where $(H, \cdot)$ is a group of odd order. The law of composition $(*)$ will be defined by $(b_i, c_i) * (b_j, c_j) = (b_i \otimes b_j, c_i c_j)$.

Quasigroups satisfying the constraint $x.xy = yx$ have been called *Stein quasigroups* by V. D. Belousov ([4], page 102) and he has called this constraint the *Stein identity* (See also section 2.1.). As Stein has pointed out, quasigroups which satisfy this identity are idempotent (we have $x.xx = xx$ and, by left cancellation, $xx = x$) and distinct elements do not commute. Also, by theorems 5.3.1 and 5.3.2, they are orthogonal to their own transposes. (See also section 12.4.) Stein has given the following examples of quasigroups of this type and it would be very interesting to find a general classification for such quasigroups.

(1) The quasigroup of order 4 with multiplication table

	1	2	3	4
1	1	3	4	2
2	4	2	1	3
3	2	4	3	1
4	3	1	2	4

(2) The groupoid denoted by $A(C_n, p, r)$ above, where $p = q^2$, $(2q + + 1)^2 \equiv 5 \bmod n$, and n is odd. (Such systems exist only when 5 is a quadratic residue of n.)

(3) The groupoid $A(GF[p^k], \alpha, \beta)$ when $5^{(p^k-1)/2} \equiv 1, \bmod p^k$.

In V. D. Belousov and A. A. Gvaramiya [1], the following interesting result has been proved: namely that, if a Stein quasigroup is isotopic to a group G, then the commutator subgroup of G is in the centre of G.

In a paper [3] which was published several years later than that discussed in the preceding pages, S. K. Stein has proved (*i*) that if a quasigroup $(Q, \cdot)$ of order n has a transitive set of n automorphisms, then there is a quasigroup $(Q, *)$ orthogonal to it: and (*ii*) that no quasigroup of order $4k + 2$ has a transitive automorphism group. It follows from (*ii*) that no counter examples to the Euler conjecture can be constructed by means of (*i*).

The properties of systems of orthogonal quasigroups have been investigated in V. D. Belousov [3]. Also, in [11], A. Sade has obtained a number of interesting and useful results concerning orthogonal groupoids, their parastrophes (conjugates) and their isotopes. In a later paper [15], which we shall discuss in detail in section 12.1, the same author has achieved the goal of S. K. Stein: namely, to construct counter-examples to the Euler conjecture. He has used a particular type of direct product of quasigroups which he calls "*produit direct singulier*" and which preserves orthogonality.

The idea of forming new quasigroups as singular direct products came to Sade as a consequence of his investigations of the related concept of singular divisors of quasigroups introduced by him earlier in A. Sade [3] and discussed further in A. Sade [6] and [8]. In the opinion of the authors, this concept is a fruitful one and has not received the attention it deserves.

We end this section with a theorem of J. R. Barra (see [1]) which we shall use on several occasions later in connection with constructions of pairs of orthogonal latin squares.

														0	4	3	2	1
$L_1 =$	0	1	2	3	4		0	1	2	3	4		0	0	1	2	3	4
	1	2	3	4	0		4	0	1	2	3		4	4	0	1	2	3
	2	3	4	0	1		3	4	0	1	2		3	3	4	0	1	2
	3	4	0	1	2		2	3	4	0	1		2	2	3	4	0	1
	4	0	1	2	3		1	2	3	4	0		1	1	2	3	4	0

														0	4	3	2	1
$L_2 =$	0	1	2	3	4		0	1	2	3	4		0	0	1	2	3	4
	2	3	4	0	1		3	4	0	1	2		4	3	4	0	1	2
	4	0	1	2	3		1	2	3	4	0		3	1	2	3	4	0
	1	2	3	4	0		4	0	1	2	3		2	4	0	1	2	3
	3	4	0	1	2		2	3	4	0	1		1	2	3	4	0	1

														0	4	3	2	1
$L_3 =$	0	1	2	3	4		0	1	2	3	4		0	0	3	1	4	2
	3	4	0	1	2		2	3	4	0	1		4	1	4	2	0	3
	1	2	3	4	0		4	0	1	2	3		3	2	0	3	1	4
	4	0	1	2	3		1	2	3	4	0		2	3	1	4	2	0
	2	3	4	0	1		3	4	0	1	2		1	4	2	0	3	1

														0	4	3	2	1
$L_4 =$	0	1	2	3	4		0	1	2	3	4		0	0	2	4	1	3
	4	0	1	2	3		1	2	3	4	0		4	2	4	1	3	0
	3	4	0	1	2		2	3	4	0	1		3	4	1	3	0	2
	2	3	4	0	1		3	4	0	1	2		2	1	3	0	2	4
	1	2	3	4	0		4	0	1	2	3		1	3	0	2	4	1

Fig. 5.3.1

Theorem 5.3.4. *From a given set of $t \leq n - 1$ mutually orthogonal quasigroups of order n, a set of t mutually orthogonal quasigroups of the same order n (but usually different from those of the original set) can be constructed of which $t - 1$ are idempotent quasigroups.*

Proof. Let the multiplication tables of the quasigroups be given by the mutually orthogonal latin squares $L_1, L_2, \ldots, L_t$ all of which are bordered in the same way. We first rearrange the rows of all the latin squares simultaneously in such a way that the entries of the leading diagonal of one square, say the square L_1, are all equal. Since the rearranged squares $L_2, L_3, \ldots, L_t$ are all orthogonal to L_1, the entries of the leading diagonal of each of them must form a transversal: that is, be all different. A relabelling of all the entries in any one of the squares does not affect its orthogonality to the remainder, so we may suppose that the squares $L_3, L_4, \ldots, L_t$ are relabelled in such a way that the entries of the leading diagonal of each become the same as those of L_2. If now each of the squares is bordered by its elements in the order in which these elements appear in the leading diagonal of L_2, the resulting Cayley tables define mutually orthogonal quasigroups $Q_1, Q_2, \ldots, Q_t$ of which all but the first are idempotent.

An example of the process is given in Fig. 5.3.1

5.4. Orthogonality in other structures related to latin squares

In this section we consider how the orthogonality concept may be generalized to apply to a number of structures related to latin squares. We consider in turn latin rectangles, permutation cubes, latin cubes and hypercubes, orthogonal arrays, and orthogonal Steiner triple systems.

Latin rectangles were defined in section 3.1. We say that two latin rectangles of the same size are *orthogonal* if, when one is superimposed on the other, each ordered pair of symbols r, s occurs in at most one cell of the superimposed pair. Also, a set of $m - 1$ latin rectangles of size $m \times n$, with $m \geq n$, is a *complete system* if each pair of the set is an orthogonal pair.

It is easy to see that the definition of orthogonality for latin squares is included as a special case of this more general definition and, by the method of theorem 5.1.5, that one cannot have more than $m - 1$ mutually orthogonal $m \times n$ latin rectangles if $m \geq n$.

In P. Quattrocchi [1], the following result has been proved:

Theorem 5.4.1. *For every prime number p and every integer q $(\geq p)$ such that each prime divisor of q is not less than p, there exists at least one complete system of mutually orthogonal $pq \times p$ latin rectangles.*

Proof. For each integer k, $k = 1, 2, \ldots, pq - 1$, we define a $pq \times p$ matrix $R_k = ||\alpha_{ij}||$, $i = 1, 2, \ldots, pq$; $j = 1, 2, \ldots, p$, by $\alpha_{ij} = $ $= \langle (i-1) + (j-1)k \rangle$, where $\langle h \rangle$ denotes the remainder on division of the integer h by pq.

It is immediate that each of the integers $0, 1, 2, \ldots, pq - 1$ occurs exactly once in each column of the matrix R_k and at most once in each of its rows. Consequently, R_k is a $pq \times p$ latin rectangle.

Let us consider two rectangles R_{k_1} and R_{k_2} of the system. Let us suppose that when R_{k_1} and R_{k_2} are placed in juxtaposition, the ordered pair (s, t) of integers appears both in the cell of the i_1th row and j_1th column and also in the cell of the i_2th row and j_2th column. Then we have

$$(i_1 - 1) + (j_1 - 1)k_1 \equiv s, \quad (i_1 - 1) + (j_1 - 1)k_2 \equiv t$$

and

$$(i_2 - 1) + (j_2 - 1)k_1 \equiv s, \quad (i_2 - 1) + (j_2 - 1)k_2 \equiv t,$$

all equalities being modulo pq. These relations imply that

$$(i_1 - i_2) + (j_1 - j_2)k_1 \equiv 0 \equiv (i_1 - i_2) + (j_1 - j_2)k_2,$$

and so

$$(j_1 - j_2)(k_1 - k_2) \equiv 0, \bmod pq.$$

Since q has no prime divisor less than p, and since j_1 and j_2 lie between 1 and p, $j_1 \neq j_2$ would imply $k_1 \equiv k_2 \bmod pq$ which is contrary to the supposition that R_{k_1} and R_{k_2} are distinct rectangles. Therefore, $j_1 = j_2$ and it follows that $i_1 = i_2$. Thus, each ordered pair of integers (s, t) occurs in at most one cell when a pair R_{k_1} and R_{k_2} of the rectangles are superimposed. This is sufficient to show that the $pq - 1$ rectangles are pairwise orthogonal.

P. Quattrocchi made use of this theorem in a construction of generalized affine spaces (equivalent to a certain type of balanced incomplete block design) from similar spaces of smaller order. The latin rectangles were used to define the incidence relation between point and line in the synthesized structure.

A latin square is a two-dimensional object, and the latin rectangle is a generalization of it in the sense that the "size" of one of these dimensions is allowed to be different from the other. A different generalization is obtained if, while retaining a fixed size, we allow the number of dimensions to be increased. If we increase the number of dimensions to three, we obtain what should properly be called a latin cube: an object having n rows, n columns and n files such that each of a set of n elements occurs exactly once in each row, once in each column, and once in each file.

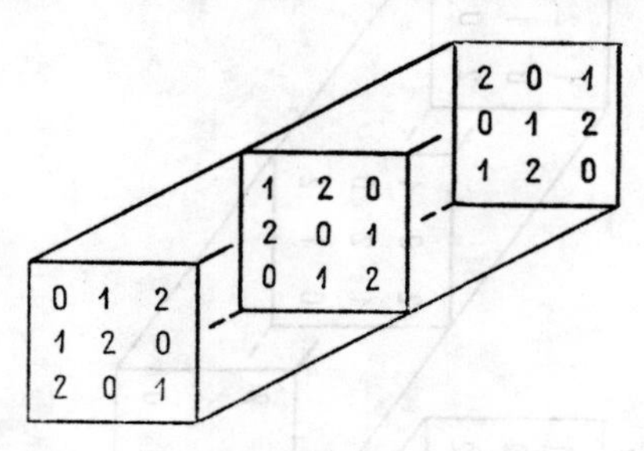

Fig. 5.4.1

An illustrative example for the case $n = 3$ is given in Fig. 5.4.1. If the number of dimensions increases still further to m say, we obtain an object which could reasonably be called an m-dimensional latin hypercube. Unfortunately, the terms latin cube and latin hypercube have been used by statisticians to denote another kind of combinatorial object which we shall describe later in this section. In view of this, we shall follow the lead given by A. Heppes and P. Révész in [1], and call the objects just introduced *permutation cubes.*

We make the following formal definition:

DEFINITION. An m-dimensional $n \times n \times \ldots \times n$ matrix, the elements of which are the integers $0, 1, 2, \ldots, n-1$, will be called an *m-dimensional permutation cube* of order n if every column (that is, every sequence of elements parallel to an edge of the cube) of the matrix contains a permutation of the integers $0, 1, 2, \ldots, n-1$. In particular, a two-dimensional permutation cube is simply a latin square of order n.

The appropriate generalization of a pair of orthogonal latin squares (or Graeco-Latin square) is an m-tuple of "orthogonal" m-dimensional permutation cubes which Heppes and Révész have called a *variational cube.* They have made the following definition(1):

(1) A quite different definition of orthogonality for permutation cubes has been introduced in P. Höhler [1] and yet another in P. D. Warrington [1].

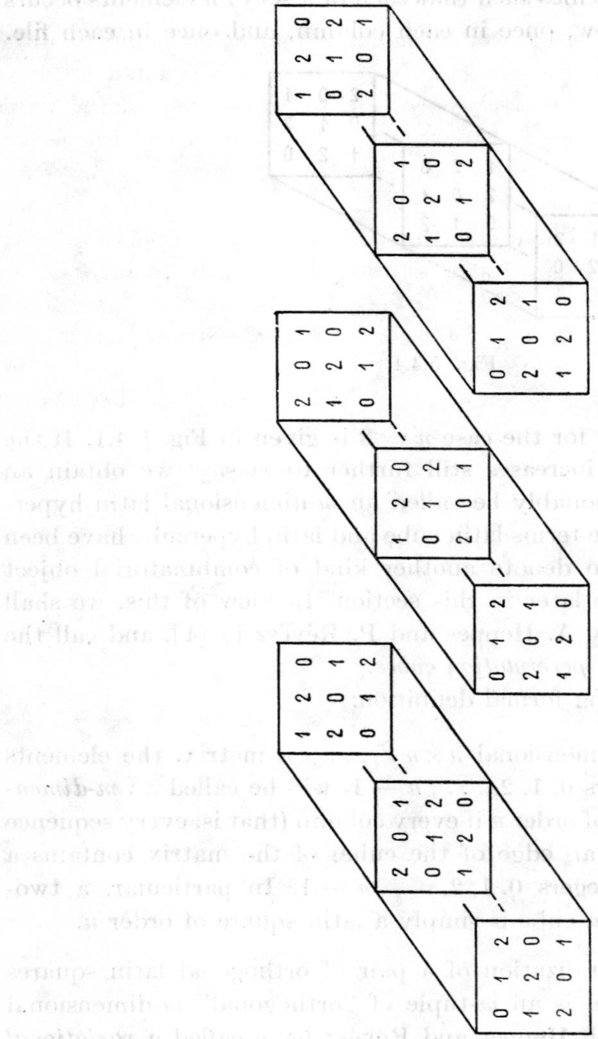

Fig. 5.4.2.

DEFINITION. The m-dimensional permutation cubes $c_1, c_2, \ldots, c_m$ constitute a variational m-tuple of cubes or, more briefly, a *variational cube* if, among the n^m m-tuples of elements chosen from corresponding cells of the m cubes, every distinct ordered m-tuple involving the integers $0, 1, 2, \ldots, n-1$ occurs exactly once.

As an example, the three 3-dimensional permutation cubes of order 3 shown in Fig. 5.4.2 form a variational cube. Each ordered triple of the integers 0, 1, 2 occurs exactly once. For example, the ordered triple (0, 0, 1) is given by the cell in the second place of the file of the second row and first column. The ordered triple (2, 1, 0) is given by the cell in the third place of the same file.

A set of mutally orthogonal latin squares is a set such that each pair of its squares form a Graeco-Latin square. In just the same way, we define a *variational set* (mutally orthogonal set) of $k(\geq m)$ m-dimensional permutation cubes of order n to be a set having the property that each subset of m cubes is a variational cube.

The four 3 dimensional permutation cubes of order 3 shown in Figs 5.4.1 and 5.4.2 form such a variational set.

In A. Heppes and P. Révész [1], the following two theorems giving a lower bound for k have been proved.

THEOREM 5.4.2. *If p is a prime and $m \leq p-1$ then a system of $p-1$ m-dimensional permutation cubes on p elements $0, 1, \ldots, p-1$ can be constructed each m of which form a variational m-tuple of cubes.*

PROOF: We prove our result by an actual construction. Let the kth m-dimensional cube of the system be denoted by C_k, $k = 1, 2, \ldots, p-1$. Let the element of C_k whose co-ordinates are $(x_1, x_2, \ldots, x_m)$; $0 \leq x_i \leq$ $\leq p-1$, be defined as the residue modulo p of the integer $\sum_{j=1}^{m} x_j k^{j-1}$. Then the difference between the elements in the hth and ith places of that column of C_k which is given by varying x_l and keeping all the other co-ordinates fixed is $(h-i)\, k^{l-1}$ if $x_l = h$ in the hth place and $x_l = i$ in the ith place. Since p is a prime and $h - i$ and k are both integers less than p, the number $(h-i)k^{l-1}$ is not congruent to zero modulo p. Consequently, the elements of this typical column are all different, so C_k is a permutation cube.

Next, let m cubes $C_{k_1}, C_{k_2}, \ldots, C_{k_m}$ be chosen arbitrarily from the system. We require to show that these cubes form a variational m-tuple of cubes. That is, we require to show that the p^m m-tuples formed by cells of $C_{k_1}, C_{k_2}, \ldots, C_{k_m}$ respectively having the same co-ordinates are all distinct. Suppose, on the contrary, that those corresponding to the sets of cells having co-ordinates $(x_1, x_2, \ldots, x_m)$ and $(x_1', x_2', \ldots, x_m')$ were the same. Then we would have

$$\sum_{j=1}^{m} x_j k_i^{j-1} = \sum_{j=1}^{m} x_j' k_i^{j-1} \text{ modulo } p, \text{ for } i = 1, 2, \ldots, m.$$

That is,

$$\sum_{j=1}^{m} (x_j - x_j')k_i^{j-1} = 0, \text{ modulo } p, \text{ for } i = 1, 2, \ldots, m.$$

We may regard these equations as being m homogeneous linear equations in m quantities $(x_j - x_j')$ not all of which are simultaneously zero, by hypothesis. The equations are consistent only if

$$\begin{vmatrix} 1 & k_1 & k_1^2 & \ldots & k_1^{m-1} \\ 1 & k_2 & k_2^2 & \ldots & k_2^{m-1} \\ \cdot & \cdot & \cdot & \cdots & \cdot \\ 1 & k_m & k_m^2 & \ldots & k_m^{m-1} \end{vmatrix} = 0.$$

The left-hand side of this equation is an alternant and can only be zero if two of the k_i are equal. Since the m-tuple of cubes under consideration are distinct by hypothesis, the equations above cannot be simultaneously satisfied and so the cubes form a variational set, as required.

THEOREM 5.4.3. *If $n = p_1^{\alpha_1} p_2^{\alpha_2} \ldots p_r^{\alpha_r}$, where the p_i are primes and the α_i are positive integers, then a system of $u = \min_i (p_i^{\alpha_i} - 1)$ m-dimensional permutation cubes on n elements can be constructed each m of which form a variational m-tuple of cubes provided that $m \leq u$.*

PROOF. As before, we prove the result by giving a construction for the set of cubes. Let $F_i = GF[p_i^{\alpha_i}]$, $i = 1, 2, \ldots, r$, denote the Galois field of $p_i^{\alpha_i}$ elements. Let $\gamma_h = (f_{h1}, f_{h2}, \ldots, f_{hr})$, where $f_{hi} \in F_i$, $i = 1, 2, \ldots, r$, and $h = 0, 1, \ldots, n-1$. Under the operations $\gamma_g + \gamma_h = (f_{g1} + f_{h1}, f_{g2} + f_{h2}, \ldots, f_{gr} + f_{hr})$ and $\gamma_g \gamma_h = (f_{g1} f_{h1}, f_{g2} f_{h2}, \ldots, f_{gr} f_{hr})$, the r-tuples γ_h form a ring R with a unit element $\gamma_1 = (f_{11}, f_{12}, \ldots, f_{1r})$, where f_{1i} is unit element of F_i. R contains divisors of zero, but every divisor of zero of R is a vector γ_h which has at least one zero component. The remaining elements of R have inverses in the ring. We take the n elements $\gamma_0, \gamma_1, \ldots, \gamma_{n-1}$ as the elements for our permutation cubes. We may also label the cells of each cube by means of co-ordinates $(x_1, x_2, \ldots, x_m)$ with $x_i \in R$, $i = 1, 2, \ldots, m$.

We may suppose without loss of generality that $u = \min (p_i^{\alpha_i} - 1) = p_1^{\alpha_1} - 1$. Assuming this, we select a set U containing u elements of R as follows:

$$\tilde{\gamma}_1 = (f_{11}, f_{12}, \ldots, f_{1r})$$
$$\tilde{\gamma}_2 = (f_{21}, f_{22}, \ldots, f_{2r})$$
$$\cdots\cdots\cdots\cdots\cdots$$
$$\tilde{\gamma}_u = (f_{u1}, f_{u2}, \ldots, f_{ur})$$

where $f_{11}, f_{21}, \ldots, f_{u1}$ are the complete set of non-zero elements of the Galois field F_1 and where $f_{1i}, f_{2i}, \ldots, f_{ui}$ are u distinct non-zero elements of the Galois field F_i, $i = 2, 3, \ldots, r$. Then none of the elements of U is a divisor of zero and neither is any difference $\tilde{\gamma}_l - \tilde{\gamma}_m$ if $\tilde{\gamma}_l$ and $\tilde{\gamma}_m$ are both elements of U.

We are now able to construct the u m-dimensional permutation cubes $C_1, C_2, \ldots, C_u$ whose existence is claimed in the theorem by defining the element of the cube C_k whose co-ordinates are $(x_1, x_2, \ldots, x_m)$, with $x_i \in R$, to be the element $\sum_{j=1}^{m} x_j \tilde{\gamma}_k^{j-1}$, for $k = 1, 2, \ldots, u$ $(\tilde{\gamma}_k \in U)$.

As in the previous theorem, we show easily from the fact that $(\tilde{\gamma}_h - \tilde{\gamma}_i)\tilde{\gamma}_k = 0$ implies $h = i$ that each of our cubes has the entries in each of its columns all different and so is a permutation cube. Also, as in the previous theorem, the set of m simultaneous equations

$$\sum_{j=1}^{m} (x_j - x_j')\tilde{\gamma}_k^{j-1} = 0, \; k = k_1, k_2, \ldots, k_m,$$

where $\tilde{\gamma}_{k_1}, \tilde{\gamma}_{k_2}, \ldots, \tilde{\gamma}_{k_m}$ are m distinct element of the set U, cannot be consistent if the set of m quantities $x_j - x_j'$ are not simultaneously all zero unless the determinant

$$\begin{vmatrix} 1 & \tilde{\gamma}_{k_1} & \tilde{\gamma}_{k_1}^2 & \ldots & \tilde{\gamma}_{k_1}^{m-1} \\ 1 & \tilde{\gamma}_{k_2} & \tilde{\gamma}_{k_2}^2 & \ldots & \tilde{\gamma}_{k_2}^{m-1} \\ . & . & . & \ldots & . \\ 1 & \tilde{\gamma}_{k_m} & \tilde{\gamma}_{k_m}^2 & & \tilde{\gamma}_{k_m}^{m-1} \end{vmatrix} \text{ is zero.}$$

This is not the case because none of the quantities $\tilde{\gamma}_{k_l} - \tilde{\gamma}_{k_m}$ is either equal to zero or is a divisor of zero in R. Consequently, every subset of m of the u permutation cubes forms a variational set.

The reader should compare theorem 5.4.3 with MacNeish's theorem given in section 11.1. As in the case of MacNeish's theorem, the lower bounds given by theorems 5.4.2 and 5.4.3 can be exceeded. Our four cubes given in Figs 5.4.1 and 5.4.2 show that the bound given in theorem 5.4.2 can be exceeded. As regards theorem 5.4.3, J. Arkin has recently constructed a variational set of three 3-dimensional permutation cubes on 10 elements. (That is, a variational cube of order 10). Here, $m = 2 \times 5$ and $u = \min(p_i^{\alpha_i} - 1) = 1$, so the lower bound of 1 is certainly exceeded. Arkin's method, which he has described in outline in [1], makes use of the existence of pairs of orthogonal latin squares of order 10.

An unsolved problem of considerable difficulty is that of finding, for given values of m and n, the maximum number $\pi_m(n)$ of permutation cubes in a variational set. The corresponding problem for latin squares ($m = 2$) has already been mentioned in section 5.2 and is discussed in greater detail in later chapters of this book.

However, as in the case of latin squares, it is not difficult to find an upper bound for $\pi_m(n)$. We have the following theorem, which was first proved in L. Humblot [1].

Theorem 5.4.4. *If $\pi_m(n)$ denotes the maximum number of m-dimensional permutation cubes of order n in a variational set, then $\pi_m(n) \leq (m-1).(n-1)$.*

Proof. The argument is similar to that of theorem 5.1.5 (which proves the result for the case $m = 2$) and could be used to provide an alternative proof of that theorem. We shall denote the cells of each m-dimensional permutation cube by means of m-tuples of co-ordinates $(x_1, x_2, \ldots, x_m)$, where each x_i takes the values $1, 2, \ldots, n$.

We may rename the symbols in all the cells of any one permutation cube simultaneously without affecting the orthogonality. By such renamings, we may arrange that the symbol occurring in the cell with co-ordinates $(i, 1, 1, \ldots, 1)$ is the symbol $i - 1$ and is the same in all the permutation cubes, for $i = 1, 2, \ldots, n$. Assuming this done, it now follows that the symbol 0 can occur in the cell $(x_1, 2, 1, 1, \ldots, 1)$ for the same value of x_1 ($x_1 \neq 1$) in at most $m - 1$ of the permutation cubes in the variational set: for if it occurred in the same place in m of the cubes, these m cubes would not form a variational cube. This is true for each of the $n - 1$ possible values of x_1 (namely, $x_1 = 2, 3, \ldots, n$) and so, since the symbol 0 must occur in some cell of the row $x_2 = 2$ in every permutation cube of the variational set, the total number of permutation cubes in the set cannot exceed $(m-1)(n-1)$.

Next, let us introduce the concepts of latin cube and latin hypercube as used by Statisticians.

DEFINITION. An $n \times n \times n$ three-dimensional matrix comprising n layers each having n rows and n columns is called a *latin cube* if it has n distinct elements each repeated n^2 times and so arranged that in each layer parallel to each of the three pairs of opposite faces of the cube all the n distinct elements appear and each is repeated exactly n times in that layer. (See R. A. Fisher [4], page 85 and K. Kishen [2], page 21.)

This concept differs from that of a permutation cube in that, in the statistician's latin cube, there may exist rows, columns or files in which some of the n elements do not occur and others are repeated. J. R. Kerr has recently distinguished the various possibilities by calling a latin cube regular (or 3-regular) if every row, column and file contains each element exactly once (implying that the concept of a regular latin cube is the same as that of a permutation cube), 2-regular if each element occurs exactly once in every row and column but not in the files (as in Fig. 5.4.3), 1-regular if each element occurs exactly once in every row but not in the columns or files and 0-regular if there are repetitions of elements in all three directions: that is, in rows, columns and files. A *latin hypercube* is the analogous concept in more than three dimensions.

Latin cubes and hypercubes were first introduced by K. Kishen in 1942 in [1] and, independently, by R. A. Fisher in 1945 in his paper [3]. Latin hypercubes of a slightly different kind were also introduced in the following year in C. R. Rao [1]. Both concepts find a use in connection with the design of statistical experiments. (For a recent paper describing such applications, see D. A. Preece, S. C. Pearce and J. R. Kerr [1].)

Both Kishen and Fisher also defined orthogonality of latin cubes, as follows:

DEFINITION. Two latin cubes are *orthogonal* if, among the n^3 pairs of elements chosen from corresponding cells of the two squares, each distinct ordered pair of the elements $0, 1, \ldots, n-1$ occurs exactly n times.

The maximum number of latin cubes in a set of pairwise orthogonal ones is $n^2 + n - 2$, as was first proved by R. L. Plackett and J. P. Burman in [1]. These authors actually obtained a more general result of which this is a special case, see below.

A method of construction of latin cubes and hypercubes and also of orthogonal sets of these of prime order p has been described in K. A. Brownlee and P. K. Loraine [1]. These authors made use of elementary abelian groups of type $(p, p, \ldots, p)$ in their constructions. Later, in 1950, a much more comprehensive paper on the same subject was written by K. Kishen (see K. Kishen [2]) and we shall give some details of its contents below. First, we illustrate by means of examples the concepts just introduced.

We exhibit a pair of orthogonal latin cubes of order 3 in Fig. 5.4.3. A set of 10 pairwise orthogonal ones of the same order 3, the maximum possible number of this order, is given in Fig. 5.4.4.

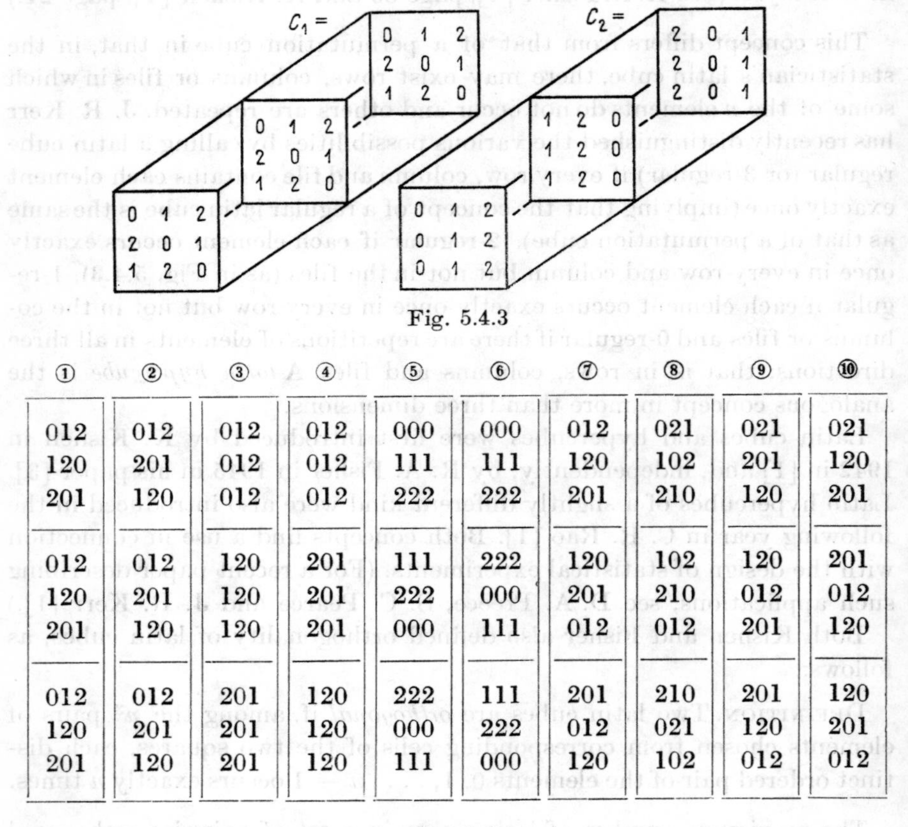

Fig. 5.4.3

①	②	③	④	⑤	⑥	⑦	⑧	⑨	⑩
012	012	012	012	000	000	012	021	021	021
120	201	012	012	111	111	120	102	201	120
201	120	012	012	222	222	201	210	120	201
012	012	120	201	111	222	120	102	120	201
120	201	120	201	222	000	201	210	012	012
201	120	120	201	000	111	012	012	201	120
012	012	201	120	222	111	201	210	201	120
120	201	201	120	000	222	012	021	120	201
201	120	201	120	111	000	120	102	012	012

Fig. 5.4.4

K. Kishen's paper [2] of 1950 on the subject of latin cubes and hypercubes was a very detailed one. In the first place, Kishen gave a general formula for the maximum number of m-dimensional latin hypercubes of order n in a set of pairwise orthogonal ones. This number is $\frac{n^m - 1}{n - 1} - m$. He also showed how a complete set of pairwise orthogonal latin cubes or hypercubes can be constructed from a finite projective space of the appropriate dimension. His method is a generalization of that given in theorem 5.2.2 for the construction of a complete set of orthogonal latin squares from a finite projective plane. In the second place, Kishen made further generalizations of the concepts both of latin hypercube and of orthogonality between pairs of these objects.

He defined an m-dimensional latin hypercube of order n and *of the* rth *class* to be an $n \times n \times \ldots \times n$ m-dimensional matrix having n^r distinct elements, each repeated n^{m-r} times, and such that each element occurs exactly n^{m-r-1} times in each of its m sets of n parallel $(m - 1)$-dimensional linear subspaces (or "layers"). Two such latin hypercubes of the same order n and class r with the property that, when one is superimposed on the other, every element of the one occurs exactly n^{m-2r} times with every element of the other, are said to be *orthogonal*. Thus the latin cubes which we have discussed above are all of the first class.

An example of a $3 \times 3 \times 3$ latin cube of the second class is exhibited in Fig. 5.4.5.

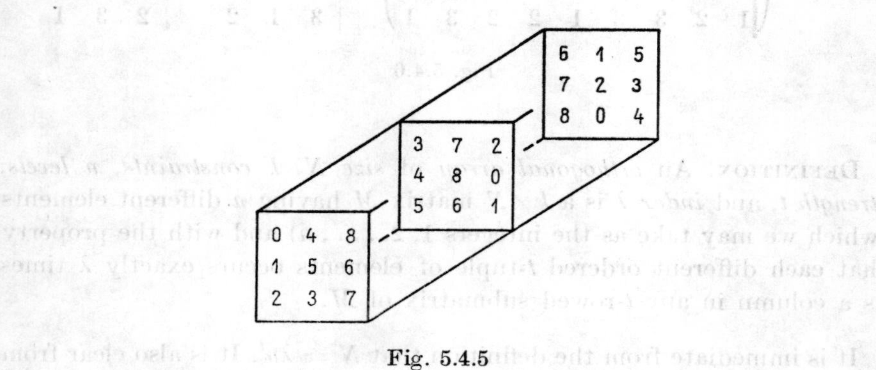

Fig. 5.4.5

Latin cubes of the second class are discussed in P. N. Saxena [3] and some methods for their construction additional to those of K. Kishen are given there.

The significance of latin cubes and hypercubes is more easily understood if we regard them as special cases of another design which arises by generalizing the concept of a set of orthogonal latin squares in a different way.

Let $L_1, L_2, \ldots, L_{k-2}$ be a set of $k-2$ mutually orthogonal latin squares of order n whose elements are the integers $1, 2, \ldots, n$, and let a $k \times n^2$ matrix M be formed in the following way. The first row of M has as its elements the integer 1 repeated n times, the integer 2 repeated n times, . . . , and finally the integer n repeated n times. The second row of M comprises the sequence of integers $1, 2, \ldots, n$ repeated n times. If $L_h = (a_{ij}^{(h)})$, the $(h+2)$th row of M has $a_{ij}^{(h)}$ as the entry in the $[(i-1)n + j]$th column: namely, the column whose first two entries are i and j. A simple example is given in Fig. 5.4.6. The matrix M so constructed has the property that each ordered pair of elements $\begin{pmatrix} i \\ j \end{pmatrix}$ appears just once as a column in any two-rowed submatrix by virtue of the facts that the squares $L_1, L_2, \ldots, L_{k-2}$ are all latin and are also orthogonal in pairs. We say that it is an *orthogonal array* of k constraints and n levels having strength 2 and index 1. More generally, we may make the following definition.

$$M = \begin{pmatrix} 1 & 1 & 1 & 2 & 2 & 2 & 3 & 3 & 3 \\ 1 & 2 & 3 & 1 & 2 & 3 & 1 & 2 & 3 \\ 1 & 2 & 3 & 2 & 3 & 1 & 3 & 1 & 2 \\ 1 & 2 & 3 & 3 & 1 & 2 & 2 & 3 & 1 \end{pmatrix}$$

1	2	3
2	3	1
3	1	2

1	2	3
3	1	2
2	3	1

Fig. 5.4.6

DEFINITION. An *orthogonal array* of *size* N, k *constraints*, n *levels*, *strength* t, and *index* λ is a $k \times N$ matrix M having n different elements (which we may take as the integers $1, 2, \ldots, n$) and with the property that each different ordered t-tuple of elements occurs exactly λ times as a column in any t-rowed submatrix of M.

It is immediate from the definition that $N = \lambda n^t$. It is also clear from the construction given above that any orthogonal array of k constraints, n levels, strength 2, and index 1 is equivalent to a set of $k-2$ mutually orthogonal latin squares of order n. We shall defer a formal proof of this fact until section 11.1. From theorem 5.1.5, we may deduce that, if M is an orthogonal array with $t = 2$ and $\lambda = 1$, then $k \leq n + 1$. (See also C. R. Rao [3].) The investigation of upper bounds for the number of constraints in other cases has been the subject of a number of papers.

We refer the reader, in particular, to R. L. Plackett and J. P. Burman [1], C. R. Rao [1] and [2], K. A. Bush [1] and [2], R. C. Bose and K. A. Bush [1], E. Seiden [2] and [3], and J. R. Blum, J. A. Schatz, and E. Seiden [1]. One result of this kind is mentioned below.

The connection between these orthogonal arrays and latin cubes and hypercubes is given by the statements: (*i*) a latin cube of order n is equivalent to an orthogonal array of 4 constraints and n levels having strength 2 and index n; (*ii*) a set of $k - 3$ mutually orthogonal latin cubes of order n is equivalent to an orthogonal array of k constraints and n levels having strength 2 and index n; (*iii*) a set of $k - m$ mutually orthogonal m-dimensional latin hypercubes of order n is equivalent to an orthogonal array of k constraints and n levels having strength 2 and index n^{m-2}.

As an illustration of statement (*ii*), we give in Fig. 5.4.7 the orthogonal array corresponding to the two orthogonal latin cubes shown in Fig. 5.4.3. Here the first, second, and third rows of the matrix M give the row, column, and file number respectively of the cell of the latin cube whose entry appears below them in the fourth or fifth row of M according to which of the two latin cubes is being considered.

$$M = \begin{pmatrix} 0&0&0&0&0&0&0&0&0&1&1&1&1&1&1&1&1&1&2&2&2&2&2&2&2&2&2 \\ 0&0&0&1&1&1&2&2&2&0&0&0&1&1&1&2&2&2&0&0&0&1&1&1&2&2&2 \\ 0&1&2&0&1&2&0&1&2&0&1&2&0&1&2&0&1&2&0&1&2&0&1&2&0&1&2 \\ 0&0&0&1&1&1&2&2&2&2&2&2&0&0&0&1&1&1&1&1&1&2&2&2&0&0&0 \\ 0&1&2&1&2&0&2&0&1&0&1&2&1&2&0&2&0&1&0&1&2&1&2&0&2&0&1 \end{pmatrix} \begin{matrix} \\ \\ \\ \leftarrow c_1 \\ \leftarrow c_2 \end{matrix}$$

Fig. 5.4.7

In R. L. Plackett and J. P. Burman [1], the authors proved that the maximum number of constraints k for an orthogonal array of n levels having strength 2 and index λ satisfies the inequality $k \leq \frac{\lambda n^2 - 1}{n - 1}$. An alternative proof of the same result was later given in R. C. Bose and K. A. Bush [1]. If we put $\lambda = n$, we get $k \leq n^2 + n + 1$ for this case, and it follows that the maximum number of latin cubes in a pairwise orthogonal set is $n^2 + n - 2$, as stated earlier.

We remark here that the general question "What is the maximum number of constraints possible in an orthogonal array of n levels, strength t and index λ?" is unsolved, though some information regarding special cases is known. For example, J. R. Blum, J. A. Schatz and E. Seiden have recently proved in [1] that for a 2-level array of strength t and odd index, $k \leq t + 1$, and that the value $k = t + 1$ is always attainable.

For further general information on the subject of orthogonal arrays, the reader is referred to the following papers and to the bibliographies contained therein: R. C. Bose and K. A. Bush [1], E. Seiden and R. Zemach [1].

Finally, we mention the concept of orthogonal Steiner triple systems. Steiner triple systems were defined in section 2.2. We say that two such systems of the same order n and defined on the same set of elements are *orthogonal* if (*i*) the systems have no triples in common, and if (*ii*) when two pairs of elements appear with the same third element in one system, then they appear with distinct third elements in the other system. As an illustration, we give two orthogonal Steiner triple systems, S_1 and S_2, of order 7 in Fig. 5.4.8 below.

S_1 has triples (1 2 6), (2 3 7), (3 4 1), (4 5 2), (5 6 3), (6 7 4), (7 1 5)

S_2 has triples (1 2 4), (2 3 5), (3 4 6), (4 5 7), (5 6 1), (6 7 2), (7 1 3)

Fig. 5.4.8

It has been conjectured by C. D. O'Shaughnessy in [1] and by R. C. Mullin and E. Németh in [2] that such orthogonal pairs of Steiner triple systems exist of all orders n congruent to 1 modulo 6. Mullin and Németh proved in [2] that if n is a prime power congruent to 1 modulo 6 then orthogonal pairs do exist and, in [4], they showed that orthogonal pairs of order 9 do not exist. The latter result suggests the further conjecture that orthogonal pairs do not exist for all orders n congruent to 3 modulo 6. However the falsity of this conjecture has been shown by A. Rosa, who has constructed a pair of orthogonal Steiner triple systems of order 27. Details have not yet been published, but see A. Rosa [1]. Recently, N. S. Mendelsohn (see [4]) has given another construction which yields orthogonal pairs of Steiner triple systems of all prime orders congruent to 1 modulo 6 and has shown that this construction can be extended to yield such pairs for all orders congruent to 1 modulo 6 except when n has odd powers of primes p which are congruent to 2 modulo 3 as factors.

A joint paper by C. C. Lindner and N. S. Mendelsohn (see [1]) written subsequently to the above paper of Mendelsohn uses a generalization of the singular direct product construction (see section 12.1) and also a new concept of *conjugate orthogonal latin squares* to obtain a sharpened version of this result.

Two quasigroups Q_1 and Q_2 and their associated latin squares L_1 and L_2 are said to be *conjugate orthogonal* if each of the five quasigroups conjugate to Q_1, as well as Q_1 itself, is orthogonal to Q_2 and also to each of the five conjugates of Q_2. (For the definition of conjugacy between quasigroups, see section 2.1.)

Orthogonal Steiner triple systems may be used to construct Room designs, as we shall explain in section 6.4. They may also be used to define perpendicular commutative quasigroups, see section 12.5.

CHAPTER 6

Connections between latin squares and magic squares

The close link that exists between magic squares and latin squares has been recognized by mathematicians ever since the by-now famous investigation into the subject of constructing magic squares which was made by L. Euler (see [1] and [2]) in the 1780's. Consequently, no further justification for devoting the first three sections of the present chapter to this subject seems necessary.

In the fourth section of the chapter, we give an account (mainly historical) of researches into the construction of Room squares since these designs too may be regarded as magic squares of a sort and recent investigations have shown that they also are inter-connected with latin squares.

6.1. Diagonal latin squares

A *magic square* of order n is an arrangement of n^2 integers (usually, but not necessarily, consecutive and usually, but not necessarily, all different) in a square such that the sums of each row, each column and each of the main diagonals are the same.

Magic squares were known in ancient times in China and India and were often engraved on metal or stone. Even to this day they are worn as amulets. The first known European writer on the subject was E. Moschopoulos, a Greek of Constantinople (c. A.D. 1300). Later, C. Agrippa (1486–1535) constructed magic squares of orders 3 to 9. A magic square of order 4 is shown in A. Dürer's picture "Melancholy".

Further general information and historical details on the subject of magic squares may be found in E. J. Armstrong [1], W. W. Rouse Ball [1], L. Euler [1], M. Frolov [3], M. Kraitchik [1], E. Maillet [7], M. M. Postnikov [1], H. Schubert [1] and M. J. Van Driel [1], [2].

In the above books and expository papers and also in J. Dénes [1] a number of results on the connection between magic squares and latin squares are given.

In this first section we shall discuss the more unusual type of magic square in which only n of the n^2 integers are distinct and each is repeated n times. Among the magic squares of this type are the so-called "diagonal latin squares".

Definition. A latin square $||a_{ij}||$ is called *left semi-diagonal*[1] if the elements $a_{11}, a_{22}, \ldots, a_{nn}$ are distinct. Similarly, if the elements a_{1n}, $a_{2\,n-1}, \ldots, a_{n1}$ are distinct, the latin square $||a_{ij}||$ is called *right semi-diagonal*. (The diagonal formed by the elements $a_{1n}, a_{2\,n-1}, \ldots, a_{n1}$ will be called the *main right-to-left diagonal.*)

A latin square is called *diagonal* if it is both left and right semi-diagonal simultaneously and, of course, every such latin square is also a magic square.

Theorem 6.1.1. *If n is any positive integer $(n \neq 2)$, there exists at least one left semi-diagonal latin square of order n.*

Proof. It is well known that, if n is any given positive integer $(n \neq 2)$ there exists an idempotent quasigroup of order n (by theorem 5.3.4 with $t = 2$) and the multiplication table of an idempotent quasigroup is a left semi-diagonal latin square.

Corollary. *If n is any positive integer $(n \neq 2)$ there exists at least one right semi-diagonal latin square of order n.*

Proof. By reversing the order of the columns of a left semi-diagonal latin square one can obtain a right semi-diagonal latin square.

Theorem 6.1.2. *If n is any odd integer which is not a multiple of 3 there exists at least one diagonal latin square of order n.*

Proof. The proof consists in first showing that if the positive integers $\alpha, \beta, \alpha + \beta, \alpha - \beta$ $(\alpha > \beta)$ are all relatively prime to n then the latin square exhibited in Fig. 6.1.1 (where the elements are taken modulo n) is a diagonal latin square. To see this, we observe that one of the main diagonals contains the elements

$$0,\ \beta + \alpha,\ 2(\beta + \alpha),\ \ldots,\ (n - 1)(\beta + \alpha)$$

and the other the elements

$$(n - 1)\alpha,\ \beta + (n - 2)\alpha,\ 2\beta + (n - 3)\alpha,\ \ldots,\ (n - 1)\beta.$$

[1] These definitions are given in A. Margossian [1] but probably date from much earlier. See, for example, G. Tarry [6] where diagonal latin squares are mentioned.

The latter elements are respectively equal to

$$n\alpha - \alpha,\ n\alpha - \alpha - (\alpha - \beta),\ n\alpha - \alpha - 2(\alpha - \beta), \ldots,$$

$$n\alpha - \alpha - (n - 1)(\alpha - \beta).$$

Since $\alpha + \beta$ and $\alpha - \beta$ are both prime to n, the elements $i(\alpha + \beta)$ $i = 0, 1, \ldots, n - 1$, are all distinct and so also are the elements $i(\alpha - \beta)$. Consequently, the displayed latin square is diagonal.

Now let us choose $\alpha = 2$ and $\beta = 1$. Then $\alpha - \beta = 1$ and $\alpha + \beta = 3$, so it is immediately obvious that if n is odd and not a multiple of three $(n > 1)$, the integers α, β, $\alpha - \beta$ and $\alpha + \beta$ are all prime to n. The existence of a diagonal latin square of order n follows.

0	α	2α	$\ldots$	$(n-1)\alpha$
β	$\beta + \alpha$	$\beta + 2\alpha$	$\ldots$	$\beta + (n-1)\alpha$
2β	$2\beta + \alpha$	$2\beta + 2\alpha$	$\ldots$	$2\beta + (n-1)\alpha$
3β	$\cdot$	$\cdot$	$\ldots$	$\cdot$
$\cdot$	$\cdot$	$\cdot$	$\ldots$	$\cdot$
$\cdot$	$\cdot$	$\cdot$	$\ldots$	$\cdot$
$(n-1)\beta$	$(n-1)\beta+\alpha$	$(n-1)\beta+2\alpha$	$\ldots$	$(n-1)\beta+(n-1)\alpha$

Fig. 6.1.1.

Remark. The above construction fails when n is even.

Theorem 6.1.3. *If $n = 2^k$, k an arbitrary positive integer, there exists at least one diagonal latin square of order n if and only if $k \neq 1$.*

Proof. Since $k = 0$ is a trivial case and it is easy to determine by trial that it is not possible to construct a diagonal latin square of order 2, we shall suppose that $k \geq 2$ holds. The cases $k = 2, 3$ are disposed of by the following examples, which show us (see Figs 6.1.2 and 6.1.3) that diagonal latin squares of orders 4 and 8 exist.

1	2	3	4
3	4	1	2
4	3	2	1
2	1	4	3

Fig. 6.1.2

For $k > 3$, we proceed by an induction argument. Let $k = r$, so that $n = 2^r$, and take as induction hypothesis that there exists at least one diagonal latin square of every order $n = 2^k$ for which $k < r$ $(k \neq 1)$.

Since $2^r = 4 . 2^{r-2}$, it follows from the induction hypothesis that 2^r can be written as the product $n_1 n_2$ of two integers n_1, n_2 such that diagonal latin squares exist both of order n_1 and of order n_2. Let us suppose now that A is a diagonal latin square of order n_1 and that $B_0, B_1, B_2, \ldots, B_{n_1-1}$

1	2	3	4	5	6	7	8
2	3	5	6	7	8	4	1
8	6	4	1	2	5	3	7
6	7	8	5	3	2	1	4
4	8	7	2	6	1	5	3
3	4	6	8	1	7	2	5
5	1	2	7	4	3	8	6
7	5	1	3	8	4	6	2

Fig. 6.1.3

are isomorphic diagonal latin squares of order n_2 such that B_i is defined on the set $\{in_2, in_2+1, \ldots, in_2+n_2-1\}$; for $i = 0, 1, \ldots, n_1-1$. If the elements $0, 1, \ldots, n_1-1$ of A are replaced by $B_0, B_1, B_2, \ldots, B_{n_1-1}$ respectively, it is easy to see that we shall obtain a diagonal latin square of order $n_1.n_2 = 2^r$. The proof can now be completed by induction on the integer k.

THEOREM 6.1.4. *If n is any integral multiple of 4, there exists at least one diagonal latin square of order n.*

PROOF. We show first that if n is even and $n = n_1n_2$ where n_1, n_2 are integers such that there exists at least one left semi-diagonal latin square of order n_1, and also at least one diagonal latin square of order n_2, and if, further, n_2 is even, then a diagonal latin square of order n can be constructed. The construction is similar to that used in the proof of theorem 6.1.3.

Let A be a diagonal latin square of order n_2. Let the elements $0, 1, \ldots, n_2 - 1$ of A, excluding those which appear in the main right-to-left diagonal, be replaced successively by $B_0, B_1, B_2, \ldots, B_{n_2-1}$ where B_i $(i = 0, 1, 2, \ldots, n_2 - 1)$ is a left semi-diagonal latin square of order n_1 and is defined on the set $\{in_1, in_1 + 1, \ldots, in_1 + n_1 - 1\}$. Further, let those elements of A which are contained in the main right-to-left diagonal be replaced successively by $B_0^*, B_1^*, B_2^*, \ldots, B_{n_2-1}^*$ where, for each i, B_i^* is defined on the same set as B_i and has the same order, but is right instead of left semi-diagonal.

Then, it is easy to see that, provided n_2 is even, the resulting square of order n_1n_2 is a diagonal latin square.

We now take the special case when $n_2 = 4$, corresponding to the case when n is an integral multiple of 4, $n = 4n_1$. By theorem 6.1.3, there exists a diagonal latin square of order 4 and, by theorem 6.1.1, there exist both left and right semi-diagonal latin squares of order n_1 so the conditions required for our construction then hold. This proves the theorem.

It is easy to see that the constructions of theorems 6.1.2, 6.1.3, and 6.1.4 do not provide the only ways of producing diagonal latin squares since none of the above methods is applicable if $n = 6$, 9 or 10 and yet diagonal squares of these orders certainly exist, as is shown by Figs 6.1.4, 6.1.7 and 6.1.8 respectively. However, it is easy to prove that no diagonal latin square of order 3 can exist, as in our next theorem.

1	2	3	4	5	6
6	5	4	3	2	1
2	4	6	1	3	5
3	1	5	2	6	4
5	3	1	6	4	2
4	6	2	5	1	3

Fig. 6.1.4

Theorem 6.1.5. *Diagonal latin squares of order 3 do not exist.*

Proof. Let us suppose that the square matrix $\|a_{ij}\|$ $(i, j = 1, 2, 3)$ is a diagonal latin square. By the diagonal property, the elements a_{11}, a_{22}, a_{33} are all different and the same is true for the elements a_{13}, a_{22}, a_{31}. This necessarily implies that either $a_{11} = a_{13}$ or $a_{11} = a_{31}$. However, if either of these equalities holds the fact that $\|a_{ij}\|$ is a latin square is contradicted.

Some examples of diagonal latin squares of orders 5, 7, 9, 10, 11, 12 and 13 are given in the following diagrams (Figs 6.1.5 to 6.1.11). Examples of orders 4, 6 and 8 have already been given.

1	2	3	4	5
4	5	1	2	3
2	3	4	5	1
5	1	2	3	4
3	4	5	1	2

Fig. 6.1.5

1	2	3	4	5	6	7
7	3	4	5	6	1	2
6	1	2	7	3	4	5
5	7	1	6	4	2	3
4	6	5	2	7	3	1
3	4	7	1	2	5	6
2	5	6	3	1	7	4

Fig. 6.1.6

1	2	3	4	5	6	7	8	9
6	3	2	8	7	5	9	1	4
7	9	8	5	6	4	3	2	1
5	7	6	9	1	8	4	3	2
9	8	5	7	4	1	2	6	3
8	1	9	6	3	2	5	4	7
3	4	7	1	2	9	6	5	8
4	5	1	2	9	3	8	7	6
2	6	4	3	8	7	1	9	5

Fig. 6.1.7

1	2	3	4	5	6	7	8	9	10
4	10	8	7	6	9	5	3	1	2
10	8	2	6	7	5	1	4	3	9
9	4	10	8	1	3	2	7	6	5
2	9	6	1	3	8	4	5	10	7
3	5	1	10	9	7	6	2	4	8
5	6	7	3	2	4	9	10	8	1
7	1	5	9	4	10	8	6	2	3
8	7	4	2	10	1	3	9	5	6
6	3	9	5	8	2	10	1	7	4

Fig. 6.1.8

1	2	3	4	5	6	7	8	9	10	1
6	3	5	1	2	8	11	4	10	9	
7	9	4	10	11	5	3	2	8	6	
2	4	6	5	3	9	1	10	7	11	
11	8	9	7	10	4	6	5	3	1	
5	6	8	9	1	7	10	11	2	3	
9	11	2	6	4	10	8	7	1	5	
10	1	11	2	7	3	4	9	6	8	
4	10	1	3	8	2	5	6	11	7	
8	5	7	11	6	1	9	3	4	2	1
3	7	10	8	9	11	2	1	5	4	

Fig. 6.1.9

1	2	3	4	5	6	7	8	9	10	11	12
2	3	1	6	4	5	9	7	8	12	10	11
3	1	2	5	6	4	8	9	7	11	12	10
7	8	9	10	11	12	1	2	3	4	5	6
8	9	7	11	12	10	3	1	2	6	4	5
9	7	8	1	10	11	2	3	1	5	6	4
10	11	12	7	8	9	4	5	6	1	2	3
12	10	11	9	7	8	5	6	4	2	3	1
11	12	10	8	9	7	6	4	5	3	1	2
4	5	6	1	2	3	10	11	12	7	8	9
6	4	5	3	1	2	12	10	11	8	9	7
5	6	4	2	3	1	11	12	10	9	7	8

Fig. 6.1.10

In a lecture given at the University of Surrey, one of the authors of the present book conjectured that diagonal latin squares exist for all positive integers n greater than 3. Since the book was prepared for printing, the truth of this conjecture has been shown by three different authors. First to give a proof was A. J. W. Hilton (see [2]) who made use of so called *cross latin squares*: that is, latin squares such that all the elements of the main left-to-right diagonal are equal and all the

1	2	3	4	5	6	7	8	9	10	11	12	13
2	3	7	8	4	9	13	10	11	12	6	1	5
5	10	4	13	12	1	9	11	8	7	2	6	3
13	12	9	5	11	8	10	7	6	3	4	2	1
11	9	8	10	6	12	5	13	4	2	1	3	7
3	4	1	2	13	7	11	5	12	9	10	8	6
10	11	12	9	2	13	8	6	7	1	3	5	4
4	5	2	3	7	11	6	9	1	8	13	10	12
7	8	13	1	9	2	12	3	10	6	5	4	11
6	1	5	7	3	4	2	12	13	11	8	9	10
8	7	10	6	1	5	3	4	2	13	12	11	9
9	6	11	12	10	3	1	2	5	4	7	13	8
12	13	6	11	8	10	4	1	3	5	9	7	2

Fig. 6.1.11

elements of the main right-to-left diagonal are equal. (The definition has to be modified slightly for squares of odd order.) Shortly afterwards, C. C. Lindner found a somewhat simpler proof (see C. C. Lindner [15]) which made use of prolongations. More recently still, a proof more elementary than either of those just mentioned has been obtained by E. Gergely (see [1]) and we shall now explain it.

(We should mention at this point that both Hilton and Lindner have used the term diagonal latin square to mean a left semi-diagonal latin square and the term *doubly diagonal latin square* to mean a diagonal latin square.)

Gergely's proof depends essentially on the following result:

THEOREM 6.1.6. *For every $n > 3$, there exist left semi-diagonal latin squares of order n which possess at least one transversal disjoint from the main left-to-right diagonal.*

PROOF. For odd n, the standard multiplication table of the cyclic group of order n, as exhibited in Fig. 6.1.12 provides an example of a latin square with the required property. Since the main diagonal is a transversal, there exist $n - 1$ further disjoint transversals by theorem 1.4.2.

1	2		n
2	3		1
.			
.			
$\frac{1}{2}(n-1)$	...	$n-2$	
$\frac{1}{2}(n+1)$	...	n	
$\frac{1}{2}(n+3)$	...	2	
.			
.			.
n			$n-1$

Fig. 6.1.12

For even n, $n \geq 8$, we first construct the standard multiplication table of the cyclic group C_{n-3} of order $n - 3$ (in the form shown in Fig. 6.1.12) and thence, by three prolongations using transversals disjoint from the main left-to-right diagonal, construct a latin square of order n having a latin subsquare in its bottom right-hand corner isomorphic to the standard multiplication table of the cyclic group C_3 (as in theorem 1.5.1, taking $k = 3$). Since $n - 3 \geq 5$, the $(n - 3)\times(n - 3)$ subsquare in the top left-hand corner of the enlarged square possesses at least one further transversal disjoint from its main left-to-right diagonal. Since the 3×3 subsquare isomorphic to C_3 has the same property, the result of the theorem follows for the enlarged square. (The construction is illustrated in Fig. 6.1.13 for the special value $n = 8$.)

1	2	3	4	5
2	3	4	5	1
3	4	5	1	2
4	5	1	2	3
5	1	2	3	4

1	6	7	8	5	2	3	4
2	3	6	7	8	4	5	1
8	4	5	6	7	1	2	3
7	8	1	2	6	3	4	5
6	7	8	3	4	5	1	2
5	2	4	1	3	6	7	8
4	1	3	5	2	7	8	6
3	5	2	4	1	8	6	7

Fig. 6.1.13

There remain the cases $n = 4$ and $n = 6$. They are dealt with by exhibiting latin squares of these orders which have the required properties, as in Fig. 6.1.14.

<table>
<tr><td>1</td><td>2</td><td>3</td><td>4</td></tr>
<tr><td>3</td><td>4</td><td>1</td><td>2</td></tr>
<tr><td>4</td><td>3</td><td>2</td><td>1</td></tr>
<tr><td>2</td><td>1</td><td>4</td><td>3</td></tr>
</table>

<table>
<tr><td>1</td><td>2</td><td>3</td><td>4</td><td>5</td><td>6</td></tr>
<tr><td>5</td><td>3</td><td>6</td><td>1</td><td>4</td><td>2</td></tr>
<tr><td>4</td><td>1</td><td>5</td><td>2</td><td>6</td><td>3</td></tr>
<tr><td>2</td><td>4</td><td>1</td><td>6</td><td>3</td><td>5</td></tr>
<tr><td>6</td><td>5</td><td>4</td><td>3</td><td>2</td><td>1</td></tr>
<tr><td>3</td><td>6</td><td>2</td><td>5</td><td>1</td><td>4</td></tr>
</table>

Fig. 6.1.14

Remark. A shorter proof of theorem 6.1.6 may be obtained by making use of the fact that, for all $n \neq 2$ or 6, there exist latin squares having orthogonal mates. (For a complete proof see chapter 11.) Such a latin square has n disjoint transversals and, by rearranging the rows, we may easily arrange that one transversal lies along the main left-to-right diagonal. We are now able to prove our main result.

Theorem 6.1.7. *For $n \geq 4$, there exists at least one diagonal latin square of order n.*

Proof. We suppose first that n is even. The case $n = 4$ is settled by the example given in Fig. 6.1.14.

For $n \geq 6$, we first construct a left semi-diagonal latin square D_1 of order $k = n/2$ on the symbols $0, 1, 2, \ldots, k - 1$ having the property described in theorem 6.1.6.

We construct a second latin square D_2 of order $k = n/2$ by reflecting D_1 in its main left-to-right diagonal and then adding k to each of its elements, the addition being modulo n.

With the aid of the squares D_1 and D_2, we construct a latin square L of order n of the form

$$L = \begin{vmatrix} D_1 & D_2 \\ D_4 & D_3 \end{vmatrix}$$

where the subsquares D_3 and D_4 are yet to be defined.

Let τ_1 denote the off-diagonal transversal of D_1 whose existence is guaranteed by theorem 6.1.6, and let τ_2 denote the corresponding transversal of D_2. We choose the elements of the main left-to-right diagonal of D_3 to be the elements of the transversal τ_2 which occur in the corresponding columns of D_2. Similarly, we choose the elements of the main right-to-left diagonal of D_4 to be the elements of the transversal τ_1 which occur in the corresponding columns of D_1. Now, let $d_{11}^{(1)} d_{22}^{(1)} \ldots d_{kk}^{(1)}$ denote the elements of the main left-to-right diagonal of D_1, and $d_{11}^{(3)}, d_{22}^{(3)}, \ldots, d_{kk}^{(3)}$ denote those of D_3.

We define a permutation π_L of the symbols of D_1 by

$$\pi_L = \begin{pmatrix} d_{11}^{(1)} & \ldots & d_{jj}^{(1)} & \ldots & d_{kk}^{(1)} \\ d_{11}^{(3)}-k & \ldots & d_{jj}^{(3)}-k & \ldots & d_{kk}^{(3)}-k \end{pmatrix}.$$

Similarly, we define a permutation π_R of the symbols of D_2 by

$$\pi_R = \begin{pmatrix} d_{1k}^{(2)} & \ldots & d_{j,k-j}^{(2)} & \ldots & d_{k1}^{(2)} \\ d_{k1}^{(4)}+k & \ldots & d_{j,k-j}^{(4)}+k & \ldots & d_{k1}^{(4)}+k \end{pmatrix},$$

where in this case the permutation is expressed in terms of the elements of the main right-to-left diagonals of D_2 and D_4.

Each element of D_3 not on the main left-to-right diagonal is defined to be the transform of the corresponding element of D_1 by the permutation π_L. Each element of D_4 not on the main right-to-left diagonal is defined to be the transform of the corresponding element of D_2 by the permutation π_R. Finally, to make the square L into a latin square, we reduce each of the elements of the transversal τ_1 of D_1 by k and at the same time increase each of the elements of the transversal τ_2 of D_2 by k. Then L is of order n and is the diagonal latin square whose existence was to be shown. We note further that the elements in the cells of the transversal τ_1 in D_1 together with the elements in the corresponding positions in D_3 form an off-diagonal transversal of L.

Using this fact, it is easy to construct from L a diagonal latin square L^* of order $n + 1$, where $n + 1$ is odd. (We illustrate the procedure below.) This completes the proof of the theorem since the case $n = 4$ $(n + 1 = 5)$ is covered by the example given in Fig. 6.1.14, which has an off-diagonal transversal.

To illustrate the construction, we shall give it in detail for the special case $n = 10$, $n + 1 = 11$.

$$D_1 = \left|\begin{array}{ccccc} 0 & \underline{1} & 2 & 3 & 4 \\ 1 & 2 & \underline{3} & 4 & 0 \\ 2 & 3 & 4 & \underline{0} & 1 \\ 3 & 4 & 0 & 1 & \underline{2} \\ \underline{4} & 0 & 1 & 2 & 3 \end{array}\right. \qquad D_2 = \left|\begin{array}{ccccc} 9 & 8 & 7 & \underline{6} & 5 \\ 5 & 9 & \underline{8} & 7 & 6 \\ 6 & \underline{5} & 9 & 8 & 7 \\ \underline{7} & 6 & 5 & 9 & 8 \\ 8 & 7 & 6 & 5 & \underline{9} \end{array}\right.$$

$$\pi_L = \begin{pmatrix} 0 & 2 & 4 & 1 & 3 \\ 7-5 & 5-5 & 8-5 & 6-5 & 9-5 \end{pmatrix} = \begin{pmatrix} 0 & 1 & 2 & 3 & 4 \\ 2 & 1 & 0 & 4 & 3 \end{pmatrix}$$

$$\pi_R = \begin{pmatrix} 5 & 7 & 9 & 6 & 8 \\ 2+5 & 0+5 & 3+5 & 1+5 & 4+5 \end{pmatrix} = \begin{pmatrix} 5 & 6 & 7 & 8 & 9 \\ 7 & 6 & 5 & 9 & 8 \end{pmatrix}$$

$$L = \left|\begin{array}{ccccc|ccccc} 0 & \underline{6} & 2 & 3 & 4 & 9 & 8 & 7 & 1 & 5 \\ 1 & 2 & \underline{8} & 4 & 0 & 5 & 9 & 3 & 7 & 6 \\ 2 & 3 & 4 & \underline{5} & 1 & 6 & 0 & 9 & 8 & 7 \\ 3 & 4 & 0 & 1 & \underline{7} & 2 & 6 & 5 & 9 & 8 \\ \underline{9} & 0 & 1 & 2 & 3 & 8 & 7 & 6 & 5 & 4 \\ \hline 8 & 9 & 5 & 6 & 2 & 7 & \underline{1} & 0 & 4 & 3 \\ 7 & 8 & 9 & 0 & 6 & 1 & 5 & \underline{4} & 3 & 2 \\ 6 & 7 & 3 & 9 & 5 & 0 & 4 & 8 & \underline{2} & 1 \\ 5 & 1 & 7 & 8 & 9 & 4 & 3 & 2 & 6 & \underline{0} \\ 4 & 5 & 6 & 7 & 8 & \underline{3} & 2 & 1 & 0 & 9 \end{array}\right.$$

$$L^* = \left|\begin{array}{ccccc|c|ccccc} 0 & \alpha & 2 & 3 & 4 & 6 & 9 & 8 & 7 & 1 & 5 \\ 1 & 2 & \alpha & 4 & 0 & 8 & 5 & 9 & 3 & 7 & 6 \\ 2 & 3 & 4 & \alpha & 1 & 5 & 6 & 0 & 9 & 8 & 7 \\ 3 & 4 & 0 & 1 & \alpha & 7 & 2 & 6 & 5 & 9 & 8 \\ \alpha & 0 & 1 & 2 & 3 & 9 & 8 & 7 & 6 & 5 & 4 \\ \hline 9 & 6 & 8 & 5 & 7 & \alpha & 3 & 1 & 4 & 2 & 0 \\ \hline 8 & 9 & 5 & 6 & 2 & 1 & 7 & \alpha & 0 & 4 & 3 \\ 7 & 8 & 9 & 0 & 6 & 4 & 1 & 5 & \alpha & 3 & 2 \\ 6 & 7 & 3 & 9 & 5 & 2 & 0 & 4 & 8 & \alpha & 1 \\ 5 & 1 & 7 & 8 & 9 & 0 & 4 & 3 & 2 & 6 & \alpha \\ 4 & 5 & 6 & 7 & 8 & 3 & \alpha & 2 & 1 & 0 & 9 \end{array}\right.$$

In [1], W. Taylor has discussed the analogue of a diagonal latin square in higher dimensions. He also remarks in his paper that he and V. Faber have together devised yet another proof of theorem 6.1.7.

As we remarked at the beginning of this section, all diagonal latin squares are magic squares. However, the more usual type of magic square contains n^2 different integers, and usually these are required to be consecutive integers. In the next section, we explain how to construct the latter type of magic square with the aid of pairs of orthogonal latin squares.

6.2. Construction of magic squares with the aid of orthogonal semi-diagonal latin squares

For our construction, we shall need to consider only pairs of orthogonal latin squares which are isotopic to the square which represents the multiplication table of the cyclic group C_n. The method which we shall describe is effective for all odd integers n. Our procedure is substantially equivalent to that of P. De la Hire but the technique of our proof involves the use of latin squares.

0	1	2	3	...	$\frac{1}{2}(n-3)$	$\frac{1}{2}(n-1)$	...	$n-2$	$n-1$
1	2	3	4	...	$\frac{1}{2}(n-1)$	$\frac{1}{2}(n+1)$	...	$n-1$	0
.	.	.	.	.	.	.	...	.	.
.	.	.	.	.	.	.	...	.	.
.	.	.	.	.	.	.	...	.	.
.	.	.	.	.	.	.	...	.	.
$n-1$	0	1	2	...	$\frac{1}{2}(n-5)$	$\frac{1}{2}(n-3)$	...	$n-3$	$n-2$

Fig. 6.2.1

Let L^* denote the latin square exhibited in Fig. 6.2.1, which is precisely the (unbordered) multiplication table of the cyclic group C_n, n odd, when represented as an additive group. By interchanging the elements $n - 1$ and $\frac{1}{2}(n - 1)$ in each of the rows of L^*, we can transform it into a latin square L whose row, column and diagonal sums are each equal to $\frac{1}{2}n(n - 1)$ and which can be characterized as follows. The main left-to-right diagonal forms a transversal of L (so L is left semi-diagonal) and each broken diagonal parallel to the main left-to-right diagonal also forms a transversal. The main right-to-left diagonal contains the element $\frac{1}{2}(n - 1)$ n times and each broken diagonal parallel to the main right-to left diagonal has all its elements equal.

A latin square L' orthogonal to L can be obtained from L by interchanging the columns of L in pairs, the rth column being replaced by the $(n - 1 - r)$th column and vice versa for $r = 0, 1, \ldots, \frac{1}{2}(n - 3)$, while the $\frac{1}{2}(n - 1)$th column is left fixed. (Here the columns are labelled from 0 to $n - 1$.) The orthogonality follows from the fact that all the elements of each broken diagonal parallel to the main left-to-right diagonal of L' are the same, whereas in L they are all different. Also, the elements of each broken diagonal parallel to the main right-to-left diagonal of L' form a transversal, whereas in L they are all the same.

We illustrate the above construction in Fig. 6.2.2 by exhibiting the squares L^*, L and L' for the case $n = 5$.

0	1	2	3	4
1	2	3	4	0
2	3	4	0	1
3	4	0	1	2
4	0	1	2	3

L^*

0	1	4	3	2
1	4	3	2	0
4	3	2	0	1
3	2	0	1	4
2	0	1	4	3

L

2	3	4	1	0
0	2	3	4	1
1	0	2	3	4
4	1	0	2	3
3	4	1	0	2

L'

Fig. 6.2.2

We can express our result in the form of a theorem as follows:

Theorem 6.2.1. *Let C_n denote the cyclic group of order n, n odd, with elements represented by the integers $0, 1, \ldots, n-1$ under addition modulo n and let its isotope under the isotopism $\varrho = (\alpha, \beta, \gamma)$ be denoted by $\varrho(C_n)$. Then, if $\varrho_1 = (\varepsilon, \varepsilon, \gamma)$ and $\varrho_2 = (\varepsilon, \beta, \gamma)$ where ε is the identity permutation and β, γ are the permutations*

$$\begin{pmatrix} 0 & 1 & \ldots & \frac{1}{2}(n-3) & \frac{1}{2}(n-1) & \ldots & n-1 \\ n-1 & n-2 & \ldots & \frac{1}{2}(n+1) & \frac{1}{2}(n-1) & \ldots & 0 \end{pmatrix}$$

and

$$\begin{pmatrix} 0 & \ldots & \frac{1}{2}(n-3) & \frac{1}{2}(n-1) & \frac{1}{2}(n+1) & \ldots & n-2 & n-1 \\ 0 & \ldots & \frac{1}{2}(n-3) & n-1 & \frac{1}{2}(n+1) & \ldots & n-2 & \frac{1}{2}(n-1) \end{pmatrix}$$

respectively, the latin squares which represent the (unbordered) multiplication tables of $\varrho_1(C_n)$ and $\varrho_2(C_n)$ are orthogonal and have the structure described above.

Theorem 6.2.2. *Let $L = \|a_{ij}\|$ and $L' = \|b_{ij}\|$ be two orthogonal latin squares of odd order n formed in the manner described in theorem 6.2.1 and having as their elements the integers $0, 1, \ldots, n-1$. Let a square matrix $M = \|c_{ij}\|$ be constructed from L and L' by putting $c_{ij} = na_{ij} + b_{ij}$. Then the sum of the elements of each row, column, and diagonal of the matrix M is equal to $\frac{1}{2}n(n^2-1)$ and the elements of M are the consecutive integers $0, 1, \ldots, n^2-1$.*

Proof. In order to prove that the row, column, and diagonal sums are each equal to $\frac{1}{2}n(n^2-1)$ $(n \neq 1)$, it is only necessary to point out the validity of the following equalities, which follow directly from the structures of the latin squares L and L' which we described in theorem 6.2.1.

$$\sum_{\substack{i \text{ fixed} \\ j=1,2,\ldots,n}} c_{ij} = \sum_{\substack{i \text{ fixed} \\ j=1,2,\ldots,n}} (na_{ij} + b_{ij}) = n \sum a_{ij} + \sum b_{ij} =$$
$$= (n+1)\left[\tfrac{1}{2}n(n-1)\right] = \tfrac{1}{2}n(n^2-1),$$

$$\sum_{\substack{i=1,2,\ldots,n \\ j=\text{fixed}}} c_{ij} = \sum_{\substack{i=1,2,\ldots,n \\ j=\text{fixed}}} (na_{ij} + b_{ij}) = (n+1)\left[\tfrac{1}{2}n(n-1)\right] = \tfrac{1}{2}n(n^2-1),$$

$$\sum_{i=1,2,\ldots,n} c_{ii} = \sum_{i=1,2,\ldots,n} (na_{ii} + b_{ii}) = (n+1)\left[\tfrac{1}{2}n(n-1)\right] = \tfrac{1}{2}n(n^2-1),$$

$$\sum_{i=1,2,\ldots,n} c_{i,n-i+1} = \sum_{i=1,2,\ldots,n} (na_{i,n-i+1} + b_{i,n-i+1}) =$$
$$= (n+1)\left[\tfrac{1}{2}n(n-1)\right] = \tfrac{1}{2}n(n^2-1).$$

To complete the proof, we have to show that no two of the elements of M are equal and that $0 \leq c_{ij} \leq n^2 - 1$ for all values of i, j. Suppose first that the two elements c_{ij} and c_{kl} of M were equal. This would imply that $n(a_{ij} - a_{kl}) + (b_{ij} - b_{kl}) = 0$. Since the elements of L' are all less than n, the latter equation could only hold if $a_{ij} = a_{kl}$ and then, since L is a latin square, we would necessarily have $i \neq k$ and $j \neq l$. It would follow that a_{ij} and a_{kl} were in the same broken diagonal parallel to the main right-to-left diagonal. However, in L' the elements of each such diagonal are all different, so $b_{ij} \neq b_{kl}$. It follows that $c_{ij} \neq c_{kl}$. Finally, since $0 \leq a_{ij} \leq n - 1$ and $0 \leq b_{ij} \leq n - 1$, the largest value that c_{ij} can take is $n(n - 1) + n - 1 = n^2 - 1$, so $0 \leq c_{ij} \leq n^2 - 1$ as required.

To demonstrate how the procedure works, we shall give the final step of the construction for the example exhibited in Fig. 6.2.2 (see Fig. 6.2.3).

0	5	20	15	10
5	20	15	10	0
20	15	10	0	5
15	10	0	5	20
10	0	5	20	15

nL

2	3	4	1	0
0	2	3	4	1
1	0	2	3	4
4	1	0	2	3
3	4	1	0	2

L'

2	8	24	16	10
5	22	18	14	1
21	15	12	3	9
19	11	0	7	23
13	4	6	20	17

$M = nL + L'$

Fig. 6.2.3

Theorem 6.2.3. *There exist orthogonal pairs of diagonal latin squares of every odd order which is not a multiple of 3.*

Proof. It follows directly from theorem 6.1.2 that, if $\alpha = 2$ and $\beta = 1$, the latin square exhibited in Fig. 6.1.1 is a diagonal latin square provided that n is odd and prime to 3. Its transpose is also a diag onal latin square and we shall show that the two squares so obtained are orthogonal, again under the condition that n is prime to 3.

Let us denote the square obtained from Fig. 6.1.1 when $\alpha = 2$ and $\beta = 1$ by L_1 and its transpose by L_2 (the squares L_1 and L_2 for the special case when $n = 5$ are exhibited in Fig. 6.2.4). We shall show first that the contents of the cells of L_1 which correspond to the cells containing the element 0 in L_2 are all different.

$L_1 =$

0	2	4	1	3
1	3	0	2	4
2	4	1	3	0
3	0	2	4	1
4	1	3	0	2

$L_2 =$

0	1	2	3	4
2	3	4	0	1
4	0	1	2	3
1	2	3	4	0
3	4	0	1	2

Fig. 6.2.4

In the rth row of L_2, $r = 0, 1, \ldots, n-1$, the element 0 appears in the cell of the $(n - 2r)$th column, where addition is modulo n. In the cell of the rth row and $(n - 2r)$th column of L_1, the element $r + 2(n - 2r)$ appears, where again the addition is modulo n. To see this, observe that the entry in the cell of the rth row and 0th column of the square L_1 is r and that for each step taken to the right along this row, the cell entry is increased by 2. Since $r + 2(n - 2r) = -3r \bmod n$ and since 3 is prime to n, the elements $-3r$ as r varies through the set $\{0, 1, 2, \ldots, n-1\}$ are all different. This proves the result.

By a similar argument, we could show that the contents of the cells of L_1 which correspond to the cells containing the element i in L_2 are all different and that this is true for each choice of i in the range $0 \leq i \leq n - 1$. It follows that L_1 and L_2 are orthogonal diagonal latin squares.

It is an obvious consequence of theorem 6.1.5 that orthogonal pairs of diagonal latin squares of order 3 do not exist. However, we can prove the following:

Theorem 6.2.4. *Orthogonal pairs of diagonal latin squares of order n can be constructed whenever n is odd or a multiple of 4 except, possibly, when n is a multiple of 3 but not of 9.*

Proof. We first point out that there exist special orthogonal pairs of diagonal latin squares of orders 4, 8, 9, and 27. Examples of such pairs corresponding to the first three of these values of n are exhibited in

Figs 6.2.5, 6.2.6 and 6.2.7 respectively and a construction which gives a pair of order 27 is described on page 192 of W. W. Rouse Ball [1]. (In Fig. 6.2.5, the diagonal squares L_1 and L_2 are first shown separately and then in juxtaposition. In the remaining cases, only the juxtaposed form is shown.)

$$L_1 = \begin{array}{|cccc} 0 & 1 & 2 & 3 \\ 2 & 3 & 0 & 1 \\ 3 & 2 & 1 & 0 \\ 1 & 0 & 3 & 2 \end{array} \qquad L_2 = \begin{array}{|cccc} 0 & 3 & 1 & 2 \\ 2 & 1 & 3 & 0 \\ 3 & 0 & 2 & 1 \\ 1 & 2 & 0 & 3 \end{array}$$

00	13	21	32
22	31	03	10
33	20	12	01
11	02	30	23

$$M = \begin{array}{|cccc|} 0 & 7 & 9 & 14 \\ 10 & 13 & 3 & 4 \\ 15 & 8 & 6 & 1 \\ 5 & 2 & 15 & 11 \end{array}$$

Fig. 6.2.5

17	50	43	04	32	75	66	21
31	76	65	22	14	53	40	07
00	47	54	13	25	62	71	36
26	61	72	35	03	44	57	10
45	02	11	56	60	27	34	73
63	24	37	70	46	01	12	55
52	15	06	41	77	30	23	64
74	33	20	67	51	16	05	42

Fig. 6.2.6

76	82	64	15	27	00	41	53	38
11	23	08	46	52	34	75	87	60
45	57	30	71	83	68	16	22	04
62	74	86	07	10	25	33	48	51
03	18	21	32	44	56	67	70	85
37	40	55	63	78	81	02	14	26
84	66	72	20	05	17	58	31	43
28	01	13	54	36	42	80	65	77
50	35	47	88	61	73	24	06	12

Fig. 6.2.7

Then, making use of theorem 6.2.3 and of the construction described in the proof of theorem 6.1.3, it is easy to deduce the truth of the present theorem.

THEOREM 6.2.5. *If n is any integer for which an orthogonal pair of diagonal latin squares of order n exists, then an $n \times n$ magic square whose entries are the consecutive integers 0 to $n^2 - 1$ can be constructed.*

PROOF. We first write the two squares in juxtaposed form, as for example in Fig. 6.2.5. Since the sums of the elements of each row, each column, and each main diagonal are all equal for each of the two squares, the same is true in the juxtaposed form. Moreover, this is true regardless of the number base selected. If we take the number base as the integer $n - 1$ we get a magic square M whose entries are the integers 0 to $n^2 - 1$ as required. Thus, to get the matrix M exhibited in Fig. 6.2.5, the number base 5 has been selected.

We may express the construction in another way by saying that the matrix M is related to the matrices L_1 and L_2 by the matrix equation $M = (n - 1)L_1 + L_2$ in the general case. (Compare the proof of theorem 6.2.2).

The constructions of magic squares by means of orthogonal pairs of diagonal latin squares or by means of orthogonal pairs of latin squares of the type given by theorem 6.2.1 have been known and used for more than two centuries. See, for example, L. Euler [2], E. Maillet [1] and [3] or [5] ([3] and [5] are two publications of the same paper), E. Barbette [1], G. Tarry [6] and [7], M. Kraichik [1] and W. W. Rouse Ball [1]. These constructions are very useful but they are not the only methods available. For example, neither of these constructions can be used to obtain a magic square of order 6 and yet magic squares of this order certainly exist, as is demonstrated by Fig. 6.2.8.

35	1	6	26	19	24
3	32	7	21	23	25
31	9	2	22	27	20
8	28	33	17	10	15
30	5	24	12	14	16
4	36	29	13	18	11

Fig. 6.2.8

Because of their importance in connection with the construction of magic squares, many authors have tried to answer the question: "For which orders n distinct from 2, 3 and 6 do there exist orthogonal pairs of diagonal latin squares?"

That such pairs exist for the case $n = 4$ has been known at least since 1723, as has been pointed out on page 190 of W. W. Rouse Ball [1].

The fact that a solution exists whenever n is odd and not a multiple of three (see theorem 6.2.3) has also been known at least since the last century. Using both these results, G. Tarry (see [7]) proved at the beginning of the present century that orthogonal pairs of diagonal latin squares exist of every order n which is a multiple of 4. Moreover, he proved in [6] that if $n = 8m$ where m is odd and not a multiple of three then it is possible to obtain pairs of orthogonal diagonal latin squares such that the magic square constructed from them by the method of theorem 6.2.5 has the additional property that the sums of the squares of the elements of each row, each column and each main diagonal are all equal.

Theorem 6.2.4 gives a further set of values of n for which orthogonal pairs of diagonal latin squares exist and comes from W. W. Rouse Ball [1]. Recently, the question has aroused new interest. In particular, C. C. Lindner has made an attempt to solve the problem by means of a construction which makes use of the singular direct product of A. Sade. (The latter concept is defined in section 12.1.) His main result is as follows: "If there are t mutually orthogonal diagonal latin squares of order $v = 2m$, t mutually orthogonal left semi-diagonal latin squares of order q each containing a diagonal latin square of order p in its top left-hand corner, and t mutually orthogonal right semi-diagonal latin squares of order $q - p$, then there are t mutually orthogonal diagonal latin squares of order $v(q - p) + p$." Taking $t = 2$, $v = 8$, $p = 1$ and $q = 5$, Lindner has deduced that there is a pair of orthogonal diagonal latin squares of order 33, which number is a multiple of 3 but not of 9. (Contrast theorem 6.2.4.) Similarly, taking $t = 6$, $v = 8$, $p = 1$ and $q = 17$, he has deduced that there are at least six mutually orthogonal diagonal latin squares of order 129. By means of the direct product construction, it is easy to show from these results that pairs of mutually orthogonal diagonal latin squares exist for an infinity of orders n which are multiples of 3 but not of 9. For the details, see C. C. Lindner [16].

By methods quite similar to those used by Lindner, A. J. W. Hilton and S. H. Scott have shown in [1] that pairs of orthogonal diagonal latin squares exist also for some orders n which are multiples of 2 but not of 4. The smallest such order for which their construction works is $n = 50$. They have also shown the existence of a pair of orthogonal diagonal latin squares of order 21. In A. J. W. Hilton [4], the author has shown that in fact it is possible to obtain a set of at least four mutually orthogonal diagonal latin squares of order 50.

As regards the maximum number $N_D(n)$ of diagonal latin squares in a pairwise orthogonal set of squares of order $n = p_1^{\alpha_1} p_2^{\alpha_2} \ldots p_r^{\alpha_r}$ (where the p_i's are distinct primes), E. Gergely has proved in [2] that

$$N_D(n) \geq \alpha(n) - 3 \text{ if } n \text{ is odd}$$

and

$$N_D(n) \geq \alpha(n) - 2 \text{ if } n \text{ is even,}$$

where $\alpha(n) = \min_{1 \leq i \leq r} p_i^{\alpha_i}$. (Compare MacNeish's theorem, page 390 of section 11.1.) Gergely has also pointed out that $N_D(n) \leq n - 3$ if n is odd ($n \geq 3$) and $N_D(n) \leq n - 2$ if n is even. (Compare theorem 5.1.5.)

In [3], A. J. W. Hilton has shown that $N_D(n) \to \infty$ as $n \to \infty$. (Compare page 174 of section 5.2.)

6.3. Additional results on magic squares

A magic square is called *pandiagonal* if the sets of elements in each of its broken diagonals have the same sum and this sum is the same as the sum of the elements of each row, column, and the two main diagonals.

E. McClintock investigated pandiagonal squares in [1] and this paper combined some of the results of P. De la Hire and extended them further.

It is well known and easy to check that, if α and β are integers which satisfy the conditions of theorem 6.1.2, then the latin square exhibited in Fig. 6.1.1 is a pandiagonal magic square. (See also W. W. R. Ball [1] and E. McClintock [1].)

Consequently, theorem 6.1.2 can be reformulated as follows.

THEOREM 6.3.1. *For every odd integer n which is not a multiple of 3 there exists at least one pandiagonal magic square of size $n \times n$.*

However, since the pandiagonal magic square exhibited in Fig. 6.1.1 is a latin square, its elements are not consecutive integers. Nevertheless, for many values of n, $n \times n$ pandiagonal magic squares whose entries are consecutive integers do exist. In Fig. 6.3.1, for example, we exhibit

10	5	49	37	32	27	15
41	29	24	19	14	2	46
16	11	6	43	38	33	28
47	42	30	25	20	8	3
22	17	12	7	44	39	34
4	48	36	31	26	21	9
35	23	18	13	1	45	40

Fig. 6.3.1

such a square of size 7×7. The sum of the elements of each of its rows, columns, and diagonals is 175.

It is a property of pandiagonal magic squares, long known and easily recognized, that a row or column which forms one of the four edges of the square can be transposed to the opposite side of the square without destroying the magic and pandiagonal attributes of the square. By reason of this remarkable property, pandiagonal squares have been given by some authors, beginning with P. De la Hire, the name *perfect* and by others, beginning with E. Lucas, the name *diabolic*.

In [1] and [2], W. W. Horner investigated a method for constructing magic squares in which not only the sum but also the product of the elements in each row, column, or main diagonal is a constant. Such a square is conveniently called an *addition-multiplication magic square*.

In [1], W. W. Horner showed how to construct addition-multiplication magic squares of any odd order and in [2] he showed how one could obtain addition-multiplication magic squares of orders 8 and 16.

Both of his methods make use of latin squares.

It would require too much space to give his constructions in detail here. Instead, we give in Figs 6.3.2 and 6.3.3 examples of addition-multiplication magic squares of orders 8 and 9 respectively, obtained by Horner's methods.

The magic sum (that is, the sum of the elements of each row, column, or diagonal) of the addition-multiplication magic square exhibited in Fig. 6.3.2 is 840 and its magic product is 2,058,068,231,856,000.

162	207	51	26	133	120	116	25
105	152	100	29	138	243	39	34
92	27	91	136	45	38	150	261
57	30	174	225	108	23	119	104
58	75	171	90	17	52	216	161
13	68	184	189	50	87	135	114
200	203	15	76	117	102	46	81
153	78	54	69	232	175	19	60

Fig. 6.3.2

200	87	95	42	99	1	46	108	170
14	44	10	184	81	85	150	261	19
138	243	17	50	116	190	56	33	5
57	125	232	9	7	66	68	230	54
4	70	22	51	115	216	171	25	174
153	23	162	76	250	58	3	35	88
145	152	75	11	6	63	270	34	92
110	2	28	135	136	69	29	114	225
27	102	207	290	38	100	55	8	21

Fig. 6.3.3

In our second example (Fig. 6.3.3) the magic sum is 848 and the magic product 5,804,807,833,440,000.

The constructions used in the proofs of theorems 6.1.2, 6.2.1 and 6.2.5, suggest the importance of orthogonal pairs of diagonal latin squares which satisfy the quadrangle criterion.

Since such squares play an important role in the construction of magic squares it is interesting to ask the question: What is the largest number of elements, chosen randomly, which can be deleted from the two members of an orthogonal pair of diagonal latin squares both of which satisfy the quadrangle criterion in such a way that the pair can be reconstructed uniquely from the remainder?

The problem is solved in theorem 6.3.2 below.

Theorem 6.3.2. *Let* $A = \| a_{ij} \|$ *and* $B = \| b_{ij} \|$ *be two orthogonal diagonal latin squares of order* $n (n \neq 4)$ *both of which satisfy the quadrangle criterion, and let the matrix of ordered pairs* (a_{ij}, b_{ij}) *be denoted by* C. *Then the maximum number of elements, chosen randomly, which can be deleted from* C *in such a way that* C *can still be uniquely reconstructed from those left is* $2n - 1$.

Proof. It follows from theorem 3.2.3 that $2n - 1$ arbitrary elements can be deleted from either one of the squares A or B separately without destroying the unique reconstructibility.

It remains to show that if some elements are deleted from A and some from B then the total number (randomly chosen) which can be deleted without destroying the unique reconstructibility still cannot exceed $2n - 1$.

However, this is easy to see. For, let us select $2n$ elements to be deleted from C of the form (a_{ij}, b_{ij}) where $a_{ij} = k$ or l (the total number of such elements is clearly $2n$). If we interchange each pair of elements of the form (k, b_{ij}), (l, b_{ij}), the square C so obtained is again a juxtaposed orthogonal pair of diagonal latin squares with the quadrangle criterion.

As a demonstration of our construction let us consider the orthogonal pair of latin squares which are exhibited in Fig. 6.2.4 and let $k = 0$ and $l = 1$. The square which we can obtain by our algorithm is exhibited in Fig. 6.3.4.

10	21	42	03	34
02	33	14	20	41
24	40	01	32	13
31	12	23	44	00
43	04	30	11	22

Fig. 6.3.4

COROLLARY. *The maximum number of elements, randomly chosen, which can be deleted from an orthogonal pair of latin squares in such a way that the pair can still be uniquely reconstructed from those left is* $2n - 1$.

The above corollary has been proved in J. Dénes [1].

We would like to end this section by pointing out that magic squares should not be regarded solely as a mathematical amusement because they also play an important role in practical applications such as experimental design in statistics and the construction of error detecting and correcting codes.

Thus, in [1] S. I. Samoilenko illustrated how generalized magic squares can be utilized for error correction. He used latin squares to obtain these generalized magic squares and then constructed variable parameter codes with the aid of the latter. We shall discuss the subject of error correcting codes more fully in section 10.1. A typical application of variable parameter codes is in the transmission of data between computers.

J. P. N. Phillips (see [1]) showed how magic squares can be applied to statistics. The magic squares which he used were obtained by means of latin squares and orthogonal pairs of latin squares.

6.4. Room squares: their construction and uses

Under the title "A new type of magic square", T. G. Room introduced in [1] what was believed at the time (1955) to be a new kind of combinatorial structure and which has subsequently become known as a *Room square* or *Room design*. As in the case of magic squares, many of these designs can be constructed with the aid of latin squares.

A *Room design* of order $2n$ comprises a square array having $2n - 1$ cells in each row and each column and such that each cell is either empty or contains an unordered pair of symbols chosen from a set of $2n$ elements. Without loss of generality, we can take these elements as the numbers $1, 2, \ldots, 2n - 1, \infty$. Each row and each column of the design contains $n - 1$ empty cells and n cells each of which contains a pair of symbols. Each row and each column contains each of the $2n$ symbols exactly once, and, further, each of the $n(2n - 1)$ possible distinct pairs of symbols is required to occur exactly once in a cell of the square. As illustrations of the concept, we exhibit two Room designs of order 8 in Fig. 6.4.1.

A Room square of side $2n - 1$ is synonymous with a Room design of order $2n$.

T. G. Room came across structures of this kind in connection with a study of Clifford matrices, a concept of Algebraic Geometry. However, what was not realised until very recently (1970) is that such designs had begun to be studied some fifty years earlier in connection with the design of tournaments for the card game known as "Bridge".

The purpose of a Duplicate Bridge tournament is to establish comparisons between every pair of players taking part. In each separate game, two pairs of players compete. We may designate each pair of players by a single symbol and, if there are $2n$ pairs, we may take the symbols denoting these to be the numbers $1, 2, \ldots, 2n - 1, \infty$ as above. The tournament consists of $2n - 1$ rounds and, in each of these rounds, all the players take part. During the course of the tournament each pair of players is required to play each other pair exactly once and also each pair of players is required to play at each of a number of different tables (or *boards*) exactly once. The arrangements of the cards at a particular table are the same for each set of players who play at that table (but are not disclosed to any pair of players until they reach that table). In this way, the desired comparisons between the play of the different pairs is effected. All these requirements can be met if and only if there exists a Room design of order $2n$, where $2n$ is the number of Bridge pairs.

To see this, let us regard the rows of the Room square as giving the rounds and the columns as giving the boards. There are $2n - 1$ of the latter, but in any particular round only n of them are in use, the others being referred to as *bye-boards*. If the cell which lies at the intersection of the rth row and sth column of the Room square is occupied, the two numbers which appear in it indicate the two pairs of players who should play the sth board in the rth round of the tournament.

The existence of such designs for use in Bridge tournaments was first investigated in 1897 by E. C. Howell, who was a Professor of Mathematics at the Massachusetts Institute of Technology at that time. According to N. S. Mendelsohn, to whom the present authors are indebted for pointing out the foregoing application of Room designs to Bridge Tournaments, Howell constructed designs for all values of n from 4 up to and including 15. In books giving instructions for the organization of Bridge tournaments, these designs are known as *Howell master sheets*. In G. W. Beynon [1], for example, master sheets for $n = 4$, 5, and 7 are listed and, in G. W. Beynon [2], master sheets for $n = 6$ and 8 as well. In M. Gruenther [1], master sheets for all values of n from 4 to 15 inclusive are given. In Fig. 6.4.1, we exhibit two master sheets (Room designs) of order 8 (that is, $n = 4$) attributed by Beynon to J. Ach and C. T. Kennedy of Cincinnati and to two authors McKennedy and Baldwin (of whom further details are lacking) respectively.

It will be noted that each of the designs shown in Fig. 6.4.1 is completely determined by its first row in the sense that each successive pair along a broken left- to- right diagonal of the square is obtained from the preceding pair in that diagonal by addition of 1, modulo 7 (or modulo $2n - 1$ in the general case), to each member of that pair. We write 7 in place of 0 for convenience. Such a Room design is called *cyclic*. Non-cyclic Room designs also exist. An example of order 8 is given in Fig. 6.4.2.

We return to the modern history of Room designs. In his paper [1], T. G. Room pointed out the non-existence of Room designs for $n = 2$ and 3 and gave an example of a non-cyclic design for $n = 4$ which we

$\infty, 1$			6, 2		5, 7	3, 4
4, 5	$\infty, 2$			7, 3		6, 1
7, 2	5, 6	$\infty, 3$			1, 4	
	1, 3	6, 7	$\infty, 4$			2, 5
		2, 4	7, 1	$\infty, 5$		
			3, 5	1, 2	$\infty, 6$	
				4, 6	2, 3	$\infty, 7$

$\infty, 1$	5, 6	2, 4		3, 7		
	$\infty, 2$	6, 7	3, 5		4, 1	
		$\infty, 3$	7, 1	4, 6		5, 2
6, 3			$\infty, 4$	1, 2	5, 7	
	7, 4			$\infty, 5$	2, 3	6, 1
7, 2		1, 5			$\infty, 6$	3, 4
4, 5	1, 3		2, 6			$\infty, 7$

Fig. 6.4.1

reproduce in Fig. 6.4.2. Here, the digits 1, 2, . . . , 7 and 8 are used, as the eighth symbol is no longer specially treated.

The next authors to write on this subject were J. W. Archbold and N. L. Johnson, who gave a geometrical construction of cyclic Room designs for all values of n of the form 4^m and made use of J. Singer's theorem (see J. Singer [1]) to enable them to express the designs they obtained in a canonical form. (See J. W. Archbold and N. L. Johnson [1]

1, 2		3, 4		5, 6		7, 8
	3, 7	2, 5			4, 8	1, 6
4, 7	1, 5			3, 8	2, 6	
			6, 8	1, 4	5, 7	2, 3
5, 8		6, 7	2, 4		1, 3	
	4, 6	1, 8	3, 5	2, 7		
3, 6	2, 8		1, 7			4, 5

Fig. 6.4.2

for the details.) These authors also showed how Room designs might be used as statistical designs for a suitable kind of experiment. (Detailed information on the subject of statistical designs is given in section 10.2.) Later, in 1960, J. W. Archbold published a further paper (J. W. Archbold [1]) in which he gave another construction for Room squares, based on difference sets, which yielded designs of orders 8, 12, 20 and 24 ($n = 4, 6, 10$ and 12). Both these kinds of design were cyclic and it is interesting to note that the design obtained for $n = 6$ is exactly the same as that published in G. W. Beynon [2] sixteen years earlier. Recently, a more detailed investigation of the effectiveness of Room squares for use in statistical designs has been carried out by K. R. Shah in [1].

In 1963, R. H. Bruck showed an interesting connection between Room designs and idempotent quasigroups, as follows:

THEOREM 6.4.1. *A Room design of order* $2n$ *is equivalent to a pair of commutative idempotent quasigroups*[1], *say* $(Q, \mathbf{r})$ *and* $(Q, \mathbf{c})$, *each of order* $2n - 1$ *and satisfying the following two orthogonality conditions: (i) if* $a \in Q$ *and if* x *and* y *are elements of* Q *such that* $x \,\mathbf{r}\, y = a$ *and* $x \,\mathbf{c}\, y = a$ *then* $x = y = a$; *and (ii) if* a *and* b *are distinct elements of* Q, *then there*

(1) this and the following pages ($\mathbf{r}$, $\mathbf{c}$, $\mathbf{r}_1$, $\mathbf{c}_1$, $\mathbf{r}_2$ and $\mathbf{c}_2$) denote quasigroup operations.

exists at most one unordered pair of elements x, y of Q such that $x \mathbf{r} y = a$ and $x \mathbf{c} y = b$.

Proof. To see the equivalence, we suppose that the given Room design has symbols $1, 2, \ldots, 2n - 1$ and ∞, as before, and we permute its rows in such a way that the ordered pair (∞, i) occurs in the ith row and then its columns in such a way that the ordered pair (∞, i) occurs also in the ith column. The quasigroups $(Q, \mathbf{r})$ and $(Q, \mathbf{c})$ are now defined by the statements that they are idempotent and that, for $x \neq y$, $x \mathbf{r} y$ and $x \mathbf{c} y$ are respectively equal to the numbers of the row and column in which the cell containing the unordered pair x, y appears in the normalized Room square.

As an example, the square due to T. G. Room exhibited in Fig. 6.4.2 takes the form shown in Fig. 6.4.3 if we carry out the above rearrangement of rows and columns after replacing the symbol 8 by the symbol ∞.

∞, 1	4, 6	2, 7			3, 5	
	∞, 2			3, 6	1, 7	4, 5
	1, 5	∞, 3	2, 6	4, 7		
2, 5	3, 7		∞, 4			1, 6
6, 7			1, 3	∞, 5	2, 4	
		1, 4	5, 7		∞, 6	2, 3
3, 4		5, 6		1, 2		∞, 7

$(\mathbf{r})$	1	2	3	4	5	6	7
1	1	7	5	6	3	4	2
2	7	2	6	5	4	3	1
3	5	6	3	7	1	2	4
4	6	5	7	4	2	1	3
5	3	4	1	2	5	7	6
6	4	3	2	1	7	6	5
7	2	1	4	3	6	5	7

Fig. 6.4.3

Following R. H. Bruck [7], we shall call this the *normalized form* of the design.

Also shown in Fig. 6.4.3 is the multiplication table of the quasigroup $(Q, \boldsymbol{r})$ which is thus defined by this square.

REMARK. The reader should note that the orthogonality conditions given in theorem 6.4.1 do not imply that the quasigroups are orthogonal in the sense defined in section 5.2. (Since both quasigroups are commutative, the equations $x \, \boldsymbol{r} \, y = a$ and $x \, \boldsymbol{c} \, y = b$ are not soluble simultaneously for all choices of a and b.) They provide our first example of a pair of perpendicular commutative quasigroups, further details of which are given in sections 12.1 and 12.5.

In [7], R. H. Bruck showed how, using this equivalence, the construction of J. W. Archbold and N. L. Johnson for a Room design of order 2^{2m+1} could be much simplified. Let Q comprise the $2^{2m+1} - 1$ non-zero elements of the Galois field $GF[2^{2m+1}]$ and define two quasigroups $(Q, \boldsymbol{r})$ and $(Q, \boldsymbol{c})$ on the set Q by the statements $x \, \boldsymbol{r} \, x = x = x \, \boldsymbol{c} \, x$ for all x in Q, $x \, \boldsymbol{r} \, y = x + y$ and $x \, \boldsymbol{c} \, y = (x^{-1} + y^{-1})^{-1}$ for all x and y in Q. Then, using the properties of the finite field, it is easy to check that the orthogonality conditions described above are satisfied for these quasigroups and so a Room design of order 2^{2m+1} can be constructed. (That is, $n = 4^m$.)

DEFINITION. A pair of idempotent quasigroups $(Q, \boldsymbol{r})$ and $(Q, \boldsymbol{c})$ which are commutative and satisfy the orthogonality conditions of theorem 6.4.1 are called a *Room pair* of quasigroups.

R. H. Bruck asserted that it was easy to see that the direct product of two Room pairs of quasigroups was itself a Room pair of quasigroups. Let $(Q_1, \boldsymbol{r}_1), (Q_1, \boldsymbol{c}_1)$ be one Room pair and $(Q_2, \boldsymbol{r}_2), (Q_2, \boldsymbol{c}_2)$ another. The direct product is defined as the pair $(Q, \boldsymbol{r}), (Q, \boldsymbol{c})$, where $Q = Q_1 \times Q_2$ and $(q_1, q_2) \, \boldsymbol{r} \, (q_1', q_2') = (q_1 \, \boldsymbol{r}_1 \, q_1', q_2 \, \boldsymbol{r}_2 \, q_2')$, $(q_1, q_2) \, \boldsymbol{c} \, (q_1', q_2') = (q_1 \, \boldsymbol{c}_1 \, q_1', q_2 \, \boldsymbol{c}_2 \, q_2')$. Later (1969), R. C. Mullin and E. Németh pointed out in [1] that such a direct product need not satisfy the second orthogonality condition and used Bruck's own construction of Room pairs of quasigroups of order $2^{2m+1} - 1$ to give an explicit counter-example. Had Bruck's assertion been true, it would have implied that from two Room designs of orders $2m$ and $2n$ respectively a Room design of order $(2m - 1)(2n - 1) + 1$ could be constructed. Recently, R. G. Stanton and J. D. Horton have shown in [1] and [2] that, although Bruck's proof of it was fallacious, the statement just made is true. We shall now give their proof.

Theorem 6.4.2. *If Room squares of sides $2m - 1$ and $2n - 1$ exist, then one can construct a Room square of side $(2m - 1)(2n - 1)$.*

Proof. Let R and S be Room squares of sides $r = 2m - 1$ and $s = 2n - 1$ whose entries are the symbols $0, 1, 2, \ldots, r$ and $0, 1, 2, \ldots, s$ respectively. Let L_1 and L_2 be a pair of (arbitrarily chosen) mutually orthogonal latin squares of order $r = 2m - 1$ whose entries are the symbols $1, 2, \ldots, r$. To construct our Room square T of side $rs = (2m - 1)(2n - 1)$, we regard T as an $s \times s$ square each of whose cells is an $r \times r$ subsquare, and we prescribe these subsquares by the following rules:

(a) If the cell (i, j) of the Room square S is empty, then the $r \times r$ subsquare t_{ij} in the corresponding cell of T is to consist entirely of empty cells.

(b) If the cell (i, j) of S is occupied by the pair $(0, k)$, then the $r \times r$ subsquare t_{ij} in the corresponding cell of T is to be the Room square obtained from R by adding kr to each of its non-zero symbols. (The zero symbol is to be left unchanged.)

(c) If the cell (i, j) of S is occupied by the pair (h, k), with $h \neq 0$ and $k \neq 0$, then the $r \times r$ subsquare t_{ij} in the corresponding cell of T is to be the square with every cell occupied by an ordered pair of symbols and which is constructed from the latin squares L_1 and L_2 in the following manner. First add hr to each of the symbols of L_1 to form a new latin square L_1^*. Similarly form a new latin square L_2^* by adding kr to each of the symbols of L_2. Finally, juxtapose L_1^* and L_2^* so as to form a square $t_{ij} = (L_1^*, L_2^*)$ whose entries are ordered pairs of symbols (l_1, l_2) with $l_1 \in \{1 + hr, 2 + hr, \ldots, r + hr\}$ and $l_2 \in \{1 + kr, 2 + kr, \ldots, r + kr\}$.

The square T so constructed has the set $0, 1 + r, 2 + r, \ldots, r + r, 1 + 2r, 2 + 2r, \ldots, r + sr$ as its set of symbols. Also, by the method of construction, each of these symbols occurs just once in each row and once in each column of T. For, we have that each of the symbols $1, 2, \ldots, r$ occurs just once in a row (column) of R and once in a row (column) of L_1 and L_2. Since each of the symbols $1, 2, \ldots, s$ occurs just once in a row (column) of S, each of the symbols $x + yr$, $1 \leq x \leq r$, $1 \leq y \leq s$ occurs just once in a row (column) of T. Since the symbol 0 occurs just once in a row (column) of S, it occurs just once in a row (column) of T.

Moreover, in each subsquare t_{ij} no unordered pair of symbols occurs more than once (an immediate consequence of the mode of formation of these subsquares), and no two subsquares t_{ij} and t_{uv} have any pair in common. For suppose that the pair $(x_1 + y_1 r, x_2 + y_2 r)$ with $x_1 \neq 0$ and

$x_2 \neq 0$ were common to t_{ij} and t_{uv}. If $y_1 \neq y_2$, it would follow that (y_1, y_2) occurred in each of the cells (i, j) and (u, v) of S, a contradiction. If $y_1 = y_2 = y$, it would follow that $(0, y)$ occurred in each of the cells (i, j) and (u, v) of S. Finally, if $(0, x + yr)$ were common to t_{ij} and t_{uv} it would again follow that $(0, y)$ occurred in each of the cells (i, j) and (u, v) of S.

We conclude that T is a Room square, as desired.

During the past two years a number of papers which investigate the question "For which values of n do Room designs exist?" have been published. The results of these investigations when taken in conjunction with theorem 6.4.2 are sufficient to show (i) that there are cyclic Room designs for all values of n from 4 to 25 inclusive and (ii) that there exist Room designs, not necessarily all cyclic, for all values of n except those for which $2n - 1$ has a Fermat prime of the form $2^{2^r} + 1$ as an unrepeated factor. The latter case remained undecided until very recently. (See the addendum to this chapter.)

The first of these results was proved by R. G. Stanton and R. C. Mullin (see [1]) with the aid of a computer. These authors considered the possibility of the existence of cyclic Room squares of side $2m + 1$ whose first rows should contain the unordered pairs

$$(\infty, 0), (1, 2m), (2, 2m - 1), \ldots, (m, m + 1),$$

not necessarily in this order. They called such squares *patterned Room squares*, and the unordered pairs just listed were said to form a *starter* for such a square. They were able to construct such patterned Room squares of all odd orders $2m + 1$ from 7 to 49 except 9. For the latter order, patterned Room squares do not exist. However, a cyclic Room square of this order ($n = 5$ in our previous notation) had previously been obtained by L. Weisner (see [2]) and, of course, Room designs of this order had also been constructed much earlier for use as Howell master sheets. The construction of Stanton and Mullin was later generalized by R. C. Mullin and E. Németh in [2]. More recently, K. Byleen has proved in [1] that patterned Room squares of side p exist for all primes p not of the form $1 + 2^s$.

The result (ii) above follows as a consequence of theorem 6.4.2 and of a construction described in R. C. Mullin and E. Németh [3] which uses a generalized form of patterned Room squares and which gives Room squares of side $2n - 1$, where $2n - 1$ is any odd prime power which is not a Fermat prime of the form $2^{2^r} + 1$.

We mentioned at the beginning of this section that Room designs can be constructed with the aid of latin squares. The following theorem is due to K. Byleen and D. W. Crowe (see [1]).

THEOREM 6.4.3. *Let L be a latin square of odd order $2n - 1$ which is orthogonal to its transpose L^T, which has elements $1, 2, \ldots, 2n - 1$, and which is standardized in such a way that these elements occur along its main left-to-right diagonal in natural order. Let the entries of the main left-to-right diagonal of L^T be all replaced by the symbol ∞ and let M denote the matrix of ordered pairs formed when L and the modified square L^T are juxtaposed. Then a Room design may be obtained from M by deletion of a selected set of n of its left-to-right (broken) diagonals provided that the n diagonals to be deleted can be chosen so that (i) every element of L appears exactly once among the remaining pairs of the first row of M, and (ii) if the diagonal which contains the cell m_{1j} of M is deleted then the diagonal which contains the cell m_{j1} is not deleted.*

PROOF. It is an immediate consequence of the fact that L and L^T are transposes that, if the cell m_{ik} $(i \neq k)$ of M contains the ordered pair (a, b), then the cell m_{ki} of M contains the ordered pair (b, a). Also, since L and L^T are orthogonal, every ordered pair of distinct elements a, b occurs just once in a cell of M. Hence, if n diagonals of M are deleted, which are so chosen that (*ii*) is satisfied, the remaining cells of M will contain each unordered pair of distinct elements a, b chosen from the set $1, 2, \ldots, 2n - 1$ just once. Moreover, the pairs $(\infty, 1)$ $(\infty, 2), \ldots,$ $(\infty, 2n - 1)$ will occur along the main left to right diagonal of M. It follows easily that if condition (*i*) is also satisfied, the remaining structure will form a Room square.

Byleen and Crowe have shown how to construct a latin square L and corresponding matrix M for which all the requirements of theorem 6.4.3 are satisfied whenever $2n - 1$ is an odd prime not of the form $1 + 2^s$. We give an example for the case $2n - 1 = 7$ in Fig. 6.4.4.

We shall discuss the general problem of constructing latin squares which are orthogonal to their own transposes in section 12.4, and we shall give a short account of Byleen and Crowe's method in that section. Full details of their construction will be found in K. Byleen and D. W. Crowe [1].

Finally, we should like to mention a construction of Room designs with the aid of a pair of orthogonal Steiner triple systems which was first given in C. D. O'Shaughnessy [1]. We remind the reader that orthogonal Steiner triple systems were defined in section 5.3.

$$
L = \begin{array}{|ccccccc}
1 & 7 & 6 & 5 & 4 & 3 & 2 \\
3 & 2 & 1 & 7 & 6 & 5 & 4 \\
5 & 4 & 3 & 2 & 1 & 7 & 6 \\
7 & 6 & 5 & 4 & 3 & 2 & 1 \\
2 & 1 & 7 & 6 & 5 & 4 & 3 \\
4 & 3 & 2 & 1 & 7 & 6 & 5 \\
6 & 5 & 4 & 3 & 2 & 1 & 7
\end{array}
$$

$1, \infty$	$7, 3$	$6, 5$		$4, 2$		
	$2, \infty$	$1, 4$	$7, 6$		$5, 3$	
		$3, \infty$	$2, 5$	$1, 7$		$6, 4$
$7, 5$			$4, \infty$	$3, 6$	$2, 1$	
	$1, 6$			$5, \infty$	$4, 7$	$3, 2$
$4, 3$		$2, 7$			$6, \infty$	$5, 1$
$6, 2$	$5, 4$		$3, 1$			$7, \infty$

Fig. 6.4.4

O'Shaughnessy's theorem is as follows:

Theorem 6.4.4. *Let S and S' be two orthogonal Steiner triple systems of the same order v* (necessarily congruent to 1 or 3 modulo 6, as shown in section 2.2) *and defined on the same set $1, 2, \ldots, v$. Then, a Room square of side v may be constructed by means of S and S' by putting the unordered pair of elements (i, j) in the cell of the kth row and k'th column of the square, where k is the third element of the triple of S which contains i and j and k' is similarly defined by S'. The square is completed by putting the ordered pair (∞, h) in the cell of the hth row and hth column. $h = 1, 2, \ldots, v$.*

Proof. The square constructed by the method just described clearly contains all unordered pairs of distinct elements obtainable from the set $1, 2, \ldots, v, \infty$. Since i occurs with k in exactly one triple of S, i

occurs exactly once in the kth row of the square. This is true for every i and every k. Similarly, i occurs exactly once in the k'th column of the square, and again this is true for every i and every k'. Hence, the proof is complete.

In C. D. O'Shaughnessy [1], the author used his method to construct Room designs of order 14 ($v = 13$) and order 20 ($v = 19$). For the interest of the reader, we give the two Steiner triple systems, S_1 and S_2, which generate the first of these designs and also the first row of the design itself (which is cyclic).

S_1 has triples $(1 + i, 4 + i, 5 + i)$ and $(1 + i, 6 + i, 12 + i)$, for $i = 0, 1, 2, \ldots, 12$, all addition being modulo 13. (We write 13 in place of 0.)

S_2 has triples $(1 + i, 2 + i, 5 + i)$ and $(1 + i, 7 + i, 12 + i)$. The Room design has first row:

$(\infty, 1)$, $(7, 9)$, —, $(6, 12)$, —, —, —, $(4, 5)$, $(10, 13)$, $(3, 8)$, —, $(2, 11)$, —.

We end this section with a question. Let us define two Room designs of the same order $2n$ to be *equivalent* if they have the same normalized form. We may ask "How many non-equivalent Room designs of order $2n$ exist?" So far as the authors are aware, the answer is unknown even for the case $n = 4$. (Very recently in C. C. Lindner [14], the author has defined a slightly more general concept than equivalence which he calls *isomorphism* of Room designs.)

Addendum to section 6.4 (added in proof).

Since the above account of Room designs was written (early in 1971) very many new results have been obtained and researches into the subject have been pursued vigorously. For a comprehensive account of the results obtained, the reader should consult part two of a recent book, W. D. Wallis, A. P. Street and J. S. Wallis [1]. Since this book was published, the existence problem for Room designs has been completely settled. It has been shown (see W. D. Wallis [6] and [9]) that these designs exist for every even order $2n$ except when $n = 2$ or 3. (That is to say, there do not exist Room squares of side 3 or 5 but, for every other odd integer $2n - 1$, squares of side $2n - 1$ do exist.)

In J. D. Horton [2], an analogue of the Room design for more than two dimensions has been introduced and investigated and an interesting connection between Room designs of order $2n$ and one-factorizations of the complete graph on $2n$ verticles has been established which Horton attributes to E. Németh.

The following is a list of papers on the subject of Room designs which have not been cited in the above account but which the reader may find it of interest to consult: E. R. Berlekamp and F. K. Hwang [1], R. J. Collens and R. C. Mullin [1], J. D. Horton [1], J. D. Horton, R. C. Mullin and R. G. Stanton [1], R. C. Mullin and E. Németh [5], C. D. O'Shaughnessy [2], E. T. Parker and A. M. Mood [1], R. G. Stanton and R. C. Mullin [1], [2], [3] and [4], W. D. Wallis [2], [3], [4], [5], [7], [8] and W. D. Wallis and R. C. Mullin [1].

CHAPTER 7

Constructions of orthogonal latin squares which involve rearrangement of rows and columns

The many known methods of constructing two or more mutually orthogonal latin squares of an assigned order n can all be put into one of two categories. On the one hand, we have methods which involve obtaining all the squares by rearrangements of the rows or columns of a single one of the set, the square in question being usually referred to as the *basis square*(1); and, on the other hand, we have methods which entail the use of previously determined sets of mutually orthogonal latin squares of smaller order, the squares of these sets being then modified or adjoined one to another in various ways to form squares of the order required. In the present chapter, we shall give an account of all those constructions which can be assigned to the first category, reserving our discussion of the second kind until chapter 11 when we shall have a knowledge of experimental designs at our disposal. We may remark at this point that the construction of R. C. Bose, S. S. Shrikhande and E. T. Parker (given in [1]) by means of which the Euler conjecture was disproved is of the second kind.

7.1. Generalized Bose construction: constructions based on abelian groups

It has been shown in A. D. Keedwell [3] and [4] that all the known constructions of the first category can be regarded as special cases of a generalization of the Bose construction which was described in theorem 5.2.4. We may formulate this generalization as follows:

THEOREM 7.1.1. *Let* $S_0 \equiv I, S_1, S_2, \ldots, S_{r-1}$ *be the permutations representing the rows of an* $r \times r$ *latin square* L_1 *as permutations of its first row*

(1) In the language of quasigroups, each of the squares of the set represents the multiplication table of a quasigroup isotopic to that whose multiplication table is represented by the basis square.

and $M_1 \equiv I, M_2, M_3, \ldots, M_h$, $h \leq r-1$, *be permutations keeping one symbol of* L_1 *fixed. Then the squares* L_i^* *whose rows are represented by the permutations* $M_i S_0, M_i S_1, M_i S_2, \ldots, M_i S_{r-1}$ *for* $i = 1, 2, \ldots, h$ *are certainly all latin and will be mutually orthogonal if, for every choice of* $i, j \leq h$, *the set of permutations*

$$S_0^{-1} M_i^{-1} M_j S_0, S_1^{-1} M_i^{-1} M_j S_1, \ldots, S_{r-1}^{-1} M_i^{-1} M_j S_{r-1}$$

is exactly simply transitive (sharply transitive) *on the symbols of* L_1.

PROOF. Let us remark first that, since each column of L_1 contains each symbol exactly once, the permutations $S_0, S_1, S_2, \ldots, S_{r-1}$ must form a sharply transitive set and that then the set of permutations $M_i S_0, M_i S_1, \ldots, M_i S_{r-1}$ will also be sharply transitive. Consequently, the columns (and, of course, the rows) of L_i will contain each symbol exactly once, so L_i will be latin.

Secondly, if $U_0, U_1, \ldots, U_{r-1}$ are permutations representing the rows of one latin square L_i as permutations of $1, 2, \ldots, n$ and if $V_0, V_1, \ldots, V_{r-1}$ are the similarly defined permutations representing the rows of another latin square L_j, then the permutations $U_0^{-1} V_0, U_1^{-1} V_1, \ldots, U_{r-1}^{-1} V_{r-1}$ map the first, second, . . . , rth rows of L_i respectively onto the first second, . . . , rth rows of L_j. When, and only when, these squares are orthogonal, each symbol of the square L_i must map exactly once onto each symbol of the square L_j since each symbol of L_i occurs in positions corresponding to those of a transversal of L_j. Thus, when and only when L_i and L_j are orthogonal, the permutations $U_0^{-1} V_0, U_1^{-1} V_1, \ldots, U_{r-1}^{-1} V_r^{-1}$ are a sharply transitive set.

The representation of a latin square by means of permutations was introduced originally by E. Schönhardt in [1], but the above two properties seem to have been observed first by H. B. Mann. (See [1].)

The requirement in the above construction that the permutations $M_1, M_2, \ldots, M_h$ be permutations keeping one symbol of L_1 fixed is equivalent to requiring that the mutually orthogonal latin squares $L_1, L_2^*, L_3^*, \ldots, L_h^*$ be standardized in such a way that one column is the same for all the squares and, as we have shown in section 5.2, such a requirement does not lead to any loss of generality. (This fact was first pointed out in H. B. Mann [1].) Notice also that the columns of any square L_i^* will always be a rearrangement of the columns of the basis square L_1 and this rearrangement will be that defined by the corresponding permutation M_i. (M_i reorders the symbols before the permutations $S_0, S_1, \ldots, S_{r-1}$ act.)

Now let us take the special case when the square L_1 is the addition table of an abelian group G. In this case, the S_i are the permutations of the Cayley representation of G (see section 9.1) and the M_i are one-to-one mappings of G onto itself. The entry in the cell of the xth row and yth column of the square L_i^* will be $xM_iS_y = xM_i + y$, where x and y belong to G and G is written in additive notation. If G is the additive group of a Galois field $\mathfrak{J}$ and the M_i effect the multiplications of $\mathfrak{J}$ so that $xM_i = xx_i$ for every x in G, then the construction of theorem 7.1.1 becomes precisely the same as that of R. C. Bose described in theorem 5.2.4.

We shall consider a number of other possibilities.

First we mention two other constructions which are applicable to the case when L_1 is the addition table of an abelian group.

(i) *The construction of D. M. Johnson, A. G. Dulmage and N. S. Mendelsohn.*

If we again take the case when the square L_1 is the addition table of an abelian group G and the S_i are the permutations of the Cayley representation of G, then the square L_i^* will be orthogonal to the square L_1 if the permutations $S_y^{-1}M_i S_y$, where y ranges through G, form a sharply transitive set. That is, if and only if

$$wS_y^{-1}M_i S_y = wS_z^{-1}M_i S_z$$

implies $y = z$ for any w in G. That is, if and only if

$$(w - y)M_i + y = (w - z)M_i + z$$

implies $y = z$. Subtracting w from each side and writing $w - y = u$, $w - z = v$, we have that L_i^* will be orthogonal to L_1 if and only if $uM_i - u = vM_i - v$ implies $u = v$.

A mapping M_i of the abelian group G onto itself which has the latter property has been called an *orthomorphism* by D. M. Johnson, A. L. Dulmage, and N. S. Mendelsohn in [2]. It will be clear to the reader that the concept of an orthomorphism is effectively the same as that of a complete mapping, which was defined in section 1.4.

Moreover, repetition of the argument leads at once to the fact that squares L_i^* and L_j^* will be orthogonal if $M_i^{-1}M_j$ is also an orthomorphism, as the above authors have shown. They have pointed out further that a one-to-one correspondence between orthomorphisms of G and transversals of the latin square representing the Cayley table of G can be established (compare section 1.4).

The entries in the cells $(x_1, y_1), (x_2, y_2), \ldots, (x_r, y_r)$, where (x_k, y_k) denotes the cell of the x_kth row and y_kth column, will form a transversal if and only if the mapping M_i defined by $x_k M_i = -y_k$ for $k = 1, 2, \ldots, r-1$ is an orthomorphism of G. For suppose firstly that we define $-y_k = x_k M_i$ for each k so that the entry in the (x_k, y_k)th cell is $x_k - x_k M_i$. Then these entries will be all distinct and form a transversal if and only if $x_h - x_h M_i = x_k - x_k M_i$ implies $x_h = x_k$; that is, if and only if M_i is an orthomorphism.

In their paper [2] already referred to above, D. M. Johnson and her co-authors have devised an algorithm for constructing orthomorphisms which is suitable for a computer search and with its aid they have found a set of four non-identity orthomorphisms of the group $C_6 \times C_2$ of order 12 suitable for the construction of five mutually orthogonal latin squares of that order. They have thus established that $N(12) \geq 5$, a result which has not been bettered up to the present. Further details will be given in chapter 13.

(ii) The construction of R. C. Bose, I. M. Chakravarti and D. E. Knuth.

The necessary and sufficient condition

$$uM_i^{-1}M_j - u = vM_i^{-1}M_j - v \Rightarrow u = v$$

that the squares L_i^* and L_j^* defined above be orthogonal may be re-written in the form

$$wM_j - wM_i = xM_j - xM_i \quad \Rightarrow \quad w = x,$$

where $w = uM_i^{-1}$ and $x = vM_i^{-1}$. In other words, the squares L_i^* and L_j^* will be orthogonal if and only if the equation $xM_j - xM_i = t$ is uniquely soluble for x.

In [1], R. C. Bose, I. M. Chakravarti, and D. E. Knuth have shown how mappings M_i having this property may be computed for abelian groups G of order $4t$ (with $4t - 1$ a prime power) and have thus obtained further sets of five mutually orthogonal latin squares of order 12. These authors have called such mappings M_i *orthogonal mappings*.

Further details of the computation will be given in chapter 13.

7.2. The automorphism method of H. B. Mann

The latin squares $L_1, L_2^*, L_3^*, \ldots, L_h^*$ of the construction described in theorem 7.1.1 can be modified by the definition $L_i = L_i^* M_i^{-1}$ for $i = 2, 3, \ldots, h$. That is to say, the xth row of the latin square L_i will be represented by the permutation $M_i S_x M_i^{-1}$. This permutation, being conjugate to the permutation S_x, is very easy to calculate when the permutation S_x is known. We note also that the squares $L_1, L_2, \ldots, L_h$ will be mutually orthogonal whenever the squares $L_1, L_2^*, \ldots, L_h^*$ are so, and that each of the squares $L_1, L_2, \ldots, L_h$ has the identity permutation as first row. Thus, these squares[1] are a *standardized set* as defined in section 5.2.

This modified form of the construction described in theorem 7.1.1 we shall call the *K-construction* and we shall refer to it several times in the present chapter.

Let the square L_1 be the addition table of a group (written in additive notation, but not necessarily abelian) and let the mappings M_i^{-1}, $i = 1, 2, \ldots, h$, represent automorphisms τ_i of G. Let the elements of G be denoted by $a, b, c, \ldots$ Then the rows of the square L_1 are represented by the permutations $S_0 \equiv I, S_a, S_b, S_c$, and so on. The sth row of the square L_i is represented by the permutation

$$M_i S_s M_i^{-1} = \begin{pmatrix} a\tau_i & b\tau_i & \ldots \\ a & b & \ldots \end{pmatrix}\begin{pmatrix} a & b & \ldots \\ a+s & b+s & \ldots \end{pmatrix}\begin{pmatrix} a & \ldots\, a+s & \ldots \\ a\tau_i & \ldots (a+s)\tau_i & \ldots \end{pmatrix} =$$

$$= \begin{pmatrix} \ldots & a\tau_i & \ldots \\ \ldots & (a+s)\tau_i & \ldots \end{pmatrix} = \begin{pmatrix} \ldots & t & \ldots \\ \ldots & t+s\tau_i & \ldots \end{pmatrix} = S_{s\tau_i},$$

since τ_i is an automorphism of G. The squares L_i and L_j will be orthogonal if I, $S_{a\tau_i}^{-1} S_{a\tau_j}$, $S_{b\tau_i}^{-1} S_{b\tau_j}$, $\ldots$, is a sharply transitive set of permutations. Since G is a group, and τ_i is an automorphism,

$$S_{a\tau_i}^{-1} S_{a\tau_j} = S_{-(a\tau_i)} S_{a\tau_j} = S_{-a\tau_i + a\tau_j}.$$

Thus, the squares will be orthogonal provided that $-s\tau_i + s\tau_j \neq -t\tau_i + t\tau_j$ for distinct elements s and t of G. That is, provided that $t\tau_i - s\tau_i \neq t\tau_j - s\tau_j$.

On writing $t - s = u$, we have that the squares L_i and L_j will be orthogonal provided that the automorphisms τ_i and τ_j have the property $u\tau_i \neq u\tau_j$ for any element u other than the identity in G. Hence we may state:

[1] The squares $L_1, L_2^*, \ldots, L_h^*$ also form a standardized set, all of them having he same first column but differing first rows.

Theorem 7.2.1. *Let G be a group and suppose that there exist h automorphisms $\tau_1, \tau_2, \ldots, \tau_h$ of G every pair of which possesses the property that $u\tau_i \neq u\tau_j$ for any element $u \in G$ except the identity element, then we shall be able to construct h mutually orthogonal latin squares based on the group G.*

Theorem 7.2.1 was first proved by H. B. Mann in [1] and, in the same paper, the author obtained an upper bound for h in terms of the number of conjugacy classes of G, see theorem 11.1.3.

7.3. The construction of pairs of orthogonal latin squares of order 10

E. T. Parker's original construction (given in [2]) of a pair of orthogonal latin squares of order 10 involved the use of orthogonal latin squares of order three. However, A. I. Ljamzin (see [1]) and L. Weisner (see [1]) subsequently produced a pair in which the rows of one square are a rearrangement of the rows of the other. Although the two authors worked independently, the pairs of squares which they obtained are equivalent. Unfortunately, neither author has given a proper account of the means by which his squares were obtained.

Using the K-construction described above, one of the authors has more recently tried to extend the Ljamzin-Weisner squares to a set of three which are mutually orthogonal, but this attempt has been unsuccessful. (For the details, see A. D. Keedwell [4].)

It is appropriate to point out at this point that, although the squares of the generalized set $L_1, L_2^*, \ldots, L_h^*$ have the property that the columns of any square L_i^* are a rearrangement of the columns of the square L_1, this is no longer necessarily true of either the rows or the columns of the set $L_1, L_2, \ldots, L_h$. We have the following theorem.

Theorem 7.3.1. *The necessary and sufficien condition that the squares $L_1, L_2, \ldots, L_h$ of the K-construction have the property that the rows of any one square L_i are the same as those of any other square L_j of the set, except that they occur in a different order, is that the operation $(\cdot)$ defined by the relation $aM_x = ax$ for each of $M_1, M_2, \ldots, M_h$, be right distributive over the operation $(+)$ defined by $aS_x = a + x$.*

Proof. It is necessary and sufficient to show that the permutations representing the rows of each square L_k are a reordering of the permutations representing the rows of the square L_1. That is, it is necessary and sufficient to have $M_k S_p M_k^{-1} = S_q$ for some q, or $M_k^{-1} S_q M_k = S_p$. Since $S_0 \equiv I$ and $M_1 \equiv I$, 0 and 1 are respective identities for $(+)$ and $(\cdot)$. Thus

$$M_k^{-1} S_q M_k = \begin{pmatrix} 0 \dots k \dots xk \dots \\ 0 \dots 1 \dots x \dots \end{pmatrix} \begin{pmatrix} 0 \dots & x & \dots \\ q \dots & x+q & \dots \end{pmatrix} \times$$

$$\times \begin{pmatrix} 0 & 1 \dots q & \dots & x+q & \dots \\ 0 & k \dots qk & \dots & (x+q)k & \dots \end{pmatrix} = \begin{pmatrix} 0 \dots & xk & \dots \\ qk \dots & (x+q)\,k & \dots \end{pmatrix}.$$

Therefore, $M_k^{-1} S_q M_k = S_p$ if and only if $p = qk$, and then

$$S_p = \begin{pmatrix} 0 \dots & xk & \dots \\ qk \dots & xk+qk & \dots \end{pmatrix}; \text{ so } (x+q)\,k = xk + qk$$

for all x (and evidently also for all q and k); $x, q = 1, 2, \dots, r-1$; $k = 1, 2, \dots, h$.

Corollary. *When the conditions of the theorem are fulfilled, the permutation M_k^{-1} represents the rearrangement of the rows of L_1 which is required to turn it into the square L_k.*

Proof. Suppose that the pth row of the square L_k is the same as the qth row of the square L_1. Then $M_k S_p M_k^{-1} = S_q$ and so $p = qk$. That is, M_k maps q into p. Thus the mapping $M_k^{-1} = \begin{pmatrix} 0 \dots p \dots \\ 0 \dots q \dots \end{pmatrix}$ represents replacement of the pth row of L_1 by its qth row; that is, it rearranges the rows of L_1 in such a way that they become the rows of L_k.

For the squares of A. I. Ljamzin mentioned above, the row permutations are as follows:

$$L_1 = \{S_0, S_1, S_2, \dots, S_8, S_9\},$$
$$L_2 = \{S_0, S_2, S_3, \dots, S_9, S_1\},$$

where

$$S_0 = I,$$
$$S_1 = (0\ 1)\,(2\ 5)\,(3\ 8\ 6\ 7\ 9\ 4),$$
$$S_2 = (0\ 2)\,(3\ 6)\,(4\ 9\ 7\ 8\ 1\ 5),$$
$$S_3 = (0\ 3)\,(4\ 7)\,(5\ 1\ 8\ 9\ 2\ 6),$$
$$S_4 = (0\ 4)\,(5\ 8)\,(6\ 2\ 9\ 1\ 3\ 7),$$
$$S_5 = (0\ 5)\,(6\ 9)\,(7\ 3\ 1\ 2\ 4\ 8),$$
$$S_6 = (0\ 6)\,(7\ 1)\,(8\ 4\ 2\ 3\ 5\ 9),$$
$$S_7 = (0\ 7)\,(8\ 2)\,(9\ 5\ 3\ 4\ 6\ 1),$$
$$S_8 = (0\ 8)\,(9\ 3)\,(1\ 6\ 4\ 5\ 7\ 2),$$
$$S_9 = (0\ 9)\,(1\ 4)\,(2\ 7\ 5\ 6\ 8\ 3).$$

The mapping $M_2 = M_x$ is the permutation $(0)(9\ 8\ 7\ 6\ 5\ 4\ 3\ 2\ 1)$, and we have, for example, $(7+5)x = 3x = 2 = 6 + 4 = 7x + 5x$. That is, $7S_5M_x = 7M_xS_{5M_x}$. So, the right-distributive law holds. Moreover, the rows of the square L_2 are obtained from the rows of the square L_1 by carrying out the permutation M_2^{-1} on those rows.

For the squares constructed by the automorphism method the row permutations are as follows:

$$L_1 = \{S_a, S_b, S_c, \ldots\}; \qquad L_i = \{S_{a\tau_i}, S_{b\tau_i}, S_{c\tau_i}, \ldots\}$$

for $i = 2, 3, \ldots, h$; and the mapping M_i is such that $xM_i = x\tau_i^{-1}$. Since τ_i is an automorphism, it is clear that the right-distributive law

$$(x+y)\tau_i^{-1} = x\tau_i^{-1} + y\tau_i^{-1}$$

holds; and, moreover, the xth row of the square L_i is the $x\tau_i$th row of the square L_1, so the permutation M_i^{-1} rearranges the rows of L_1 in such a way that they become the rows of L_i.

A further illustration of the theorem is provided, for example, by the complete sets of mutually orthogonal latin squares which correspond to the Veblen–Wedderburn–Hall translation planes. (See section 8.4.)

We notice finally that, when the square L_1 is the addition table of a group G and the conditions of theorem 7.3.1 are satisfied, each permutation M_x defines an automorphism of G: for the validity of the right-distributive law $(a + b)x = ax + bx$ implies that the mapping $a \to ax$ is an automorphism of G.

7.4. The column method

This method is another specialization of the K-construction in which once again the latin square L_1 is taken to be the multiplication table of a (not-necessarily abelian) group. The method has been used by its author to construct some triads of mutually orthogonal latin squares of order 15 (see A. D. Keedwell [4]). Hence, $N(15) \geq 3$, a result which is an improvement on that obtained by R. C. Bose, S. S. Shrikhande, and E. T. Parker in [1], see section 11.4. (Recently, the result $N(15) \geq 3$ has been obtained again by A. Hedayat in [1].)

The validity of the method is a consequence of the following theorem.

THEOREM 7.4.1. *Let G be a finite group of order r whose elements are denoted by $e, a_1, a_2, \ldots, a_{r-1}$ and suppose that these elements, excluding* *the identity element e of G, can be arranged in a row array* $b_1, b_2, \ldots, b_{2r-2}$ *in such a way that* $b_1, b_3, \ldots, b_{2r-3}$ *are all different (implying that each non-identity element of G occurs once in an "odd" position), that* $b_2, b_4, \ldots, b_{2r-2}$ *are all different (implying that each non-identity element of G occurs once in an "even" position), and that* $b_{2p-1}\, b_{2p} = b_{2p+1}$ *for* $p = 1, 2, \ldots, r-1$, *where* $b_{2r-1} \equiv b_1$. *Then there exist at least two orthogonal latin squares of order r.*

If the products $b_2 b_4, b_4 b_6, \ldots, b_{2r-2}\, b_2$ *are also all different, there exist at least three mutually orthogonal latin squares of order r; and, more generally, if the sets of products of adjacent k-tuples of elements of the set* $b_2, b_4, \ldots, b_{2r-2}$ *are all different for* $k = 2, 3, \ldots, h-1$, *then there exist at least h mutually orthogonal latin squares of order r.*

Proof. The proof of the first part consists in showing that, when the conditions of the theorem are satisfied, the permutations

$$\begin{array}{lcl} M_x & = & (e)\quad (b_1\quad b_3\quad b_5\quad \cdots\quad b_{2r-1}) \\ S_1^{-1} M_x S_1 & = & (a_1)\quad (a_1b_1\quad a_1b_3\quad a_1b_5\quad \cdots\quad a_1b_{2r-1}) \\ \cdots & & \cdots \\ \cdots & & \cdots \\ S_{r-1}^{-1} M_x S_{r-1} & = & (a_{r-1})\,(a_{r-1}\,b_1\quad a_{r-1}\,b_3\quad a_{r-1}\,b_5\quad \cdots\quad a_{r-1}\,b_{2r-1}) \end{array}$$

are a sharply transitive set and form a latin square by means of which the permutations S_i may be defined. The latter fact is obvious since the given display represents a Cayley table for G.

To see that these permutations are a sharply transitive set, we have only to show that each element of G occurs as immediate successor to any other element of G exactly once. This is ensured if

$$a_s\, b_{2t-1} = a_u\, b_{2v-1} \Rightarrow a_s\, b_{2t+1} \neq a_u\, b_{2v+1}$$

where $a_0 \equiv e$, $b_{2r-1} \equiv b_1$) since each element of G occurs exactly once in each row. That is, if

$$a_s\, b_{2t-1}\, b_{2t} = a_u\, b_{2v-1}\, b_{2t} \Rightarrow a_s\, b_{2t+1} \neq a_u\, b_{2v+1}:$$

namely, if

$$a_s\, b_{2t+1} = a_u\, b_{2v-1}\, b_{2t} \Rightarrow a_s\, b_{2t+1} \neq a_u\, b_{2v-1}\, b_{2v}.$$

Since, by the hypotheses of the theorem, $b_{2t} \neq b_{2v}$, this is certainly true. The squares L_1, L_2 whose rows are defined by the permutations

$$L_1 = \{I, S_1, S_2, \ldots, S_{r-1}\},$$
$$L_2 = \{I, M_x S_1 M_x^{-1}, M_x S_2 M_x^{-1}, \ldots, M_x S_{r-1} M_x^{-1}\}$$

will then be orthogonal, as has already been proved above.

For the second part of the theorem, we observe that the conditions stated are necessary and sufficient to ensure that the sets of permutations

$$S_0^{-1} M_x^{-m} M_x^n S_0, S_1^{-1} M_x^{-m} M_x^n S_1, \ldots, S_{r-1}^{-1} M_x^{-m} M_x^n S_{r-1}$$

be sharply transitive for $m, n = 0, 1, \ldots, h-1$, and $m < n$. Consequently, when the conditions are satisfied, the latin squares

$$L_i = \{I, M_x^i S_1 M_x^{-i}, M_x^i S_2 M_x^{-i}, \ldots, M_x^i S_{r-1} M_x^{-i}\}$$

for $i = 0, 1, 2, \ldots, h-1$, will be mutually orthogonal.

We note that the latin square $L_1 = \{I, S_1, S_2, \ldots, S_{r-1}\}$ of theorem 7.4.1 is the multiplication table of the group G. Also, it follows from theorems 1.4.1 and 1.4.2 that a necessary and sufficient condition that there exists a latin square orthogonal to that formed by the multiplication table of a group G is that G possesses a "complete mapping". L. J. Paige has shown further that a necessary condition that G possesses a complete mapping is that the product of its elements in some order be equal to the identity (see theorem 1.4.5). Hence we conclude that a necessary condition that the requirements of theorem 7.4.1 be satisfied in a given group G is that the product of the elements of G in some order be equal to the identity element of G. However, this result can be deduced directly from theorem 7.4.1. With the notation of that theorem, we have $b_{2p-1} b_{2p} = b_{2p+1}$ for $p = 1, 2, \ldots, r-1$. It follows that $b_1 b_2 b_4 b_6 \ldots b_{2r-2} = b_{2r-1} = b_1$, and so $b_2 b_4 b_6 \ldots b_{2r-2} = e$. Since $b_2, b_4, b_6, \ldots, b_{2r-2}$ are all distinct, they are the elements of G in some order.

By theorem 1.4.7 another necessary condition that there exists a latin square orthogonal to that formed by the multiplication table of a group G is that the Sylow 2-subgroups of G be non-cyclic. It follows at once that, if G is a group of order $2p$, where p is an odd prime, the column method for the construction of orthogonal latin squares must fail. In fact it turns out that the smallest value of r for which the method leads to results of serious interest is the value $r = 15$. For this value of r, the method allows the construction of triads of mutually orthogonal squares, as mentioned above.

We append one example of an array $b_1, b_2, \ldots, b_{28}$ suitable for this purpose, where each b_j is an element of the cyclic group $G = \{x/x^{15} = e\}$ (which is the only group of order 15 that exists). Also, as an illustration of the applicability of theorem 7.4.1 to non-abelian groups G, we give a sharply transitive set of permutations derived from the dihedral group D_4 of order 8, by means of which a pair of orthogonal latin squares of order 8 can be constructed.

For $G = \{x/x^{15} = e\}$, the array is as follows, where numbers represent powers of x:

$${}^{1}{}_{1}{}^{2}{}_{4}{}^{6}{}_{12}{}^{3}{}_{11}{}^{14}{}_{10}{}^{9}{}_{3}{}^{12}{}_{14}{}^{11}{}_{8}{}^{4}{}_{6}{}^{10}{}_{13}{}^{8}{}_{5}{}^{13}{}_{7}{}^{5}{}_{2}{}^{7}{}_{9}$$

For $G = D_4 = \{a, b/a^4 = e, b^2 = e, ab = ba^3\}$, the permutations are as follows:

$$\begin{array}{lcl}
M_x & = & (e)\,(ba^2{}_{a^2}\;{}^{b}{}_{a}\;{}^{ba}{}_{b}\;{}^{a^3}{}_{ba^2}\;{}^{ba^3}{}_{ba}\;{}^{a^2}{}_{a^3}\;{}^{a}{}_{ba^3}) \\
S_1^{-1} M_x S_1 & = & (a)\,(ba \quad ba^3 \quad b \quad e \quad ba^2 \quad a^3 \quad a^2\,) \\
S_2^{-1} M_x S_2 & = & (a^2)\,(b \quad ba^2 \quad ba^3 \quad a \quad ba \quad e \quad a^3\,) \\
S_3^{-1} M_x S_3 & = & (a^3)\,(ba^3 \quad ba \quad ba^2 \quad a^2 \quad b \quad a \quad e\,) \\
S_4^{-1} M_x S_4 & = & (b)\,(a^2 \quad e \quad a \quad ba^3 \quad a^3 \quad ba^2 \quad ba\,) \\
S_5^{-1} M_x S_5 & = & (ba)\,(a \quad a^3 \quad e \quad b \quad a^2 \quad ba^3 \quad ba^2\,) \\
S_6^{-1} M_x S_6 & = & (ba^2)\,(e \quad a^2 \quad a^3 \quad ba \quad a \quad b \quad ba^3\,) \\
S_7^{-1} M_x S_7 & = & (ba^3)\,(a^3 \quad a \quad a^2 \quad ba^2 \quad e \quad ba \quad b\,)
\end{array}$$

7.5. The diagonal method

For this construction it is not necessary that the latin square L_1 be the multiplication table of a group and, mainly for this reason, the method is effective in obtaining either two or a complete set of mutually orthogonal latin squares (according as r is not or is a prime power) for every order r except one, two and six up to at least the value $r = 20$. It is therefore of considerable interest to see why the method fails when $r = 6$ so as to obtain some explanation for the peculiarity of that integer in regard to the theory.

In order to motivate the construction, let us look at the set-up given by the K-construction in the case when we have a set of mutually orthogonal latin squares based on a Galois field GF $[r]$. In that case, the multiplications are effected by the elements $1, x, x^2, \ldots,$ x^{r-2} of a cyclic group of order $r-1$ and the corresponding multiplication permutations are

$$M_1 \equiv I, \; M_x = (0)(1 \; x \; x^2 \; \ldots \; x^{r-2}), \; M_x^2, \; M_x^3, \ldots, M_x^{r-2}.$$

The addition permutations are

$$S_0 \equiv I, S_1, S_x, \ldots, S_{x^{r-2}}$$

and, because the latin squares L_1 and L_x are orthogonal, the permutations

$$\begin{array}{llllll}
M_x = & (0)\,(& 1 & x & \ldots & x^{r-2} \quad) \\
S_{x^{r-2}}^{-1} M_x S_{x^{r-2}} = & (x^{r-2})\,(& 1+x^{r-2} & x+x^{r-2} & \ldots & x^{r-2}+x^{r-2}) \\
S_{x^{r-3}}^{-1} M_x S_{x^{r-3}} = & (x^{r-3})\,(& 1+x^{r-3} & x+x^{r-3} & \ldots & x^{r-2}+x^{r-3}) \\
\cdots & \cdots & \cdots & \cdots & \cdots & \cdots \\
\cdots & \cdots & \cdots & \cdots & \cdots & \cdots \\
S_x^{-1} M_x S_x = & (x)\,(& 1+x & x+x & \ldots & x^{r-2}+x \quad) \\
S_1^{-1} M_x S_1 = & (1)\,(& 1+1 & x+1 & \ldots & x^{r-2}+1 \quad)
\end{array}$$

are a sharply transitive set.

We note that, if the first row and column are disregarded, the quotients of the elements in corresponding places of any two adjacent secondary[(1)] diagonals are constant. Moreover, in each such diagonal, the element of the pth row and qth column is x times the element of the $(p+1)$th row and $(q-1)$th column. In consequence of this fact and the Galois field relationship between addition and multiplication, it is clear that each of the elements $1, x, x^2, \ldots, x^{r-2}$ occurs exactly once in each secondary diagonal. These properties will be exploited in the method of construction which we are about to explain.

We shall suppose that all elements except 0 are expressed as powers of x and, in that case, we may disregard x and write indices only. We shall also write $r-1$ in place of the element 0. Let the indices (natural numbers) $0, 1, 2, \ldots, r-3$, be ordered in such a way that the dif-

(1) By a *secondary diagonal* of an $m \times m$ matrix $A = \| a_{rs} \|$, where $r, s = 0, 1, \ldots, (m-1)$, is meant a set of elements

$$a_{0p}, a_{1p-1}, \ldots, a_{p0}, a_{p+1\,m-1}\; a_{p+2\,m-2}, \ldots, a_{m-1\,p+1}.$$

The secondary diagonal given by $p = m-1$ will be called the *main secondary diagonal*.

ferences between adjacent numbers are all different, taken modulo $(r - 1)$, and so that no difference is equal to 1. As an example, take the case $r = 8$. Then a solution is 4 0 2 1 5 3, the differences being 3, 2, 6, 4, 5. We set up an array whose main secondary diagonal consists entirely of 7's and such that all other secondary diagonals consist of the indices 0, 1, 2, . . . , 6 in descending order, columns being taken cyclically, column 0 = column 7, and so on. The result is shown in Fig. 7.5.1.

4	0	2	1	5	3	7
6	1	0	4	2	7	3
0	6	3	1	7	2	5
5	2	0	7	1	4	6
1	6	7	0	3	5	4
5	7	6	2	4	3	0
7	5	1	3	2	6	4

Fig. 7.5.1

It is clear from the method of construction that, if the entries of the first column of this array are all different, so are the entries of each other column. We seek arrays A_8^* such that this is the case. Fig. 7.5.2, with first row and column deleted, provides an example of such an array. In fact, Fig. 7.5.2 is derived from the array A_8^* by bordering the array with a 0th row and 0th column whose entries are those missing from the appropriate column or row of the array A_8^*. Thus, the complete square shown in Fig. 7.5.2 is a latin square.

$$
\begin{aligned}
M_x \quad &= (7)(0\ 1\ 2\ 3\ 4\ 5\ 6) \\
S_{x^6}^{-1} M_x S_{x^6} &= (6)(2\ 5\ 0\ 4\ 3\ 1\ 7) \\
S_{x^5}^{-1} M_x S_{x^5} &= (5)(4\ 6\ 3\ 2\ 0\ 7\ 1) \\
S_{x^4}^{-1} M_x S_{x^4} &= (4)(5\ 2\ 1\ 6\ 7\ 0\ 3) \\
S_{x^3}^{-1} M_x S_{x^3} &= (3)(1\ 0\ 5\ 7\ 6\ 2\ 4) \\
S_{x^2}^{-1} M_x S_{x^2} &= (2)(6\ 4\ 7\ 5\ 1\ 3\ 0) \\
S_x^{-1} M_x S_x &= (1)(3\ 7\ 4\ 0\ 2\ 6\ 5) \\
S_1^{-1} M_x S_1 &= (0)(7\ 3\ 6\ 1\ 5\ 4\ 2)
\end{aligned}
$$

Fig. 7.5.2

Since the differences between adjacent entries of the first row of the array A_8^* are all different, the same is true of each other row provided that each such row is regarded as starting and ending at the entry 7. Moreover, the disposition of the 7's is such that every other integer follows and precedes an entry 7 exactly once. Thus, Fig. 7.5.2 provides a sharply transitive set of permutations from which we can derive permutations

$$S_0 = I, S_1 = S_{x^0}, S_{x^1}, S_{x^2}, \ldots, S_{x^6}$$

representing the rows of a latin square L_1 and with the property that the latin square L_x derived from L_1 by means of the multiplication permutation M_x will be orthogonal to L_1.

In fact, the sharply transitive set of permutations shown in Fig. 7.5.2 arises from the Galois field GF[8]: for, in this field, $x^8 - x = 0$ and a primitive root x satisfies $x^3 + x + 1 = 0$ (or $x^3 = x + 1$, since $-1 \equiv 1$ in GF[8]). So, if $M_x = (0)(1\ x\ x^2\ x^3\ x^4\ x^5\ x^6)$, we have

$$S_{x^6}^{-1} M_x S_{x^6} = (x^6)(1 + x^6\quad x + x^6\ \ldots\ x^5 + x^6\quad x^6 + x^6) =$$
$$= (x^6)(x^2\ x^5\ 1\ x^4\ x^3\ x\ 0),$$

which is equivalent to the expression given in Fig. 7.5.2.

However, arrays of type A_r^* exist when r is not a prime power. For example, when $r = 10$, we get just two possible arrays A_{10}^*. One of these is shown (bordered by the appropriate 0th row and 0th column) in Fig. 7.5.3 and leads to one of the pairs of orthogonal latin squares of order 10 obtained by L. Weisner and displayed in Fig. 2 of his paper [1].

(9)(0 1 2 3 4 5 6 7 8)
(8)(5 3 0 2 7 6 1 4 9)
(7)(2 8 1 6 5 0 3 9 4)
(6)(7 0 5 4 8 2 9 3 1)
(5)(8 4 3 7 1 9 2 0 6)
(4)(3 2 6 0 9 1 8 5 7)
(3)(1 5 8 9 0 7 4 6 2)
(2)(4 7 9 8 6 3 5 1 0)
(1)(6 9 7 5 2 4 0 8 3)
(0)(9 6 4 1 3 8 7 2 5)

Fig. 7.5.3

The discussion above may be summarized into the following theorem:

THEOREM 7.5.1. *If r is an integer for which an array A_r^* exists, then there exist at least two mutually orthogonal latin squares of order r. When r is a power of a prime, there exists at least one array A_r^* which can be used to generate a complete set of mutually orthogonal latin squares of order r, representing the desarguesian projective plane of that order.*

Following A. D. Keedwell [4], we shall now give two simple criteria for the existence of an array A_r^* corresponding to a given integer r.

THEOREM 7.5.2. *A necessary and sufficient condition that an array A_r^* exists for a given integer r is that the residues $2, 3, \ldots, (r-2)$, modulo $(r-1)$, can be arranged in a row array P_r in such a way that the partial sums of the first one, two, $\ldots, (r-3)$, are all distinct and non-zero modulo $(r-1)$ and so that, in addition, when each element of the array is reduced by 1, the new array P_r' has the same property.*

PROOF. Suppose firstly that an array A_r^*, such as is given in Fig. 7.5.2, exists. The differences between successive entries of the first row of A_r^*, excluding the last element $(r-1)$, form an array P_r of the type specified in the theorem, since, if this were not the case, the entries of that first row would not be all distinct.

Moreover, if we write

$$A_r^* = \begin{array}{|llll} a_{11} & a_{12} & a_{13} & \cdots \\ a_{21} & a_{22} & a_{23} & \cdots \\ a_{31} & a_{32} & a_{33} & \cdots \\ \cdots & \cdots & \cdots & \cdots, \end{array}$$

we have $a_{ij} = a_{i-1 j+1} - 1$. Consequently,

$$a_{21} - a_{11} = (a_{12} - 1) - a_{11} = (a_{12} - a_{11}) - 1,$$

$$a_{31} - a_{21} = (a_{22} - 1) - a_{21} = (a_{22} - a_{21}) - 1 = (a_{13} - a_{12}) - 1,$$

and generally,

$$a_{i+11} - a_{i1} = (a_{i2} - 1) - a_{i1} = (a_{i2} - a_{i1}) - 1 = (a_{1i+1} - a_{1i}) - 1,$$

so that the differences between successive entries of the first column of A_r^*, excluding the last element $(r-1)$, form an array of the type specified in the theorem, with each element reduced by one from the corresponding element of P_r: for, if this were not the case, the entries of the first column of A_r^* would not be all distinct.

Conversely, suppose that the residues $2, 3, \ldots, (r-2)$, modulo $(r-1)$, can be arranged in the manner described in the theorem. Then an array A_r^* exists. We shall find it easiest to illustrate this by means of an example. We take the case $r = 8$.

$$P_8 = 3,\ 2,\ 4,\ 6,\ 5; \quad P_8' = 2,\ 1,\ 3,\ 5,\ 4.$$

Here, $3 = 3$, $3 + 2 = 5$, $3 + 2 + 4 = 2$, $3 + 2 + 4 + 6 = 1$, $3 + 2 + 4 + 6 + 5 = 6$, so the entries of the first row of A_8^* are

$$x,\ x+3,\ x+5,\ x+2,\ x+1,\ x+6,\ r-1=7.$$

Since the entry $r - 2 = 6$ is not to appear in the first row, we must have $x + 4 = 6$, that is $x = 2$. Then A_8^* is as shown in Fig. 7.5.2, the entries in the first row and column being all distinct in virtue of the properties of the row array P_8.

Corollary 1. *With each array A_r^* occurs a dual array $A_r^{(d)}$, obtained from A_r^* by replacing each entry q in the corresponding row arrays P_r, P_r', by its complement $(r-1)-q$, taken modulo $(r-1)$, to obtain a dual row array $P_r^{(d)}$ and hence a dual array $A_r^{(d)}$.*

Proof. If P_r and P_r' are transformed in the manner specified, the entries of the transform of P_r' become one greater than the corresponding entries of the transform of P_r. It is easy to see, therefore, that the transform of P_r' must have the same properties as P_r.

For example, if P_8 is as given in the theorem above, we have

$$P_8^{(d)} = 5,\ 6,\ 4,\ 2,\ 3; \quad P_8'^{(d)} = 4,\ 5,\ 3,\ 1,\ 2.$$

Corollary 2. *With each array A_r^* occurs a mirror image array $A_r^{(m)}$ (which may coincide with $A_r^{(d)}$), obtained from A_r^* by first constructing P_r, reversing the order of its entries to obtain a row array $P_r^{(m)}$, and then constructing the corresponding array $A_r^{(m)}$ in the manner described in the theorem.*

Proof. We have only to show that the row array obtained by reversing the order of the entries of the row array P_r has the same properties as P_r. Let

$$P_r = d_1, d_2, \ldots, d_{r-3}; \quad P_r' = d_1 - 1,\ d_2 - 1,\ \ldots,\ d_{r-3} - 1.$$

The property possessed by P_r is that

$$e_1 = d_1, \quad e_2 = d_1 + d_2, \quad \ldots, \quad e_{r-3} = \sum_{i=1}^{r-3} d_i$$

are all distinct and non-zero modulo $(r-1)$. Writing

$$P_r^{(m)} = d_{r-3}, d_{r-4}, \ldots, d_2, d_1,$$

we have

$$d_{r-3} = e_{r-3} - e_{r-4}, \; d_{r-3} + d_{r-4} = e_{r-3} - e_{r-5}, \ldots,$$

$$\sum_{i=r-3}^{i=2} d_i = e_{r-3} - e_1, \quad \sum_{i=r-3}^{i=1} d_i = e_{r-3} - 0,$$

and these are evidently all distinct and non-zero, since $e_{r-3}, e_{r-4}, \ldots, e_2, e_1, 0$ are all distinct, and $e_{r-3} \neq 0$.

Consequently, $P_r^{(m)}$ has the same property as P_r. The same argument applied to P'_r shows that $P_r'^{(m)}$ has the same property. Thus, $P_r^{(m)}$ has all the same properties as P_r, as required.

For our second criterion for the existence of an array A_r^* of the type described above, we shall need the concept of a *neofield*.

Definition. A set $\mathrm{J} = \{a, b, c, \ldots\}$ on which are defined two binary operations $(+)$ and $(\cdot)$ such that J is a loop with respect to the operation $(+)$ with identity element 0 say, $\mathrm{J} \setminus 0$ is a group with respect to the operation $(\cdot)$ and the distributive laws $a(b+c) = ab + ac$ and $(b+c)a = ba + ca$ hold, is called a *neofield*. The neofield is *commutative* if the loop $(\mathrm{J}, +)$ is commutative.

Neofields were first introduced by L. J. Paige in his paper [2], and he derived a number of their principal properties in that paper. They had not subsequently been used until the following theorem was proved in A. D. Keedwell [4]:

Theorem 7.5.3. *A necessary and sufficient condition that an array A_r^* exists for a given integer r is that there exists a neofield of r elements whose multiplicative group is cyclic of order $(r-1)$ and which possesses the property that*

$$(1 + x^t)/(1 + x^{t-1}) = (1 + x^u)/(1 + x^{u-1})$$

implies $t = u$, where x is any generating element of the multiplicative group.

Proof. Suppose first that such a neofield of order r exists. There exists a unique element x^s of the additive loop such that[1] $1 + x^s = 0$ Then $x^t + x^{s+t} = 0$ for all integers t. The addition table of the neofield will consequently be of the form shown in Fig. 7.5.4. Here, if the 0th row and 0th column be disregarded, the main secondary diagonal consists entirely of zeros. The remaining secondary diagonals comprise elements which can be represented as powers of x in descending natural order.

Moreover, the differences between the indices of adjacent elements of the first row of the addition table are all distinct and non-zero modulo $(r - 1)$: for

$$(x^{r-2} + x^{s+p})/(x^{r-2} + x^{s+p-1}) = (x^{r-2} + x^{s+q})/(x^{r-2} + x^{s+q-1})$$

would imply

$$(1 + x^{s+p+1})/(1 + x^{s+p}) = (1 + x^{s+q+1})/(1 + x^{s+q})$$

with $p \neq q$, contrary to hypothesis[2]. Likewise, the differences between the indices of adjacent elements of the first column of the addition table are all distinct and non-zero modulo $(r - 1)$: for

$$(x^{r-p} + x^{s})/(x^{r-p+1} + x^{s}) = (x^{r-q} + x^{s})/(x^{r-q+1} + x^{s})$$

would imply

$$(1 + x^{s+p-1})/(1 + x^{s+p-2}) = (1 + x^{s+q-1})/(1 + x^{s+q-2})$$

with $p \neq q$, contrary to hypothesis. Consequently, if we replace the non-zero elements of the addition table by their indices when represented as powers of x, and the zeros by $(r - 1)$, we shall obtain an array A_r^*.

Conversely, suppose that an array A_r^* is given. We border the array with a 0th row and 0th column in such a way that the bordered array forms an $r \times r$ latin square. Upon replacing each element $t \neq r - 1$ by x^t and each element $(r - 1)$ by zero, and then identifying the square with that given in Fig. 7.5.4 for a suitable choice of s (determined by the position of the 1 in the 0th row), we shall define the addition table of a neofield of the type specified in the theorem.

(1) Note that the property $(1 + x^t)/(1 + x^{t-1}) = (1 + x^u)/(1 + x^{u-1}) \Rightarrow t = u$ certainly holds if one of t, $t - 1$, u, $u - 1$ is equal to s.

(2) Moreover, no index is 1 since $(1 + x^t)/(1 + x^{t-1}) = x$ would imply $1 + x^t = x + x^t$ which is impossible for a loop.

The property that $(1+x^t)/(1+x^{t-1}) = (1+x^u)/(1+x^{u-1})$ holds if and only if $t = u$ in a neofield with a cyclic multiplicative group has been called the ***divisibility property*** and a neofield of the kind specified in theorem 7.5.3 has been called a neofield with property D or, more briefly, a *D-neofield*.

In A. D. Keedwell [4], the author has shown by a consideration of the addition table that a D-neofield is commutative if and only if the row arrays $P_r^{(d)}$ and $P_r^{(m)}$ associated with it, and defined as in the corollaries to theorem 7.5.2, coincide.

A quite different proof of this result will be found in A. D. Keedwell [5]. In the latter paper, the conditions under which two D-neofields of a given order are isomorphic or anti-isomorphic have been analysed and, with the aid of a computer, all isomorphically distinct D-neofields of orders less than or equal to 17 have been catalogued. Some examples of such neofields have also been given for the orders 18, 19 and 20. It appears clear from the results that the number of D-neofields of an assigned order r increases rapidly with r and it is conjectured by the author that D-neofields exist for all orders r except 6. (We note at this point that, in particular, every Galois field is a D-neofield.)

As regards the explanation for the non-existence of a D-neofield of order 6 (and consequent non-existence of a pair of orthogonal latin squares of that order constructible by the above method), we easily see that, when $r = 6$, a row array P_r having the properties of theorem 7.5.2 cannot exist: for the integers 2, 3, 4 and 2—1, 3—1, 4—1 cannot be simultaneously reordered so that their partial sums taken modulo 5 are all distinct and non-zero. Essentially, this is due to the fact that the integers 2, 3 occur in each triad and must occur consecutively in one or the other. This observation suggests that the non-existence of orthogonal latin squares of order 6 may be a consequence of nothing more profound than the paucity of combinatorial rearrangements of the integers 0 to 4.

A number of unsolved problems concerning D-neofields have been listed in A. D. Keedwell [6] and these also appear as problems 7.3, 7.4 and 7.5 of the present book.

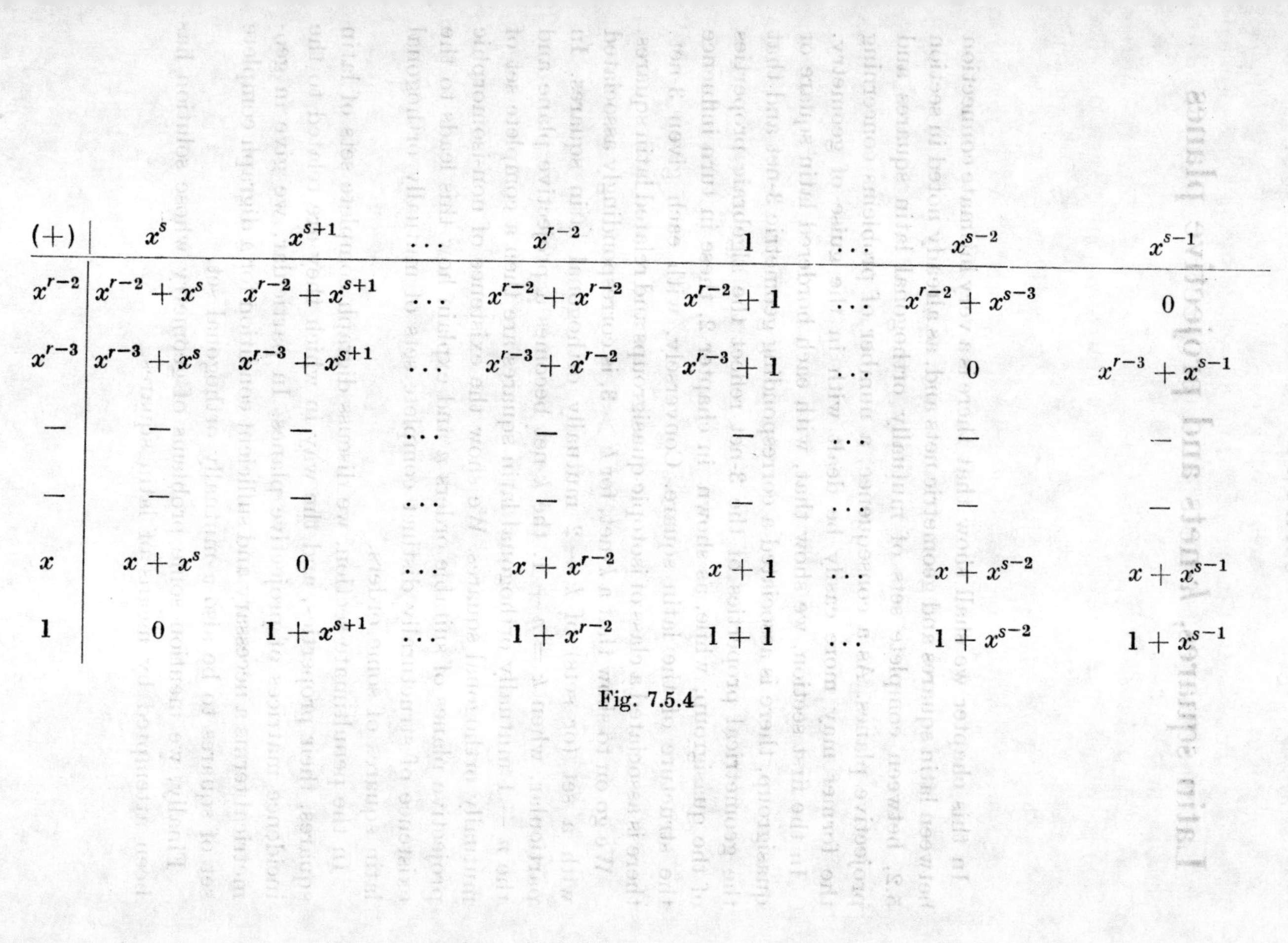

$(+)$	x^s	x^{s+1}	$\dots$	x^{r-2}	1	$\dots$	x^{s-2}	x^{s-1}
x^{r-2}	$x^{r-2}+x^s$	$x^{r-2}+x^{s+1}$	$\dots$	$x^{r-2}+x^{r-2}$	$x^{r-2}+1$	$\dots$	$x^{r-2}+x^{s-3}$	0
x^{r-3}	$x^{r-3}+x^s$	$x^{r-3}+x^{s+1}$	$\dots$	$x^{r-3}+x^{r-2}$	$x^{r-3}+1$	$\dots$	0	$x^{r-3}+x^{s-1}$
—	—	—	$\dots$	—	—	$\dots$	—	—
—	—	—	$\dots$	—	—	$\dots$	—	—
x	$x+x^s$	0	$\dots$	$x+x^{r-2}$	$x+1$	$\dots$	$x+x^{s-2}$	$x+x^{s-1}$
1	0	$1+x^{s+1}$	$\dots$	$1+x^{r-2}$	$1+1$	$\dots$	$1+x^{s-2}$	$1+x^{s-1}$

Fig. 7.5.4

CHAPTER 8

Latin squares, *k*-nets and projective planes

In this chapter we shall show that there is a very intimate connection between latin squares and geometric nets and, as already noted in section 5.2, between complete sets of mutually orthogonal latin squares and projective planes. As a consequence, a number of problems concerning the former may more easily be dealt with in the guise of geometry.

In the first section, we show that, with each bordered latin square or quasigroup, there is associated a corresponding geometric 3-net and that the geometrical properties of the 3-net reflect the algebraic properties of the quasigroup; while, as shown in chapter 2, these in turn influence the structure of the latin square. Conversely, with each given 3-net, there is associated a class of isotopic quasigroups and related latin squares.

We go on to show that a k-net, for $k > 3$, is correspondingly associated with a set (or sets) of $k - 2$ mutually orthogonal latin squares. In particular, when $k = n + 1$, the k-net becomes a projective plane and the $n - 1$ mutually orthogonal latin squares are then a complete set of mutually orthogonal squares. We show the existence of non-isomorphic projective planes of suitable orders n and explain how this leads to the existence of structurally distinct complete sets of mutually orthogonal latin squares of some orders.

In the penultimate section, we discuss digraph complete sets of latin squares, their properties, and the way in which they are related to the incidence matrices of projective planes. In particular, we give in geometrical terms a necessary and sufficient condition for a digraph complete set of squares to be also a mutually orthogonal set.

Finally, we mention some problems of geometry whose solution has been attempted by means of latin squares.

8.1. Quasigroups and 3-nets

The concept of a geometric net occurs naturally in connection with the problem of assigning co-ordinates to the points and lines of an affine or projective plane. Historically, the concept arose also in the study of certain topological problems of differential geometry. Among early papers on the subject are R. Baer [1] and [2], W. Blaschke [1], W. Blaschke and G. Bol [1], G. Bol [1], K. Reidemeister [1] and G. Thomsen [1]. Later papers concerning the connections between quasigroups, geometric nets, and projective planes are R. H. Bruck [3], T. G. Ostrom [5] and G. Pickert [1]. Extensive bibliographies of the more recent papers on the subject will be found in J. Aczél's survey paper [1] or [2], and in R. H. Bruck [5], G. Pickert [2], and chapter 11 of V. D. Belousov [7]. Hungarian readers may also like to consult F. Kárteszi [2].

Very recently, V. D. Belousov [11] and [12] have been published which are complete books devoted to the connection between nets and quasigroups.

Let us begin by giving a general definition of a net as used in geometry. We should mention here that the "lines" of our definition may be curves of the real plane in the applications to differential geometry or to nomograms.

Definition. A *geometric net* is a set of objects called "points" together with certain designated subsets called "lines". The lines occur in classes, called "parallel classes", such that (*a*) each point belongs to exactly one line of each parallel class; (*b*) if l_1 and l_2 are lines of different parallel classes, then l_1 and l_2 have exactly one point in common; (*c*) there are at least three parallel classes and at least two points on a line. (If the number of parallel classes is one or two and the remaining conditions are fulfilled, the system may be called a *trivial net*.)

A net possessing k parallel classes is called a *k-net*.

If the net is finite, then it is characterized by a parameter n, called the *order* of the net, such that (*i*) each line contains exactly n points; (*ii*) each parallel class consists of exactly n lines; and (*iii*) the total number of points is n^2.

Statements (*i*) and (*ii*) follow at once from conditions (*a*) and (*b*) of the definition of a net as soon as we postulate that one line of one parallel class has n points or that one parallel class has n lines. Then, since the lines of any one of the parallel classes contain all the points and since each of the n lines of this class has n points, statement (*iii*) follows.

An *affine plane* π^* comprises a set of objects called "points" together with certain distinguished subsets called "lines" such that (a) two distinct points belong to (are incident with) exactly one line; (b) every point exterior to a given line l of π^* is incident with exactly one line which has no point in common with l; and (c) there are at least three points not all incident with one line. The connection between affine planes and geometric nets follows immediately from the statement that:

THEOREM 8.1.1. *An affine plane is a geometric net in which each pair of points is incident with a line.* (Equivalently, we may say that *an affine plane of order n is a net of order n with $n + 1$ parallel classes.*)

PROOF. Let l be a line of the net containing n points and let P a point not on l. Then there are exactly $n + 1$ lines through P if each pair of points of the net is connected by a line: for the joins of P to the points of l give n lines and there is also one line through P belonging to the parallel class of l. These $n + 1$ lines through P necessarily belong to distinct parallel classes, so there are $n + 1$ parallel classes.

A *projective plane* may be regarded as an affine plane to which one extra line l_∞ has been adjoined, each of the $n + 1$ points of l_∞ being a point of concurrence (point at infinity) for the $n + 1$ lines of a parallel class. (Such a point on l_∞ is sometimes called the *vertex* of the parallel class.) Thus, a projective plane has a total of $n^2 + n + 1$ lines and $n^2 + n + 1$ points. When one line and the points on it are deleted, we get a geometric net of order n having $n + 1$ parallel classes.

For the purpose of introducing co-ordinates for the points of an affine plane or geometric net, we may assign arbitrary symbols to the lines of just two of its parallel classes, C_1, and C_2, and then assign the co-ordinate pair (x, y) to the point through which pass the line x of class C_1 and the line y of class C_2. There will be no loss of generality in using symbols from the same set Q (of cardinal n) for each of the two parallel classes C_1 and C_2. If the same set Q of symbols is used to label the lines of a third parallel class C_3 and if the line w of this class is incident with the point (x, y), then we can define a binary operation $(*)$ on the set Q by the statement $x*y = w$. The properties of the geometric net then ensure that each equation $x*y = w$ is uniquely soluble for x, y or w in Q when the other two variables are specified, and so $(Q, *)$ is a quasigroup. Conversely, we may prove:

THEOREM 8.1.2. *If a bordered latin square of order n is given, representing the multiplication table of a quasigroup* $(Q, *)$, *then there can be associated with it a geometric net of order n having exactly three parallel classes and such that the lines* x, y *of the parallel classes* C_1 *and* C_2 *are incident with the line* w *of the parallel class* C_3 *if and only if* $x*y = w$.

PROOF. As the n^2 points of our 3-net we take the n^2 ordered pairs (x, y) of symbols of our quasigroup $(Q, *)$. For each fixed choice of x, the n points (X, y), $y \in Q$, form a line $l_X^{(1)}$ of the parallel class C_1. For each fixed choice of y the n points (x, Y), $x \in Q$, form a line $l_Y^{(2)}$ of the parallel class C_2. For each fixed choice of w, the set of all points (x, y) such that $x*y = W$ form a line $l_W^{(3)}$ of the parallel class C_3. Since the multiplication table of $(Q,*)$ is a latin square, each line of C_3 has exactly n points.

It is immediate from the definitions that no two lines of C_1 or of C_2 or of C_3 have a point in common and that each point belongs to exactly one line of each parallel class. Also, two lines belonging to distinct parallel classes have exactly one point in common. The lines $l_X^{(1)}$ and $l_Y^{(2)}$ have the point (X, Y) in common. The lines $l_X^{(1)}$ and $l_W^{(3)}$ have the point $(X, \bar{y})$ in common, where $\bar{y}$ is the unique solution of the equation $X*\bar{y} = W$. The lines $l_Y^{(2)}$ and $l_W^{(3)}$ have the point $(\bar{x}, Y)$ in common, where $\bar{x}$ is the unique solution of the equation $\bar{x}*Y = W$. This completes the proof.

As an example, we give in Fig. 8.1.1 below a quasigroup of order 4 and its associated 3-net N.

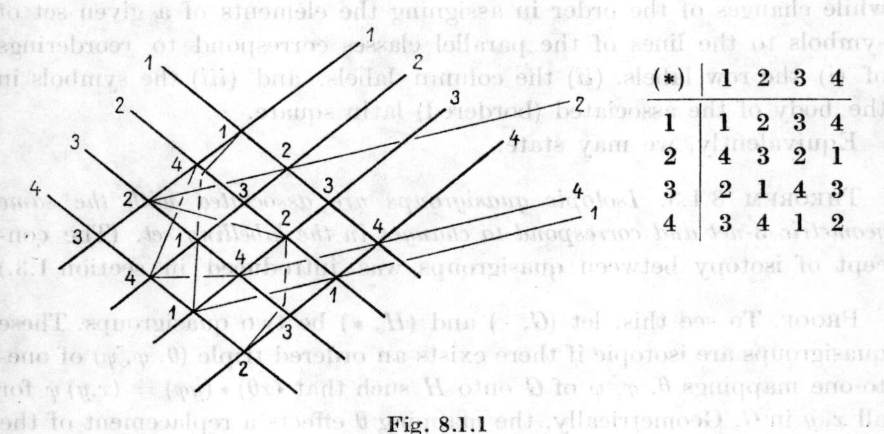

(*)	1	2	3	4
1	1	2	3	4
2	4	3	2	1
3	2	1	4	3
4	3	4	1	2

Fig. 8.1.1

We shall show presently that the algebraic properties of the quasigroup $(Q, *)$ are reflected in the geometrical properties of its associated geometric 3-net.

In co-ordinatizing an affine (or projective) plane, it is usual to introduce two binary operations denoted by $(+)$ and $(\cdot)$ respectively, the first being associated with a 3-net whose three "parallel" classes have their vertices collinear on the special line l_∞ (that is, the lines of each parallel class are "parallel" in the sense used with reference to an affine plane), and the second being associated with a 3-net such that two of its "parallel" classes coincide with two of those of the addition net and have vertices on l_∞ while the third one comprises the pencil of lines through a finite point. The finite point (vertex of the parallel class) is regarded as having been deleted when treating the system as a net.

In the case of the real affine plane, these binary operations coincide with addition and multiplication of real numbers if the four parallel classes in question comprise the lines parallel to the x-axis, the lines parallel to the y-axis, the lines of gradient 1 and the lines through the origin 0 of cartesian coordinates.

We shall discuss the co-ordinatization of affine and projective planes in more detail in section 8.3.

Let us consider further the relation between latin squares and geometric 3-nets. In the first place, let us observe that if a geometric 3-net is given, the choice of co-ordinatizing set Q and the procedure for assigning its elements one-to-one to the lines of each parallel class are not unique. Indeed, it is easy to see that change of the co-ordinatizing set of a 3-net N corresponds to a change of symbols in the associated latin square L, while changes of the order in assigning the elements of a given set of symbols to the lines of the parallel classes correspond to reorderings of (*i*) the row labels, (*ii*) the column labels, and (*iii*) the symbols in the body of the associated (bordered) latin square.

Equivalently, we may state:

Theorem 8.1.3. *Isotopic quasigroups are associated with the same geometric 3-net and correspond to changes in the labelling set.* (The concept of isotopy between quasigroups was introduced in section 1.3.)

Proof. To see this, let $(G, \cdot)$ and $(H, *)$ be two quasigroups. These quasigroups are isotopic if there exists an ordered triple (θ, φ, ψ) of one-to-one mappings θ, φ, ψ of G onto H such that $(x\theta) * (y\varphi) = (x.y)\psi$ for all x, y in G. Geometrically, the mapping θ effects a replacement of the symbols of the set G which are assigned to the lines of one parallel class C_1 of a geometric net by symbols of the set H. The mappings φ and ψ respectively effect similar relabellings of the lines of the parallel classes C_2 and C_3. In the relabelled net, the lines of the classes C_1 and C_2 labelled $x\theta$ and $y\varphi$ are incident with the line $(x.y)\psi$ when $(x\theta)*(y\varphi) = (x.y)\psi$: that is, when the lines of the classes C_1 and C_2 which were originally labelled x and y are incident with the line of C_3 which was originally labelled $x.y$. The latter lines are the same as the former.

Corollary. *Isotopisms of a quasigroup $(G, \cdot)$ onto itself correspond to changes of order in the labelling set G of the associated 3-net.*

Proof. The mappings θ, φ, ψ are now mappings of the set G onto itself. That is, they represent permutations of the labelling set.

In chapter 1, we proved that every isotope $(H, *)$ of a quasigroup $(G, \cdot)$ is isomorphic to a principal isotope of the quasigroup (theorem 1.3.2) and that, among the principal isotopes of a quasigroup $(G, \cdot)$ there always exist loops (theorem 1.3.3). These facts imply the truth of the following theorem:

Theorem 8.1.4. *If $(G, \cdot)$ is a quasigroup associated with a given geometric 3-net, then there are as many other isomorphically distinct quasigroups associated with that net as there are principal isotopes of $(G, \cdot)$. Moreover, among these quasigroups there always exist loops. Consequently, any geometric 3-net has loops among its co-ordinate systems.*

We can interpret theorem 1.3.3 of chapter 1 geometrically as follows:

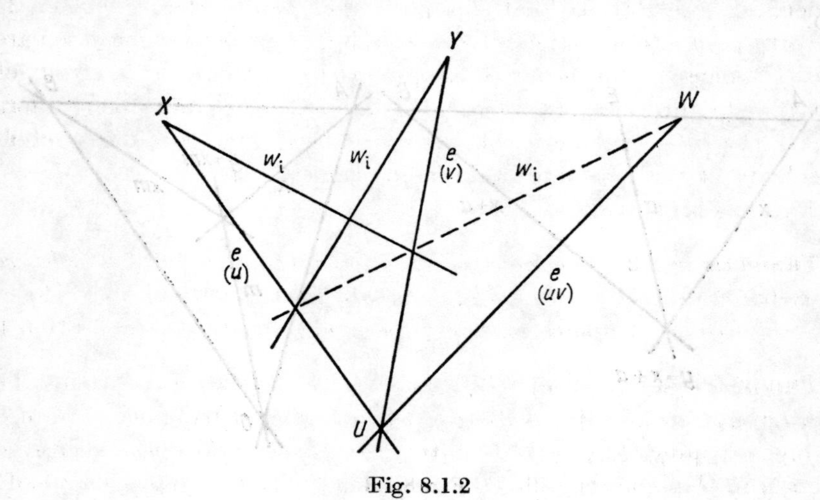

Fig. 8.1.2

In the 3-net whose parallel classes are the lines through the points X, Y, W, we may select the particular lines of the (X) and (Y) parallel classes which carry the labels u, v of $(G, \cdot)$ as the ones to be relabelled as the identity lines e for the principal isotope $(G, \otimes)$. The point of intersection U of these particular lines will be called unit point. We now relabel the lines of the parallel classes (X) and (Y) in such a way that, if w_i is any line of the parallel class (W) other than the line WU, then the line of the parallel class (X) through the point $w_i \cap YU$ will be labelled w_i and the line of the parallel class (Y) through the point $w_i \cap XU$ will also be labelled w_i. In the quasigroup $(G, \otimes)$ thus defined, $e \otimes w_i = w_i = w_i \otimes e$. Moreover, if we now consider how we should relabel the line WU, we see that it should carry the same label e as XU and YU. So, the identity element e of $(G, \otimes)$ is the element uv of $(G, \cdot)$ as in theorem 1.3.3. (See Fig. 8.1.2)

As a corollary to theorem 1.3.4 of chapter 1, we proved that if a loop is isotopic to a group, then the loop is a group isomorphic to the given group. One implication of this theorem is that, if a quasigroup has loops isotopic to it which satisfy the associative law, then so do all loops isotopic to it. We are entitled to expect, therefore, that the property of associativity corresponds to some geometrical property of the associated 3-net and that this will have significance for the theory of projective planes. This is indeed the case, as we shall shortly show. We remind the reader at this point that, in theorem 1.2.1, we interpreted the property of associativity in terms of the quadrangle criterion for the latin square representing the multiplication table of the quasigroup or loop.

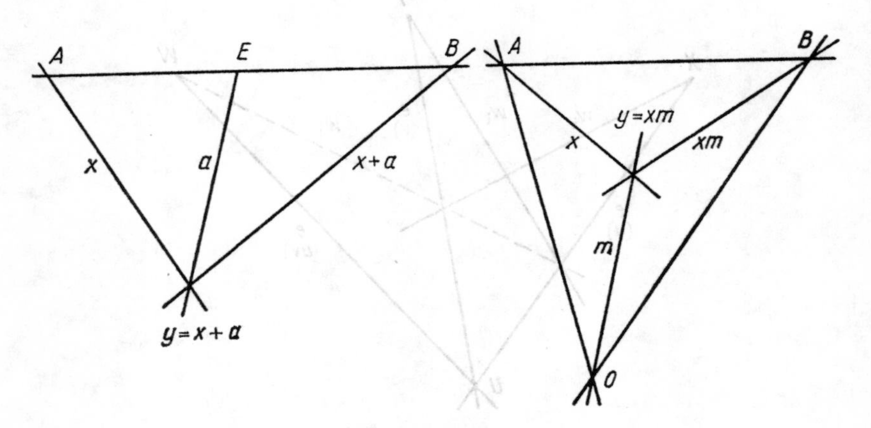

Fig. 8.1.3

Before developing these ideas further, it is desirable to mention again that, in plane projective geometry, two distinct types of 3-net arise, *affine* nets and *triangular* nets. In the first, the points A, E, B of l_∞ take the roles of X, Y, W and the parallel classes are the pencils with the collinear vertices A, E, B. For such nets, the quasigroup is usually written additively. Lines through E have "equations" of the form $y = x + a$. For triangular nets on the other hand, the vertices A, O, B of a proper triangle are the vertices of three pencils of lines representing the "parallel classes" (the lines of the pencil with vertex O being no longer parallel in the geometrical sense), and, for such nets, the associated quasigroup is usually written multiplicatively. Lines through O have "equations" of the form $y = xm$. (See Fig. 8.1.3) If the elements of either the additive or the multiplicative quasigroup satisfy some algebraic identity, this corresponds to closure of a certain geometrical configuration.

We shall now examine the most important of these configurations. The reader new to the subject may find it helpful to read section 8.3 on the introduction of co-ordinates into a projective plane before proceeding further with the present discussion.

Let us consider first the algebraic effect of requiring closure of the Pappus configuration. The assertion that the Pappus configuration has incidence closure is usually called *Pappus' theorem*. This may be stated as follows: if P_1, P_2, P_3 and Q_1, Q_2, Q_3 are two triads of collinear points, then the triad of points R_1, R_2, R_3, where $R_i = P_j Q_k \cap P_k Q_j$ $(i, j, k = 1, 2, 3;\ i \neq j \neq k)$ are collinear. We have

Theorem 8.1.5. *If the parallel classes of the net comprise the lines through the points X, Y, W and if the lines through X are denoted by $x_1, x_2, \ldots$, and the lines through Y are denoted by $y_1, y_2, \ldots$, then Pappus' theorem implies the algebraic relation*

$$x_1 y_2 = x_2 y_1 \quad \text{and} \quad x_1 y_3 = x_3 y_1 \ \Rightarrow\ x_2 y_3 = x_3 y_2$$

for the quasigroup $(G, \cdot)$ where the x_i and y_i are the symbols of G, and where we identify the points P_i, Q_i, R_i with the points shown in Fig. 8.1.4.

Proof. If we label the various lines of the configuration as in Fig. 8.1.4, the result becomes obvious.

To interpret the above result in terms of latin squares we may suppose without loss of generality that the elements of the quasigroup $(G, \cdot)$

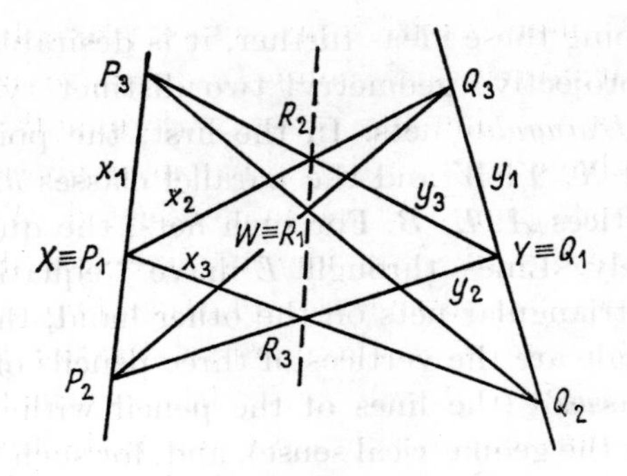

Fig. 8.1.4

are labelled in such a way that $ab'_i = b_i a'$ for $i = 1, 2, 3, \ldots$. Then if Pappus' theorem closes in the associated net, we shall have $b_i b'_j = b_j b'_i$ for $i, j = 1, 2, 3, \ldots$; $i \neq j$. This follows from the fact that $ab'_i = b_i a'$ and $ab'_j = b_j a'$ together imply $b_i b'_j = b_j b'_i$. Thus, the multiplication table of $(G, \cdot)$ will be a bordered latin square of the form shown in Fig. 8.1.5.

$(\cdot)$	$\ldots a'$	b'_1	b'_2	b'_3	$b'_4 \ldots$
$\vdots$					
a	v	w	x	y	z
b_1	w	$\cdot$	p	q	r
b_2	x	p	$\cdot$	s	t
b_3	y	q	s	$\cdot$	u
b_4	z	r	t	u	$\cdot$
$\vdots$					

Fig. 8.1.5

In the application to projective planes, we regard the three parallel classes as comprising the lines through the co-ordinatizing points A, O, B. Lines through A have symbols $x_1, x_2, \ldots$, and lines through O have symbols $m_1, m_2, \ldots$, representing their "gradients". Then, upon identifying A with the point X above and O with the point Y above, we get the relation

$$x_1 m_2 = x_2 m_1 \quad \text{and} \quad x_1 m_3 = x_3 m_1 \Rightarrow x_2 m_3 = x_3 m_2 .$$

It is usual to co-ordinatize a projective plane by means of a loop rather than an arbitrary quasigroup. To do so, we select one particular line through O as an "identity" line with equation $Y = X$. That is, we require it to have unit gradient. Let us suppose that the line with gradient m_1 and equation $y = xm_1$ is so selected. We further select one particular line through A as an identity line, say the line carrying the symbol x_1. The point of intersection of these two identity lines is then to be unit point. That is, in the co-ordinatization by means of a loop with identity element e, it is to have the co-ordinates (e, e) relative to A and B as co-ordinatizing points. We now make the change of co-ordinates given by the mappings

$$x \to X = x\sigma^{-1} \quad \text{where} \quad x\sigma^{-1} = xm_1,$$
$$m \to M = m\tau^{-1} \quad \text{where} \quad m\tau^{-1} = x_1 m,$$
$$y \to y,$$

and hence, by defining $x \otimes m = (x\sigma).(m\tau)$, obtain a loop principal isotope $(G, \otimes)$ of the quasigroup $(G, \cdot)$ with identity element $e = x_1 m_1$ as in theorem 1.3.3.

In the new co-ordinate system, the line $x = x_i$ becomes $X = x_i m_1$, since $X = x\sigma^{-1}$, and the line $y = xm_i$ becomes $y = (xm_1) \otimes (x_1 m_i)$ or, say, $y = X \otimes (x_1 m_i)$, where $x_1 m_i$ is the new gradient. (Since y co-ordinates remain unaltered, the lines with equations $x = x_i$ and $y = xm_i$ in the old system, or $X = x_i m_1$ and $y = X \otimes (x_1 m_i)$ in the new system, necessarily meet at the point for which $y = x_i m_i = x_i m_1 \otimes (x_1 m_i)$ in both systems.) In particular, the line $x = x_1$ becomes the line $X = e$ and the line $y = xm_1$ becomes unit line $y = X \otimes e$ or $y = X$ as required.

Since $(G, \cdot)$ is a quasigroup, for given elements x_1, m_1, y_2 and y_3 of G, there always exist elements m_2, m_3, x_2, and x_3 such that $x_1 m_2 = x_2 m_1 = y_2$ and $x_1 m_3 = x_3 m_1 = y_3$. Then, if and only if Pappus' theorem holds, these equations imply that $x_2 m_3 = x_3 m_2 = y_1$, say. So, since $y_2 \otimes y_3 = (y_2\sigma)(y_3\tau) = x_2 m_3$ and $y_3 \otimes y_2 = (y_3\sigma)(y_2\tau) = x_3 m_2$, we have that $y_2 \otimes y_3 = y_3 \otimes y_2 = y_1$. Since the original choices of the elements x_1 and m_1 of the quasigroup G were arbitrary, we may assert the following:

THEOREM 8.1.6. *If and only if Pappus' theorem holds for all lines through the points A, B, and O, then every LP-isotope of the multiplicative quasigroup defined by the* 3*-net whose parallel classes are the pencils of lines with vertices A, O, and B is commutative.*

PROOF. By theorem 1.3.3, every LP-isotope of the quasigroup $(G, \cdot)$ is obtained by mappings σ, τ of the type defined above, so the above analysis is valid for all such LP-isotopes. On the other hand, the single LP-isotope of $(G, \cdot)$ obtained by means of the mappings σ, τ corresponding to fixed choices of x_1 and m_1 is commutative if and only if Pappus' theorem holds with respect to the fixed lines $x = x_1$ through A, $y = xm_1$ through O and all lines through B. This may easily be seen by inspection of Fig. 8.1.6.

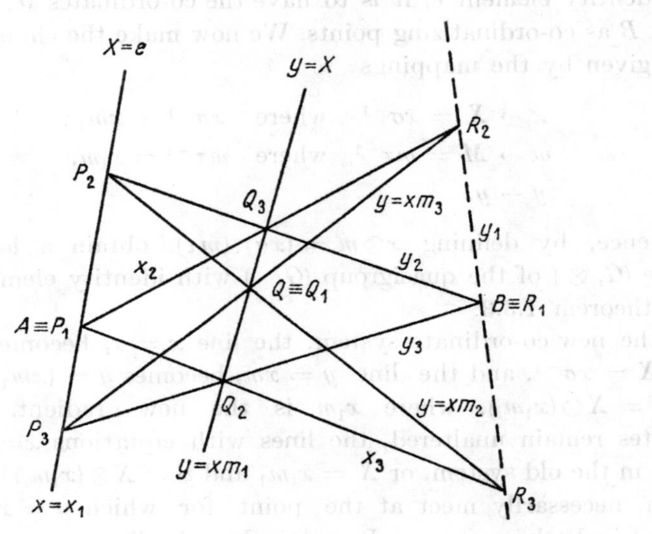

Fig. 8.1.6

We may illustrate the above result by showing how an LP-isotope $(G, \otimes)$ is obtained from the quasigroup $(G, \cdot)$ whose multiplication table is given in Fig. 8.1.5. Let us choose the point of intersection of the lines carrying the labels a and a' as unit point. Then the mappings σ and τ are defined by $x\sigma^{-1} = xa'$ and $x\tau^{-1} = ax$. If $h = b_i b_j'$ in $(G, \cdot)$, then $h = (b_i\sigma^{-1}) \otimes (b_j\tau^{-1})$ in $(G, \otimes)$. That is, in Fig. 8.1.5, $p = b_1 b_2'$ implies $p = (b_1 a') \otimes (ab_2') = w \otimes x$, $q = b_1 b_3'$ implies $q = (b_1 a') \otimes (ab_3') = w \otimes y$, and so on. Hence, the multiplication table of $(G, \otimes)$ takes the form shown in Fig. 8.1.7 and is a commutative loop with $v = aa'$ as identity element.

We now digress slightly to prove the interesting and important fact that, if all LP-isotopes of a quasigroup $(G, \cdot)$ are commutative, they are all associative too. (See R. Baer [1] and G. Bol [1].) However, from

$(\otimes)$	$\cdot$	v	w	x	y	z	$\cdot$
$\vdots$							
v		v	w	x	y	z	
w		w	$\cdot$	p	q	r	
x		x	p	$\cdot$	s	t	
y		y	q	s	$\cdot$	u	
z		z	r	t	u	$\cdot$	
$\vdots$							

Fig. 8.1.7

the commutativity of a single LP-isotope, its associativity does not necessarily follow.

THEOREM 8.1.7. *If all the loop-principal isotopes of the quasigroup* $(G, \cdot)$ *are commutative, they are all associative too.*

PROOF. Let $x \otimes y = (x\sigma).(y\tau)$ where $x\sigma^{-1} = xv$ and $y\tau^{-1} = uy$. By the hypothesis of the theorem, the loop so defined may be assumed commutative. Now let a, b, c be arbitrary elements of G. Then since $(G,\cdot)$ is a quasigroup, there are unique elements x, y, w in G such that $a = xv$, $b = uy$, $c = wv$. Define a second loop-principal isotope of $(G,\cdot)$ by $x * y = (x\sigma).(y\theta)$ where $x\sigma^{-1} = xv$ as before and $y\theta^{-1} = wy$. This loop also is commutative by hypothesis. We find that $a \otimes b = xy = (xv) * (wy) =$ $= a * (wv \otimes uy) = a * (c \otimes b)$. So, $(a \otimes c) \otimes b = (c \otimes a) \otimes b = (c \otimes a) * (c \otimes b) =$ $= (c \otimes b) * (c \otimes a) = (c \otimes b) \otimes a = a \otimes (c \otimes b)$, which proves the theorem.

COROLLARY 1. *Under the hypotheses of the theorem (that is, in a net for which Pappus' theorem is satisfied relative to the parallel classes with vertices* A, O, B*), every loop isotopic to* $(G,\cdot)$ *is an abelian group.*

PROOF. As every isotope of a quasigroup is isomorphic to a principal isotope by theorem 1.3.2, it follows that every loop isotopic to $(G,\cdot)$ is isomorphic to an LP-isotope. The latter are commutative and associative, so they are abelian groups.

COROLLARY 2. *If Pappus' theorem holds relative to the vertices* A, B, O *for all choices of lines with one line fixed through* O*, then it holds for all choices of lines through* O.

Proof. In the argument above, the mapping σ is kept fixed as the LP-isotope varies. Geometrically, this corresponds to using the same line $y = xv$ through O throughout.

If, instead of the triangular net considered above, we have to deal with an affine net whose parallel classes have collinear vertices A, E and B, then the whole of the above discussion remains valid if the Pappus configuration is replaced by the Thomsen configuration (see G. Thomsen [1]) which is illustrated in Fig. 8.1.8. This may be regarded as the configuration obtained from the Pappus configuration when $O \to E$ on l_∞. The assertion that the Thomsen configuration has incidence closure is often called the *axial minor theorem of Pappus*. This may be stated as follows: If P_1, P_2, P_3, Q_2, Q_3 are five given points such that P_1, P_2, P_3 are collinear, and if R_1 is the point $P_2Q_3 \cap P_3Q_2$ and P_1R_1 meets Q_2Q_3 at Q_1, then the points $R_1, R_2 \equiv P_1Q_3 \cap P_3Q_1$ and $R_3 \equiv P_1Q_2 \cap P_2Q_1$ are collinear.

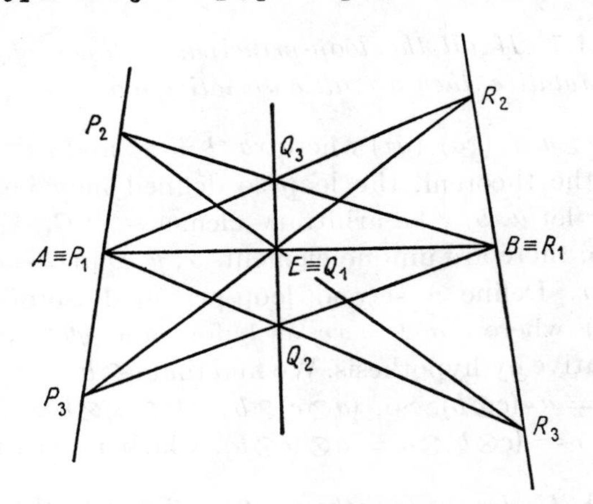

Fig. 8.1.8

In the application to projective planes, closure of the Thomsen configuration is the condition that the additive loop of the co-ordinate system be commutative and closure of the Pappus configuration the condition that the multiplicative loop be commutative.

We consider next the algebraic consequence for a triangular net of requiring closure of the large Reidemeister configuration (see K. Reidemeister [1]). This asserts that, if the quadrangles $P_1P_2P_3P_4$ and $P_1'P_2'P_3'P_4'$ have the joins of vertices P_1P_2, P_3P_4, $P_1'P_2'$, $P_3'P_4'$ concurrent in a point B, P_2P_3, P_1P_4, $P_2'P_3'$, $P_1'P_4'$ concurrent in a point A, and $P_1P'_1$, $P_2P'_2$, $P_3P'_3$ concurrent in a point O, then the join $P_4P'_4$ passes through the same point (Fig. 8.1.9). In B. I. Argunov [1], this configurational proposition has been denoted by the symbol $mA(11; 11, 12)$.

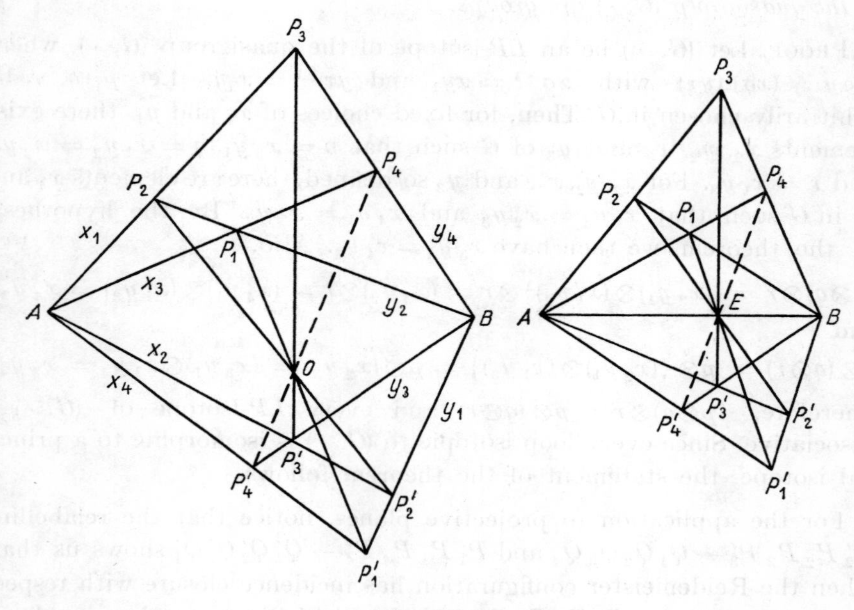

Fig. 8.1.9

We have the following important result.

Theorem 8.1.8. *If the lines through the point A have symbols $x_1, x_2, \ldots$, and the lines through the point B have symbols $y_1, y_2, \ldots$, then closure of the large Reidemeister configuration implies satisfaction of the algebraic relation*

$$x_1 y_2 = x_2 y_1,\ x_1 y_4 = x_2 y_3 \quad and \quad x_3 y_2 = x_4 y_1 \Rightarrow x_3 y_4 = x_4 y_3$$

in the quasigroup $(G, \cdot)$ associated with the net whose parallel classes are the pencils of lines through the points A, B, and O.

Proof. If we label the various lines of the configuration as in Fig. 8.1.9 the result becomes obvious.

Let us observe that the algebraic relation given in theorem 8.1.8 is precisely the quadrangle criterion which was first introduced in theorem

1.2.1. This observation makes it clear that our next result is only to be expected.

THEOREM 8.1.9. *Closure of the large Reidemeister configuration in the 3-net associated with the quasigroup $(G, \cdot)$ implies that all loops isotopic to the quasigroup $(G, \cdot)$ are groups.*

PROOF. Let $(G, \otimes)$ be an *LP*-isotope of the quasigroup $(G, \cdot)$, where $x \otimes y = (x\sigma).(y\tau)$ with $x\sigma^{-1} = xy_1$ and $y\tau^{-1} = x_1 y$. Let p, q, r be arbitrarily chosen in G. Then, for fixed choices of x_1 and y_1, there exist elements x_2, y_2, x_3 and y_3 of G such that $p = x_3 y_1$, $q = x_1 y_2 = x_2 y_1$, and $r = x_1 y_3$. For x_2, y_2, x_3, and y_3 so defined, there are elements x_4 and y_4 in G such that $x_1 y_4 = x_2 y_3$ and $x_4 y_1 = x_3 y_2$. By the hypothesis of the theorem, we then have $x_3 y_4 = x_4 y_3$. Also,

$$(p \otimes q) \otimes r = \{(x_3 y_1) \otimes (x_1 y_2)\} \otimes r = (x_3 y_2) \otimes r = (x_4 y_1) \otimes (x_1 y_3) = x_4 y_3 ;$$

and

$$p \otimes (q \otimes r) = p \otimes \{(x_2 y_1) \otimes (x_1 y_3)\} = p \otimes (x_2 y_3) = (x_3 y_1) \otimes x_1 y_4 = x_3 y_4 .$$

Therefore, $(p \otimes q) \otimes r = p \otimes (q \otimes r)$ and every *LP*-isotope of $(G, \cdot)$ is associative. Since every loop isotopic to $(G, \cdot)$ is isomorphic to a principal isotope, the statement of the theorem follows.

For the application to projective planes, notice that the relabelling $P_2' P_2 P_3 P_3' \rightarrow Q_1 Q_2 Q_3 Q_4$ and $P_1' P_1 P_4 P_4' \rightarrow Q_1' Q_2' Q_3' Q_4'$ shows us that, when the Reidemeister configuration has incidence closure with respect to the quadrangles $P_1 P_2 P_3 P_4$ and $P_1' P_2' P_3' P_4'$, it also has incidence closure with respect to the quadrangles $Q_1 Q_2 Q_3 Q_4$ and $Q_1' Q_2' Q_3' Q_4'$. For the latter quadrangles, the roles of the vertices O and B are interchanged so, if the lines through O carry symbols $m_1, m_2, \ldots$, representing their gradients, we have

$$x_1 m_2 = x_2 m_1,\ x_1 m_4 = x_2 m_3 \text{ and } x_3 m_2 = x_4 m_1 \Rightarrow x_3 m_4 = x_4 m_3$$

where $(\cdot)$ now represents the multiplicative operation in the co-ordinatization of the projective plane. By the same type of argument as that used in the discussion of the Pappus' configuration, it now follows from the preceding theorem that closure of the large Reidemeister configuration is the condition that the multiplicative loop of the co-ordinate system of the projective plane is associative.

In the first corollary to theorem 8.1.7, we showed that, in a triangular net for which Pappus' theorem is satisfied relative to the parallel classes with vertices A, O and B, every co-ordinatizing loop is an abelian group. This means, a fortiori, that every such loop is associative and so the large

Reidemeister configuration closes (since $(p \otimes q) \otimes r = p \otimes (q \otimes r)$ implies that $x_4 y_3 = x_3 y_4$ in theorem 8.1.9). Thus, the geometrical equivalent of theorem 8.1.7 is the following:

THEOREM 8.1.10. *Closure of the Pappus' configuration relative to the parallel classes with vertices A, O, B in a projective plane implies closure of the large Reidemeister configuration relative to these same parallel classes.*

A wholly geometrical proof of this result is possible. See G. Hessenburg [1] and W. Klingenberg [1] and [2].

When $O \to E$ (a point on AB), our triangular net is replaced by an affine net and the large Reidemeister configuration is replaced by the small Reidemeister configuration, denoted by the symbol $aA(10; 11, 13)$ in B. I. Argunov [1](1) and shown in Fig. 8.1.9.

Closure of the latter configuration is the geometrical condition that the additive loop of the co-ordinate system of the projective plane is associative. The geometrical equivalent of theorem 8.1.7 for such an affine net is

THEOREM 8.1.11. *Closure of the Thomsen configuration* (or validity of the axial minor theorem of Pappus) *relative to the parallel classes with vertices A, E, B in a projective plane implies closure of the configuration $aA(10;11,13)$ relative to those same parallel classes.*

A purely geometrical proof of this result has been given in A. D. Keedwell [2].

Among the further infinite number of configurational propositions whose algebraic interpretations relative to a given 3-net we might consider, the so-called Bol configurations (first introduced by G. Bol in [1]), the central minor theorem of Pappus, and the hexagon configuration have, up to the present, shown themselves to be of most importance.(2) However, it would be inappropriate in the present text to discuss the algebraic implications of each of these configurations in full detail,

(1) In Argunov's notation, which makes some attempt to be systematic, the leading small letter m or a stands for "multiplication" or "addition", the capital A stands for "associativity" and the three numbers following, say $(R; P, L)$ are respectively the rank, the number of points, and the number of lines of the configuration. The *rank* R is defined by the formula $R = 2(P + L) - I$, where I is the number of incidences in the unclosed configuration.

(2) The configuration of Desargues is of crucial importance in the theory of 4-nets and projective planes but not so much so in the theory of 3-nets.

and so we shall content ourselves with a summary of the most importan results.[1]

The Bol configurations arise from the Reidemeister configuration when P_1 lies on P_3P_3' (see Fig. 8.1.10) and are denoted by $mA(10;11,11$ and $aA(9;\ 11,\ 12)$ respectively in Argunov's notation. From th manner of the derivation of these configurations from the Reidemeiste

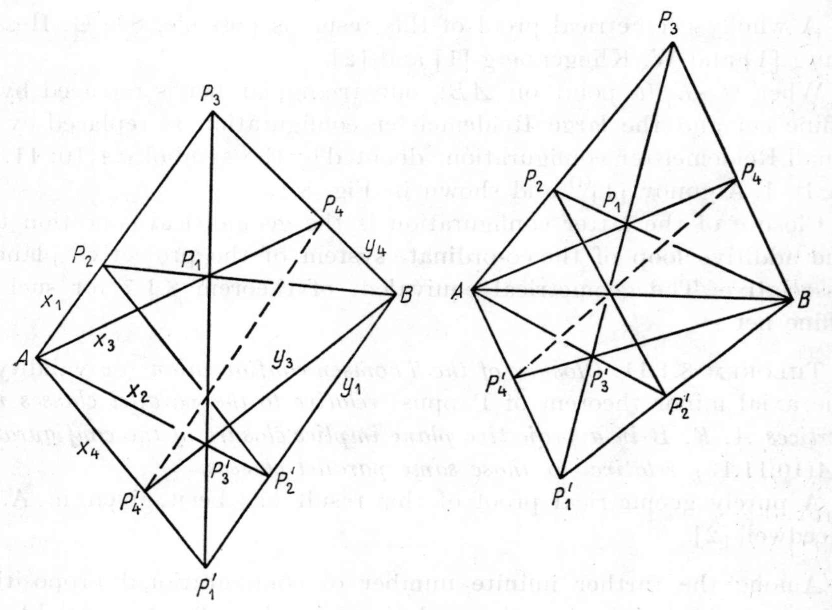

Fig. 8.1.10.

configurations, it is immediately clear that the roles of the vertice A, O, B are no longer interchangeable. (Closure of the Pappus and Reidemeister configurations relative to the parallel classes with vertice A, O, B taken in any one order implies their closure relative to these parallel classes taken in any other order. In particular, we showed

[1] For a detailed and comprehensive account of the subject of the inter-relation between geometrical configurations and quasigroup identities treated from the geometrical point of view, the reader is referred to B. I. Argunov's paper [1] and to G. Pickert's book [2] on projective planes. Further information on the subject will be found also in J. Aczél's paper [1] or [2], in V. D. Belousov [11] and in chapter 8 of V. D. Belousov [7].

explicitly on page 264 that the roles of O and B were interchangeable for the large Reidemeister configuration.)

If the vertices A, O, B are as in Fig. 8.1.10, closure of the configuration $mA(10;11,11)$ implies the condition

$$x_1 y_2 = x_2 y_1 \text{ and } x_1 y_4 = x_2 y_3 = x_3 y_2 = x_4 y_1 \Rightarrow x_3 y_4 = x_4 y_3$$

(condition B_3).

If O and A are interchanged and the lines are suitably labelled (P_2P_2', P_1P_1', P_4P_4' as x_1, x_2, x_4 and $P_3'P_4'$, $P_1'P_2'$, P_3P_4, P_1P_2 as y_1, y_2, y_3, y_4 respectively), then closure of the configuration implies the condition

$$x_1 y_2 = x_2 y_1,\ x_1 y_4 = x_2 y_3, \text{ and } x_2 y_2 = x_4 y_1 \Rightarrow x_2 y_4 = x_4 y_3$$

(condition B_1).

Finally, if O and B are interchanged instead and the lines are suitably labelled, then closure of the configuration implies the condition

$$x_1 y_2 = x_2 y_1,\ x_1 y_4 = x_2 y_2 \text{ and } x_3 y_2 = x_4 y_1 \Rightarrow x_3 y_4 = x_4 y_2$$

(condition B_2).

J. Aczél (in [1] or its English equivalent [2]) has shown algebraically that conditions B_1 and B_2 together imply condition B_3 (a result originally proved in G. Bol [1], and he has also demonstrated the followings results:

(1) condition B_1 implies that all the LP-isotopes of the quasigroup $(G, \cdot)$ are M_1-loops: that is, loops with the property $y(z.yx) = (y.zy)x$ for all x, y, z in the loop;

(2) condition B_2 implies that all the LP-isotopes of the quasigroup $(G, \cdot)$ are M_2-loops: that is, loops with the property $(xy.z)y = x(yz.y)$ for all x, y, z in the loop;

(3) if a loop is an M_1-loop and an M_2-loop, then it is a Moufang loop: that is, it has the property $(xy.z)y = x(y.zy)$;

(4) if a loop is an M_1-loop, its associated net satisfies condition B_1; if a loop is an M_2-loop, its associated net satisfies condition B_2; if a loop is Moufang, its associated net satisfies all of the conditions B_1, B_2, B_3.

An equivalent set of results has also been given by R. H. Bruck in [7]. One implication of these results is that the multiplicative loop of any co-ordinate system of a projective plane that is based on the co-ordinatizing points A, B, O, E will be Moufang if and only if the configuration $mA(10;11,11)$ has incidence closure relative to the "parallel classes" with vertices A, B, and O.

The *central minor theorem of Pappus* is the geometric dual of the axial minor theorem of Pappus discussed earlier. It is illustrated in Fig. 8.1.11. For a triangular net, its closure implies the condition

$$x_1 y_2 = x_2 y_1 \quad \text{and} \quad x_1 y_3 = x_2 y_2 = x_3 y_1 \;\Rightarrow\; x_2 y_3 = x_3 y_2$$

(condition H)

For an affine net, the same condition is implied by the *third minor theorem of Pappus*: that is, by closure of the "hexagon configuration" which arises from the central minor theorem when $O \to E$ on AB. The third minor theorem of Pappus may be stated as follows: if P_1, P_2, Q_1, Q_2 are four points which form a proper quadrangle and if R_3 is the point $P_1Q_2 \cap P_2Q_1$, S is the point $P_1P_2 \cap Q_1Q_2$, R_3S intersects P_1Q_1 at R_1 and $P_2R_1 \cap Q_1Q_2 \equiv Q_3$, $Q_2R_1 \cap P_1P_2 \equiv P_3$, then the point $R_2 \equiv P_1Q_3 \cap P_3Q_1$ lies on R_1R_2 (see Fig. 8.1.11).

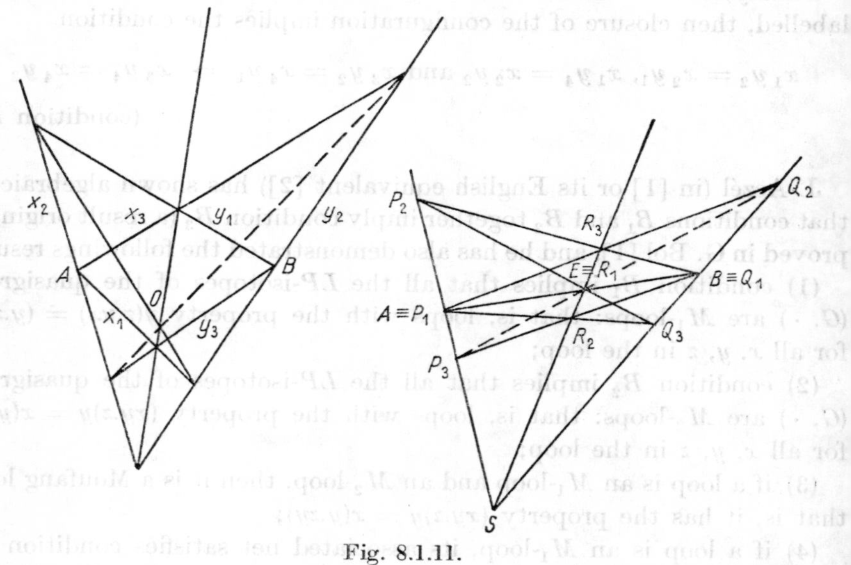

Fig. 8.1.11.

The following facts are well known:

(5) Condition H implies and is implied by the statement that all loops isotopic to $(G, \cdot)$ are power associative.

(6) Condition H implies and is implied by the statement that, in each loop isotopic to $(G, \cdot)$ each element has a two-sided inverse.

With the aid of the results (5) and (6) it is quite easy to prove the following important result:

THEOREM 8.1.12. *If all loops isotopic to the quasigroup* $(G, \cdot)$ *are power associative, then they are all strongly power associative too. That is, if in any loop isotope* $(G, \otimes)$ *of* $(G, \cdot)$ *we define* x^n *by* $x^0 = e$ *and* $x^n = x \otimes x^{n-1}$ *for* $n > 0$, x^{-1} *by* $x^{-1} \otimes x = e$, *and* $x^n = (x^{-1})^{-n}$ *for* $n < 0$, *then* $x^m \otimes x^n = x^{m+n}$ *for all integers* m *and* n.

The interested reader will find proofs of all these results in J. Aczél [1] and [2].

We may summarize the results of this section by saying that every quasigroup $(G, \cdot)$ (that is, every bordered latin square) determines a 3-net of a particular geometrical type which it co-ordinatizes. Conversely, to each such particular 3-net there corresponds a family of isotopic quasigroups any one of which co-ordinatizes it, and this family includes loops. An identity such as $(p \otimes p) \otimes p = p \otimes (p \otimes p)$ (validity of power-associativity) which is valid in all loops isotopic to the quasigroup $(G, \cdot)$ corresponds to closure of a certain geometrical configuration in the associated 3-net. This leads to the following definition:

DEFINITION. An identity in a loop $(G, \otimes)$ is called *universal* if it is satisfied in every loop isotopic to $(G, \otimes)$.

In [1], V. D. Belousov and V. V. Ryzkov have given an algorithm for constructing the closure figure corresponding to a given universal identity and, as an illustration, have applied it to the Moufang identity $x \otimes [y \otimes (x \otimes z)] = [(x \otimes y) \otimes x] \otimes z$.

Some further discussion of connections between quasigroups and geometric nets will be found in M. Neumann [1] and [2], A. Sade [10] and V. D. Belousov [9].

8.2. Orthogonal latin squares and k-nets

In the previous section, we defined a k-net and showed that every 3-net N defines a class of corresponding latin squares (representing the multiplication tables of a set of isotopic quasigroups Q_N) and that the geometric properties of the 3-net are reflected in the algebraic properties of the quasigroups associated with these squares. Also, in section 5.2, we showed that each $(n + 1)$-net of order n (that is to say, each affine plane) defines a set of $n - 1$ mutually orthogonal latin squares, and conversely. Both of these results are special cases of the general statement that each geometric k-net N of order n defines and is defined by a corresponding set of $k - 2$ mutually orthogonal latin squares of order n.

Moreover, the discussion of the previous section makes it evident that the algebraic properties of the quasigroups whose multiplication tables are represented by these latin squares reflect geometrical properties of the various corresponding sub-3-nets of the net N.

A slight modification of the argument of theorem 8.1.2 is all that is necessary to show that our general statement is true.

In view of its importance, we formulate it as a theorem:

THEOREM 8.2.1. *Each geometric k-net N of order n defines, and is defined by, a corresponding set of $k-2$ mutually orthogonal latin squares of order n.*

PROOF. We designate the various parallel classes of the k-net N by script letters $\mathcal{A}$, $\mathcal{E}$, $\mathcal{B}_1$, $\mathcal{B}_2, \ldots, \mathcal{B}_{k-2}$. Let the lines of these parallel classes be labelled as follows: $a_1, a_2, \ldots, a_n$ are the lines of the class $\mathcal{A}$; $e_1, e_2, \ldots, e_n$ are the lines of the class $\mathcal{E}$; $b_{j1}, b_{j2}, \ldots, b_{jn}$ are the lines of the class $\mathcal{B}_j$. Every point $P(h, i)$ of N can then be identified with a set of k numbers $(h, i, l_1, l_2, \ldots, l_{k-2})$ describing the k lines e_h, a_i, b_{1l_1}, $b_{2l_2}, \ldots, b_{k-2\, l_{k-2}}$ with which it is incident, one from each of the k parallel classes, and a set of $k-2$ mutually orthogonal latin squares can be formed in the following way: In the jth square, put l_j in the (h, i)th place. Each square is latin since, as h varies with i fixed so does l_j; and, as i varies with h fixed, so does l_j. Each two squares L_p and L_q are orthogonal: for if not, we would have two lines belonging to distinct parallel classes with more than one point in common.

Conversely, from a given set of $k-2$ mutually orthogonal latin squares, we may construct a k-net N. We define a set of n^2 points (h, i), $h = 1, 2, \ldots, n$; $i = 1, 2, \ldots, n$; where the point (h, i) is to be identified with the k-tuple of numbers $(h, i, l_1, l_2, \ldots, l_{k-2})$, l_j being the entry in the hth row and ith column of the jth latin square L_j. We form kn lines b_{jl}, $j = -1, 0, 1, 2, \ldots, k-2$; $l = 1, 2, \ldots, n$; where b_{jl} is the set of all points whose $(j+2)$th entry is l and $b_{-1l} \equiv e_l$, $b_{0l} \equiv a_l$. Thus, we obtain k sets of n parallel lines. (Two lines are *parallel* if they have no point in common.) Also, from the orthogonality of the latin squares, it follows that two lines of distinct parallel classes intersect in one and only one point, so we have a k-net.

In [3], R. H. Bruck has used the representation of a set of $k-2$ mutually orthogonal latin squares by means of a k-net to obtain an interesting criterion for such a set of squares to have a common transversal, and hence has obtained a simple necessary (but not sufficient) condition for such a set of squares to be extendible to a larger set (of the same order). In a subsequent paper [6] and again using the net representation, he has obtained a sufficient condition in terms of the relative sizes of k and n for a set of mutually orthogonal latin squares to be extendible to a complete set. We shall give a detailed account of these important results in the next chapter.

8.3. Co-ordinatization of a k-net or projective plane

In his well-known paper [1], M. Hall has shown that co-ordinates may be introduced into an arbitrary projective plane of order r in the following way:

We select any two points A, B in the given projective plane π. If from π we remove the line l_∞ joining A, B and the points on l_∞, the remaining plane π^* is an affine plane. We use A and B as centres of perspectivities to introduce co-ordinates for the points of π^*. No ambiguity will arise if, as we shall sometimes find convenient, we speak of π and π^* as the same plane as long as l_∞ is fixed.

Let the lines of the pencil through A in π^* be denoted by $x = x_0$, $x = x_1, \ldots, x = x_{r-1}$ where r is their cardinal number and $x_0, x_1, \ldots, x_{r-1}$ are r different symbols. Similarly, denote the lines of the pencil through B by $y = y_0$, $y = y_1, \ldots, y = y_{r-1}$. (The cardinal number of the y's is necessarily the same as that of the x's.) Through every point P of π^*, there is exactly one line $x = x_i$ and one line $y = y_j$. Denote P by (x_i, y_j). On an arbitrary line L of π^*, not through A or B, there are points (x_i, y_j) where each x and each y occurs exactly once. Henceforward, suppose that the symbols $x_0, x_1, \ldots, x_{r-1}$ are the same as $y_0, y_1, \ldots, y_{r-1}$, though not necessarily in the same order. Then, an arbitrary line L, not through A or B, is associated with the permutation $\begin{pmatrix} \ldots x_i \ldots \\ \ldots y_i \ldots \end{pmatrix}$ where the (x_i, y_i) are the points of L. This expresses the fact that L determines a one-to-one correspondence between the lines of the pencil through A and the lines of the pencil through B. Without confusion we may write $L = \begin{pmatrix} \ldots x_i \ldots \\ \ldots y_i \ldots \end{pmatrix}$ since distinct lines contain at most one point in common and hence are associated with distinct permutations.

It is evident that the same method of co-ordinatization may be used for a k-net of order r.

We now choose two particular points O and I which are joined by a line but are not collinear with A or B and assign them the co-ordinates (0,0) and (1,1) respectively. (In the case of a projective plane, every two points are joined by a line and so the choice of O and I is entirely arbitrary.) This assignment of co-ordinates is equivalent to assigning the labels 0, 1 to two of the symbols $x_0, x_1, \ldots, x_{r-1}$, say $x_0 = 0$, $x_1 = 1$.

Let OI meet AB at E and let the remaining points of AB be denoted by $B_2, B_3, \ldots, B_{k-2}$, $k \leq r + 1$. If the line OB_m meets the line AI at the point $(1, m)$, denote the point B_m by the symbol (m). In particular denote the point E $(\equiv B_1)$ by the symbol (1). We are now able to define operations $(+)$ and $(\wedge)$ on the symbols $x_0, x_1, \ldots, x_{r-1}$ by means of which the lines of the pencils with vertices E and O may be assigned equations. Each line through E has a permutation representation of the form $\begin{pmatrix} 0 \ldots x_i \ldots \\ a \ldots y_i \ldots \end{pmatrix}$. We define $x_i + a = y_i$ for $i = 0, 1, 2, \ldots, r - 1$ and say that the line has equation $y = x + a$. This defines the result of the operation $(+)$ on every ordered pair of the r symbols. Each line through O has a permutation representation of the form $\begin{pmatrix} 0 & 1 \ldots x_i \ldots \\ 0 & m \ldots y_i \ldots \end{pmatrix}$. We define $x_i \wedge m = y_i$ for $i = 0, 1, 2, \ldots, r - 1$ and say that the line has equation $y = x \wedge m$. This defines the result of the operation $(\wedge)$ for all choices of x and for $k - 1$ finite values of m.

We can provide equations for the remaining lines of the plane with the aid of a ternary operation defined on the set of symbols $x_0, x_1, \ldots x_{r-1}$ in the following way. If x, m, a are any three of the symbols, but with m restricted to $k - 1$ finite values in the case of a k-net, we define $T(x, m, a) = y$, where y is the second co-ordinate of the point on the line joining the points (m) and $(0, a)$ whose first co-ordinate is x. We can then say that $y = T(x, m, a)$ is the "equation" of the line joining the points (m) and $(0, a)$. The connection between this ternary operation on the symbols $x_0, x_1, \ldots, x_{r-1}$ and the binary operations $(+)$ and $(\wedge)$ previously introduced is given by the relations $a + b = T(a, 1, b)$ and $a \wedge b = T(a, b, 0)$. We may also observe that, both in the case of a complete projective plane and in the case of a k-net with $k < r + 1$, the lines which pass through the point E are those which are represented in the permutation representation by permutations which displace all the symbols and by the identity permutation. We shall denote this subset of the set S of permutations representing the lines by the symbol $\bar{S}$.

Our purpose in the present section is to establish a connection between the above representation of a projective plane or k-net by means of permutations and associated natural domain[1] operations $(+)$ and $(\wedge)$ and the representation of the same projective plane or k-net by means of a set of $k-2$ mutually orthogonal $r \times r$ latin squares (explained in the previous section) and associated operations $(+)$ and $(\cdot)$ defined by $xS_b = x + b$ and $xM_i = x.x_i$, where the S_i are the permutations representing the rows of one latin square L_1 as permutations of its first row and the M_i are the permutations defining the remaining latin squares $L_2^*, L_3^*, \ldots, L_{k-2}^*$ in the manner explained in section 7.1.

Let us identify the points $A, E(\equiv B_1), B, B_2, B_3, \ldots, B_{k-2}$ of the above discussion with the vertices of the parallel classes $\mathcal{A}, \mathcal{E}, \mathcal{B}_1, \mathcal{B}_2, \ldots, \mathcal{B}_{k-2}$ of the k-net in the notation of section 8.2. In the Hall co-ordinatization, the lines through the point A are denoted by $x_0, x_1, \ldots, x_{r-1}$ and the lines through the point B by $y_0, y_1, \ldots, y_{r-1}$. We may set up a one-to-one correspondence between that symbolism and the symbolism of section 8.2 by denoting the lines of the parallel class $\mathcal{A}$ by $a_{x_0}, a_{x_1}, \ldots, a_{x_{r-1}}$ and the lines of the parallel class $\mathcal{B}_1$ by $b_{1y_0}, b_{1y_1}, \ldots, b_{1y_{r-1}}$. The finite point with co-ordinates (x, y) will then be the point of intersection of the lines a_x of the parallel class $\mathcal{A}$ and b_{1y} of the parallel class $\mathcal{B}_1$. The line e_z of the parallel class $\mathcal{E}$ (line through the point E) which is incident with the point $P(x, y)$ will be associated with the permutation $\begin{pmatrix} \ldots x \ldots \\ \ldots y \ldots \end{pmatrix}$ of the set $\bar{S}$.

Now let $e_z = \begin{pmatrix} 0 & \ldots & i & j & k & \ldots \\ z & \ldots & p & q & r & \ldots \end{pmatrix}$ be an arbitrary permutation of the set $\bar{S}$. Then the pairs of lines a_0, b_{1z}; a_i, b_{1p}; a_j, b_{1q}; a_k, b_{1r} are incident on the line e_z. That is, in the notation of section 8.2, the points $(z, 0)$, (z, i), (z, j), (z, k) are respectively incident with the lines $b_{1z}, b_{1p}, b_{1q}, b_{1r}$ of the parallel class $\mathcal{B}_1$ (consisting of lines through the point B). Thus, in the latin square L_1 the entry in the zth row and 0th column will be z, the entry in the zth row and ith column will be p, the entry in the zth row and jth column will be q, and so on. We deduce that the entries in the zth row of the latin square L_1 provide just that transposition of symbols from the natural order $0, 1, \ldots, i, j, k, \ldots$, which is defined by the permutation e_z of the set $\bar{S}$. Also, if, as here, the notation for the lines through E is so chosen that the line e_z through E is the one represented

[1] as they have been called in M. Hall [1].

by the permutation which transforms 0 into z, then it is the entries in the zth row of the latin square L_1 (the row corresponding to the suffix) which are arranged according to this same permutation of the symbols from their natural order. We note too that permutation from natural order is permutation from 0th row order, since the set $\bar{S}$ always contains the identity permutation.

Thus we have proved that, in this representation by means of latin squares, the permutations $S_0 \equiv I, S_1, S_2, \ldots, S_{r-1}$, representing the rows of the latin square L_1 as permutations of its first row, may be identified with the permutations of the set $\bar{S}$ in the M. Hall co-ordinatization. It follows at once that the addition defined by $aS_b = a + b$ then coincides with natural domain addition as defined by M. Hall.

We shall formulate our result as a theorem.

THEOREM 8.3.1. *In the representation of a projective plane or a k-net by means of a set of mutually orthogonal latin squares the permutations* $S_0 \equiv I, S_1, S_2, \ldots, S_{r-1}$ *representing the rows of one latin square* L_1 *as permutations of its first row may be identified with the permutations of the set* $\bar{S}$ *in a suitably chosen Marshall Hall type co-ordinatization. The addition defined by* $aS_b = a + b$ *then coincides with natural domain addition as defined by Marshall Hall.*

As regards the connection between latin square multiplication $(\cdot)$, defined by $aM_i = a.x_i$, and natural domain multiplication $(\wedge)$, we have the following theorem (which was first given in A. D. Keedwell [4]):

THEOREM 8.3.2. *If* $L_i = \{M_i S_0 M_i^{-1}, M_i S_1 M_i^{-1}, \ldots, M_i S_{r-1} M_i^{-1}\}$ *for* $i = 1, 2, \ldots, k-2$ *are a standardized set of mutually orthogonal latin squares (as introduced in section 7.4) and* $(\wedge)$ *denotes multiplication in the natural domain of the associated projective plane or k-net, then the statement* $u \wedge p = v$ $(p \neq 1)$ *is equivalent to the statement that there exists a permutation* M_p *such that* $1M_p S_g M_p^{-1} = 0 = uM_p S_h M_p^{-1}$; *where* g, h *are defined by the relations* $1S_g = p$, $uS_h = v$.

Further, if, in a complete projective plane, the operation $(\cdot)$ *is left-distributive over the operation* $(+)$, *then the operations* $(\cdot)$ *and* $(\wedge)$ *are equivalent, provided that the notation is chosen so that* $0M_i = 0$ *and* $1M_i = x_i$ *for* $i = 1, 2, \ldots, r-1$,[1] *with* $x_1 = 1$.

[1] This is equivalent to requiring that the symbol 1 be a left-identity for the operation $(\cdot)$. In effect, if $1M_i = x_i$, we define $xx_i = xM_i$. Since $M_1 \equiv I$, the symbol 1 is already required to be a right-identity.

PROOF. With the notation used above, $a_0, a_1, \ldots, a_{r-1}$ are the lines of the parallel class $\mathscr{A}$, $e_0, e_1, \ldots, e_{r-1}$ are the lines of the parallel class $\mathscr{E}$, $b_{j0}, b_{j1}, \ldots, b_{jr-1}$ are the lines of the parallel class $\mathscr{B}_j$ for $j = 1, 2, \ldots, k-2$.

Since the set of squares $L_1, L_2, \ldots, L_{k-2}$ are a standardized set, the element 0 occurs in the 0th row and column of each square. Consequently, the lines b_{j0} are incident, for every j, with the point of intersection of the lines e_0, a_0. It follows that the line b_{j0} passes through the point (0, 0) relative to A, B_1 as co-ordinatizing points, for $j = 2, 3, \ldots, k-2$. Consequently, each such line b_{p0} must have an equation of the form $y = x \wedge p$ for some $p \neq 0, 1$. Also the line e_0 has equation $y = x \wedge 1$. (That is, $y = x$.)

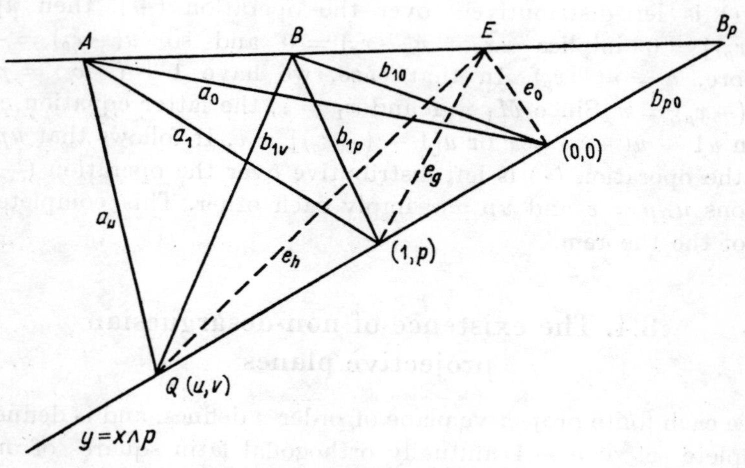

Fig. 8.3.1.

If the line b_{p0} has equation $y = x \wedge p$, then the lines a_1, b_{1p} are incident on b_{p0}. If then e_g is the line of the pencil vertex E which passes through the point of intersection $(1, p)$ of the lines a_1, b_{1p}, we have that p is the entry in the gth row and first column of the square L_1, and that 0 is the entry in the corresponding row and column of the square L_p. So, $1M_p\, S_g\, M_p^{-1} = 0$, where $1S_g = p$. Conversely, when the latter condition holds, the lines b_{p0} and b_{1p} pass through the point $e_g \cap a_1$ and so the line b_{p0} has equation $y = x \wedge p$ relative to A, B as co-ordinatizing points.

Now, let $Q(u, v)$ be any other point on the line b_{p0}. Then we shall have $u \wedge p = v$. This implies that the lines a_u and b_{1v} pass through Q. If the line EQ is the line e_h, we have that v is the entry in the hth row and uth column of the square L_1 and that 0 is the entry in the corresponding row and column of the square L_p. So, $uM_p S_h M_p^{-1} = 0$, where $uS_h = v$. Conversely, when this condition holds, the line b_{p0} contains the point $Q(u, v)$.

Thus, when $1M_p S_g M_p^{-1} = 0$, where $1S_g = p$; and $uM_p S_h M_p^{-1} = 0$, where $uS_h = v$; the line b_{p0} has equation $y = x \wedge p$ and contains Q. Therefore, $u \wedge p = v$ if and only if these conditions hold.

If we suppose the notation chosen so that $0M_i = 0$ and $1M_i = x_i$ for $i = 1, 2, \ldots, r - 1$ (in a complete projective plane) and with $x_1 = 1$, we may restate this first part of the theorem in the form: *the statement that $u \wedge p = v$ is equivalent to the statement that, if g, h are defined by the equations $1 + g = p$ and $u + h = v$, then $x_p + g = 0$ and $ux_p + h = 0$.*

For the second part, let us denote the unique[1] solution x of the equation $cS_x = 0$ by $-c$. Then $g = -x_p$ and $h = -ux_p$. If the operation $(\cdot)$ is left-distributive[2] over the operation $(+)$, then $u[x_p + (-x_p)] = 0$ implies $ux_p + u(-x_p) = 0$ and so $u(-x_p) = -ux_p$. Therefore, $h = u(-x_p)$. In that case, we have $1 + (-x_p) = p$ and $u + u(-x_p) = v$. Since $M_1 = I$ and $x_1 = 1$, the latter equation can be written $u1 + u(-x_p) = v$ or $u[1 + (-x_p)] = v$. It follows that $up = v$. So, if the operation $(\cdot)$ is left-distributive over the operation $(+)$, the equations $u \wedge p = v$ and $up = v$ imply each other. This completes the proof of the theorem.

8.4. The existence of non-desarguesian projective planes

Since each finite projective plane of order n defines, and is defined by, a complete set of $n - 1$ mutually orthogonal latin squares of order n (see theorem 5.2.2), it is evident that the answer to the question "How many non-equivalent complete sets of mutually orthogonal latin squares of order n exist?" is equivalent to the question "How many geometrically distinct projective planes of order n exist?". We have already shown that there exists a Galois or desarguesian[3] plane of every positive

[1] necessarily unique because L_1 is a latin square.

[2] Notice that the assumption of the left-distributive law involves the assumption that the product $a.b$ is defined for all the symbols b which occur in the cells of the latin squares. In other words, it involves the assumption that this set of squares is a complete set, representing a complete projective plane. Consequently, our corollary is not necessarily valid for k-nets with $k < n + 1$.

[3] A Galois plane is called a desarguesian plane because in such a plane the well-known configurational theorem of Desargues concerning perspective triangles is universally valid. (It is conjectured that every plane of prime order is desarguesian.)

integral order n that is a power of a prime number (theorem 5.2.3) and we wish now to demonstrate the existence of non-desarguesian planes.

The simplest type of non-desarguesian projective planes to describe are the so-called *translation planes*.

We may specify such a plane by the nature of its co-ordinate system. We suppose co-ordinates to have been introduced into the plane in the manner described at the beginning of the preceding section, so that there is a special line $AB = l_\infty$ whose points are represented by symbols (m) and so that all other points are represented by co-ordinate pairs (x, y), where m, x, y belong to a co-ordinate set Σ. Also lines through A have equations of the type $x = x_r$, lines through B have equations of the form $y = y_r$, and all other lines have equations of the form $y = T(x, m, a)$, where T is a ternary operation on the set Σ. By means of T, binary operations $(+)$ and $(\wedge)$ may be defined on Σ, where $a + b = T(a, 1, b)$ and $a \wedge b = T(a, b, 0)$, 0 and 1 being two selected symbols of the set Σ. If then the algebraic system $(\Sigma, +, \wedge)$ has the properties (i) $(\Sigma, +)$ is an abelian group with 0 as its identity element, (ii) $(\Sigma - 0, \wedge)$ is a loop with 1 as its identity element, (iii) $(a + b) \wedge m = a \wedge m + b \wedge m$ for all a, b, m in Σ, and (iv) if $r \neq s$, the equation $x \wedge r = x \wedge s + t$ has a unique solution x in Σ when r, s, t are in Σ, we say that it is a *Veblen–Wedderburn system* or a *right quasifield*. If, further, the ternary operation T on Σ satisfies the condition $T(a, m, b) = a \wedge m + b$ (which is consistent with the relations describing the binary operations $(+)$ and $(\wedge)$ in terms of T previously given), then we may easily verify that the axioms for a projective plane are satisfied and we call the resulting plane a *translation plane*.

In the first place, we have to check that there is a unique line joining two given points (x_r, y_r) and (x_s, y_s). If $x_r = x_s$, the required unique line is that with equation $x = x_r$. If $y_r = y_s$, the required line is $y = y_s$. If $x_r \neq x_s$ and $y_r \neq y_s$, the required line is that with equation $y = x \wedge m + b$ where m and b satisfy the equations $x_r \wedge m + b = y_r$ and $x_s \wedge m + b = y_s$: that is, m is the unique solution of the equation $(x_r - x_s) \wedge m = y_r - y_s$ and b is then determined uniquely by the requirement that $x_r \wedge m + b = y_r$. In the second place, there is a unique point common to each two lines. The lines $x = x_r$, $x = x_s$ have the point A in common. Similarly, the lines $y = y_r$, $y = y_s$ have the point B in common. The lines $x = x_r$, $y = y_s$ have the point (x_r, x_s) in common. The lines $x = x_r$ and $y = x \wedge m + b$ have the point $(x_r, x_r \wedge m + b)$ in common. The lines $y = y_s$, $y = x \wedge m + b$ have the point (x_s, y_s) in common, where x_s is the unique solution of the equation $x_s \wedge m = y_s - b$.

Finally, the lines $y = x \wedge m_1 + b_1$ and $y = x \wedge m_2 + b_2$ intersect in the point whose x co-ordinate is the unique solution of the equation $x \wedge m_1 = x \wedge m_2 + (b_2 - b_1)$ and whose y co-ordinate can then be obtained from either one of the equations of the two lines.

It is easy to see that, in particular, every Galois field is a right quasifield. Moreover, in the case when the right quasifield is a Galois field, homogeneous co-ordinates can be introduced in a manner analogous to that used in elementary geometry and it is then quite simple to show that the plane is isomorphic to the Galois plane of the same order. For the details, the reader is referred to books on projective planes. (See, for example, G. Pickert [2].) We deduce that the existence of finite translation planes which are geometrically distinct from the Galois planes is dependent upon the existence of finite right quasifields which are not fields. The following theorem, due to M. Hall (see [1]), provides one method of constructing such quasifields.

Theorem 8.4.1. *Let F be a given commutative field and let $f(x) = x^2 - rx - s$ be a quadratic polynomial irreducible over F. Then the set of elements $a = a_1 + ua_2$, $a_1, a_2 \in F$, is a Veblen–Wedderburn system J under the following rules of addition and multiplication:*

$$(1)\ (a_1 + ua_2) + (b_1 + ub_2) = (a_1 + b_1) + u(a_2 + b_2);$$

$$(2)\ (a_1 + ua_2)\,b_1 = a_1b_1 + ua_2b_1 \quad \text{if} \quad b_1 \in F;$$

$$(3)\ (a_1 + ua_2)\,(b_1 + ub_2) = sA_2 + b(A_1 + rA_2)$$

(where $A_1 = a_1 - a_2b_1b_2^{-1}$, $A_2 = a_2b_2^{-1}$)

$$= sa_2b_2^{-1} + a_1b_1 - a_2b_1^2b_2^{-1} + ra_2b_1b_2^{-1} + u(a_1b_2 - a_2b_1 + ra_2),$$

provided $b_2 \neq 0$. Here, $b = b_1 + ub_2$ and $a = a_1 + ua_2 = A_1 + bA_2$.

Each element of J which is not in F satisfies the equation $f(x) = 0$; and, if $x, y \in J$, $c \in F$, we have $cx = xc$, $c(xy) = (cx)y = (xc)y$.

Proof. We note firstly that the rule (3) for multiplication by b when $b \notin F$ is chosen so that each element of J not in F shall satisfy the equation $f(x) = 0$. We have $ab = sA_2 + b(A_1 + rA_2)$ where $a = A_1 + bA_2$. Thus, when $a = b$, we have $A_1 = 0$, $A_2 = 1$, and then $b^2 = s + rb$. That is, $f(b) = 0$. We also note that, if $a_2 = 0$, rule (3) gives $a_1(b_1 + ub_2) = a_1b_1 + ua_1b_2 = b_1a_1 + ub_2a_1 = (b_1 + ub_2)a_1$ by rule (2).

Thus, a product of two elements $a, b \in J$ commutes if either or both belong to F, but not when both belong to $J \setminus F$ (by rule (3)).

It follows easily from (1) that the elements of J form an abelian group under addition. Since, in (2) and (3), each term of the product involves one of a_1, a_2 to the first degree and the other not at all, the right-distributive law $(a^{(1)} + a^{(2)})\, b = a^{(1)} b + a^{(2)} b$ holds.

To show that multiplication is a loop, we need to prove the unique solubility of each of the equations $xb = c$ and $ay = c$. The solubility of $xb = c$ is easily shown again using the fact that, in (2) and (3), each term of the product xb involves one of x_1, x_2 to the first degree and the other not at all. As regards the solubility of $ay = c$, suppose firstly that $a_1 c_2 = a_2 c_1$. Then $y_1 = c_1/a_1 = c_2/a_2$, $y_2 = 0$ is a solution by rule (2). If $a_1 c_2 \neq a_2 c_1$, we try to find a solution with $y_2 \neq 0$ using rule (3). We have $a = A_1 + yA_2$ and $c = C_1 + yC_2$, where $A_1 = a_1 - a_2 y_1 y_2^{-1}$, $A_2 = a_2 y_2^{-1}$, $C_1 = c_1 - c_2 y_1 y_2^{-1}$, $C_2 = c_2 y_2^{-1}$. Therefore $ay = c$ implies $sA_2 + y(A_1 + rA_2) = C_1 + yC_2$, which requires $C_1 = sA_2$ and $C_2 = A_1 + rA_2$. Therefore, $c_1 y_2 - c_2 y_1 = sa_2$ and $a_1 y_2 - a_2 y_1 = c_2 - ra_2$. These two equations are soluble for y_1, y_2 and we have

$$y_2 = \frac{c_2^2 - ra_2 c_2 - sa_2^2}{a_1 c_2 - a_2 c_1} = \frac{a_2^2 f(c_2/a_2)}{a_1 c_2 - a_2 c_1} \neq 0$$

unless $a_2 = 0$, since $f(x)$ is irreducible over F and $a_1 c_2 \neq a_2 c_1$ by hypothesis. If $a_2 = 0$, $ay = ya$ and we have already shown that $ya = c$ is soluble in J.

Finally, to show that, when $m \neq n$, the equation $xm = xn + h$ has a unique solution, it is sufficient to find a solution $x^{(1)}$ since, if $x^{(2)}$ were another solution, we should have $(x^{(1)} - x^{(2)})\, m = (x^{(1)} - x^{(2)})\, n$ in contradiction to the loop property of multiplication just proved. If m, n both belong to F, the products xm, xn each commute so, using the right-distributive law, we deduce that the unique solution of the equation $(m - n)x = h$ solves our equation $xm = xn + h$. If $m \notin F$, $n \in F$, suppose that $x = x_1 + ux_2 = X_1 + mX_2$, $n = n_1 \in F$, $h = h_1 + uh_2 = H_1 + mH_2$. Then, $xm = xn_1 + h$ implies $sX_2 + m(X_1 + rX_2) = X_1 n_1 + mX_2 n_1 + H_1 + mH_2$ and so a solution exists if the equations

$$\left.\begin{aligned} sX_2 - n_1 X_1 &= H_1 \\ (r - n_1)X_2 + X_1 &= H_2 \end{aligned}\right\} \text{ are soluble for } X_1,\ X_2 \text{ in } F. \text{ Since}$$

$\{s + (r - n_1)\, n_1\}\, X_1 = sH_2 - (r - n_1)\, H_1$, $\{s + (r - n_1)\, n_1\}\, X_2 = H_1 + n_1 H_2$, and $s + (r - n_1)\, n_1 = -f(n_1) \neq 0$ for $n_1 \in F$, solutions exist. If $m \in F$, $n \notin F$, we consider similarly the equation $xn = xm - h$. Finally, to deal with the case $m \notin F$, $n \notin F$, we put $x = X_1 + mX_2$, $h = H_1 + mH_2$, and $n = N_1 + mN_2$. We observe that, by rule (2), $m(ab) = (ma)b$ if a, b both belong to F, and so

$$x = X_1 + mX_2 = X_1 - N_1N_2^{-1}X_2 + (N_1 + mN_2)(N_2^{-1}X_2),$$

where we have also used the right-distributive law. Hence, since $x = X_1 - N_1N_2^{-1}X_2 + n(N_2^{-1}X_2)$, we find that

$$xn = sN_2^{-1}X_2 + n\{(X_1 - N_1N_2^{-1}X_2) + rN_2^{-1}X_2\} =$$

$$= sN_2^{-1}X_2 + (N_1 + mN_2)\{(X_1 - N_1N_2^{-1}X_2) + rN_2^{-1}X_2\} =$$

$$= sN_2^{-1}X_2 + N_1X_1 - N_1^2N_2^{-1}X_2 + rN_1N_2^{-1}X_2 + m(N_2X_1 - N_1X_2 + rX_2).$$

We also have $xm = sX_2 + m(X_1 + rX_2)$ as before, and so a solution of the equation $xm = xn + h$ exists if the equations

$$\left.\begin{aligned} -N_1X_1 + (N_1^2N_2^{-1} - rN_1N_2^{-1} - sN_2^{-1} + s)X_2 &= H_1 \\ (1 - N_2)X_1 + N_1X_2 &= H_2 \end{aligned}\right\} \text{ are soluble for}$$

X_1, X_2 in J. Since we have

$$(1 - N_2)(N_1^2N_2^{-1} - rN_1N_2^{-1} - sN_2^{-1} + s) + N_1^2 =$$

$$= N_1^2N_2^{-1} - r(1 - N_2)N_1N_2^{-1} - s(1 - N_2)^2N_2^{-1} =$$

$$= (1 - N_2)^2 N_2^{-1} f\left(\frac{N_1}{1 - N_2}\right)$$

and $N_2 \neq 0$, $f\left(\frac{N_1}{1 - N_2}\right) \neq 0$, solutions exist.

The smallest order for which the above construction gives a right quasifield which is not a field is the order 9. Each of the quadratic polynomials $x^2 + 1$, $x^2 - x - 1$, and $x^2 + x - 1$, which are irreducible over $GF[3]$, gives rise to a proper right quasifield. Although these three right quasifields are not isomorphic, they all give rise to the same non-desarguesian projective plane of order 9; the unique proper translation plane of that order. Two further non-desarguesian planes of order 9 are known. One is the dual of the translation plane. The other is a member of another class of non-desarguesian planes known as the Hughes planes (see D. R. Hughes [2]), although that of order 9 was first constructed by Veblen and Wedderburn in 1907 (see O. Veblen and J. H. M. Wedderburn [1]).

As an illustration, we shall describe the right quasifield constructed over $GF[3]$ by means of the polynomial $x^2 - x - 1$. Its elements are of the form $a_1 + ua_2$ with a_1, a_2 in $GF[3]$. Addition is defined by

$(a_1 + ua_2) + (b_1 + ub_2) = (a_1 + b_1) + u(a_2 + b_2)$ and the multiplication table is as given in Fig. 8.4.1. Every element not in $GF[3]$ satisfies the equation $x^2 - x - 1 = 0$, which is irreducible over $GF[3]$.

	1	-1	u	$1+u$	$-1+u$	$-u$	$1-u$	$-1-u$
1	1	-1	u	$1+u$	$-1+u$	$-u$	$1-u$	$-1-u$
-1	-1	1	$-u$	$-1-u$	$1-u$	u	$-1+u$	$1+u$
u	u	$-u$	$1+u$	1	$-1-u$	$-1+u$	-1	$1-u$
$1+u$	$1+u$	$-1-u$	$1-u$	$-1+u$	1	-1	$-u$	u
$-1+u$	$-1+u$	$1-u$	1	$-u$	u	$-1-u$	$1+u$	-1
$-u$	$-u$	u	$-1-u$	-1	$1+u$	$1-u$	1	$-1+u$
$1-u$	$1-u$	$-1+u$	-1	u	$-u$	$1+u$	$-1-u$	1
$-1-u$	$-1-u$	$1+u$	$-1+u$	$1-u$	-1	1	u	$-u$

Fig. 8.4.1

When the complete set of mutually orthogonal latin squares which are defined by a desarguesian plane, by a translation plane, or by the dual of a translation plane, are constructed and put into standardized form (see page 159), it is found that the rows of any one square L_k of the set are the same as those of any other square L_h of the set, except that they occur in a different order. That this is necessarily so was proved for the case of desarguesian planes by R. C. Bose and K. R. Nair in [2]. However, there exist other types of projective plane for which the squares do not have this property as the same authors pointed out, among them being the Hughes planes mentioned above. From the point of view of the theory of latin squares it is therefore important to have a criterion for distinguishing the two cases. Such a criterion has been given by D. R. Hughes (see [1]) and in a more geometrical form by one of the present authors, who has proved

Theorem 8.4.2. *A necessary and sufficient condition that, in a standardized complete set of mutually orthogonal latin squares, the rows of the square* L_k *be the same as those of the square* L_h *except that they occur in a different order is that the squares represent the incidence structure of a projective plane in which the first minor theorem of Desargues*[(1)] *holds affinely with E as a vertex of perspective and* A, B_h, B_k *as meets of corresponding sides of the two triangles.* (The notation is that of the previous section.)

Proof. Let z_h be an arbitrary row of the square L_h and suppose that this is identical with the z_kth row of the square L_k. Then the entry in the x_rth column of the z_hth row of the square L_h and in the x_rth column of the z_kth row of the square L_k are the same, say y_r. Therefore the line denoted by b_{hy_r} is incident with the point of intersection P of the lines e_{z_h}, a_{x_r}; and the line denoted by b_{ky_r} is incident with the point of intersection Q of the lines e_{z_k}, a_{x_r}. Similarly, the entries in the x_sth column of the z_hth row of the square L_h and of the z_kth row of the square L_k are the same, say y_s. Therefore, the line denoted by b_{hy_s} is incident with the point of intersection P' of the lines e_{z_h}, a_{x_s}; and the line denoted by b_{ky} is incident with the point of intersection Q' of the lines e_{z_k}, a_{x_s}.

Now, the entry y_r occurs in some column of the first row of the square L_h: namely in the y_rth column since the squares are in standardized form, the entry y_r occurs also in the y_rth column of the first row of the square L_k. Therefore, the lines b_{hy_r}, b_{ky_r} are both incident with the point of intersection of the lines e_1, a_{yr}. That is, the lines b_{hy_r}, b_{ky_r} intersect at some point R of the line e_1. Similarly, the lines b_{hy_s}, b_{ky_s} intersect at some point R' of the line e_1. The triangles PQR, $P'Q'R'$ are in perspective with E as centre and AB_hB_k as axis. Since z_h, x_r, x_s were arbitrary so also were the points P, P' and so the first minor theorem of Desargues is satisfied affinely with ε as vertex of perspective and A, B_h, B_k as meets of corresponding sides of the two triangles.

Moreover, it is clear that when the minor theorem of Desargues is so satisfied for arbitrary choice of P, P' on an arbitrary line e_{z_h} through

[(1)] The first minor theorem of Desargues is the special case of Desargues' theorem in which the vertex of perspectivity of the two triangles lies on the axis of perspectivity. It is said to be satisfied affinely when this axis is the special line l_∞.

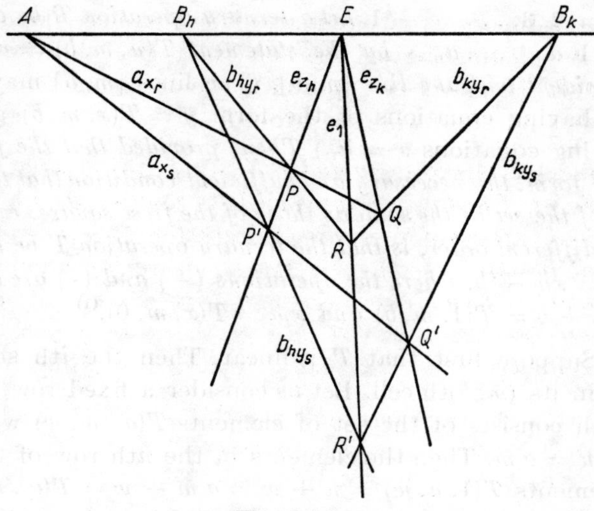

Fig. 8.4.2

E, then each row z_h of square L_h will be identical with some row z_k of square L_k.

It is a well known fact in the theory of projective planes that, if a given plane is co-ordinatized in the Marshall Hall manner, the ternary operation which is defined on the co-ordinatizing set Σ will be linear (that is, we shall have the equality $T(a, m, b) = a \wedge m + b)$ if and only if the minor theorem of Desargues holds affinely for suitable choices of vertex on l_∞. This explains the connection between the above theorem and the criterion of D. R. Hughes, which reads

Theorem 8.4.3. *Let a projective plane π be constructed from a given complete set of mutually orthogonal latin squares in the following way. The points of π are symbols (∞), (0), (1), (a_i), $i = 2, 3, \ldots, n-1$, and ordered pairs of symbols (x, y) where x and y belong to the set $0 = a_0$, $1 = a_1$, $a_2, \ldots, a_{n-1}$ of symbols occurring in the latin squares. The lines of π are symbols $[\infty]$, $[0]$, $[1]$, $[a_i]$, $i = 2, 3, \ldots, n-1$, and ordered pairs of symbols $[m, b]$ where m and b belong to the set $0, 1, a_2, \ldots, a_{n-1}$. Incidences between points and lines are defined as follows: (i) the line $[m, b]$ is incident with the point (a_i, y), $a_i \neq 0$, if y occurs in the (m, b)th cell of the ith latin square; (ii) the line $[m, b]$ is incident with the point $(0, y)$ if $y = b$; (iii) the line $[m, b]$ is incident with the point (a_i) if $m = a_i$; (iv) the line $[a_i]$ is incident with the point (a_i, y) for all choices of y and is also incident with the point (∞); (v) the line $[\infty]$ is incident with all of the points (∞), (0),*

(1), (a_i), $i = 2, 3, \ldots, n-1$. *Let a ternary operation T be defined on the symbols* $0, 1, a_2, \ldots, a_{n-1}$ *by the statement* $T(u, m, b) = v$ *if the point* (u, v) *is incident with the line* $[m, b]$. (The lines $[m, b]$ may then be regarded as having equations of the form $y = T(x, m, b)$ and the lines $[a_i]$ as having equations $x = a_i$.) *Then, provided that the first square is in standard form, the necessary and sufficient condition that the rows of the* ith *square of the set be the same as those of the first square, except that they occur in a different order, is that the ternary operation T be linear: that is* $T(x, m, b) = xm + b$, *where the operations* $(+)$ *and* $(\cdot)$ *are defined by the relations* $m + b = T(1, m, b)$ *and* $xm = T(x, m, 0)$.[1]

PROOF. Suppose first that T is linear. Then the ith square L_i has $T(a_i, m, b)$ in its (m, b)th cell. Let us consider a fixed row, say the mth in L_i, which consists of the set of elements $T(a_i, m, w)$ with a_i and m fixed. Let $n = a_i m$. Then the elements in the nth row of the square L_1 are the elements $T(1, n, w) = n + w = a_i m + w = T(a_i, m, w)$, so the nth row of L_1 is the same as the mth row of L_i.

Conversely, suppose that the rows of each square L_i, $i = 2, 3, \ldots, n-1$, are the same as the rows of the square L_1 except that they occur in a different order. That is, suppose that the mth row of L_i is the same as the $(m\sigma)$th row of L_1. Then $T(a_i, m, w) = T(1, m\sigma, w)$ for all columns w. In particular, $a_i m = T(a_i, m, 0) = T(1, m\sigma, 0) = m\sigma$.[2] This defines the mapping σ. Then, $T(a_i, m, b) = T(1, a_i m, b) = a_i m + b$, and so T is linear.

As an example (of smallest possible order) of a standardized complete set of mutually orthogonal latin squares for which the rows of the squares $L_2, L_3, \ldots, L_{n-1}$ are not a reordering of those of the square L_1, we give in Fig. 8.4.3 the eight squares corresponding to the so-called Hughes plane of order 9. (These have also been given in L. J. Paige and C. Wexler [1].)

For further information concerning the various kinds of projective plane that may exist and their properties, the reader is referred to the textbooks on this subject, especially G. Pickert [2] and P. Dembowski [1] and, for Hungarian readers, F. Kárteszi [2]. For a concise account written from the point of view of the student of latin squares, the reader is recommended to consult chapter two of R. Guérin [4].

[1] It is easy to check that the symbols 0 and 1 act as two-sided identity elements for the operations $(+)$ and $(\cdot)$ respectively if and only if the set of latin squares is in standardized form.

[2] $T(1, k, 0) = k$ provided that the first square L_1 is in standard form.

L_1

1	2	3	4	5	6	7	8	9
2	5	6	9	8	7	3	1	4
3	7	9	8	6	5	2	4	1
4	8	2	7	1	9	5	3	6
5	4	7	1	2	3	6	9	8
6	9	4	5	3	8	1	7	2
7	3	8	6	9	1	4	2	5
8	1	5	3	4	2	9	6	7
9	6	1	2	7	4	8	5	3

L_2

1	2	3	4	5	6	7	8	9
8	4	7	1	6	2	5	9	3
6	9	4	5	3	8	1	7	2
2	3	6	8	7	5	9	1	4
3	5	1	6	9	4	8	2	7
9	6	8	7	2	1	4	3	5
5	7	9	2	8	3	6	4	1
4	8	2	9	1	7	3	5	6
7	1	5	3	4	9	2	6	8

L_3

1	2	3	4	5	6	7	8	9
4	8	9	6	7	5	2	3	1
8	1	5	3	4	2	9	6	7
3	6	1	9	8	7	4	5	2
7	9	8	2	6	1	3	4	5
5	3	7	1	9	4	6	2	8
6	4	2	8	1	9	5	7	3
9	5	6	7	2	3	8	1	4
2	7	4	5	3	8	1	9	6

L_4

1	2	3	4	5	6	7	8	9
7	6	8	3	4	1	9	2	5
5	4	7	1	2	3	6	9	8
6	7	9	5	3	2	8	4	1
9	3	2	7	1	8	4	5	6
3	8	1	2	7	9	5	6	4
8	1	5	9	6	4	2	3	7
2	9	4	6	8	5	1	7	3
4	5	6	8	9	7	3	1	2

L_5

1	2	3	4	5	6	7	8	9
6	7	4	8	9	3	1	5	2
4	8	2	7	1	9	5	3	6
9	5	7	1	2	4	3	6	8
8	6	9	3	4	5	2	7	1
7	1	5	6	8	2	9	4	3
2	9	6	5	3	7	8	1	4
3	4	1	2	7	8	6	9	5
5	3	8	9	6	1	4	2	7

L_6

1	2	3	4	5	6	7	8	9
5	3	2	7	1	9	6	4	8
9	6	1	2	7	4	8	5	3
7	9	4	3	6	8	1	2	5
2	1	5	8	3	7	9	6	4
8	5	6	9	4	3	2	1	7
4	8	7	1	2	5	3	9	6
6	7	8	5	9	1	4	3	2
3	4	9	6	8	2	5	7	1

L_7

1	2	3	4	5	6	7	8	9
3	9	1	5	2	4	8	7	6
7	3	8	6	9	1	4	2	5
8	1	5	2	4	3	6	9	7
4	7	6	9	8	2	5	1	3
2	4	9	8	6	7	3	5	1
9	5	4	3	7	8	1	6	2
5	6	7	1	3	9	2	4	8
6	8	2	7	1	5	9	3	4

L_8

1	2	3	4	5	6	7	8	9
9	1	5	2	3	8	4	6	7
2	5	6	9	8	7	3	1	4
5	4	8	6	9	1	2	7	3
6	8	4	5	7	9	1	3	2
4	7	2	3	1	5	8	9	6
3	6	1	7	4	2	9	5	8
7	3	9	8	6	4	5	2	1
8	9	7	1	2	3	6	4	5

Fig. 8.4.3

8.5. Digraph-complete sets of latin squares

In connection with an investigation of the relation between a complete set of mutually orthogonal latin squares and the incidence matrix of the projective plane represented by those squares, L. J. Paige and C. Wexler have introduced the related concept of a digraph-complete set of latin squares. We shall now describe this concept and its connections with that of a complete set of mutually orthogonal squares. We shall follow quite closely the discussion given in Paige and Wexler's paper [1].

The incidence matrix of a finite projective plane of order n is an $(n^2 + n + 1)\times(n^2 + n + 1)$ matrix obtained as follows. Number the points and lines arbitrarily from 1 to $n^2 + n + 1$ and form an incidence matrix in which the rows are identified with the lines and the columns are identified with the points. The entry in the (i, j)th position is either 1 or 0 according as the line l_i is incident with the point P_j or not.

It is not easy to obtain the incidence matrix of a projective plane directly from the set of orthogonal latin squares representing it, but the construction of a digraph complete set of latin squares as an intermediate step provides a means of doing so.

The incidence matrix is not unique but depends upon the choice of numbering for the points and lines. Let us specialize the incidence matrix by so numbering the points and lines of our projective plane π that (property (i))

$P_1, P_2, \ldots, P_{n+1}$ are the points of l_1;

$l_1, l_2, \ldots, l_{n+1}$ are the lines through P_1;

for $k = 1, 2, \ldots, n$, $P_{kn+2}, P_{kn+3}, \ldots, P_{kn+n+1}$ are incident with l_{k+1} (which also contains the point P_1);

for $k = 1, 2, \ldots, n$, $l_{kn+2}, l_{kn+3}, \ldots, l_{kn+n+1}$ are incident with P_{k+1} (which also lies on the line l_1).

With this choice of labelling, the incidence matrix takes the form shown in Fig. 8.5.1.

The points P_i and the lines l_i with $i > n + 1$ are now conveniently separated into groups of n and the incidence matrix, excluding the first $n + 1$ rows and columns, may be considered as a matrix whose elements are $n\times n$ matrices C_{ij} $(i, j = 1, 2, \ldots, n)$ as shown in Fig. 8.5.1.

We note that in any matrix C_{ij} no row may contain more than one entry 1: for otherwise two lines would intersect in two distinct points.

lines \ points	1	2	3	$n+1$	$n+2$	$2n+1$	$2n+2$	$3n+1$		n^2+2	n^2+n+1
1	1	1	1	1	0	0	0	0		0	0
2	1	0	0	0	1	1	0	0		0	0
3	1	0	0	0	0	0	1	1		0	0
$n+1$	1	0	0	0	0	0	0	0		1	1
$n+2$	0	1	0	0							
					C_{11}		C_{12}			C_{1n}	
$2n+1$	0	1	0	0							
$2n+2$	0	0	1	0							
					C_{21}		C_{22}			C_{2n}	
$3n+1$	0	0	1	0							
n^2+2	0	0	0	1							
					C_{n1}		C_{n2}			C_{nn}	
n^2+n+1	0	0	0	1							

Fig. 8.5.1

On the other hand, each line of π contains $n+1$ points and so there must be exactly one entry 1 in each row of C_{ij}.

A dual argument (interchanging the words point and line) shows that each column of C_{ij} contains exactly one entry 1. Thus, the matrices C_{ij} are permutation matrices of order n. That is, they are matrices obtained from the identity matrix of order n by permuting its rows or columns.

We define a *canonical incidence matrix* of the projective plane π as an incidence matrix satisfying property (i) which has the additional property that the matrices C_{1j} and C_{i1} $(i, j = 1, 2, \ldots, n)$ are $n \times n$ identity matrices. We call the matrices C_{ij} the *kernel* of the canonical incidence matrix.

The following properties of the matrices C_{ij} of a canonical incidence matrix are an immediate consequence of the nature of C_{1j} and C_{i1} and of the facts that two lines intersect in a unique point and two points lie on a unique line.

(*ii*) C_{ij} with i and j greater than 1, contains no entry 1 on its main diagonal;

(*iii*) the kth rows of C_{is} and C_{ir} for $s \neq r$, $i > 1$, are distinct;

(*iv*) the kth rows of C_{is} and C_{ir} with $s \neq r$ and $i > 1$, $(k = 1, 2, \ldots, n)$ cannot be simultaneously identical to the pth rows of C_{ms} and C_{mr} for any p and for $m \neq i > 1$.

From a canonical incidence matrix we can construct $n - 1$ square arrays D_{i-1}, $i = 2, 3, \ldots, n$, by defining the jth column of the array D_{i-1} to be the column matrix $C_{ij}(1\ 2\ \ldots\ n)^T$ for $j = 1, 2, \ldots, n$, where (T) denotes transpose. Then, clearly, every column of each D_{i-1} contains the numbers $1, 2, \ldots, n$. If the number j occurred twice in the kth row of D_{i-1}, this would imply that C_{is} and C_{ir} would have identical kth rows contrary to property (*iii*) above. Hence, the numbers $1, 2, \ldots, n$ occur in each row of D_{i-1}, and we conclude that the D_{i-1} $(i=2, 3, \ldots, n)$ form $n - 1$ latin squares. In general, these latin squares will not be orthogonal, but they always possess the following property:

(*v*) for any given pair of columns, say the rth and sth, $s \neq r$, the $n(n - 1)$ number pairs (h, k) obtained by picking out the entries which occur in these columns for each row of each of the $n - 1$ latin squares in turn are all distinct and, since each square is latin, the pairs (h, h), $h = 1, 2, \ldots, n$, do not occur among them.

We shall call the number pairs defined in property (*v*) above *digraphs*. The validity of property (*v*) is an immediate consequence of property (*iv*).

If we are given a set of $n - 1$ latin squares D_i which possess property (*v*) (such a set of squares will be called a *digraph complete set*), we may reorder the rows of each one separately (without affecting property (*v*)) so as to get the first column of each to be $1, 2, \ldots, n$ in natural order. We may then define permutation matrices C_{ij} by the requirement that the matrix C_{ij} effect that permutation of the first column of the square D_{i-1} which will transform it into the jth column of D_{i-1}. The validity of property (*v*) then assures us that these permutation matrices may be used to construct the kernel of a canonical incidence matrix of a finite projective plane π.

We may summarize our argument in the form of a theorem.

THEOREM 8.5.1. *A necessary and sufficient condition for the existence of a finite projective plane with $n^2 + n + 1$ points and lines is that there exist a digraph complete set of latin squares of order n.*

It should be pointed out that we have used the rows of the kernel of the canonical incidence matrix to form our digraph complete set of latin squares; however, we could have used the columns instead and obtained a second set.

Next we will show how to construct a digraph complete set of latin squares from a set of $n - 1$ mutually orthogonal latin squares. We can then construct quite simply an incidence matrix corresponding to the finite projective plane π whose existence is implied by the mutually orthogonal latin squares.

Theorem 8.5.2. *Every complete set of mutually orthogonal latin squares defines a digraph complete set of latin squares of the same order, and conversely.*

Proof. Let $L_1, L_2, \ldots, L_{n-1}$ be a set of mutually orthogonal latin squares of order n and assume without loss of generality that L_1 has been normalized in the first row and first column to be $1, 2, \ldots, n$ in natural order and that the remaining squares have been normalized in the first row to be $1, 2, \ldots, n$.

We proceed to construct the digraph complete set of latin squares D_i.

Let the first column of D_i $(i = 1, 2, \ldots, n - 1)$ be $(1\ 2 \ldots n)^T$.

Let P_{kj} $(j, k \neq 1)$ be the permutation taking the kth row of L_1 into the kth row of L_j.

We now define the jth column of D_{k-1} to be the permutation P_{kj}, considered as a permutation of the first column of D_{k-1}. For the nth column of D_{k-1} we let P_{kn} be the permutation taking the kth row of L_1 into the first row of L_1.

Before we prove that the set D_i are digraph complete let us observe the relationship between the digraphs of the D_i and the orthogonal latin squares L_i.

If $i < j < n$, the digraph (r, s) occurring in the ith and jth columns and in the hth row of D_u is precisely the number pair occurring in the orthogonal pair of latin squares (L_i, L_j) in the position occupied by h in the $(u + 1)$th row of L_1. (In the case $i = 1$ we see, of course, that $r = h$.)

The previous statement is made clear by means of Fig. 8.5.2:

The ith column of D_u is merely the permutation that, in particular, takes h of L_1 into r of L_i, and, for the jth column, h of L_1 goes into s of L_j.

We must consider the case $j = n$ separately. Here again a diagram proves helpful and we see from Fig. 8.5.3 that the digraph (r, s) which occurs in the ith and nth columns and in the hth row of D_u corresponds to the element appearing in the $(u+1)$th row of L_i under the column headed by s in L_1 (or L_i, since the first rows have been normalized).

Each column of D_u contains the numbers $1, 2, \ldots, n$. The arguments above assure us that each row of D_u contains $1, 2, \ldots, n$ so that D_u is a latin square.

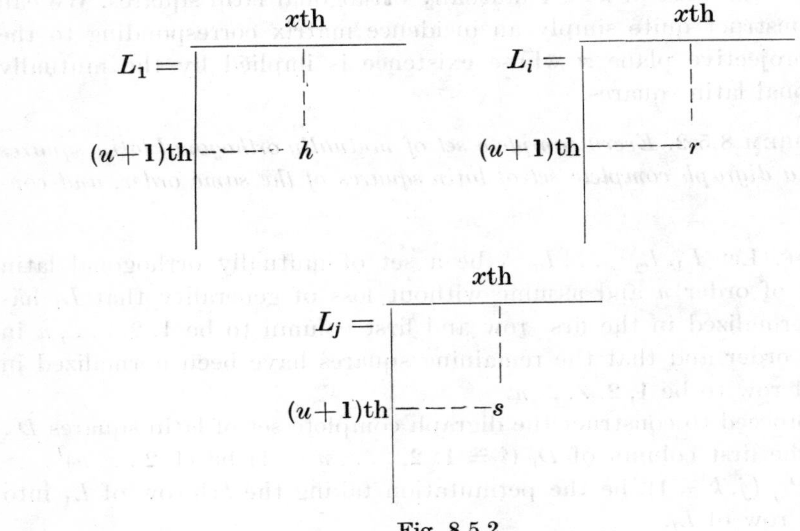

Fig. 8.5.2

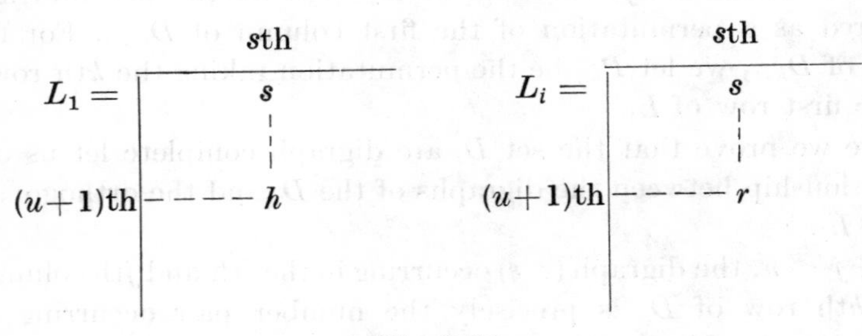

Fig. 8.5.3

We are now prepared to show that the set of latin squares D_u are digraph complete. Since D_u is a latin square, the digraph (r, r) cannot appear and our proof is completed if we show that for the same two columns no digraph is repeated in all of the D_u's.

First, no digraph can repeat in the same two columns of D_u since D_u is a latin square. Hence, let us consider the case that the digraph (r, s) occurs in the ith and jth columns of D_u and D_v $(i < j < n$ and $u \neq v)$.

From our previous discussion we see that this would imply that the couple (r, s) would occur in the $(u + 1)$th and $(v + 1)$th rows of the orthogonal latin squares (L_i, L_j). This contradiction shows that the digraphs (r, s) are all distinct.

Again the case $j = n$ deserves special attention. Referring again to the relationship between the digraphs and the latin squares L_i, we see that a repeated digraph (r, s) involving the nth column of D_u and D_v would imply that the number r is repeated in the sth column of L_i, and hence L_i would not be a latin square.

Thus the latin squares D_i form a digraph complete set and we may proceed to construct the canonical incidence matrix in the way we explained earlier.

The converse problem of constructing a mutually orthogonal set of latin squares from a digraph complete set of latin squares, and hence from the canonical incidence matrix of a finite projective plane, is perfectly straightforward.

Let $D_1, D_2, \ldots, D_{n-1}$ be a digraph complete set of latin squares which have been normalized so as to have $(1\ 2\ \ldots\ n)^T$ as their first column. This corresponds to making the matrices C_{i1} of the canonical incidence matrix equal to the identity matrix.

Let the first row of L_i $(i = 1, 2, \ldots, n - 1)$ be $(1\ 2\ \ldots\ n)$.

Let the kth row $(k > 1)$ of L_1 be the inverse of the permutation taking the first column of D_{k-1} into the last column of D_{k-1}.

Let the $(k + 1)$th row of L_j $(j > 1)$ be the row obtained from the $(k + 1)$th row of L_1 after applying the permutation carrying the first column of D_k into the jth column of D_k. (Of course, the nth column is not used here.)

The proof that these L_i are orthogonal latin squares follows from the reversibility of the construction given above for the construction of the digraph complete set of latin squares D_i.

As an illustration of the above procedures, we give in Fig. 8.5.4 the incidence matrix of the finite projective plane of order 4 using the notation of section 5.2 for its points and with the lines suitably reordered so that the incidence matrix takes canonical form. For each symbol x the line x is the line whose line co-ordinates are the same as the point co-ordinates of the point x. Thus, for example, the line e_1 has line co-ordinates $[\alpha, 1, 1]$. Its point equation is $\alpha x + y + z = 0$ and it contains the five points $b_1 = (0, 1, 1)$, $f_2 = (1, \alpha^2, 1)$, $f_3 = (1, 1, \alpha^2)$, $c_3 = (1, \alpha, 0)$, $d_2 = (1, 0, \alpha)$.

From the matrices $C_{11}, C_{12}, \ldots, C_{44}$ of the kernel of the incidence matrix, we derive the digraph complete set of latin squares D_1, D_2, D_3 and hence the mutually orthogonal latin squares L_1, L_2, L_3. The process is easily reversed.

	b_1	u	a_1	f_1	e_1	b_2	b_3	w	v	f_2	f_3	c_3	d_2	e_2	e_3	c_2	d_3	a_2	a_3	c_1	d_1
b_1	1	1	1	1	1																
u	1					1	1	1	1												
e_1	1									1	1	1	1								
f_1	1													1	1	1	1				
a_1	1																	1	1	1	1
b_2		1				1				1				1				1			
b_3		1					1				1				1				1		
w		1						1				1				1				1	
v		1							1				1				1				1
a_2			1			1							1			1			1		
a_3			1				1					1					1	1			
c_1			1					1			1			1							1
d_1			1						1	1					1					1	
e_2				1		1					1						1			1	
e_3				1			1			1						1					1
c_2				1				1					1		1			1			
d_3				1					1			1		1					1		
f_2					1	1						1			1						1
f_3					1		1						1	1						1	
c_3					1			1		1							1		1		
d_2					1				1		1					1		1			

$$D_1 = \begin{array}{|cccc} 1 & 4 & 3 & 2 \\ 2 & 3 & 4 & 1 \\ 3 & 2 & 1 & 4 \\ 4 & 1 & 2 & 3 \end{array} \quad D_2 = \begin{array}{|cccc} 1 & 2 & 4 & 3 \\ 2 & 1 & 3 & 4 \\ 3 & 4 & 2 & 1 \\ 4 & 3 & 1 & 2 \end{array} \quad D_3 = \begin{array}{|cccc} 1 & 3 & 2 & 4 \\ 2 & 4 & 1 & 3 \\ 3 & 1 & 4 & 2 \\ 4 & 2 & 3 & 1 \end{array}$$

$$L_1 = \begin{array}{|cccc} 1 & 2 & 3 & 4 \\ 2 & 1 & 4 & 3 \\ 3 & 4 & 1 & 2 \\ 4 & 3 & 2 & 1 \end{array} \quad L_2 = \begin{array}{|cccc} 1 & 2 & 3 & 4 \\ 3 & 4 & 1 & 2 \\ 4 & 3 & 2 & 1 \\ 2 & 1 & 4 & 3 \end{array} \quad L_3 = \begin{array}{|cccc} 1 & 2 & 3 & 4 \\ 4 & 3 & 2 & 1 \\ 2 & 1 & 4 & 3 \\ 3 & 4 & 1 & 2 \end{array}$$

Fig. 8.5.4

It will be noticed that the digraph complete set of latin squares D_1, D_2, D_3 of the above example are themselves mutually orthogonal. However, not all digraph complete sets need be orthogonal as was remarked by Paige and Wexler themselves.

A criterion for determining under what circumstances a digraph complete sets of squares will also be mutually orthogonal has been given by R. B. Killgrove in [3].

Killgrove began by proving the following theorem:

THEOREM 8.5.3. *From a given digraph complete set of latin squares* $D_1, D_2, \ldots, D_{n-1}$ *an affine plane can be constructed directly as follows: For the points of the plane, we take the* n^2 *ordered pairs* (h, k) *where* h *and* k *are chosen from the symbols occurring in the latin squares and where, without loss of generality, we may suppose that these symbols are the integers* $0, 1, 2, \ldots, n-1$. *We also suppose that these same integers are used to label the rows and columns of the latin squares. For the lines of the plane we take certain subsets of the points: namely, the line* $[r, c]$ *of gradient* r *is to comprise the points* $(0, c)$ *and* $(j, a_r(c, j))$ *for* $j = 1, 2, \ldots, n-1$, *where* $a_r(c, j)$ *denotes the element of the* cth *row and* jth *column of the latin square* D_r. *This defines* $(n-1)n$ *lines. The line* $[0, c]$ *of gradient* 0 *is to comprise the point* $(0, c)$ *and the points* (j, c) *for* $j = 1, 2, \ldots, n-1$. *The line* $[\infty, c]$ *is to comprise the points* (c, i) *for* $i = 0, 1, \ldots, n-1$. *This defines* $2n$ *further lines and completes the affine plane.*

PROOF. There is a unique line joining two given points (j, α) and (k, β), $j \neq k$ and $\alpha \neq \beta$, since the ordered pair α, β occurs as a row pair of the jth and kth columns in exactly one of the squares D_i by virtue of the fact that they are digraph complete. Let this square be the square D_r and suppose that the row in question is the cth. Then the unique line joining the points (j, α) and (k, β) is the line $[r, c]$. If $\alpha = \beta$, the unique line joining the points (j, α) and (k, β) is the line $[0, \alpha]$. If $j = k$, the unique line joining the points (j, α) and (k, β) is the line $[\infty, j]$.

Two given lines $[r, c]$ and $[r, d]$ having the same gradient r have no point in common. For, suppose that (l, γ) were a common point. Then, for $r \neq 0$ or ∞, we would have $\gamma = a_r(c, l) = a_r(d, l)$ which contradicts the fact that the lth column of the square D_r contains each symbol only once. If $r = 0$ or ∞, the result is obvious.

There exist n lines of gradient r each of which contains n points and no two of which have any point in common, so, between them, the lines of gradient r contain all n^2 points. It follows that any line of gradient s must meet one or more of the lines of gradient r. A line of gradient

s cannot intersect a line of gradient r in two points otherwise there would be two lines joining these two points. It follows that each line of gradient s meets each line of gradient r in exactly one point. Thus, the n^2 points and $n^2 + n$ lines form an affine plane, as required.

It is worth pointing out here (see the next theorem) that the affine plane constructed by Killgrove's method as above and the projective plane derived from the same digraph complete set of latin squares by the Paige-Wexler method are connected. It is not clear to the authors whether Killgrove himself appreciated this connection.

THEOREM 8.5.4. *Let π be the projective plane derived from the digraph complete set of latin squares $D_1, D_2, \ldots, D_{n-1}$ by the Paige-Wexler method, and let π^* be the affine plane obtained from π by deleting the line l_1 and the points $P_1, P_2, \ldots, P_{n+1}$. Then the affine plane constructed by the method of theorem* 8.5.3 *is always isomorphic to π^*.*

PROOF. To see this, we label the points $P_{kn+2}, P_{kn+3}, \ldots, P_{kn+n+1}$ by the co-ordinate pairs $(k-1, 0), (k-1, 1), \ldots, (k-1, n-1)$ respectively, for $k = 1, 2, \ldots, n$ and the lines $l_{kn+2}, l_{kn+3}, \ldots, l_{kn+n+1}$ by the co-ordinate pairs $[k-1, 0], [k-1, 1], \ldots, [k-1, n-1]$ respectively, for $k = 1, 2, \ldots, n$. We label the lines $l_2, l_3, \ldots, l_{n+1}$ by the co-ordinate pairs $[\infty, 0], [\infty, 1], \ldots, [\infty, n-1]$. From Fig. 8.5.1, we see at once that the line $[\infty, c]$ comprises the points (c, i) for $i = 0, 1, 2, \ldots, n-1$. Also, since each of the matrices $C_{11}, C_{12}, \ldots, C_{1n}$ is the identity matrix, the line $[0, c]$ comprises the points (j, c) for $j = 0, 1, \ldots, n-1$ as in Killgrove's definition. Finally, consider the line l_{kn+c+2} with co-ordinates $[k-1, c]$. If $a_{k-1}(c, j)$ is the entry in the cth row and jth column of the latin square D_{k-1} ($c = 0, 1, \ldots, n-1$ and $j = 0, 1, \ldots, n-1$) then a 1 occurs in the cth row and $a_{k-1}(c, j)$th column of the matrix $C_{k,j+1}$. That is, the line $[k-1, c]$ is incident with the points $(j, a_{k-1}(c, j))$ for $j = 0, 1, \ldots, n-1$.

Suppose now that the latin squares D_r and D_s are not orthogonal. Then, for some two ordered pairs c_1, j_1 and c_2, j_2 of the integers $0, 1, \ldots, n-1$, we have $a_r(c_1, j_1) = a_r(c_2, j_2) = \alpha$ and $a_s(c_1, j_1) = a_s(c_2, j_2) = \beta$. This implies that the points (j_1, α), (j_2, α), (j_1, β), and (j_2, β) lie respectively on the lines $[r, c_1]$, $[r, c_2]$, $[s, c_1]$ and $[s, c_2]$. Hence, in the diagram of Fig. 8.5.5, the points C_1 and C_2 must be collinear with A. Killgrove has called the configuration of Fig. 8.5.5 with collinearity of the points C_1, C_2, A, *the K-configuration*. Thus, as in R. B. Killgrove [3], we get

THEOREM 8.5.5. *The necessary and sufficient condition that a given digraph complete set of latin squares be a mutually orthogonal set is that no K-configurations exist in the associated projective plane* π (as defined by the Paige-Wexler incidence matrix).

Killgrove has shown that there do exist K-configurations in the Hughes plane of order 9 (see section 8.4 for details of this plane) but that such configurations cannot exist in any desarguesian plane. (See R. B. Killgrove [4].) Nor can any which have A, B, S, R on l_∞ exist in a translation plane. (The concept of translation plane is explained in section 8.4.) The latter remarks are easily justified as follows: In Fig. 8.5.5, the triangles FDC_1 and GEC_2 are in axial perspective from the line RSA and the joins FG and DE meet at B. Consequently, if the minor theorem of Desargues is satisfied with l_∞ as axis, the join C_1C_2 must pass through B. In that case, it cannot also pass through A.

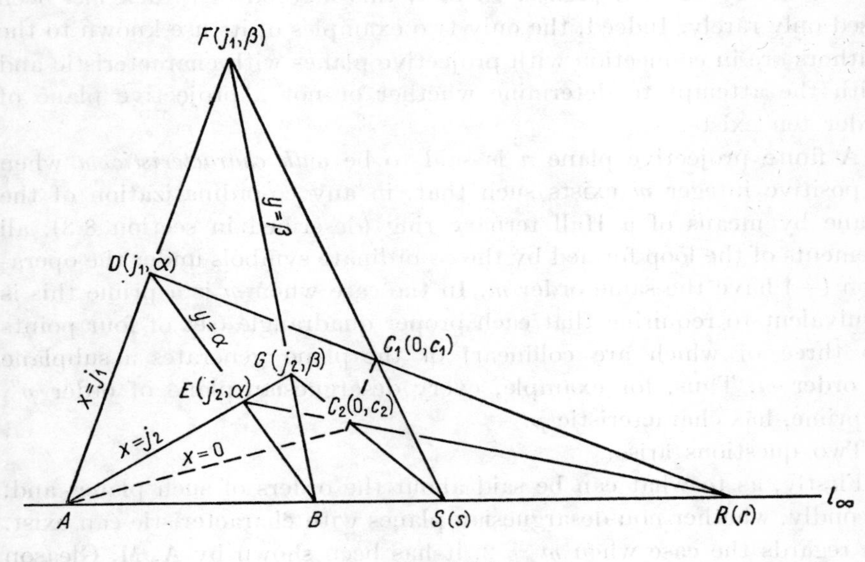

Fig. 8.5.5

In his paper [3], Killgrove has also considered the possibility that K-configurations may exist in infinite projective planes and has shown that, in particular, the Moulton plane contains such configurations. (For the concept of a Moulton plane, the reader is referred to books on projective planes, especially G. Pickert [2] and P. Dembowski [1].)

We end this section by remarking that a digraph complete set of latin squares provides a solution (unfortunately not an economic one) to the problem, raised in section 2.2, of the complete counterbalancing for both immediate and remote effects of a sequential experiment. However, interaction between those remote effects themselves is not allowed for. That is, we get a solution of the first of the two kinds described on page 83.

8.6. Some further geometrical problems involving latin squares

In consequence of the fact that every finite projective plane is equivalent to a complete set of mutually orthogonal latin squares (as we showed in section 5.2), it is natural to expect that latin squares would be used directly in attempting to solve some of the existence problems arising in the theory of such planes. In fact, this method of attack has been used only rarely. Indeed, the only two examples of its use known to the authors are in connection with projective planes with characteristic and with the attempt to determine whether or not a projective plane of order ten exists.

A finite projective plane π is said to be *with characteristic m* when a positive integer m exists such that, in any co-ordinatization of the plane by means of a Hall ternary ring (described in section 8.3), all elements of the loop formed by the co-ordinate symbols under the operation $(+)$ have the same order m. In the case when m is a prime this is equivalent to requiring that each proper quadrangle (set of four points no three of which are collinear) of the plane generates a subplane of order m. Thus, for example, every desarguesian plane of order p^n, p prime, has characteristic p.

Two questions arise:

Firstly, as to what can be said about the orders of such planes and, secondly, whether non-desarguesian planes with characteristic can exist. As regards the case when $m = 2$, it has been shown by A. M. Gleason in [1] that every finite projective plane of characteristic two is desarguesian and consequently that it has order equal to a power of two. For the case $m = 3$, it has been shown in A. D. Keedwell [1] that, under an additional restriction, the order must be a power of three.(1) In the latter

(1) Recently, the same author has shown that, if $m = p$, p prime, then, under the same restriction, the order of the plane is a power of p. (See A. D. Keedwell [7].)

author's attempt to deal with the second question for this case, it proved effective to use the latin square representation (see A. D. Keedwell [3]) and two orthogonal latin squares suitable for the construction of a non-desarguesian projective plane in which affine quadrangles would generate subplanes of order three were quite easily constructed. However, the question as to the existence or non-existence of such planes remains unanswered.

The problem of determining whether or not a projective plane of order ten exists is probably the one whose solution has been attempted in recent years more often than any other in the theory of projective planes. Ten is the smallest integer which is neither a prime power nor is excluded as the possible order of a projective plane by the Bruck-Ryser theorem (theorem 5.2.5). Here, the latin square representation gives an obvious method of attack. It is known that pairs of orthogonal latin squares of order ten exist (see, for example, sections 7.3 and 11.2) and so an obvious next step would be to obtain a triad of such squares and then to examine their structural relationship. Attempts to construct such a triad have been made by many authors, as was mentioned in chapter 5, but none has so far been successful. Some more details are given in chapter 13.

ADDED IN PROOF. The attention of the authors has been drawn to several recent papers which investigate problems concerning the structure of projective planes by means of their representational latin squares. See L. J. Haranen [1], L. I. Istomina [1], A. D. Lumpov [1], A. E. Malih [1] and L. I. Panteleeva [1].

CHAPTER 9

The application of graph theory to the solution of latin square problems

In the first section of this chapter, we discuss a number of connections between the theory of graphs and that of latin squares.

In the following two sections, we show how, making use of k-nets and graph theory, R. H. Bruck has obtained some powerful and interesting results concerning the conditions under which a given set of mutually orthogonal latin squares may be extended to a larger set. The main results are as follows:

(A) Let G be a finite loop of order n. If G contains a normal subloop H of odd order such that the quotient loop G/H is a cyclic group of even order, then the latin square L_G representing the multiplication table of G does not possess a transversal (and consequently L_G has no orthogonal mate).

(B) Let S be a set of $k-2$ mutually orthogonal latin squares of order n and let $d = n + 1 - k$. Then, (1) whenever $n > (d-1)^2$, if S can be completed to a complete set of mutually orthogonal latin squares at all, it can be so completed in only one way; (2) a sufficient (but not necessary) condition that such a set S can be augmented to a complete set of mutually orthogonal squares in at least one way is that $n > \mathscr{P}(d-1)$, where $\mathscr{P}(x) = \frac{1}{2}x^4 + x^3 + x^2 + \frac{3}{2}x$.

As regards the result (A), notice that the requirement that G be a loop instead of a quasigroup is equivalent to the requirement that L_G be in standard form (see section 5.1) and does not entail any loss of generality. If G has order $n = m.2^t$ where m is odd and $t \geq 1$, then the result (A) implies that a sufficient condition that L_G have no transversal is that G possess a normal subloop H of order m and that G/H be cyclic of order 2^t. When G is a group, its Sylow 2-subgroups have order 2^t. Moreover, when G is abelian we have $G = H \times S$ and $G/H \cong S$, where S is the unique Sylow 2-subgroup of G. Consequently, we may make an interesting comparison between this result of Bruck's and the conjecture of M. Hall and L. J. Paige which we discussed in section 1.4; namely that a finite group G will possess a complete mapping (and consequently a transversal) when and only when its Sylow 2-subgroups are non-cyclic. (See theorems 1.4.7, 1.4.8, 1.4.9 and 1.4.10.)

The results of (B) were originally stated and proved by R. H. Bruck in the language of nets, a set of $k - 2$ mutually orthogonal latin squares of order n being equivalent to a k-net of order n as we showed in section 8.2; and we too shall find this language convenient when we come to give the detailed proofs of these results in section 9.3.

9.1. Miscellaneous connections between latin squares and graphs

We begin this section by discussing the close relationship that exists between the existence of horizontally complete latin squares and latin rectangles and the existence of decompositions of complete graphs on n vertices into disjoint circuits or disjoint Hamiltonian paths.

We remind the reader that an $n \times n$ latin square is called *horizontally complete* if every ordered pair of its elements appears exactly once as a pair of adjacent elements in some row of the square. A latin rectangle of $n/2$ rows and n columns (n necessarily even) is *horizontally complete* if every pair of its elements appears exactly once as an unordered pair of adjacent elements in some row of the square. These concepts were introduced in sections 2.3 and 3.1 respectively.

We first prove:

THEOREM 9.1.1. *If a horizontally complete latin square of order n exists, then (i) the complete directed graph on n vertices can be separated into n disjoint Hamiltonian paths, and (ii) the complete directed graph on $n + 1$ vertices can be separated into n disjoint circuits each of length $n + 1$.*

PROOF. For (*i*), we associate a directed graph with the given horizontally complete latin square in such a way that the vertices of the graph correspond to the n distinct elements of the latin square and that an edge of the graph directed from the vertex x to the vertex y exists when and only when the elements x and y appear as an ordered pair of adjacent elements in some row of the latin square. Then, because the latin square is horizontally complete, the graph obtained will be the complete directed graph on n vertices. Moreover, it is immediate to see that the rows of the latin square define n disjoint Hamiltonian paths into which the graph can be decomposed.

For (ii) we adjoin an extra column to the given horizontally complete latin square L, the elements of which are equal but distinct from the n elements of the latin square. We associate a directed graph with the $n \times (n+1)$ matrix so formed in the same way as before but this time treating each row cyclically so that if the rth row ends with the element x and begins with the element y then the associated graph has an edge directed from the vertex labelled x to the vertex labelled y. Because each element of L appears just once in its last column and just once in its first row, the directed graph on $n+1$ vertices which we obtain is complete. Also, the rows of the $n \times (n+1)$ matrix define n disjoint circuits into which it can be decomposed.

The relationship described in theorem 9.1.1 (i) has been pointed out in J. Dénes and E. Török [1] and also in N. S. Mendelsohn [1]. That described in theorem 9.1.1(ii) does not seem to have been noticed until now.

Illustrative examples of these decompositions are given in Figs 9.1.1 and 9.1.2. Fig. 9.1.1 shows the decomposition of the complete directed graph with four vertices into disjoint Hamiltonian paths with the aid of the 4×4 complete latin square previously displayed in Fig. 2.3.1. Fig. 9.1.2 shows the decomposition of the complete directed graph with five vertices into disjoint circuits of length five with the aid of the 4×5 matrix obtained by augmenting this same 4×4 complete latin square in the way described in theorem 9.1.1.

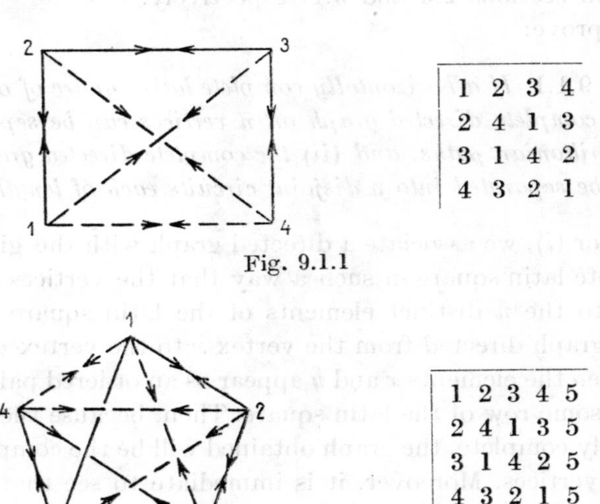

1	2	3	4
2	4	1	3
3	1	4	2
4	3	2	1

Fig. 9.1.1

1	2	3	4	5
2	4	1	3	5
3	1	4	2	5
4	3	2	1	5

Fig. 9.1.2

By virtue of theorem 2.3.1, decompositions of the above kind always exist when n is even. It is also of interest to note that when the decomposition is effected with the aid of a horizontally complete latin square formed in the manner described in theorem 2.3.1, one half of the paths obtained are the same as the other half but described in the opposite direction.

E. G. Strauss has posed the question whether a complete directed graph with an odd number n of vertices can likewise be separated into n disjoint Hamiltonian paths. By virtue of the fact, already mentioned in section 2.3, that N. S. Mendelsohn has published several examples of complete latin squares of order 21, the question is answered in the affirmative for this value of n but the general question of the existence of such decompositions for odd values of n remains an open one.

We next prove an analogue of theorem 9.1.1 for complete undirected graphs. This time our questions are only meaningful in the case when n is even. We are also able to show a connection with the construction of an Eulerian line.

THEOREM 9.1.2. *Horizontally complete $m \times 2m$ latin rectangles exist for every positive integer m and, correspondingly, (i) for every positive integer m, the complete undirected graph on $2m$ vertices has a decomposition into m disjoint Hamiltonian paths. Also, (ii) the complete undirected graph on $2m + 1$ vertices has a decomposition into m disjoint circuits each of length $2m + 1$. Moreover, (iii) every complete undirected graph on an odd number $2m + 1$ of vertices has an Eulerian line with the property that, when a certain vertex and all the edges through it are deleted, the remaining portions of the Eulerian line are Hamiltonian paths of the residual graph on $2m$ vertices.*

PROOF. Consider the horizontally complete latin square formed by taking as first row the integers

$$0, 1, 2m - 1, 2, 2m - 2, \ldots, k, 2m - k, \ldots, m + 1, m$$

(as in theorem 2.3.1) and whose subsequent rows are obtained, each from its predecessor, by adding 1 modulo $2m$ to its elements (Fig. 9.1.3). The last m rows are the same as the first but in reverse order. Thus, in the latin rectangle formed by the first m rows every unordered pair of elements occurs as a consecutive pair just once in some row. The first m rows define the required Hamiltonian decomposition of the undirected graph on $2m$ vertices. This proves (*i*).

0	1	$2m-1$	2	$2m-2$	$\dots$	$m+2$	$m-1$	$m+1$	m
1	2	0	3	$2m-1$	$\dots$	$m+3$	m	$m+2$	$m+1$
$\cdot$	$\cdot$	$\cdot$	$\cdot$	$\cdot$	$\cdot$	$\cdot$	$\cdot$	$\cdot$	$\cdot$
$m-1$	m	$m-2$	$m+1$	$m-3$	$\dots$	1	$2m-2$	0	$2m-1$
m	$m+1$	$m-1$	$m+2$	$m-2$	$\dots$	2	$2m-1$	1	0
$\cdot$	$\cdot$	$\cdot$	$\cdot$	$\cdot$	$\cdot$	$\cdot$	$\cdot$	$\cdot$	$\cdot$
$2m-1$	0	$2m-2$	1	$2m-3$	$\dots$	$m+1$	$m-2$	m	$m-1$

Fig. 9.1.3

Now let an additional column be adjoined to the latin rectangle formed by the first m rows of L. Since the leading entries of these m rows together with their final entries exactly cover the set $0, 1, \dots, 2m-1$, the m augmented rows may be read cyclically (see Fig. 9.1.4) to define an Eulerian path of the complete undirected graph on $2m+1$ vertices which has the properties described in statement (*iii*) of the theorem. Finally, to obtain the disjoint cycle decomposition whose existence is claimed in statement (*ii*) it is only necessary to regard each separate row of the augmented rectangle as defining one such cycle in the manner described in theorem 9.1.1.

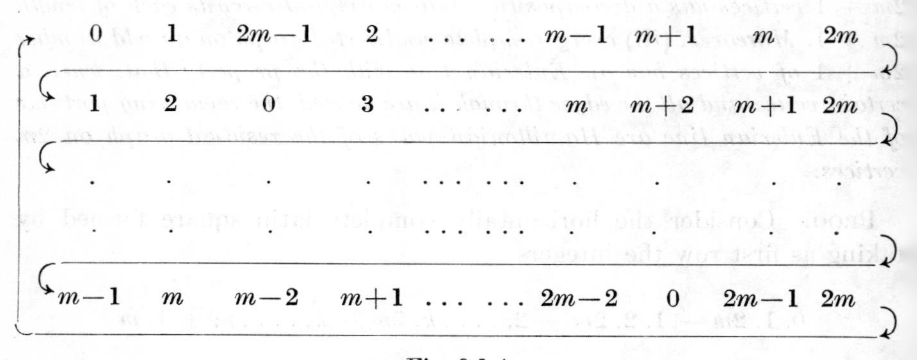

Fig. 9.1.4

Another graph problem of a similar kind to that just discussed and which has been shown by A. Kotzig to have connections with a particular type of latin square concerns the more general problem of the decomposition of a complete undirected graph into a set of disjoint closed paths of arbitrary lengths.

In [1], Kotzig gave the name *P-groupoid* (partition groupoid) to a groupoid $(V, \cdot)$ which has the following properties: (*i*) $a.a = a$ for all $a \in V$; (*ii*) $a \neq b$ implies $a \neq a.b \neq b$ for all $a, b \in V$; (*iii*) $a.b = c$ implies and is implied by $c.b = a$ for all $a, b, c \in V$.

He showed that there exists a one-to-one correspondence between P-groupoids of n elements and decompositions of complete undirected graphs of n vertices into disjoint closed paths. This correspondence is established by labelling the vertices of the graph with the elements of the P-groupoid and prescribing that the edges (a, b) and (b, c) shall belong to the same closed path of the graph if and only if $a.b = c, a \neq b$. We illustrate this relationship in Fig. 9.1.5 and from it we easily deduce:

THEOREM 9.1.3. *In any P-groupoid* $(V, \cdot)$ *we have (i) the number of elements is necessarily odd, and (ii) the equation* $x.b = c$ *is uniquely soluble for* x.

PROOF. The result (*i*) is deduced by using the correspondence between P-groupoids and graphs just described. Since for a complete undirected graph which separates into disjoint closed paths the number of edges which pass through each vertex must clearly be even, any such complete undirected graph must have an odd number of vertices all together. This is because each vertex has to be joined to an even number of others. The number of elements of a P-groupoid is equal to the number of vertices in its associated graph.

The result (*ii*) is a consequence of the definition of a groupoid and the fact that $x.b = c$ implies $c.b = x$.

COROLLARY. *The multiplication table of a P-groupoid is a column latin square.* (See section 3.1 for the definition of this concept.)

	1	2	3	4	5
1	1	5	5	5	3
2	4	2	4	3	4
3	5	4	3	2	1
4	2	3	2	4	2
5	3	1	1	1	5

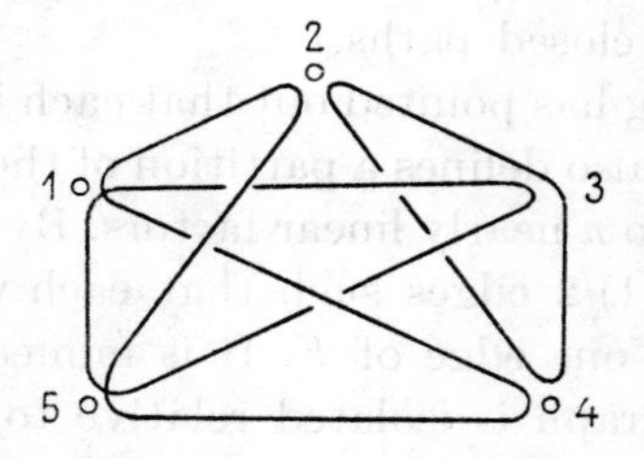

Fig. 9.1.5

The example given in Fig. 9.1.5 illustrates the fact that there exist P-groupoids which are not quasigroups. This observation leads us to make the following definition.

Definition. A P-groupoid which is also a quasigroup will be called a *P-quasigroup*.

Theorem 9.1.4. *Let $(V, \cdot)$ be a P-quasigroup and let a groupoid $(V, *)$ be defined by the statement that $a.(a * b) = b$ holds for all $a, b \in V$. Then $(V, *)$ is an idempotent and commutative quasigroup. Moreover, with any given idempotent commutative quasigroup $(V, *)$ there is associated a P-quasigroup $(V, \cdot)$ related to $(V, *)$ by the correspondence $a.b = c \Leftrightarrow a * c = b$.*

Proof. In a P-quasigroup, the equation $a.x = b$ is uniquely soluble for x, so the binary operation $(*)$ is well-defined. Also, $a.x = b$ implies $b.x = a$ so $b * a = a * b$; that is $(V, *)$ is commutative. The equation $a.x = a$ has the solution $x = a$, so $a * a = a$ and $(V, *)$ is idempotent. If the equation $a * y = c$ or the equation $y * a = c$ had two solutions, the groupoid property of $(V, \cdot)$ would be contradicted. Hence, $(V, *)$ is a quasigroup.

The second statement of the theorem may be justified similarly by defining the operation $(\cdot)$ in terms of the operation $(*)$ by the statement that $a*(a.b) = b$ for all $a, b \in V$.

The multiplication table of the quasigroup $(V, *)$ is an *idempotent symmetric latin square*. Thus, a consequence of theorems 9.1.3 and 9.1.4 above is that idempotent symmetric latin squares exist only for odd orders n, a result which we may contrast with the fact that unipotent symmetric latin squares exist only when n is even. (The latter result has been proved in B. Elspas, R. C. Minnick and R. A. Short [1]. See also theorem 10.1.7.)

It also follows from theorem 9.1.4 that each idempotent symmetric latin square of (necessarily odd) order n defines a P-quasigroup of order n and hence a decomposition of the complete undirected graph on n vertices into disjoint closed paths.

Kotzig has pointed out that each idempotent symmetric latin square of order n also defines a partition of the complete undirected graph on n vertices into n nearly linear factors. By a *nearly linear factor* is meant a set F of $(n - 1)/2$ edges such that each vertex of the graph is incident with at most one edge of F. It is immediately clear that exactly one vertex of the graph is isolated relative to a given nearly linear factor F.

As an illustration of this concept, let us point out that the three edges (1, 2), (3, 7), (4, 6) of the complete undirected graph on seven vertices form a nearly linear factor. Likewise, the three edges (1, 3), (4, 7), (5, 6) form another nearly linear factor of the same graph. (See also Fig. 9.1.6.) We shall formulate this result of Kotzig's as a theorem.

Theorem 9.1.5. *To each idempotent symmetric latin square of order n there corresponds a partition of the complete undirected graph on n vertices into n nearly linear factors, and conversely.*

Proof. The correspondence is established as follows. Let k be a fixed element of the idempotent commutative quasigroup $(V, *)$ defined by the given symmetric latin square. Then there exist $(n-1)/2$ unordered pairs (a_i, b_i) of elements of V such that $a_i * b_i = k (= b_i * a_i)$. These $(n-1)/2$ pairs define the $(n-1)/2$ edges (a_i, b_i) of a nearly linear factor of the complete undirected graph G_n whose n vertices are labelled by the elements of V. The truth of this statement follows immediately from the fact that $(V, *)$ is an idempotent and commutative quasigroup and that the element k consequently occurs $(n-1)/2$ times in that part of its multiplication table which lies above the main left-to-right diagonal and at most once in each row and at most once in each column.

The converse is established by defining a quasigroup $(V, *)$ by means of a given decomposition of G into nearly linear factors according to the following rules: (*i*) if (a_i, b_i) is an edge of the nearly linear factor whose isolated vertex is labelled k, then $a_i * b_i = k = b_i * a_i$ and (*ii*) for each symbol $k \in V$, $k * k = k$.

Kotzig has asserted further that if $n = 2k - 1$ and either n or k is prime then there exists a partition of the corresponding complete graph on n vertices into n nearly linear factors with the property that the union of every two of them is a Hamiltonian path of the graph. (See A. Kotzig [1].) He has not published a proof of this assertion.

As an example of this situation we give in Fig. 9.1.6 below an idempotent and commutative quasigroup $(V, *)$ of order 7 which defines such a decomposition of the corresponding complete graph G_7 on seven vertices into nearly linear factors whose unions in pairs are Hamiltonian paths of G_7.

(*)	1	2	3	4	5	6	7
1	1	5	2	6	3	7	4
2	5	2	6	3	7	4	1
3	2	6	3	7	4	1	5
4	6	3	7	4	1	5	2
5	3	7	4	1	5	2	6
6	7	4	1	5	2	6	3
7	4	1	5	2	6	3	7

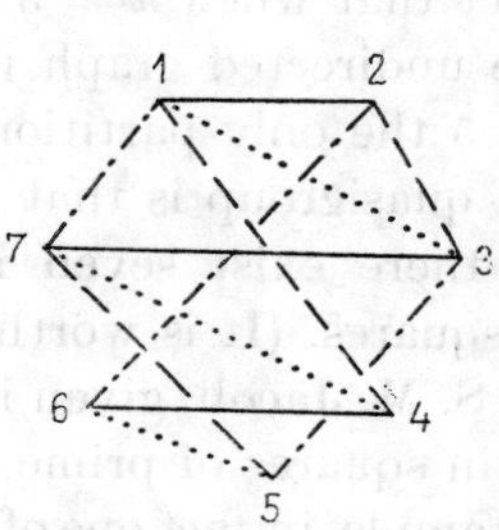

F_1 (2, 7), (3, 6), (4, 5)
F_2 (1, 3), (4, 7), (5, 6) . . . .
F_3 (1, 5), (2, 4), (6, 7)
F_4 (1, 7), (2, 6), (3, 4) —·—·—
F_5 (1, 2), (3, 7), (4, 6) ———
F_6 (1, 4), (2, 3), (5, 7) - - - -
F_7 (1, 6), (2, 5), (3, 4)

Fig. 9.1.6

Kotzig has proposed as a conjecture that a partition of the kind just described may exist for all odd values of n.

In Fig. 9.1.7 we give the P-quasigroup $(V, \cdot)$ which defines and is defined by the idempotent commutative quasigroup $(V, *)$ of Fig. 9.1.6 and show the associated partitions of the graph G_7 into closed paths.

$(\cdot)$	1	2	3	4	5	6	7
1	1	3	5	7	2	4	6
2	7	2	4	6	1	3	5
3	6	1	3	5	7	2	4
4	5	7	2	4	6	1	3
5	4	6	1	3	5	7	2
6	3	5	7	2	4	6	1
7	2	4	6	1	3	5	7

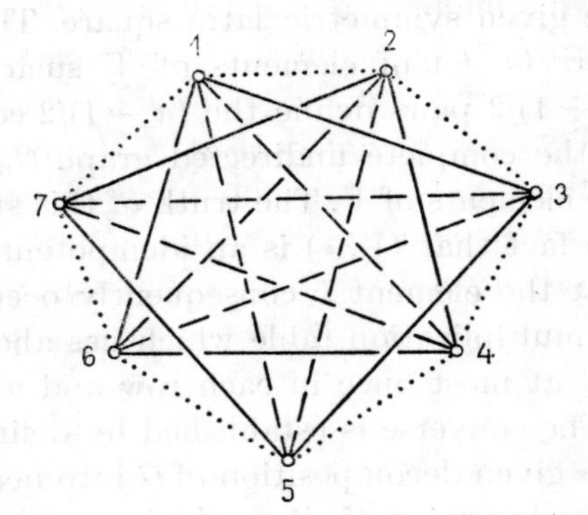

Fig. 9.1.7

With the aid of the two graph theoretical interpretations of a given idempotent symmetric latin square thus established, Kotzig has enumerated all such symmetric latin squares of orders 3, 5 and 7. For each of the orders 3 and 5, there is up to isomorphism just one such square (see Fig. 9.1.8).

1	3	2
3	2	1
2	1	3

1	4	2	5	3
4	2	5	3	1
2	5	3	1	4
5	3	1	4	2
3	1	4	2	5

Fig. 9.1.8

This corresponds to the fact that when $n = 3$ the only partition into closed paths of the complete undirected graph is into a single path of length 3 and that when $n = 5$ the only partition which corresponds to a P-groupoid which is also a quasigroup is that into two paths each of length 5. For the order 7, there exist seven isomorphically distinct idempotent symmetric latin squares. (It is worth noticing that this disproves an early conjecture of S. M. Jacob, given in [1], to the effect that all idempotent symmetric latin squares of prime order satisfy the quadrangle criterion. A counter-example, in fact one of order 7, was first given in S. M. Kerewala [4].)

Kotzig has raised the further question "For what values of n does a P-quasigroup exist which defines a partition of G_n into a single closed path?" (Such a single closed path which contains every edge of G_n is usually called as *Eulerian path.*) He has shown that such a P-quasigroup exists when $n = 3$ or 7 but not when $n = 5$. We may ask "What are the distinguishing features of such a P-quasigroup? What are the distinguishing features of the corresponding symmetric latin square?" No effective answers to these questions have so far been given.[1]

We show in Fig. 9.1.9 the P-quasigroup of order 7 which defines a partition of G_7 into a single closed Eulerian path. In Fig. 9.1.10 we give the corresponding symmetric latin square. This is, of course, isomorphically distinct from that displayed in Fig. 9.1.6.

$(\cdot)$	1	2	3	4	5	6	7
1	1	3	5	6	7	2	4
2	6	2	4	7	3	1	5
3	7	1	3	5	2	4	6
4	5	7	2	4	6	3	1
5	4	6	1	3	5	7	2
6	2	5	7	1	4	6	3
7	3	4	6	2	1	5	7

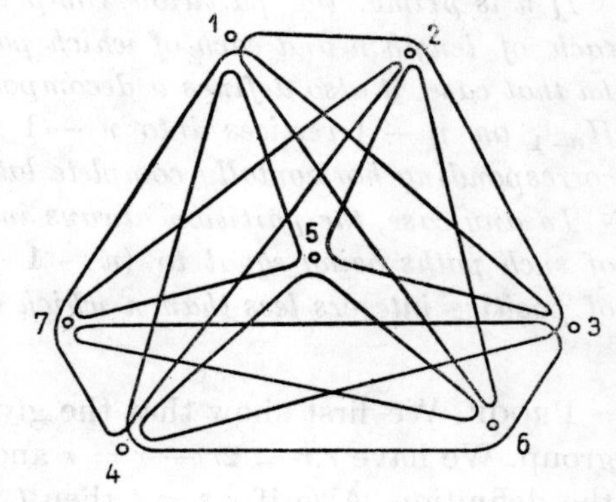

Fig. 9.1.9

$(*)$	1	2	3	4	5	6	7
1	1	6	2	7	3	4	5
2	6	2	5	3	7	1	4
3	2	5	3	6	4	7	1
4	7	3	6	4	1	5	2
5	3	7	4	1	5	2	6
6	4	1	7	5	2	6	3
7	5	4	1	2	6	3	7

Fig. 9.1.10

[1] One of the present authors has proved recently (Autumn 1973) that P-quasigroups of the abovementioned kind exist whenever n is an integer of the form $4r + 3$ such that $r \not\equiv 2 \bmod 5$ and $r \not\equiv 1 \bmod 6$. They also exist when $n = 4r + 3$ and $0 \leq r \leq 7$.

As one of the present authors has observed, the P-quasigroup of order 7 exhibited in Fig. 9.1.7 and also those which can be defined by means of the symmetric latin squares displayed in Fig. 9.1.8 belong to an infinite class of similar kind, the members of which have several interesting properties. We have the following theorem. (See also J. Dénes and A. D. Keedwell [1].)

Theorem 9.1.6. *If n is any odd positive integer, the set $V = \{1, 2, \ldots, n\}$ forms a P-quasigroup under the operation $(\cdot)$ defined by $r.s = 2s - r \pmod n$ and this P-quasigroup defines a partition of the complete undirected graph G_n on n vertices into disjoint closed paths whose lengths are divisors of n and each of which passes through each vertex of G_n at most once.*

If n is prime, the partition comprises $(n-1)/2$ disjoint closed paths each of length n and each of which passes through all the vertices of G_n. In that case, it also defines a decomposition of the complete directed graph H_{n-1} on $n-1$ vertices into $n-1$ disjoint Hamiltonian paths and a corresponding horizontally complete latin square.

In any case, the partition always includes paths of length n, the number of such paths being equal to $(n-1-d)/2$ where d denotes the number of positive integers less than n which are not prime to n.

Proof. We first show that the given operation $(\cdot)$ defines a P-quasigroup. We have $r.r = 2r - r = r$ and $r \neq r.s \neq s$ if $r \neq s$ directly from the definition. Also if $r.s = t$ then $t.s = (2s - r).s = 2s - (2s - r) = r$ as required.

Next, let r and s be arbitrary integers $1 \leq r, s \leq n$. We wish to find the closed path which contains the edge (r, s). We have $r.s = 2s - r \pmod n$. Then, $s.(2s - r) = 2(2s - r) - s = 3s - 2r$, $(2s - r).(3s - 2r) = 2(3s - 2r) - (2s - r) = 4s - 3r, \ldots$ Hence, each of the edges (r, s), $(s, 2s - r)$, $(2s - r, 3s - 2r)$, $(3s - 2r, 4s - 3r), \ldots, (hs - \overline{h-1}\,r, \overline{h+1}\,s - hr), \ldots$, belong to the same path. Now, $(hs - \overline{h-1}\,r, \overline{h+1}\,s - hr) = (r, s)$ implies $hs \equiv hr \pmod n$ and so if $s - r$ is prime to n, we have $h = n$. The path is then of length n and passes through each vertex of G_n just once since, if two vertices of the above path were the same, there would be integers k and l both less than n such that $ls - \overline{l-1}\,r \equiv ks - \overline{k-1}\,r \pmod n$ which would imply $(l - k)(s - r) \equiv 0 \pmod n$ and so $l \equiv k \pmod n$ if $s - r$ is prime to n. On the other hand, if $s - r$ is not prime to n, the smallest positive solution h^* of the equation $hs \equiv hr \pmod n$ is a divisor of n and the length of the path which contains the edge (r, s) is then equal to h^*. In this case too, it is easy to see that the path does not pass through any vertex of G_n more than once, since $ls - \overline{l-1}\,r \equiv ks - \overline{k-1}\,r \pmod{n}$ implies that $l - k$ is a multiple of h^*. This proves the first statement of the theorem.

Next, let us consider the paths which contain the vertex labelled 1. If $s - 1$ is not prime to n, the length of the path which contains the edge $(1, s)$ is less than n and a divisor of n. If $s - 1$ is prime to n, the length of the path is equal to n. Thus, the number of paths of length n (each of which contains all the vertices of G_n and, in particular, the vertex labelled 1) is equal to half the number of the integers $s - 1$ for $s = 2, 3, \ldots, n$ which are prime to n. (Note that each path contains two edges passing through the vertex 1.) This number is $(n - 1 - d)/2$ where d denotes the number of integers less than n which are not prime to n. When n is prime, $d = 0$ and, in that case, the partition comprises $(n - 1)/2$ disjoint closed paths each of length n and each of which passes through each vertex of G_n just once.

In the latter case (n prime), let one vertex of G_n and all the edges which pass through it be deleted. The $(n - 1)/2$ disjoint closed paths of length n of G_n then become Hamiltonian paths of G_{n-1}. If we suppose that the vertex which is deleted is that labelled $n (\equiv 0)$, these Hamiltonian paths are the following:

$$\boxed{(0, k)}, (k, 2k), (2k, 3k), \ldots, \ldots, (\overline{n-2}\,k, \overline{n-1}\,k), \boxed{(\overline{n-1}\,k, 0)}$$

where k takes the values $1, 2, \ldots, (n - 1)/2$.

If we now suppose the graph to be directed and count each path in both senses, we get $n - 1$ directed Hamiltonian paths whose vertices and edges are given by the $n - 1$ rows of the following horizontally complete latin square, where all entries are to be reduced modulo n.

1	2	3	$\ldots$	r	$\ldots$	$n-2$	$n-1$
2	4	6	$\ldots$	$2r$	$\ldots$	$2(n-2)$	$2(n-1)$
$\cdot$	$\cdot$	$\cdot$	$\ldots$	$\cdot$	$\ldots$	$\cdot$	
k	$2k$	$3k$	$\ldots$	kr	$\ldots$	$k(n-2)$	$k(n-1)$
$\cdot$	$\cdot$	$\cdot$	$\cdot$	$\cdot$	$\cdot$	$\cdot$	$\cdot$
$n-1$	$2(n-1)$	$3(n-1)$	$\ldots$	$(n-1)r$	$\ldots$	$(n-1)(n-2)$	$(n-1)^2$

The fact that the square is latin follows from the fact that, for n prime, the integers $r, 2r, \ldots, kr, \ldots, (n-1)r$ are all distinct modulo n and that each vertex occurs exactly once on each path. This completes the proof of the theorem.

The complete latin square exhibited in Fig. 2.3.1 and whose corresponding directed graph is exhibited in Fig. 9.1.1 is of the type constructed in theorem 9.1.6.

Next, let $v_1, v_2, v_3 \ldots v_h$ be the vertices of a closed path of a complete undirected graph G_n belonging to a partition of G_n defined by a P-quasigroup $(V, \cdot)$ and which is such that the vertices $v_1, v_2, \ldots, v_h$ are all distinct. (The paths defined by the partitions obtained in theorem 9.1.6 above are of this kind.) Then we have the relations $v_i.v_{i+1} = v_{i+2}$ for $i = 1, 2, \ldots, h-2$ and $v_{h-1}.v_h = v_1$, $v_h.v_1 = v_2$. Since the vertices $v_1, v_2, \ldots, v_h$ are all distinct, the cells $v_1.v_2, v_2.v_3, v_3.v_4, \ldots, v_{h-1}.v_h$, $v_h.v_1$ lie one in each row and one in each column of the multiplication table of $(V, \cdot)$ and so the entries $v_3, v_4, \ldots, v_h, v_1, v_2$ in these h cells form a partial transversal of the latin square formed by this multiplication table. (See section 3.1 for the definition of a partial transversal.) We call a partial transversal of this kind a *cyclic partial transversal* of the quasigroup $(V, \cdot)$. If V has n elements, a cyclic partial transversal of the maximum length n is called a *cyclic transversal*. The question has been raised "For what odd integers n do P-quasigroups with cyclic transversals exist?" We see that theorem 9.1.6 allows us to give a complete answer to this question as follows:

Theorem 9.1.7. *A P-quasigroup $(V, \cdot)$ which is isotopic to the abelian group Z_n formed by the integers $0, 1, \ldots, n-1$ under addition modulo n and which possesses cyclic transversals exists for every odd order n. If n is prime, the multiplication table of $(V, \cdot)$ can be separated into n disjoint cyclic transversals (the case $n = 5$ is illustrated in* Fig. 9.1.11). *If n is composite, the multiplication table contains both cyclic transversals and also cyclic partial transversals, at least one of each length which is a divisor of n (the case $n = 9$ is illustrated in* Fig. 9.1.12).

Proof. The P-quasigroup $(V, \cdot)$ with the properties stated is that defined on the set $V = \{1, 2, \ldots, n\}$ by the operation $(\cdot)$ given by $r.s = 2s - r \pmod{n}$.

To see that this is isotopic to the group Z_n formed by the integers $0, 1, 2, \ldots, n-1$ under addition modulo n, we define one-to-one mappings θ, φ of V onto these integers by the relations $r\theta = -r$ for $r = 1$,

$2, \ldots, n-1$ and $n\theta = 0$, $s\varphi = 2s \pmod n$ for $s = 1, 2, \ldots, n-1$ and $n\varphi = 0$. Then $rs = t$ in $(V, \cdot)$ implies and is implied by $r\theta + s\varphi = t \pmod n$ in the abelian group Z_n.

The remaining statements of the theorem are immediate consequences of theorem 9.1.6 above.

(·)	1	2	3	4	5
1	1	3	5	2	4
2	5	2	4	1	3
3	4	1	3	5	2
4	3	5	2	4	1
5	2	4	1	3	5

1.2=3	1.3=5	1.4=2	1.5=4
2.3=4	3.5=2	4.2=5	5.4=3
3.4=5	5.2=4	2.5=3	4.3=2
4.5=1	2.4=1	5.3=1	3.2=1
5.1=2	4.1=3	3.1=4	2.1=5

Fig. 9.1.11

(·)	1	2	3	4	5	6	7	8	9
1	1	3	5	7	9	2	4	6	8
2	9	2	4	6	8	1	3	5	7
3	8	1	3	5	7	9	2	4	6
4	7	9	2	4	6	8	1	3	5
5	6	8	1	3	5	7	9	2	4
6	5	7	9	2	4	6	8	1	3
7	4	6	8	1	3	5	7	9	2
8	3	5	7	9	2	4	6	8	1
9	2	4	6	8	1	3	5	7	9

1.3=5	1.4=7
3.5=7	4.7=1
5.7=9	7.1=4
7.9=2	
9.2=4	
2.4=6	
4.6=8	2.8=5
6.8=1	8.5=2
8.1=3	5.2=8

Fig. 9.1.12

An investigation which has some connections with the result of theorem 9.1.7 has recently been published by H. P. Yap in [1]. Yap has shown that if $n = 6m$ or $n = 2^\alpha 3^\beta m$ (≥ 3) where $(6, m) = 1$, $\alpha = 0$, 2 or 3 and $\beta = 0, 1$ or 2 then there exists a finite abelian group (of order $> n$) which has a cyclic partial transversal of length n. He has used the term *simple recurrence relation* for the concept which we have called a cyclic partial transversal.

A relation between transversals of a latin square and linear factors of a bicoloured graph has been established in F. Harary [1]. Harary has denoted a bicoloured graph which has r vertices of one colour and n vertices of another by $K_{r,n}$. He has shown that each latin square of order n may be associated with a bicoloured graph $K_{n,n}$ having $2n$ vertices.

In the Harary graph $K_{n,n}$ of a latin square of order n there are n vertices (say $v_1, v_2, \ldots, v_n$) of one colour which correspond to the n rows of the latin square and n vertices (say $u_1, u_2, \ldots, u_n$) of another colour which correspond to the columns. Each v vertex is joined to each u vertex by an edge and each edge is coloured with one of n colours corresponding to the n symbols which occur in the latin square in such a way that the colour assigned to the edge $v_i u_j$ is that associated with the symbol which occurs in the cell of the ith row and jth column of the square. It follows that each vertex of the graph is incident with exactly n edges and that each of these edges has a different colour.

A *linear factor* or 1-*factor* of a graph Γ is a subgraph which has the same number of vertices as Γ, is without isolated vertices, and has only one edge incident with each vertex. A graph is said to be 1-*factorizable* if it can be decomposed into such 1-factors. Clearly, a latin square has a transversal if and only if the corresponding Harary graph contains a 1-factor whose edges are labelled by n distinct colours. Further, a latin square can be split into n disjoint transversals if and only if the corresponding Harary graph can be factored into 1-factors of this kind.(1)

Another graph representation of a latin square has been introduced by R. C. Bose in connection with a study of graphical representations of statistical designs which show their association schemes (see page 415 of R. C. Bose [2]), and is as follows:

Let $L = \|a_{ij}\|$ be a latin square of order n and let P_L be a graph with n^2 vertices, each vertex being labelled by one of the n^2 ordered triples (a_{i1}, a_{1j}, a_{ij}) $(i, j = 1, 2, \ldots, n)$ of elements of L. Two vertices of P_L are joined by an edge if and only if the triples by which they are labelled have one of their three components in common.

Such a graph P_L has been called a *latin square graph* by F. C. Bussemaker and J. J. Seidel. In [1], these authors have used the latin square

(1) E. Németh and J. D. Horton have introduced the concept of orthogonal 1-factorizations of a graph. Two distinct 1-factorizations of a 1-factorizable graph are called *orthogonal* if any two 1-factors, one from each factorization have at most one edge in common. A Room design of order n (for the definition, see section 6.4) is equivalent to a pair of orthogonal 1-factorizations of a complete graph with n vertices. For the details, see J. D. Horton [2].

graphs of the twelve main classes (see section 4.2) of 6×6 latin squares for the construction of regular symmetric Hadamard matrices of order 36.

We turn now to the particular case of latin squares which satisfy the quadrangle criterion (see section 1.2): that is, to latin squares which represent multiplication tables of groups.

Any finite group G of order n can be represented by a graph by means of the so-called *Cayley representation*. (See A. Cayley [1] and [3].) According to Cayley's representation theorem, an isomorphism may be set up between G and a subgroup of the symmetric group S_n of degree n in such a way that the element a_i of G $(i = 1, 2, \ldots, n)$ corresponds to the permutation

$$\begin{pmatrix} a_1 & a_2 & \ldots & a_n \\ a_1 a_i & a_2 a_i & \ldots & a_n a_i \end{pmatrix}.$$

Then, to each element a_i of G, a directed graph Γ_{a_i} with n labelled vertices and n labelled edges can be made to correspond as follows. The vertices of the graph are labelled by the n elements $a_1, a_2, \ldots, a_n$ of G. Secondly from each vertex a_j of the graph an edge is directed to the vertex whose label is the element $a_j a_i$ of Γ_{a_i} and this edge is given the label a_i. Thus, the graph has n edges all of which are given the same label a_i and which are such that one is directed away from each vertex and one towards each vertex. We shall call Γ_{a_i} the *Cayley subgraph of the element* a_i.

By superimposing the n distinct Cayley subgraphs Γ_{a_i} which correspond to the n distinct elements of G, we obtain a complete directed graph with labelled edges in which the edge directed from the vertex a_i to the vertex a_j is labelled $a_i^{-1} a_j$. Because Cayley originally used colours for labelling the edges, this type of graph is called the *Cayley colour graph* of G and we shall denote it by Γ_G.

If only the subgraphs corresponding to a set S of generating elements of G are superimposed, we get a reduced graph (with the same number of vertices as Γ_G but a smaller number of edges) which we shall show still determines G, see theorem 9.1.9 below. Such a subgraph of G is necessarily connected[1]: for let a_i and a_j be the labels of two vertices which

[1] A directed graph is said to be *connected* if every two distinct vertices of it are connected by an undirected path (a series of consecutive edges not necessarily all directed in the same sense). If, on the other hand, each vertex v_i is connected to each other vertex v_j by a series of consecutive edges all directed in the same sense (that is, in the sense from v_i to v_j), the graph is said to be *strongly connected*. In the present context, strong connection is not required.

are not joined by an edge in the subgraph. Then, since $a_i^{-1}a_j$ is an element of G, it is equal to a product of generators of G and their inverses, say $h_1^{\varepsilon_1} h_2^{\varepsilon_2} \ldots h_k^{\varepsilon_k}$ where each $\varepsilon_i = \pm 1$ and each $h_i \in S$. This implies that a path from a_i to a_j is provided by one of the edges joining the vertices a_i and $a_i h_1^{\varepsilon_1}$, one of the edges joining $a_i h_1^{\varepsilon_1}$ to $a_i h_1^{\varepsilon_1} h_2^{\varepsilon_2}, \ldots$, and finally one of the edges joining $a_i h_1^{\varepsilon_1} h_2^{\varepsilon_2} \ldots h_{k-1}^{\varepsilon_{k-1}}$ to $a_i h_1^{\varepsilon_1} h_2^{\varepsilon_2} \ldots h_k^{\varepsilon_k} = a_j$ all of which belong to the reduced graph. For, if $\varepsilon_1 = +1$, the directed edge from a_i to $a_i h_1^{\varepsilon_1}$ belongs to the reduced graph; while if $\varepsilon_1 = -1$, the directed edge from $a_i h_1^{\varepsilon_1}$ to a_i belongs to it. Similarly, one of the two edges joining each other pair of vertices listed belongs to the reduced graph.

Conversely, we have (c.f. H. S. M. Coxeter and W. O. J. Moser [1]):

Theorem 9.1.8. *If, from the Cayley colour graph of a given finite group G, all the edges labelled by a particular element a_1 are deleted, leaving a Cayley subgraph $\Gamma_1 = \Gamma_G - \Gamma_{a_1}$; then all the edges labelled by a particular element a_2 are deleted, leaving a Cayley subgraph $\Gamma_2 = \Gamma_G - \Gamma_{a_1} - \Gamma_{a_2}$; and so on, until the deletion of the set of edges labelled by one further element a_{i+1} would leave a Cayley subgraph Γ_{i+1} which is no longer connected, then the set S of elements labelling the edges of the Cayley subgraph Γ_i left at the last stage at which there is still connection form a generating set for G.*

Proof. Let a_j be an element not included in S. Since Γ_i is connected, there exists a path from the vertex e representing the identity element of G to the vertex a_j and, if the edges of this path taken in order are labelled $h_1, h_2, \ldots, h_k$, then we have $a_j = h_1^{\varepsilon_1} h_2^{\varepsilon_2} \ldots h_k^{\varepsilon_k}$ where $\varepsilon_i = +1$ or -1 according as the edge on the connecting path labelled h_i is directed towards or away from the vertex a_j. Thus, each element of G can be expressed as a finite product of elements from the set S and their inverses.

The following result, mentioned in H. S. M. Coxeter and W. O. J. Moser [1], is also true, as we stated above.

Theorem 9.1.9. *Let $\Gamma_G^* = \Gamma_{a_1} \cup \Gamma_{a_2} \cup \Gamma_{a_3} \cup \ldots \cup \Gamma_{a_r}$ be the union of the Cayley subgraphs of the elements $a_1, a_2, \ldots, a_r$ in the group G. Then, if Γ_G^* is a connected graph, it determines the group G and hence the complete Cayley colour graph Γ_G of G uniquely up to isomorphism.*

Proof. We first show that each element $g \in G$ can be expressed in at least one way as a product of elements from the set $S = \{a_1, a_2, \ldots a_r\}$ and their inverses. Let e be the vertex[1] representing the identity of

[1] This vertex may be chosen arbitrarily since all vertices of Γ_G^* are "alike".

G and let g be any other element of G. There exists an undirected path joining the vertex e to the vertex labelled g since Γ_G^* is connected. Let the labels of the edges of this path taken in order be $a_{i_1}, a_{i_2}, \ldots, a_{i_k}$. Then $g = e^{-1}g = a_{i_1}^{\varepsilon_1} a_{i_2}^{\varepsilon_2} \ldots a_{i_k}^{\varepsilon_k}$, where $\varepsilon_j = +1$ or -1 according as the edge on the connecting path which is labelled a_{i_j} is or is not directed towards the vertex g. In this way, each element $g \in G$ can be expressed in at least one way as a product of elements taken from the set S and their inverses. Consequently, S is a generating set for G.

Next we show that each product $g_p g_q$ of elements of G is uniquely defined by means of Γ_G^*. Let $g_q = a_{h_1}^{\varepsilon_1} a_{h_2}^{\varepsilon_2} \ldots a_{h_l}^{\varepsilon_l}$, where the a_{h_j} belong to S and $\varepsilon_j = \pm 1$ as before. Then there exists an edge labelled a_{h_1} joining the vertex labelled g_p to another vertex $g_{p_1} = g_p a_{h_1}^{\varepsilon_1}$ (as in Fig. 9.1.13). There exists an edge labelled a_{h_2} joining the vertex labelled g_{p_1} to another vertex $g_{p_2} = g_{p_1} a_{h_2}^{\varepsilon_2} = g_p a_{h_1}^{\varepsilon_1} a_{h_2}^{\varepsilon_2}$, and so on. Hence, finally, we reach a vertex $g_{p_l} = g_p a_{h_1}^{\varepsilon_1} a_{h_2}^{\varepsilon_2} \ldots a_{h_l}^{\varepsilon_l} = g_p g_q$ and the product $g_p g_q$ is thus defined.

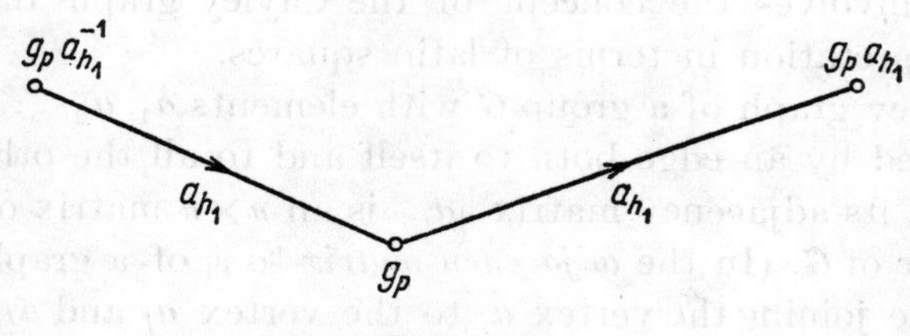

Fig. 9.1.13

Let us note here that the question of the unique reconstructibility of Γ_G from a subgraph has some relevance to the subject matter of section 3.2 in which we considered under what circumstances a latin square could be reconstructed uniquely if some of its entries had been deleted.

In this connection as well as others, it is to be observed that the representation by means of a colour graph in which the edge directed from the vertex a_i to the vertex a_j is labelled by the unique solution x

	e	a	b	c	d
e	e	a	b	c	d
a	a	b	e	d	c
b	b	c	d	e	a
c	c	d	a	b	e
d	d	e	c	a	b

Fig. 9.1.14

of the equation $a_i x = a_j$ can be useful even in the case of quasigroups.[1] For example, it is immediately clear from Fig. 9.1.15 that the loop of order 5 whose multiplication table is exhibited in Fig. 9.1.14 has three elements of "left order" five and one of "left order" three.

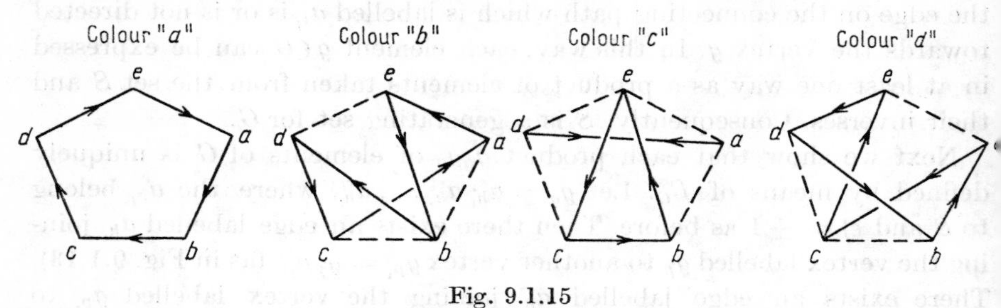

Fig. 9.1.15

We digress here to explain two combinatorial interpretations of the Feit-Thompson theorem to the effect that all groups of odd order are soluble. One involves the concept of the Cayley graph and the other a direct interpretation in terms of latin squares.

In the Cayley graph of a group G with elements $a_1, a_2, \ldots, a_n$, every vertex is joined by an edge both to itself and to all the other vertices. Consequently, its adjacency matrix $\|a_{ij}\|$ is an $n \times n$ matrix of 1's, where n is the order of G. (In the *adjacency matrix* $\|a_{ij}\|$ of a graph, $a_{ij} = 1$ if there is an edge joining the vertex a_i to the vertex a_j and $a_{ij} = 0$ otherwise.)

The edge joining a_i to a_j is labelled $a_i^{-1} a_j = a_k$ say. The adjacency matrix can be written as the sum of permutation matrices $P_1, P_2, \ldots, P_n$, where P_k has an entry 1 in the cell of the ith row and jth column if and only if $a_i^{-1} a_j = a_k$. Thus,

$$\begin{pmatrix} 1 & 1 & \ldots & 1 \\ 1 & 1 & \ldots & 1 \\ . & . & \ldots & . \\ 1 & 1 & \ldots & 1 \end{pmatrix} = \sum_{k=1}^{n} P_k$$

In the multiplication table of G with row border $a_1^{-1}, a_2^{-1}, \ldots, a_n^{-1}$ and column border $a_1, a_2, \ldots, a_n$, the cells which contain a_k are exactly those which contain the entry 1 in the matrix P_k.

Let $S_{hl} = \{j_1, j_2, \ldots, j_l\}$ be a subset of the integers $1, 2, \ldots, n$. There exist $\binom{n}{l}$ such subsets of l integers, and we denote them by $S_{1l}, S_{2l}, \ldots$, and

[1] The idea is exploited in S. C. Shee [1] and [2] and in S. C. Shee and H. H. Teh [1].

so on. Let $\sigma_{S_{hl}} = \sum_{i=1}^{l} P_{j_i}$, $j_i \in S_{hl}$, and write $M = \sum_{l=1}^{n} (-1)^{n-l} \sum_{h=1}^{h=\binom{n}{l}} (\sigma_{S_{hl}})^n$. By a theorem proved in J. Dénes and É. Török [1] (theorem 2.5, page 268), if the group G is a P-group in the sense defined on page 35 of section 1.4, then the necessary and sufficient condition that G be not perfect (that is, not equal to its derived group) is that the $n \times n$ matrix M has at least one zero entry. Using this and the fact that A. Rhemtulla has proved that every finite soluble group is a P-group (see section 1.4), one of the present authors has deduced that the following two statements would together imply the Feit–Thompson theorem: (*i*) every finite group of odd order is a P-group, (*ii*) if G is a group of odd order, the matrix M defined above contains at least one zero entry.

The same author has also interpreted the Feit–Thompson theorem in terms of latin squares as follows: Every latin square of composite odd order n which satisfies the quadrangle criterion can be decomposed into disjoint latin subsquares all of the same size k in such a way that the array of latin subsquares is itself a latin square whose multiplication table is that of an abelian group and so that the number of different latin subsquares is a divisor of n.

He has suggested that these combinatorial interpretations might lead to an easier combinatorial proof of this important theorem.

Let us end this section by mentioning two further connections between latin squares and graphs which are somewhat indirect. We remember that every latin square is the multiplication table of a quasigroup and reflects the algebraic nature of that quasigroup.

In a series of papers ([1], [2] and [3]) A. C. Choudhury has devised a means of defining an ordering relation on any quasigroup Q and with the aid of this definition has associated a directed graph Γ with Q. The properties of Γ serve to illuminate those of Q.

A method of representing the identities satisfied by a given quasigroup by means of graphs has been described in I. M. H. Etherington [1]. Etherington has made use of bifurcating rooted trees: that is, trees every vertex of which is of degree two (except the end-points) and in which one particular vertex is considered as the root. If a, b, c are three elements of a quasigroup Q such that $ab = c$ then this relationship is exhibited as in Fig. 9.1.16. By repeated use of this idea, an identity such as $ab.cd = ef$ is represented in the manner shown in Fig. 9.1.17.

Etherington's main purpose in introducing this representation was to show pictorially what other identities are deducible from a given one

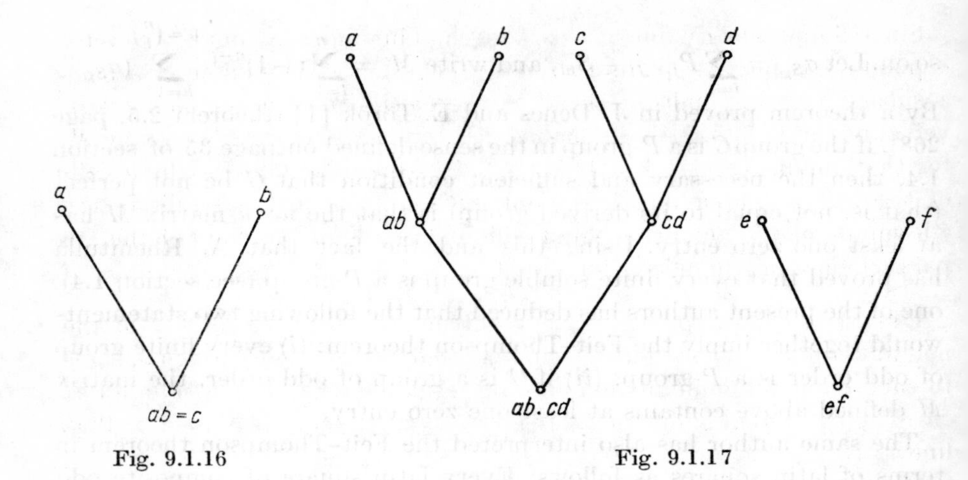

Fig. 9.1.16 Fig. 9.1.17

in a totally symmetric medial quasigroup. (For the definition of this concept, the reader is referred to chapter 2.) He has proved the validity of the following two rules for obtaining new identities from a given one in such a quasigroup:

(*i*) Two factors in a product may be interchanged if their altitudes in the tree have the same parity: and

(*ii*) two factors on opposite sides of the equality sign in an identity may be interchanged if their altitudes have opposite parities.

Here "altitudes" are counted upwards from the root, which has zero altitude. Thus, in the identity $ab.cd = ef$, a and c or ab and cd may be interchanged and, for example, a and e or $ab.cd$ and f can also be interchanged. For further details, the reader is referred to I. M. H. Etherington [1] and [2].

9.2. R. H. Bruck's sufficient condition for the non-existence of a transversal

In his paper [3], R. H. Bruck obtained the result which we have called (A) as a corollary to some results of a more general nature concerning the extendibility of nets. These are of considerable interest in themselves and we shall now explain them. We begin with two definitions.

Definition. If S is a subset of the points of a k-net N such that each line of N contains exactly one point of S, we shall say that S can be *adjoined as a line* to N.

Since S meets every line of a parallel class just once, S contains exactly n points and no two of these are collinear in N. The existence of such a subset S corresponds precisely to the existence of a common transversal of the $k - 2$ mutually orthogonal latin squares determined by N: for the "points" (h_l, i_l), $l = 1, 2, \ldots, n$, of the subset S are the n cells of the common transversal, the entries appearing in these cells in the jth square of the set being the n different labels of the lines of the jth parallel class of the net N, for $j = 1, 2, \ldots, k - 2$.

Definition. Let N be a finite k-net and let f be a mapping of the net into the set of integers (positive, negative and zero) which assigns to each point a unique integer. We say that the integer m is *represented on N by f* if, for each line of N, the sum of the integers assigned to that line is m.

If, in addition, f takes only non-negative values, m is said to be *represented positively* on N by f.

We denote the least positive integer represented on N by $\varphi(N)$. (Note that $\varphi(N)$ is not required to be represented positively on N.)

For example, if f assigns 0 to every point of N, we see that 0 is represented positively on N by f. Again, if f_1 assigns 1 to every point of N, we see that n is represented positively on N by f_1.

Theorem 9.2.1. *Let N be a finite k-net of order n. Then:*

(i) a necessary and sufficient condition that a line can be adjoined to N is that 1 *be positively represented on N;*

(ii) n is positively represented on N;

(iii) $k - 1$ is represented on N;

(iv) $\varphi(N)$ is the positive greatest common divisor of the integers represented on N and so it divides the greatest common divisor of n and $k - 1$;

(v) if N is an affine plane (that is, if $k = n + 1$), then $\varphi(N) = n$;

(vi) with at most a finite number of exceptions, every positive integer divisible by $\varphi(N)$ is positively represented on N.

Proof. (*i*) If S can be adjoined as a line to N, define $f(P) = 0$ or 1 according as the point P is or is not in S. Then, since each line of N contains exactly one point of S, 1 is positively represented on N by f. Conversely, if 1 is positively represented on N by some f, let S be the set of points P for which $f(P) \neq 0$. Clearly, there is exactly one such point on each line of N and for this point we must have $f(P) = 1$. Thus, the set S so defined can be adjoined as a line to N.

(*ii*) If $f_1(P) = 1$ for every point P of N, then f_1 represents n positively on N.

(*iii*) Select an arbitrary point C of the net N and define a mapping h as follows: $h(C) = k - n$, $h(P) = 1$ if P is distinct from but collinear with C, $h(P) = 0$ otherwise. If l is a line of N through C, h sums over the points of l to $k - n + (n - 1) = k - 1$. If l is a line of N not through C, the $k - 1$ lines through C which are not in the same parallel class as l meet l in $k - 1$ distinct points to each of which h assigns the integer 1; hence h sums to $k - 1$ in this case also. Therefore, h represents $k - 1$ on N.

(*iv*) If P_i, for $i = 1$ to n^2, are the points of N, let $f(P_i) = n_i$ be that mapping of these points into the set of integers which gives rise to the least positive integer $\varphi(N)$ which is represented on N. Then, for each line l of N, $\sum_{P_i \in l} f(P_i) = \varphi(N)$. Let $f_1(P_i) = m_i$ be another mapping by means of which an integer m is represented on N, so that $\sum_{P_i \in l} f_1(P_i) = m$ for each line l of N. If $m < 0$, the mapping $f_1^*(P_i) = -m_i$ will represent $-m > 0$ on N, so we may assume that m is positive, $m \geq \varphi(N)$. In that case, the mapping $f_2(P_i) = m_i - qn_i$ will represent the integer $m - q\varphi(N)$ on N. Choose q so that $0 \leq m - q\varphi(N) < \varphi(N)$, as is always possible for $m > \varphi(N) > 0$. If $m - q\varphi(N) = r \neq 0$, the mapping f_2 represents a positive integer less then $\varphi(N)$ on N. Consequently, $r = 0$ and m is an integer multiple of $\varphi(N)$.

Thus, $\varphi(N)$ divides every integer which can be represented on N and any multiple $h\varphi(N)$ of $\varphi(N)$ can be represented on N by a suitable function $f^*(P_i) = hn_i$. Let d be the greatest common divisor of the integers represented on N. If $\varphi(N) \neq d$, we have $d = s\varphi(N)$ for some integer s. Choose an integer t prime to s. Then $t\varphi(N)$ is represented on N. Consequently, d divides $t\varphi(N)$. But this implies that s divides t. Therefore, $s = 1$ and $\varphi(N) = d$.

(*v*) Let $\varphi(N)$ be represented by f on the affine plane N and suppose that $f(P_i) = n_i$ as in (*iv*) above. Let s be the sum of the values of f over the n^2 points of N. That is, $\sum_{i=1}^{n^2} f(P_i) = s$. If we consider the sum of the sums $\sum_{P_i \in l} f(P_i) = \varphi(N)$ over the n lines of some parallel class of lines, we see that $n\varphi(N) = s$. On the other hand, if C is a point of N, every point of N (other then C) lies on exactly one of the $n + 1$ lines through C. Considering the sum of the sums $\sum_{P_i \in l} f(P_i) = \varphi(N)$ over the $n + 1$ lines through C, we find that $nf(C) + s = (n + 1)\varphi(N)$. Since

$s = n\varphi(N)$, $nf(C) = \varphi(N)$. Consequently, n divides $\varphi(N)$. But, by (*iv*) above, $\varphi(N)$ divides n. Therefore, $\varphi(N) = n$. It follows incidently that $f(C) = 1$ for every point C of n.

(*vi*) In virtue of (*ii*), every positive integral multiple of n is positively represented on N. Let r be an integer divisible by $\varphi(N)$ in the range $0 < r < n$. Certainly r is represented on N by some function f, say $f(P_i) = r_i$. Let m be the least value assumed by f. Clearly, $m < 1$, since $f_1(P_i) = 1$ for all i represents n. Then, if h is any integer satisfying $h \geq -m$, the integer $r + hn$ is positively represented on N by the function $f + hf_1$, since $m + h \geq 0$, where $f_1(P_i) = 1$ for all P_i as before. Thus, in the residue class $[r]$ of integers taken modulo n, all the positive integers are positively represented if $m \geq 0$ and all except the positive integers $r, r + n, r + 2n, \ldots, r + (-m - 1)n$ are positively represented if $m < 0$. Since there are only n residue classes of integers modulo n, this means that a finite number at most of positive integers which are multiples of $\varphi(N)$ cannot be represented positively on N.

Corollary. *A necessary condition that a line can be adjoined to N is that $\varphi(N) = 1$.*

Proof. We know from (*i*) that the integer 1 can be positively represented on N if a line can be adjoined to N. Since there is no positive integer less than 1, it is clear that this means that the smallest positive integer which can be represented on N, positively or otherwise, is 1: that is, whenever N has the property that a line can be adjoined to it, $\varphi(N) = 1$.

For Bruck's second theorem, we shall need to introduce co-ordinates for our k-net N of order n in much the same way as we did for a projective plane in section 8.3. We designate the various parallel classes of the k-net N by script letters $\mathcal{B}_1, \mathcal{B}_2, \ldots, \mathcal{B}_k$. We label the lines of each class by means of n symbols $x_1, x_2, \ldots, x_n$ in some arbitrary order. We then define k point-functions I_i, which we call *the indicators*, by the statement $I_i(P) = x_j$ if the line of the parallel class $\mathcal{B}_i$ which passes through P is the one labelled x_j.

If f is a mapping from the set $\{x_1, x_2, \ldots, x_n\}$ into the set of integers (positive, negative or zero) which assigns a unique integer to each x_i, we shall write f^σ to denote the sum $f(x_1) + f(x_2) + \ldots + f(x_n)$.

Theorem 9.2.2. *Let N be a k-net of order n. Then*:

(*i*) *a necessary and sufficient condition that the integer m be represented on N is that m be represented modulo n on N;*

(ii) *if the symbols of the labelling set* $\{x_1, x_2, \ldots, x_n\}$ *are assigned to the lines of the parallel classes of the net* N *in such a way that the lines labelled* x_1 *all pass through the same point* C (which we shall refer to as the *centre* C of the net) *and if* $f_1, f_2, \ldots, f_k$ *are mappings from the set* $\{x_1, x_2, \ldots, x_n\}$ *into the set of integers* (like f above) *which satisfy the conditions*

$$(a)\ f_i(x_1) \equiv 0 \ mod\ n\ for\ i = 1, 2, \ldots, n,\ and$$

$$(b)\ \sum_{i=1}^{k} f_i\{I_i(P)\} \equiv 0\ mod\ n\ for\ each\ P\ of\ N,$$

then $\varphi(N)$ *is the smallest positive integer* s *which satisfies the equation* $sf_1^\sigma \equiv 0$ *mod* n.

PROOF. Let $a_1, a_2, \ldots, a_{kn}$ be the kn lines and $P_1, P_2, \ldots, P_{n^2}$ be the n^2 points of N, arbitrary ordered. We define the line-point *incidence matrix* A of N as the $kn \times n^2$ matrix which has 1 or 0 in the (u, v)th place according as the line a_u is or is not incident with the point P_v. Let U represent a $kn \times 1$ column matrix with every entry equal to 1, and let X be an $n^2 \times 1$ column matrix. Then, the (arbitrary) integer m is represented on N if and only if the matrix equation $AX = mU$ has a solution for the matrix X such that every entry of X is an integer. Certainly, if X_1 is the $n^2 \times 1$ column matrix with every entry equal to 1, we have $AX_1 = nU$ because exactly n points of N are incident with each line and so each row of the matrix A has n non-zero entries. It follows that the matrix equation $AX = mU$ always has the rational solution in which every entry of the matrix X is equal to m/n.

If rank $A = r$, there exist unimodular matrices T and Q with integer elements such that

$$TAQ = \begin{pmatrix} D_r & 0 \\ 0 & 0 \end{pmatrix}$$

where $D_r = \text{diag}\,(e_1, e_2, \ldots, e_r)$ and where the integers e_i are the elementary divisors of the matrix A, so that each e_i divides its successor e_{i+1}. If we define the matrices V, Y by the equations $V = TU$ and $Y = Q^{-1}X$ we find that $AX = mU$ implies

$$\begin{pmatrix} D_r & 0 \\ 0 & 0 \end{pmatrix} Y = TAQY = TAX = T(mU) = mTU = mV,$$

whence $e_j y_j = m v_j$ for $j = 1, 2, \ldots, r$. Here, the v_j and e_j are integers and, since $X = QY$, the matrix equation $AX = mU$ has an integral solution for X if and only if the solutions for $y_1, y_2, \ldots, y_r$ of the equations $e_j y_j = m v_j$ are all integers. (Evidently, $y_{r+1}, y_{r+2}, \ldots, y_n$ may be assigned arbitrarily without affecting the equality $A(QY) = mU$ and so we may take them as integers.) If d_j denotes the greatest common divisor of e_j and v_j so that $e_j = h_j d_j$ and $v_j = k_j d_j$ then we have $h_j y_j = m k_j$ with h_j and k_j coprime, whence h_j divides m if y_j is an integer. It follows that when the equation $AX = mU$ has an integral solution for m, every h_j divides m and so the positive lowest common multiple h^* of $h_1, h_2, \ldots, h_r$ divides m. Also, when $m = h^*$, the y_j are all integers and so the equation $AX = h^*U$ has an integral solution for X. We conclude that h^* is the smallest positive integer which is represented on N. That is, $\varphi(N) = h^*$.

Next, let u be any integer divisible by e_r (and hence by each e_j). Clearly, m is represented modulo u on N if and only if $AX \equiv mU \pmod{u}$ has a solution for the matrix X with integer entries or, equivalently, if and only if $e_j y_j \equiv m v_j \pmod{u}$ for integer values of $y_1, y_2, \ldots, y_r$. That is, if and only if $e_j y_j = m v_j + pu$ for integer values of the variables. Since e_j divides u, this equality holds if and only if e_j divides $m v_j$; in other words, if and only if h_j divides $m k_j$ or h_j divides m for each j. That is, m is represented modulo u on N when and only when $\varphi(N)$ divides m. But, when $\varphi(N)$ divides m, then m is certainly represented on N since $\varphi(N)$ is the positive greatest common divisor of the integers represented on N (theorem 9.2.1). Thus, we shall have proved statement (*i*) of theorem 9.2.2 when we show that e_r divides n.

For $i = 1, 2, \ldots, k$, let R_i denote the submatrix of A comprising the n rows of A which correspond to the n lines of the ith parallel class $\mathcal{B}_i$. Since each point of N lies on exactly one line of the parallel class $\mathcal{B}_i$, the sum of the n rows of the matrix R_i is a row consisting entirely of 1's. Now suppose that the tth row of the matrix R_i $(i > 1)$ corresponds to the line x_1 through the centre C of the net N. Add all the rows of R_i (except the tth) to the tth row of R_i and subtract the sum of the rows of R_1 from the new tth row of R_i. This replaces the tth row of R_i $(i > 1)$ by a row of zeros and is an elementary row operation on A effected by premultiplication of A by a product of elementary matrices each of determinant unity. By a succession of such premultiplications, we obtain a matrix B which differs from A in having the rows corresponding to the lines of the parallel classes $\mathcal{B}_2, \mathcal{B}_3, \ldots, \mathcal{B}_k$ of N which pass through the centre C of the net replaced by zeros. Hence we get the relation

$SA = B$ where S is a unimodular[1] matrix effecting a succession of elementary row operation on A. Hence, B has the same rank and the same elementary divisors as A. It follows that, for the relation

$$TAQ = \begin{pmatrix} D_r & 0 \\ 0 & 0 \end{pmatrix}$$

we can substitute one in which the matrix T (effecting row operations on A) is replaced by a matrix T^* which has (in particular, in its first r rows) zeros in the columns matching with the $k-1$ rows of A which are replaced in B by rows of zeros, since the reduction to elementary divisor form can be effected without using the latter rows.

Let τ_j denote the jth row ($j = 1, 2, \ldots, r$) of the matrix T^*. Since Q is unimodular, it follows from the mode of construction of D_r that e_j is the greatest common divisor of the elements of the matrix $\tau_j A$. Let $g_i(x_l)$ denote the element of the jth row τ_j of T^* which occurs in the column corresponding to the line labelled x_l in the parallel class $\mathscr{B}_i$. Then we have just shown that without loss of generality we may take it that $g_i(x_1) = 0$ for $i > 1$. In the $1 \times n^2$ matrix $\tau_j A$, the entry in the column corresponding to the point P is $\sum_{i=1}^{k} g_i\{I_i(P)\}$ so[2] this is congruent to 0 modulo e_j. In the particular case when $P = C$, the relation

$$\sum_{i=1}^{k} g_i\{I_i(P)\} \equiv 0 \pmod{e_j}$$

reduces to $g_1(x_1) \equiv 0 \pmod{e_j}$, since we already have $g_i(x_1) = 0$ for $i > 1$. Thus, we certainly have $g_i(x_1) \equiv 0 \pmod{e_j}$ for $i = 1, 2, \ldots, k$. If we select a fixed line x_l of the parallel class $\mathscr{B}_i$ and sum the congruence

$$\sum_{i=1}^{k} g_i\{I_i(P)\} \equiv 0 \pmod{e_j}$$

over the n points of this line, we get

$$\sum_{h \neq i} [g_h(x_1) + g_h(x_2) + \ldots + g_h(x_n)] + ng_i(x_l) \equiv 0 \pmod{e_j}$$

[1] Since none of the operations effected by S involve multiplication of the rows of A by a scalar other than 1, and since S can be obtained by carrying out the same sequence of elementary row operations on the identity matrix, S has determinant unity and all its elements are integers: that is, S is unimodular.

[2] If $I_i(P) = x_l$, the column corresponding to the point P in the submatrix R_i of A has a 1 in the row corresponding to the line x_l of the parallel class $\mathscr{B}_i$ and zeros elsewhere.

since $I_h(P)$ for $h \neq i$ ranges over the whole set $\{x_1, x_2, \ldots, x_n\}$ as P describes the fixed line, but $I_i(P) = x_l$ for all P on the chosen line. Therefore, $ng_i(x_l)$ is independent of x_l since $\sum_{h\neq i} g_h^\sigma$ is independent of x_l; whence $ng_i(x_l) = ng_i(x_1) \equiv 0 \pmod{e_j}$ because $g_i(x_1) \equiv 0 \pmod{e_j}$ for each i. So, if d is the greatest common divisor of n and e_j and $e_j = de^*$, we have $g_i(x_l) \equiv 0 \pmod{e^*}$ for all i and x_l. But, since T^* is unimodular and the greatest common divisor of the elements of its jth row τ_j divides det T^*, the greatest common divisor is unity and so $e^* = 1$. That is, e_j divides n. Since this is true for every j, we have in particular that e_r divides n. This completes the proof of part (*i*) of theorem 9.2.2.

To prove part (*ii*), let $g_i(x_l)$ denote the (integer) element of the jth row τ_j of the matrix T^* which occurs in the column corresponding to the line labelled x_l of the parallel class $\mathcal{B}_i$, as above, and let $f_i(x_l) = n_j g_i(x_l)$ where $n_j = n/e_j$. Then $g_i(x_1) \equiv 0 \pmod{e_j}$ implies $f_i(x_1) \equiv 0 \pmod{n}$ which is condition (*a*) of (*ii*), and

$$\sum_{i=1}^{k} g_i\{I_i(P)\} \equiv 0 \pmod{e_j}$$

implies

$$\sum_{i=1}^{k} f_i\{I_i(P)\} \equiv 0 \pmod{n}$$

which is condition (*b*) of (*ii*). Further, if $V = T^*U$ as before, we have $\tau_j U = v_j$ and so $\sum_{h=1}^{k} f_h^\sigma = n_j v_j$. Since $e_j y_j = mv_j$, on multiplying this equation through by n_j we have $ny_j = mn_j v_j$. Therefore, m is represented on N if and only if $mn_j v_j \equiv 0 \pmod{n}$, for $j = 1, 2, \ldots, r$. (We shall call this "result A") Since

$$\sum_{h\neq i} [g_h(x_1) + g_h(x_2) + \ldots + g_h(x_n)] + ng_i(x_l) \equiv 0 \pmod{e_j},$$

we have (on multiplication of this equation by n_j) that $\sum_{h\neq i} f_h^\sigma \equiv 0 \pmod{n}$ and so, since $\sum_{h=1}^{k} f_h^\sigma = n_j v_j$, $f_i^\sigma \equiv n_j v_j \pmod{n}$. This is true for each value of i from 1 to k, as can be seen by allowing the parallel class $\mathcal{B}_i$ to vary in the proof of the relation

$$\sum_{h\neq i} g_h^\sigma + ng_i(x_l) \equiv 0 \pmod{e_j}$$

given earlier. In particular, $f_1^\sigma \equiv n_j v_j \pmod{n}$. If s is defined as the least positive integer such that $sf_1^\sigma \equiv 0 \pmod{n}$, we have $sn_j v_j \equiv 0$

(mod n), whence s is certainly represented on N (by result A), and so $\varphi(N)$ divides s.

To complete the proof, we need to show that s divides $\varphi(N)$. Certainly, since $\varphi(N)$ is represented on N, we have $AX = \varphi(N)U$ for a matrix X with integer elements. Let $f_1, f_2, \ldots, f_k$ be arbitrary integer valued functions satisfying conditions $(ii)(a)$ and $(ii)(b)$ in the statement of theorem 9.2.2, and let τ be the row-vector which has $f_i(x_l)$ in the column corresponding to the line labelled x_l in the parallel class $\mathscr{B}_i$. Then the matrix τA has $\sum_{i=1}^{k} f_i\{I_i(P)\}$ as entry in the column corresponding to the point P, while the 1×1 matrix τU is equal to $\sum_{i=1}^{k} f_i^\sigma$. Therefore, the equation $\tau AX = \varphi(N)\tau U$ together with the congruences of condition $(ii)(b)$ imply that $0 \equiv \varphi(N) \sum_{i=1}^{k} f_i^\sigma \pmod{n}$. (We shall call this "result B".) If we select a fixed line x_l of the parallel class $\mathscr{B}_i$ and sum the congruence $\sum_{h=1}^{k} f_h\{I_h(P)\} \equiv 0 \pmod{n}$ over the n points of this line, we get $\sum_{h \neq i} f_h^\sigma + nf_i(x_l) \equiv 0 \pmod{n}$ since $I_h(P)$ for $h \neq i$ ranges over the whole set $\{x_1, x_2, \ldots, x_n\}$ as P describes the fixed line but $I_i(P) = x_l$ for all P on the chosen line. Therefore, $\sum_{h \neq i} f_h^\sigma \equiv 0 \pmod{n}$.

From result B, we deduce that $\varphi(N)f_i^\sigma \equiv 0 \pmod{n}$ for each i. In particular, $\varphi(N)f_1^\sigma \equiv 0 \pmod{n}$. Since s is the least positive integer with the property that $sf_1^\sigma \equiv 0 \pmod{n}$, we conclude that s divides $\varphi(N)$. Therefore, $\varphi(N) = s$ as required.

For the proof of Bruck's third theorem, we shall need to make use of the concept of a normal subloop of a loop. This, like the corresponding concept for a group, is closely related to the concept of a homomorphic mapping.

DEFINITIONS. A subloop H of a loop G is called a *normal* subloop of G if it satisfies each of the following relations for all x, y in G:

$$xH = Hx, \ (Hx)y = H(xy), \ y(xH) = (yx)H.$$

A single-valued mapping θ of a loop $(G, \cdot)$ into or onto a loop $(K, *)$ is called a *homomorphism* of G into or onto K if and only if $(xy)\theta = (x\theta) * (y\theta)$ for all x, y in G. The *kernel* H of the homomorphism θ is the set of all h in G which map onto the identity element e^* of K.

We may easily show that *the kernel of a homomorphism from G into or onto a subloop K is a normal subloop of G.* For simplicity, we shall use the same operation symbol for both G and K.

Let θ be the homomorphism and H its kernel. Since $xy = z$ in G implies $(x\theta)(y\theta) = z\theta$ and since $h\theta = e^*$ for every h in H, if any two of the elements x, y, z belong to H so does the third. Also, $e\theta = e^*$, where e is the identity element of G, so H is a subloop of G.

If h is in H, x in G, there exists an element k of G such that $kx = xh$. Then $(k\theta)(x\theta) = (x\theta)e^* = x\theta$ so $k\theta = e^*$ and k is in H. Thus, $xH = Hx$ for all x in G.

If x, y, z belong to G, there exist unique elements p, q, r, s in G such that $z = (px)y = q(xy) = x(yr) = (xy)s$ and then $z\theta = [(p\theta)(x\theta)](y\theta) = (q\theta)[(x\theta)(y\theta)] = (x\theta)[(y\theta)(r\theta)] = [(x\theta)(y\theta)](s\theta)$. If one of p, q, r, s belongs to H, we have $z\theta = (x\theta)(y\theta)$ and then all of p, q, r, s are in H. Therefore, $(Hx)y = H(xy) = x(yH) = (xy)H$ for all x, y in G. This completes the proof.

Again as for groups, *if H is a normal subloop of G, the cosets Hx of H partition G and themselves form a loop under the operation $(Hx)(Hy) = H(xy)$, called the quotient loop G/H.*

For the proof, let H be a normal subloop of a loop G with identity element e. Since e belongs to H, x is in Hx for every x of G. If y belongs to Hx, then $y = hx$ for some h in H. Hence, $Hy = H(hx) = (Hh)x = Hx$. It follows that the relation $\sim$ defined by the statement $y \sim x$ if and only if y belongs to Hx is an equivalence relation on G: for $y \sim x$ implies and is implied by $Hy = Hx$, whence it is reflexive, symmetric, and transitive. We can conclude, therefore, that the right cosets Hx of G, modulo H, partition G. Furthermore, using the defining relations for a normal subloop H, we have

$$(Hx)(Hy) = [(Hx)H]y = [H(Hx)]y = [(HH)x]y = (Hx)y = H(xy).$$

Since y in Hx implies $Hy = Hx$, the mapping φ of G defined by $x\varphi = Hx$ is a single-valued mapping of G onto the set G/H of right-cosets which maps each coset into itself. In virtue of the relation $(Hx)(Hy) = H(xy)$, φ is a homomorphism of G upon the groupoid G/H. The identity element of G/H is the coset H. If $H(xy) = H(xz)$ then, since $x(Hy) = x(yH) = (xy)H = H(xy)$, we have $x(Hy) = x(Hz)$ whence $Hy = Hz$. If $H(yx) = H(zx)$ then $(Hy)x = (Hz)x$ and so $Hy = Hz$. Therefore, G/H is a cancellation regular groupoid. The equations $(Ha)(Hx) = Hb$ and $(Hy)(Ha) = Hb$ have the solutions Hx and Hy defined by the equations

$ax = b$ and $ya = b$. In virtue of the cancellation regularity, these solutions are necessarily unique. Therefore, G/H is a loop as required.

If H is the kernel of a homomorphism θ of G onto a loop G^{θ}, the mapping ψ of G/H defined by $(Hx)\psi = x\theta$ is an isomorphism (induced by θ) of G/H onto G^{θ}. So, if H is a normal subloop of a loop G then H defines a natural homomorphism $x \to Hx$ of G onto the quotient loop G/H. Any homomorphism θ of a loop G onto a loop G^{θ} may be regarded as the product of a canonical homomorphism $x \to Hx$ of G onto the quotient loop G/H, where H is kernel of the homomorphism, and an isomorphism $G/H \to G^{\theta}$.

With the aid of the above concepts we may now prove

Theorem 9.2.3. *Let G be a finite loop of order n and let N be the 3-net defined by G. If G contains a normal subloop H of odd order such that the quotient loop G/H is a cyclic group of even order, then $\varphi(N) = 2$. In all other cases, $\varphi(N) = 1$.*

Necessary and sufficient conditions that $\varphi(N) = 2$ are that $n = m.2^t$ for m odd and $t \geq 1$, that G contains a normal subloop K of order m, and that G/K be the cyclic group of order 2^t.

If $n = 4m + 2$, then $\varphi(N) = 2$ if and only if G contains a subloop of order $2m+1$. In particular, if G is a group of order $4m+2$, then $\varphi(N)=2$.

Proof. The labelling set for N comprises the n elements $x_1 = e, x_2, x_3, \ldots, x_n$ of the loop G. We suppose that the lines through the centre C of the net are those labelled e and that the indicators are such that for an arbitrary point P we have $I_1(P) = x_i$, $I_2(P) = x_j$, and $I_3(P) = x_i \cdot x_j$ where $(\cdot)$ is the loop operation. N is then the 3-net defined by the loop G. For this net N, the conditions (*ii*) (*a*) and (*ii*) (*b*) of theorem 9.2.2 become

(*a*) $f_1(e) \equiv f_2(e) \equiv f_3(e) \equiv 0 \pmod{n}$, and

(*b*) $f_1(x_i) + f_2(x_j) + f_3(x_i \cdot x_j) \equiv 0 \pmod{n}$ for each pair of elements x_i and x_j of G. In particular, on putting $x_i = e$ and $x_j = e$ alternately in (*b*) and using (*a*), we find that $-f_3(x_i) \equiv f_2(x_i) \equiv f_1(x_i) \pmod{n}$ for each x_i in G. Hence, from (*b*),

$$f_1(x_i \cdot x_j) = f_1(x_i) + f_1(x_j) \pmod{n}.$$

This means that the mapping $x_i \to f_1(x_i)$ is a homomorphism of G onto some subgroup Z of the additive group of the integers taken modulo n. Thus, Z is necessarily a cyclic group.

Conversely, if G can be mapped homomorphically onto a cyclic group $Z \cong G/H$ of order n^*, there will be no loss of generality in assuming that Z is a subgroup of the additive group of integers taken modulo n and that the homomorphic mapping f_1 satisfies $f_1(x_i \cdot x_j) = f_1(x_i) + f_1(x_j)$ (mod n). Also $n = mn^*$, where m is the order of the kernel H and exactly m elements of G map onto each integer of Z. If t denotes the sum of the elements of Z, it follows that

$$f_1^\sigma = f_1(x_1) + f_1(x_2) + \dots + f_1(x_n) \equiv mt \pmod{n}.$$

By theorem 9.2.2, $\varphi(N)$ is the least positive integer s which satisfies the equation $sf_1^\sigma \equiv 0 \bmod n$. Now, if Z has odd order, $t \equiv 0 \bmod n$ since each positive integer h of Z, excepting the identity element 0, can be paired with its additive inverse $-h$. If Z has even order, it has a unique element $n/2$ of order 2 and t is equal to this element since, excepting 0, each other integer can be paired with its inverse as before. We conclude that if Z has odd order, the least positive integer s such that $smt \equiv 0 \bmod n$ is 1, since we have $t \equiv 0 \bmod n$. Thus, $\varphi(N) = 1$ in this case. If Z has even order and m is even, the same is true, since then $mt \equiv 0$ modulo n. If Z has even order and m is odd, the least positive integer s such that $smt \not\equiv 0 \bmod n$ is 2, since in that case $2mt \equiv 0 \bmod n$ but $mt \not\equiv 0 \bmod n$. Therefore, $\varphi(N) = 2$ for this case. This completes the proof of the first statement of theorem 9.2.3.

If G has order $n = m_1 m_2 2^t$ where $m_1 m_2$ is odd and $t \geq 1$ then necessary conditions that $\varphi(N) = 2$ are that either G contains a normal subloop H of order $m_1 m_2$ such that G/H is the cyclic group of order 2^t or that G contains a normal subloop H_i of order m_i ($i = 1$ or 2) such that G/H_i is the cyclic group of order $m_j 2^t$ ($j \neq i$, $j = 1$ or 2). Both these conditions are sufficient. We shall show that the second condition implies the first and the second statement of theorem 9.2.3 will then follow. There is no loss of generality in assuming that $i = 1$. Thus, G has a normal subloop H_1 of order m_1 such that G/H_1 is the cyclic group of order $m_2 2^t$. In that case, G/H_1 has a normal subgroup H_2 of order m_2 and there exists a homomorphism ψ of G/H_1 onto $(G/H_1)/H_2$ with kernel $H_2 = \{H_1g_1, H_1g_2, \dots, H_1g_r\}$ say, where $r = m_2$. Let φ denote the canonical homomorphism $G \to G/H_1$. Then $\varphi\psi$ is a homomorphism from G onto $(G/H_1)/H_2$ and its kernel consists of all the elements of the cosets H_1g_1, $H_1g_2, \dots, H_1g_r$ of H_2. Thus, the kernel has order $m_1 m_2$, whence G has a normal subloop H of order $m_1 m_2$ and $G/H \cong (G/H_1)/H_2$ is the cyclic group of order 2^t.

The last statement of theorem 9.2.3 is an immediate consequence of the second and provides a second proof of the fact that if a latin square L of order $4m + 2$ represents the Cayley table of a quasigroup which contains a subquasigroup of order $2m + 1$ then L has no orthogonal mate (see theorem 5.1.4).

From theorem 9.2.3, and the corollary to theorem 9.2.1 the result (A) stated at the beginning of this chapter follows immediately: namely, if G is a finite loop of order n and G contains a normal subloop H of odd order such that the quotient loop G/H is a cyclic group of even order, then the latin square L_G representing the multiplication table of G does not possess a transversal. (The result (A) can also be deduced from theorems 1.5.2 and 12.3.1).

9.3. A sufficient condition for a given set of mutually orthogonal latin squares to be extendible to a complete set

In contrast to the above sufficient condition for a single latin square not to possess any transversal, S. S. Shrikhande has proved in [2] that, for $n > 4$, the squares of any set of $n - 3$ mutually orthogonal latin squares always possess sufficient transversals to allow completion of the set to a set of $n - 1$ mutually orthogonal squares. In the language of nets, this means that in any $(n - 1)$-net N of order n it is always possible to find a sufficient number of subsets of points which can be adjoined as lines to N to make up two further parallel classes of lines and thus to complete N to an affine plane. Shrikhande's original proof was expressed in statistical language and used as its instrument the concept of a partially balanced incomplete block design (PBIBD), which we shall define in section 10.2. As we shall show in that section, there is a one-to-one correspondence between PBIBD's of a particular kind with two associate classes and geometric nets.

Shrikhande's result implies that, if k is sufficiently large compared with n, a k-net of order n can always be embedded in an affine plane, or, equivalently, that a set of $k - 2$ mutually orthogonal latin squares of order n can be extended to a complete set of $n - 1$ such squares whenever k is sufficiently large relative to n.

DEFINITION. The number $d = (n - 1) - (k - 2) = (n + 1) - k$ by which a given set of $k - 2$ mutually orthogonal latin squares of order n fall short of being a complete set, equal to the number of parallel classes of lines by which a given k-net falls short of being an $(n+1)$-net, has been called by R. H. Bruck (see [6]) the *deficiency*.

Amongst a number of other more specialized results concerning this quantity, Bruck has proved the following;

(1) whenever $n > (d-1)^2$, a k-net which can be completed to an affine plane can be so completed in only one way;

(2) a sufficient (but not necessary) condition that a k-net can be completed to an affine plane is that $n > \mathscr{F}(d-1)$, where

$$\mathscr{F}(x) = \tfrac{1}{2}x^4 + x^3 + x^2 + \tfrac{3}{2}x .$$

These statements are equivalent to those introduced as result (B) at the beginning of this chapter.

Since $\mathscr{F}(1) = 4$, the statement (2) includes Shrikhande's result as a special case. (We have $k-2 = n-3$ when $d=2$.) As regards (1), it is important to observe that there is no implication that, for $n > (d-1)^2$, every k-net can be completed to an $(n+1)$-net. Indeed, for the case $n=4$, $d=2$, $k=3$, the 3-net which represents the latin square given by the multiplication table of the cyclic group C_4 provides a counter-example, since that latin square has no transversals. (Let us note here that, in any representation of the unique projective plane of order 4 by three orthogonal latin squares of order 4, the squares are the multiplication table of the group $C_2 \times C_2$ and/or of isotopes of that group.) On the other hand, the result (1) is "best possible" in the sense that, when $n = (d-1)^2$, there exist k-nets of order n which can be completed to affine planes in more than one way as has been shown by T. G. Ostrom (see [2] and [1]). We shall now give the proofs of statements (1) and (2).

DEFINITIONS. If P, Q are two points of a given geometrical net N, we shall say that P and Q are *joined* in N if there is a line of N (necessarily unique) which contains both P and Q: if no such line exists, we shall say that P and Q are *not joined* in N. A non-empty set S of points of N with the property that no two of its members are joined in N will be called a *partial transversal* of N. If N has order n a partial transversal having n points will be called a *transversal*.[1] Thus, in the language of the preceding section, a transversal of N is a subset S of n points which can be *adjoined as a line* to N.

[1] The reader will observe that a transversal as defined here corresponds to a common transversal of the set of mutually orthogonal latin squares which represent the net in the manner described in theorem 8.2.1. The connection is made more explicit in theorem 9.3.3.

Theorem 9.3.1. *Let N be a non-trivial finite k-net of order n and deficiency d. Then,*

(i) if S is a partial transversal of N, we have $|S| \leq n$, where $|S|$ denotes the cardinal of set S;

(ii) if P is a point of N then, of the $n^2 - 1$ points of N distinct from P, $n_1 = k(n - 1)$ are joined to P in N and $n_2 = d(n - 1)$ are not joined to P in N;

(iii) if P is a point of N and L a line of N not containing P, then P is joined to $k - 1$ points of L and is not joined to d points of L;

(iii)′ if P is a point of N and T a transversal of N not containing P, then P is not joined to $d - 1$ points of T and is joined to k points of T;

(iv) if P and Q are distinct points joined in N, then, of the remaining $n^2 - 2$ points, $p_{11}^1 = n - 2 + (k - 1)(k - 2)$ are joined to both P and Q, $p_{12}^1 = (k - 1)d$ are joined to P and not joined to Q, $p_{21}^1 = (k - 1)d$ are not joined to P and joined to Q, $p_{22}^1 = d(d - 1)$ are not joined to either P or Q;

(iv)′ if P and Q are distinct points not joined in N, then, of the remaining $n^2 - 2$ points $p_{22}^2 = (n - 2) + (d - 1)(d - 2)$ are not joined to either P or Q, $p_{21}^2 = (d - 1)k$ are not joined to P and are joined to Q, $p_{12}^2 = (d - 1)k$ are joined to P and not joined to Q, $p_{11}^2 = k(k - 1)$ are joined to both P and Q.

Proof. (*i*) is obvious. Since P is incident with just one line of each of the k parallel classes of N and since each line through P has $n - 1$ points distinct from P, there are just $k(n - 1)$ points joined to P and $(n^2 - 1) - k(n - 1) = d(n - 1)$ points not joined to P. This proves (*ii*).

For (*iii*), since each two lines of N from distinct parallel classes have just one point in common, each of the $k - 1$ lines through P which do not belong to the same parallel class as L intersects L in a point, giving a total of $k - 1$ points of L which are joined to P and $n - (k - 1) = d$ points of L which are not joined to P.

A transversal T of N has n points and these necessarily lie one each on each of the n lines of a parallel class of N. Just k lines of N pass through P and each of these contains one point of T, so P is joined to k points of T and is not joined to $n - k = d - 1$ points of T. This gives (*iii*)′.

For (*iv*), we are given that P and Q lie on a line PQ of N. This line PQ has $n - 2$ points distinct from P and Q all of which are joined to both P and Q. Also, each of the $k - 1$ lines through P distinct from PQ is met by the $k - 1$ lines through Q distinct from QP in $k - 2$ distinct points, since each one of the latter lines is parallel to some one of the former and so does not meet it. This gives a total of $p_{11}^1 = n - 2 + (k - 1)(k - 2)$ points which are joined in N to both P and Q. Since $p_{11}^1 + p_{12}^1$ is equal to the total number of points other than Q which are joined to P in N, we have $p_{11}^1 + p_{12}^1 + 1 = n_1 = k(n - 1)$. Therefore, $p_{12}^1 = [(k - 1)(n - 1) + (n - 1)] - 1 - [(n - 2) + (k - 1)(k - 2)] = (k - 1)d$. By symmetry, we also have $p_{21}^1 = (k - 1)d$. Again, since $p_{21}^1 + p_{22}^1$ is equal to the total number of points which are not joined to P in N, we have $p_{21}^1 + p_{22}^1 = n_2 = (n - 1)d$. Hence, $p_{22}^1 = (d - 1)d$.

Finally, for $(iv)'$ we have that P and Q are not joined in N. Hence, none of the k lines through P contains Q and each of these lines is met by the $k - 1$ lines through Q which are not parallel to it in $k - 1$ points. This gives a total of $p_{11}^2 = k(k - 1)$ points which are joined to both P and Q. The sum $p_{11}^2 + p_{12}^2$ is equal to the number $n_1 = k(n - 1)$ of lines joined to P, so $p_{12}^2 = (d - 1)k$. By symmetry, $p_{21}^2 = p_{12}^2$. Lastly, the sum $p_{22}^2 + p_{21}^2$ gives the number of points distinct from Q which are not joined to P, so $p_{22}^2 + p_{21}^2 = n_2 - 1 = d(n - 1) - 1$ and hence $p_{22}^2 = (n - 2) + (d - 1)(d - 2)$.

Theorem 9.3.2. *Let N be a non-trivial finite k-net of order n, deficiency $d > 0$. Let T be a transversal of N and let S be a partial transversal of N not contained in T but containing at least two points of T. Then $|S \cap T| \leq d - 1$ and $|S| \leq (d - 1)^2$.*

Proof. By hypothesis, S contains at least one point R which is not in T and, by $(iii)'$ of the previous theorem, there are precisely $d - 1$ points of T not joined to R. Among these $d - 1$ points must be the points of $S \cap T$, since R is joined to no other points of S. Therefore, $|S \cap T| \leq d - 1$.

Again, by hypothesis, $S \cap T$ contains at least two distinct points P and Q. By $(iv)'$ of the previous theorem, there are precisely $p_{22}^2 = (n - 2) + (d - 1)(d - 2)$ points joined neither to P nor to Q, and the points of $(S \cup T) \setminus \{P, Q\}$ must be among these p_{22}^2 points. Therefore, $|S \cup T| \leq p_{22}^2 + 2$. From this and our inequality for $|S \cap T|$ we get

$$|S| + |T| = |S \cap T| + |S \cup T| \leq (d - 1) + n + (d - 1)(d - 2) = n + (d - 1)^2.$$

Since $|T| = n$ because T is a transversal, this yields $|S| \leq (d - 1)^2$.

Corollary. *If N is a non-trivial finite k-net of order n with deficiency $d > 0$ and if $n > (d - 1)^2$ then two distinct transversals of N can have at most one common point.*

PROOF. Suppose that S as well as T is a transversal of N. Then our result becomes $n \leq (d-1)^2$ whenever S and T are two distinct transversals with at least two points in common. Therefore, for $n > (d-1)^2$, two distinct transversals can have at most one point in common.

In attempting to extend a given k-net N to an affine plane, we wish to adjoin transversals of N as lines. Since no two distinct lines of an affine plane or geometric net can intersect in more than one point, we need to adjoin transversals having at most one point in common. This shows the importance of the above corollary in proving Bruck's results (1) and (2). In fact, the result (1) comes immediately from our next theorem.

THEOREM 9.3.3. *Let N be a non-trivial finite k-net of order n with deficiency d satisfying $n > (d-1)^2$. Let N^* be the system whose points are the points of N and whose lines are the lines of N together with the transversals of N, and whose incidence relation is the natural one. Then,*

(*i*) *if t is the total number of distinct transversals of N, we have $t \leq dn$;*

(*ii*) *a necessary and sufficient condition*[1] *that N be embeddable in an affine plane of order n is that $t = dn$;*

(*iii*) *if N is embeddable in an affine plane N_1 of order n, then N_1 is isomorphic to N^*. That is, N^* is the only possibility for an affine plane of order n containing the net N.*

PROOF. By the corollary to theorem 9.3.2., two distinct transversals of N have at most one common point. Moreover, two distinct lines of N have at most one common point (by definition of a geometric net), and a line and a transversal of N (by definition of the latter) have exactly one common point. Consequently, two distinct lines of N^* have at most one common point.

For each point P of N (and N^*), let $t(P)$ denote the number of distinct transversals of N which contain P. Thus the number of distinct lines of N^* containing P is exactly $k + t(P)$. Two such lines have only the point P in common. Therefore the number of points, distinct from P, to which P can be joined in N^* is $[k + t(P)]\,(n-1) \leq n^2 - 1$. Since $n > 1$, we deduce that $k + t(P) \leq n + 1 = k + d$ and so that $t(P) \leq d$ for every point P of N. Moreover, for any fixed P, $t(P)$ is equal to d precisely when P can be joined (in N^*) to every other point. By summing the

[1] For a sharper result applicable to the case $n=10$, see A. Bruen and J. C. Fisher [1].

relation $t(P) \leq d$ over the n^2 points P of N and using the fact that every transversal has exactly n points, we see that $t \leq dn$ and that equality holds precisely when every two distinct points are joined in N^*.

If N is embeddable in an affine plane N_1 of order n, then (when N is considered as a subsystem of N_1) every line of N_1 is either a line of N or a transversal of N. Hence, every line of N_1 is a line of N^*. Since every two distinct points are joined in N_1, we must conclude that the equality $t = dn$ holds. This proves (*i*) and the necessity part of (*ii*).

Suppose, conversely, that the equality $t = dn$ holds. Then also from our argument above it follows that $t(P) = d$ for every point P, and so every two distinct points are joined in N^*. We consider a transversal T and a line L of N and note that T and L have a unique common point Q. Let P be any point of L distinct from Q. Then P is not in T. Hence, by theorem 9.3.1 (*iii*)′, there are exactly $d - 1$ distinct points of T not joined in N to P. Each of these is joined to P by a unique line of N^*, giving a total of $d - 1$ distinct transversals of N which contain P and intersect T. Since $t(P) = d$, there remains a unique transversal which contains P and is parallel to T. As P varies over the $n - 1$ points of L distinct from Q, we get in this way $n - 1$ distinct transversals parallel to T. No two of these transversals intersect, for a common point R would lie on two distinct transversals parallel to T. Consequently, when we include T, we get a set of n distinct and mutually parallel transversals. These must contain all the points of N: namely n points on each of n transversals. It now follows that the $t = dn$ transversals of N form d distinct parallel classes of lines of N^*, distinct from the kn lines of N. Therefore, N^* is a net of order n having $k + d = n + 1$ parallel classes. That is, N^* is an affine plane. This proves the sufficiency part of (*ii*) and also, because of the preceding paragraph, we see that, when $t = dn$, N_1 is isomorphic to N^*. Thus, (*iii*) is true.

In order to obtain Bruck's result (2), we shall need to introduce the concept of a *net-graph*. The graph G_1 of a k-net N of finite order n has n^2 vertices corresponding to the n^2 points of N and two vertices of G_1 are joined by an edge of G_1 if and only if the points of N which they represent are joined in N. We see that each vertex of G_1 lies on exactly $n_1 = k(n - 1)$ edges of G_1 and so G_1 is a *regular* graph. Since the edges are unordered, G_1 is also *symmetric*. The *complementary* graph G_2 of G_1 is the graph which has the same vertices as G_1 and also has the property that two vertices are joined by an edge in G_2 if and only if they are not so joined in G_1. It follows that G_2 also is regular and symmetric and that each vertex of G_2 lies on exactly $n_2 = d(n-1)$ edges. However, as a consequence of the properties (*iii*), (*iii*)′, (*iv*), and (*iv*)′ of theorem 9.3.1, both G_1 and G_2 have further regularity of structure. For example, each two distinct vertices P and Q of G_2 which are joined in G_2 are separately joined to $p^2_{22} = n - 2 + (d-1)(d-2)$ other vertices of G_2. From these observations, we are led to the concept of a *pseudo net graph* as being a graph whose regularity of structure mirrors that of a net graph or its complement. We make the following formal definition:

DEFINITION. By a *pseudo net graph* G of order n, degree d, and deficiency k, where n, d, k are non-negative integers related by $d + k = n + 1$, is meant a symmetric graph with n^2 vertices such that (*i*) each vertex of G is joined by an edge of G to exactly $n_1 = d(n-1)$ other vertices of G; (*ii*) two distinct vertices P and Q of G which are joined in G are both joined to exactly $p^1_{11} = n - 2 + (d-1)(d-2)$ other vertices of G; two distinct vertices P and Q of G which are not joined in G are both joined to exactly $p^2_{11} = d(d-1)$ other vertices of G.

Our objective is to find sets of conditions on a pseudo net graph which will ensure that it will actually be the graph of a d-net and thus provide the complement necessary for completing a given k-net to an affine plane.

We shall need some further concepts. By a *clique* of a graph G we mean a subgraph of G every two of whose vertices are joined in the subgraph. That is, a clique is a complete subgraph. If G is a pseudo net graph of order n, we define a *line of* G to be a clique with exactly n vertices. When G is the complementary graph G_2 of a net N, the cliques of G are the partial transversals of N, and the lines of G are the transversals of N. In this case, by theorem 9.3.1 (*i*), no clique of G has more than n elements. The same fact is true for pseudo net graphs, but requires a different proof. We shall obtain the result as a corollary to our next theorem, but for the proof we shall first need a set theoretical result of quite general applicability.

Let A and B be non-empty sets and let φ be a finite subset of the cartesian product set $A \times B$. For each a in A, let $a\varphi$ denote the subset of B consisting of all b in B such that (a, b) is in φ; and, for each b in B, let φb denote the subset of A consisting of all a in A such that (a, b) is in φ. Then,

$$\sum_{a \in A} |a\varphi| = \sum_{b \in B} |\varphi b|$$

The proof is simple. For each a in A, the set $(a, a\varphi)$, consisting of all pairs (a, b) with b in $a\varphi$, contains precisely $|a\varphi|$ elements of φ. Also, the sets $(a, a\varphi)$, as a ranges over A, partition φ, provided we ignore the empty sets which may arise. Hence the sum $\sum_{a \in A} |a\varphi|$ is equal to $|\varphi|$. An exactly similar argument shows that the sum $\sum_{b \in B} |\varphi b|$ is also equal to $|\varphi|$, so the result stated holds.

Theorem 9.3.4. *Let G be a pseudo net graph of order n and degree d, and let L be a line of G. Then (i) each vertex of G which is not in L is joined in G to exactly $d - 1$ distinct vertices of L; and (ii) L is a maximal clique of G.*

Corollary. *No clique of G has more than n vertices.*

Proof. Let L' be the set consisting of the $n^2 - n$ vertices of G which are not in L. For each integer x in the range $0 \leq x \leq n$, let $g(x)$ denote the number of vertices in L' which are joined in G to exactly x distinct vertices in L. Then, by means of our set theory result, we shall show that $\Sigma g(x) = n^2 - n$, $\Sigma x g(x) = (d-1)(n^2 - n)$, and $\Sigma x^2 g(x) = (d-1)^2(n^2 - n)$, where summation is over the range of x.

Since each vertex in L' is joined to x vertices of L for some x, the first formula is evident. For the second, let L be the set A, L' be the set B, and φ be the set of all pairs (a, b) such that $\{a, b\}$ is an edge. Then $xg(x)$ is the number of elements $(\varphi b, b)$ of φ with b joined to exactly x edges. Hence,

$$\Sigma x g(x) = \sum_{b \in B} |\varphi b| = |\varphi|.$$

On the other hand, since L has n members and each vertex of L lies on $d(n-1)$ edges (including $n - 1$ joining it to the remaining points of L), we have

$$\sum_{a \in A} |a\varphi| = n\,[d(n-1) - (n-1)] = (d-1)(n^2 - n).$$

Therefore,

$$\Sigma x g(x) = (d-1)(n^2 - n).$$

For the third formula, we take the set A to be the set of $n(n-1)$ ordered pairs of distinct vertices of L, the set B to be L', and φ to be the subset of $A \times B$ consisting of all pairs (a, b) such that there are edges of G joining b to both of the vertices making up a. Then $x(x-1)\,g(x)$ is that part of the sum $\sum_{b \in B} |\varphi b|$ which is contributed by the vertices b which are joined to exactly x vertices of L. Thus,

$$\Sigma x(x-1)\, g(x) = \sum_{b \in B} |\varphi b|.$$

By the definition of a pseudo net graph, each ordered pair of vertices of the set A is joined to $p_{11}^1 = n - 2 + (d-1)(d-2)$ other vertices of G and among these vertices are $n - 2$ vertices of L. Thus, for each a in A, $|a\varphi| = (d-1)(d-2)$. Therefore,

$$\sum_{a \in A} |a\varphi| = (n^2 - n)(d-1)(d-2).$$

By our set theoretical result proved above

$$\Sigma x(x-1) g(x) = (n^2 - n)(d-1)(d-2).$$

Since $\Sigma x g(x) = (d-1)(n^2 - n)$, we deduce that

$$\Sigma\, x^2 g(x) = (n^2 - n)(d-1)^2,$$

as required.

From the formulae for $\Sigma g(x)$, $\Sigma x g(x)$, and $\Sigma x^2 g(x)$, we deduce that

$$\Sigma\, (d-1-x)^2\, g(x) = (d-1)^2 \Sigma g(x) - 2(d-1)\Sigma x g(x) + \Sigma x^2 g(x) =$$

$$= (n^2 - n)\,[(d-1)^2 - 2(d-1)^2 + (d-1)^2] = 0$$

and so $g(x) = 0$ for $x \neq d - 1$. Since $\Sigma\, g(x) = n^2 - n$, we deduce that $g(d-1) = n^2 - n$ and statement (*i*) of the theorem follows. From (*i*) and the fact that n exceeds $d - 1$, we see that for every vertex P in L', the set $L \cup \{P\}$ is not a clique since P is joined to only $d - 1$ vertices and so is not joined to all the vertices in L. This means that L is a maximal clique, and the corollary follows at once.

Our intention is to show that when n is sufficiently large relative to d, the lines of a pseudo net graph have the properties required of them to represent lines of a net. For this purpose we shall need the concepts of a *major clique*, a *grand clique*, and a *claw* of G.

Definition. A *major clique* K is a clique such that $|K| \geq n - (d-1)^2(d-2)$. A *grand clique* is a major clique which is also a maximal clique.

We note that the definition of a line of G as a clique with n vertices, together with the fact that a clique cannot have more than n vertices, shows that a line is a maximal clique with $|K| = n$. That is to say, when lines exist in G, they are grand cliques. Conversely, when $d = 1$ or 2, major cliques and grand cliques are lines of G by virtue of the theorem just proved and the fact that a major clique K has $|K| = n$ if $d = 1$ or 2. We shall show that, when n is sufficiently large relative to d, grand cliques are always lines.

Definition. By a *claw*, (P, S) of a pseudo net graph G we shall mean an ordered pair consisting of a vertex P, the *vertex* of the claw, and a non-empty set S of vertices distinct from P such that every vertex in S is joined to P in G but no vertices in S are joined in G. By the *order* of a claw (P, S) we shall mean the number $|S|$ of vertices in S.

When G is the complementary graph of a net N of deficiency d, it is easy to see that a claw (P, S) of order d exists for every vertex P. Indeed, let L be any line of N not containing P, and let S consist of the d distinct points of L not joined to P in N; then every two points of S are joined in N. Hence, in G, (P, S) is a claw of order d with vertex P.

Theorem 9.3.5. *Let G be a pseudo net graph of order n and degree d, and let K, L be two distinct cliques of G. Then, (i) if $K \cup L$ is not a clique, $|K \cap L| \leq d(d-1)$; (ii) if $K \cap L$ has at least two vertices, $|K \cup L| \leq$ $\leq n + (d-1)(d-2)$; (iii) if both the preceding formulae hold, $|K| +$ $+ |L| \leq n + 2(d-1)^2$.*

Proof. If $K \cup L$ is not a clique, there must exist a vertex P in $K \setminus L$ and a vertex Q in $L \setminus K$ such that P and Q are not joined in G. Then P and Q are both joined to exactly $p_{11}^2 = d(d-1)$ other vertices (by definition of a pseudo net graph) and these must include the vertices of $K \cap L$, which proves (*i*).

If $K \cap L$ contains two distinct vertices R and S, then R and S are joined in G and so (from the definition of a pseudo net graph) are joined to exactly $p_{11}^1 = n - 2 + (d-1)(d-2)$ other vertices. Among these are certainly included the vertices of $K \cup L \setminus \{R, S\}$. From this, (*ii*) follows. (*iii*) is an immediate consequence of the equality $|K| + |L| =$ $= |K \cup L| + |K \cap L|$.

Corollary. *If G is a pseudo net graph of order n and degree d and if $n > 2(d-1)^3$, then two distinct grand cliques of G have at most one common vertex.*

Proof. Suppose that K and L are two distinct maximal cliques with at least two common vertices. Then (*ii*) holds. Moreover, $K \cup L$ cannot be a clique, so (*i*) holds. Hence, (*iii*) also holds. If K and L are also both major cliques, we have

$$2[n - (d-1)^2(d-2)] \leq |K| + |L| \leq n + 2(d-1)^2$$

and so

$$n \leq 2(d-1)^3.$$

Consequently, two distinct grand cliques cannot have as many as two common vertices if $n > 2(d-1)^3$. This is the result stated.

THEOREM 9.3.6. *Let G be a pseudo net graph of order n and degree d, and let (P, S) be a claw of G of order s. Let T be the set of all vertices of G other than P and those in S. For each integer x in the range $0 \leq x \leq s$, let $f(x)$ denote the number of vertices in T which are joined to exactly x vertices of S and also to P. Then the formulae* (*i*) $\sum_{x=0}^{s} f(x) = d(n-1) - s$, (*ii*) $f(0) - \sum_{x=2}^{s} (x-1)f(x) = (d-s)(n-1) - s(d-1)(d-2)$, *and* (*iii*) $2f(0) + \sum_{x=3}^{s} (x-1)(x-2)f(x) = \alpha_s + 2(d-s)(n-1) - 2s(d-1)(d-2)$ *hold, where*

(*a*) *if $s \leq 2$, the summation on the left hand side of* (*iii*) *should be omitted,*

(*b*) *if $s = 1$, the summation on the left hand side of* (*ii*) *should be omitted,*

(*c*) *α_s is an integer such that $0 \leq \alpha_s \leq s(s-1)(d^2 - d - 1)$ and the upper bound is attained precisely when every vertex of T which is joined to at least two distinct vertices of S is also joined to P.*

PROOF. The left hand side of (*i*) is the number of vertices of T which are joined to P. Since P is joined in G to exactly $d(n-1)$ distinct vertices (by definition of a pseudo net graph) of which s are in S and the rest in T, there are $d(n-1) - s$ vertices of T which are joined to P. Thus (*i*) holds.

To obtain (*ii*), we first prove that

$$\sum_{x=1}^{s} xf(x) = s[n - 2 + (d-1)(d-2)]$$

using the set theory result of page 336. We take the set A to be the set of all vertices in T which are joined to P, the set B to be S, and φ to be the subset of $A \times B$ comprising all (a, b) with a in A and b in B such that $\{a, b\}$ is an edge of G. For any $x \geq 1$, $xf(x)$ is the sum of the integers

$|a\varphi|$ as a ranges over the vertices in A which are joined to exactly x vertices in B. Hence,

$$\sum_{x=1}^{s} xf(x) = \sum_{a \in A} |a\varphi| = |\varphi| .$$

For any vertex b in S ($\equiv B$), since P and b are joined, there are exactly $p_{11}^1 = n - 2 + (d-1)(d-2)$ vertices joined to both P and b; and these are in A. Thus, for each b in B, $|\varphi b| = p_{11}^1$. Since $B \equiv S$ has s members, $\sum_{b \in B} |\varphi b| = sp_{11}^1$. Hence, from the result of page 336, $\sum_{x=1}^{s} xf(x) = \sum_{a \in A} |a\varphi| = \sum_{b \in B} |\varphi b| = s[n - 2 + (d-1)(d-2)]$. On subtracting the formula *(i)* from this, we get *(ii)*.

Next, for every ordered pair U, V of distinct vertices in S, we define $f_1(U, V)$ to be the number of vertices in T which are joined to U, V and also to P, $f_0(U, V)$ to be the number of vertices in T which are joined to U, V but not to P. For each such pair U, V, there are exactly $p_{11}^2 = d(d-1)$ vertices in G joined to both U and V. One of these vertices is P and the rest are in T, so $f_1(U, V) + f_0(U, V) = d(d-1) - 1$. We define α_s by the statement that $\alpha_s = \Sigma f_1(U, V)$ where the summation is over the $s(s-1)$ ordered pairs of vertices U, V in S. Then $\alpha_s + \Sigma f_0(U, V) = s(s-1)(d^2 - d - 1)$. Since $\Sigma f_0(U, V)$ is a non-negative integer, α_s satisfies $0 \leq \alpha_s \leq s(s-1)(d^2 - d - 1)$ and attains its upper bound under the conditions stated in the theorem.

To obtain the formula *(iii)*, we again make use of the set theory result of page 336 taking the set A to be the set of all vertices in T which are joined to P as before, the set B to be the set of all ordered pairs U, V of distinct vertices of S, and φ to be the subset of $A \times B$ comprising all triples (a, U, V) with a joined to both U and V. By definition of α_s, $|\varphi| = \sum_{b \in B} |\varphi b| = \alpha_s$. Also, for each $x \geq 2$, $x(x-1)f(x)$ is the sum of $|a\varphi|$ over all those a in A which are joined to exactly x elements of S. Therefore, $\sum_{x=2}^{s} x(x-1) f(x) = \sum_{a \in A} |a\varphi| = \sum_{b \in B} |\varphi b| = \alpha_s$. On multiplying the formula *(ii)* by 2 and adding the result to the formula just derived, we get the formula *(iii)* of our theorem.

THEOREM 9.3.7. *Let G be a pseudo net graph of order n and degree d. Then, if $n > \mathcal{S}(d-1)$ where $\mathcal{S}(x) = \frac{1}{2}x^4 + x^3 + x^2 + \frac{3}{2}x$, G has no claws*[1] *of order $d + 1$.*

[1] Note that the non-existence of claws of order $d + 1$ is a necessary requirement if G is to be the graph of a d-net.

PROOF. Let us suppose, on the contrary, that G has a claw of order $d + 1$. Then, from formula (*iii*) of theorem 9.3.6,

$$2f(0) + \sum_{x=3}^{d+1} (x-1)(x-2)f(x) = \alpha_{d+1} - 2(n-1) - \\ - 2(d+1)(d-1)(d-2).$$

The left hand side of this formula is a non-negative integer so certainly if we replace α_{d+1} by its upper bound $(d+1)d(d^2-d-1)$, the right-hand side will be non-negative. This gives

$$(d+1)d(d^2-d-1) - 2(n-1) - 2(d+1)(d-1)(d-2) \geq 0,$$

whence

$$\begin{aligned} 2(n-1) &\leq (d+1)\,[d(d^2-d-1) - 2(d-1)(d-2)] = \\ &= (d+1)\,[\{(d-1)+1\}\{(d-1)^2+(d-1)-1\} - \\ &\quad - 2(d-1)\{(d-1)-1\}] = \\ &= [(d-1)+2][(d-1)^3+2(d-1)-1] = 2[\mathscr{S}(d-1)-1], \end{aligned}$$

and so $n \leq \mathscr{S}(d-1)$. Thus, for $n > \mathscr{S}(d-1)$, G can have no claws of order $d+1$.

THEOREM 9.3.8. *Let G be a pseudo net graph of order n and degree d such that $n - 1 > (d-1)^2(d-2)$. Then (i) to every pair P, Q of distinct vertices joined in G there corresponds at least one claw (P, S) of order d such that S contains Q; (ii) if G has no claws of order $d+1$, every edge of G is contained in at least one grand clique of G.*

PROOF. We begin by noting that $(P, \{Q\})$ is a claw of order one. If $d = 1$, the proof of (*i*) is complete. Consequently, we consider the case $d > 1$ and may assume inductively that there exists a claw (P, S) of order s such that S contains Q and $1 \leq s \leq d-1$, since this is true for $s = 1$. Since $s \leq d-1$, the right hand side of formula (*ii*) of theorem 9.3.6 is at least $n-1-(d-1)^2(d-2)$; and, since the sum $\sum_{x=2}^{s}(x-1)f(x)$ is non-negative, we deduce that $f(0) \geq n-1-(d-1)^2(d-2) > 0$, the last inequality being a consequence of the hypothesis of the present theorem. If R is any one of the $f(0)$ vertices in T which are joined to P but to no vertex in S, then $(P, S \cup \{R\})$ is a claw of order $s+1$. The result (*i*) now follows by induction.

Next, let $\{P, Q\}$ be any edge of G. By result (i), there exists at least one claw (P, S') of order d such that S' contains the vertex Q. Let us define S by $S' = S \cup \{Q\}$ where S does not contain Q. Define T to be the set of all vertices of G other than P and those in S, and let H be the subset of all elements of T which are joined to P but to no element of S. Certainly H contains Q and also $|H| = f(0)$, where we have already shown that $f(0) \geq n - 1 - (d-1)^2(d-2) > 0$. Therefore, if $K = H \cup \{P\}$, $|K| \geq n - (d-1)^2(d-2)$. We claim that K is a clique. Indeed, every element of H is joined to P. So if K contains two distinct vertices A and B not joined in G, then $(P, S \cup \{A, B\})$ is a claw of order $d+1$, contrary to hypothesis. In consequence of the inequality satisfied by $|K|$, K is a major clique. Consequently, if K' is any maximal clique containing K, then K' is a grand clique containing the edge $\{P, Q\}$. This proves result (ii) of the theorem.

THEOREM 9.3.9. *Let G be a pseudo net graph of order n and degree d subject to the following three conditions:* (i) *G has no claws of order $d+1$;* (ii) *two distinct grand cliques of G have at most one common point; and* (iii) *$n > \mathcal{R}(d-1)$, where $\mathcal{R}(x) = 2x^3 - x^2 - x + 1$. Then every vertex of G lies in exactly d distinct grand cliques, and every grand clique of G is a line of G.*

PROOF. We observe first that

$$\mathcal{R}(d-1) - 1 = (d-1)^2(d-2) + d(d-1)(d-2)$$

and so the inequality $n > \mathcal{R}(d-1)$ implies the inequality $n - 1 > (d-1)^2(d-2)$ which occurs in the hypothesis of theorem 9.3.8. So, by that theorem, if P is any vertex of G, there exists at least one claw (P, S) of order d with vertex P. We denote the d vertices of S by $A_1, A_2, \ldots, A_d$. For each i in the range $1 \leq i \leq d$, we denote by H_i the set of vertices of G, distinct from P and the A_j for $j \neq i$, which are joined to P but to no vertex A_j for $j \neq i$. By the argument of theorem 9.3.8 (ii), taking $Q = A_i$, $\{P\} \cup H_i$ is, for each i, a major clique containing the vertices P and A_i. We denote by K_i a grand clique containing $\{P\} \cup H_i$. Since, for $i \neq j$, H_i and H_j have no common elements (otherwise such a common vertex would form with (P, S) a claw of order $d+1$), it follows from hypothesis (ii) of the present theorem that the only common element of K_i and K_j is P. We wish to show that the d grand cliques $K_1, K_2, \ldots, K_d$ are the only grand cliques containing P.

We begin by recalling that (P, S) is a claw of order d and that, if T denotes the set of all vertices of G other than P and those in S, the set

$H = H_1 \cup H_2 \cup \ldots \cup H_d$ consists of S (since H_i contains A_i for $i = 1, 2, \ldots, d$) and of all vertices of T which are joined to exactly one of the vertices $A_1, A_2, \ldots, A_d$ of S and are also joined to P. That is, in the notation of theorem 9.3.6, $|H| = f(1) + d$. Also, since G has no cliques of order $d + 1$, we have $f(0) = 0$. Hence, the formulae (*i*) and (*ii*) of theorem 9.3.6 (with $s = d$) can be rewritten as $|H| + \sum_{x=2}^{d} f(x) = d(n-1)$ and $\sum_{x=2}^{d} (x-1) f(x) = d(d-1)(d-2)$ respectively. In the case $d = 1$, the summation disappears in the first of these formulae, giving $|H| = d(n-1)$, whence the inequality

$$d[n-1-(d-1)(d-2)] \leq |H| \leq d[n-1-(d-2)]$$

holds trivially. In the case $d > 1$, we have $1 \leq x - 1 \leq d - 1$ for each x in the range $2 \leq x \leq d$ and so the second formula gives

$$\sum_{x=2}^{d} f(x) \leq d(d-1)(d-2) \leq (d-1) \sum_{x=2}^{d} f(x) \, .$$

From this,

$$-d(d-1)(d-2) \leq -\sum_{x=2}^{d} f(x) \leq -d(d-2)$$

and this inequality combined with the formula

$$|H| + \sum_{x=2}^{d} f(x) = d(n-1)$$

again gives

$$d[n-1-(d-1)(d-2)] \leq |H| \leq d[n-1-(d-2)].$$

Now suppose that P is contained in at least one grand clique K distinct from $K_1, K_2, \ldots, K_d$. Then each of the $d(n-1)$ vertices of G (distinct from P) which are joined to P, is contained in at most one of the $d + 1$ grand cliques by hypothesis (*ii*) of our theorem. Moreover, $K_1, K_2, \ldots, K_d$ together contain at least $|H|$ of these vertices, and K, being a grand clique, contains at least $n - 1 - (d-1)^2(d-2)$ more. So by the inequality for $|H|$ just obtained,

$$d[n-1-(d-1)(d-2)] + [n-1-(d-1)^2(d-2)] \leq d(n-1)$$

and hence

$$n - 1 \leq (d-1)^2(d-2) + d(d-1)(d-2) = \mathfrak{R}(d-1) - 1,$$

implying $n \leq \mathfrak{R}(d-1)$. Therefore, if $n > \mathfrak{R}(d-1)$, every vertex of G lies in exactly d distinct grand cliques.

Suppose that Q were a vertex joined to P but not in any of the grand cliques $K_1, K_2, \ldots, K_d$. Then, by theorem 9.3.8(*ii*), there would exist a grand clique K, distinct from $K_1, K_2, \ldots, K_d$ containing P and Q. We conclude that each of the $d(n-1)$ vertices joined to P lies in one (and only one, by hypothesis (*ii*) of our theorem) of $K_1, K_2, \ldots, K_d$. By theorem 9.3.4, no maximal clique can have more than n elements and so each of $K_1, K_2, \ldots, K_d$ must contain exactly n vertices. That is, each K_i is a line of G. Thus, all grand cliques containing an arbitrary vertex P (there are just d of them) are lines.

If K is any given grand clique of G, we fix attention on a vertex P contained in K and use the fact, just proved, that every grand clique containing P is a line; in particular, K is a line. This completes the proof of theorem 9.3.9.

THEOREM 9.3.10. *Let G be a pseudo net graph of order n and degree d with $n \geq d \geq 1$. Then, if any of the following three sets of conditions hold, G is the graph of one, and only one, d-net of order n. The sets of conditions are:*

(*a*) (*i*) *G has no claws of order $d+1$; (ii) two distinct grand cliques of G have at most one common vertex; (iii) $n > \mathcal{R}(d-1)$, where $\mathcal{R}(x) = 2x^3 - x^2 - x + 1$;*

(*b*) (*i*) *G has no claws of order $d+1$; (ii) $n > 2(d-1)^3 \geq 0$ (and in the case that $d = 1$, also $n > 1$);*

(*c*) *$n > \mathcal{S}(d-1)$, where $\mathcal{S}(x) = \frac{1}{2}x^4 + x^3 + x^2 + \frac{3}{2}x$, and either $d = 1$, $n > 1$, or $d > 1$.*

PROOF. The conditions (*a*) are the same as the hypotheses of theorem 9.3.9. Hence we see that every grand clique of G is a line and that each vertex of G lies on exactly d distinct lines of G. Let P be a vertex of G and let L be a line of G not containing P. By theorem 9.3.4 (*i*), P is joined to exactly $d-1$ distinct vertices of L. Since $n > \mathcal{R}(d-1)$ implies $n - 1 > (d-1)^2(d-2)$ (see theorem 9.3.9), the hypotheses of theorem 9.3.8 are satisfied and so each edge of G containing P is contained in at least one grand clique of G. From hypothesis (*ii*) of the present theorem it then follows that the $d-1$ distinct vertices of L to which P is joined in G lie one each on $d-1$ lines through P. Consequently, there is one and only one line L' through P which is parallel to L (that is, has no vertex in common with L). If we choose any line M which meets L, the same argument shows that through each vertex in M but not in L there passes a unique line parallel to L. Since there exists just one line parallel to L through each vertex not contained in L, no two of these parallels can intersect. Since M contains exactly $n - 1$ points distinct from its intersection with L, we see that L determines a parallel class consisting of n lines, including L itself, each two of which are parallel. This, together with the facts that each grand clique of G is a line, that two such lines have at most one common point (hypothesis (*ii*) above), and that there is a line of each parallel class through each point P (implied by the above argument) is sufficient to show that the vertices of G (considered as points) and the lines of G constitute a net of order n and degree d. Since two distinct vertices of G are joined in G if and only if they lie on a common line of G (by theorem 9.3.8(*ii*)), we see that G is the graph of the net.

As regards the conditions (*b*), condition (*b*)(*i*) is the same as (*a*)(*i*). By the corollary to theorem 9.3.5, condition (*b*)(*ii*) implies (*a*)(*ii*) Also, since $2(d-1)^3 - \mathcal{R}(d-1) = (d-1)^2 + (d-1) - 1 > 0$ for $d > 1$ and since $\mathcal{R}(0) = 1$, condition (*b*)(*ii*) implies (*a*)(*iii*). Thus, conditions (*b*) imply conditions (*a*) and so the theorem is true when the former hold.

By theorem 9.3.7, condition (*c*) implies (*b*)(*i*). Also, since $\mathcal{S}(d-1) - 2(d-1)^3 = \frac{1}{2}[(d-1)^4 - 2(d-1)^3 + 2(d-1)^2 + 3(d-1)] = \frac{1}{2}[(d-1)^3(d-3) + 2(d-1)^2 + 3(d-1)] > 0$ if $d > 1$, condition (*c*) implies (*b*)(*ii*). Thus, condition (*c*) implies conditions (*b*), and so the theorem is true when the former hold.

Bruck's result (2) can be deduced immediately from theorem 9.3.10 (*c*). For suppose that N is a k-net of order n with $n > \mathcal{S}(d-1)$ and $d = (n+1) - k$, having a net graph G_1. The complementary graph G_2 is a pseudo net graph of order n, degree d, and deficiency k, where $n > \mathcal{S}(d-1)$. From theorem 9.3.10(*c*), it follows that G_2 is the net graph of a uniquely determined d-net N' of order n and, by definition of G_2 as complementary graph of G_1, each two points of N which are not joined by a line of N are joined by a line of N'. The lines of N' form d parallel classes distinct from the k parallel classes of N, and so the union of N and N' is an $(n+1)$-net; that is, an affine plane.

It is easy to show from the above result that, if $N(n) < n - 1$, then necessarily $N(n) < n - 1 - (2n)^{1/4}$ where $N(n)$ denotes the maximum number of mutually orthogonal latin squares of order n which can be constructed (see section 5.2). For, by the definition of d, we have $d = (n-1) - N(n)$ when a set of $N(n)$ mutually orthogonal latin squares exists. By definition of $N(n)$, this number of squares cannot be extended to a complete set since otherwise $N(n)$ would not be maximal. Consequently, we must have $n \leq \mathcal{S}(d-1)$. But $\frac{1}{2}(x+1)^4 = \frac{1}{2}x^4 + 2x^3 +$

$+ 3x^2 + 2x + \frac{1}{2} = \mathscr{S}(x) + (x^3 + 2x^2 + \frac{1}{2}x + \frac{1}{2})$ and so $\mathscr{S}(d-1) < \frac{1}{2}d^4$ for $d > 1$. Also, $\mathscr{S}(d-1) < \frac{1}{2}d^4$ for $d = 1$ trivially, since $\mathscr{S}(0) = 0$. Therefore, $n < \frac{1}{2}[(n-1) - N(n)]^4$ if $N(n) < n - 1$. This inequality yields $(2n)^{1/4} < (n-1) - N(n)$, whence $N(n) < n - 1 - (2n)^{1/4}$, as required.

In his paper [1], L. Zhang has given a construction for obtaining an $(n-1)$th square orthogonal to all the squares of a given set of $(n-2)$ mutually orthogonal latin squares, thus making up a complete set. The authors have not had access to the contents of this paper and consequently have been unable to ascertain whether this construction is more direct than that implicit in S. S. Shrikhande's paper [2] (in which he proved that a set of $n-3$ mutually orthogonal latin squares can always be extended to a complete set provided $n \neq 4$).

If N is a given k-net of order n we say that the net N has *critical* deficiency if $n = (d-1)^2$. It follows from theorem 9.3.3 that, when d is less than the critical deficiency, then (1) N can be extended to an affine plane in at most one way, (2) the number of distinct transversals is less than or equal to dn, and (3) N can be embedded in an affine plane of order n when and only when equality holds in (2). By methods similar in kind to those used by R. H. Bruck in obtaining these results, T. G. Ostrom has shown in [2] that, when d is critical, then (1)′ N can be extended to an affine plane of order n in at most two ways, (2)′ the number of distinct transversals is less than or equal to $2dn$, and (3)′ N can be extended to a plane in two different ways if equality holds in (2)′. The same author has subsequently used similar methods both to obtain a new class of projective planes and also to unify many of the known ways of constructing finite projective planes by making an analysis of these constructions in terms of net structure (see T. G. Ostrom [3], [4] and [5]).

In R. C. Bose [2], the concepts of a *partial geometry* and a *pseudo geometric net* have been introduced as generalizations of those of a *net* and a *pseudo net graph* respectively and used to illuminate the study of partially balanced incomplete block designs. (For a description of the latter, see section 10.2). In the same paper, a number of R. H. Bruck's theorems given above have been generalized so as to make them applicable to this more general situation.

CHAPTER 10

Practical applications of latin squares

Up to the present time, there have been two main practical applications of the latin square concept. The first is to the construction of error correcting telegraph codes and we devote the first section to a detailed account of this.

The second application is to the design of statistical experiments, one aspect of which has already been discussed in detail in section 2.3.

In section two of this chapter, we consider statistical designs more generally. We explain the role of orthogonal latin squares in statistics and then we introduce the concept of balanced incomplete block design and its generalizations. Among other things, we show that a partially balanced incomplete block design with a particular kind of association scheme is equivalent to a set of mutually orthogonal latin squares. However, the introduction of the block design is necessary for another reason. Namely, it is one of the tools used in the disproof of Euler's conjecture and will be needed for this purpose in the next chapter.

10.1. Error detecting and correcting codes

Suppose that we wish to transmit a sequence of digits $a_1, a_2, \ldots$, across a noisy channel: for example, through a telegraph cable. Occasionally, the channel noise will cause a transmitted digit a_i to be mistakenly interpreted as a different digit a_j with the result that the message received at the destination will differ from that which was transmitted.

Although it is not possible to prevent the channel from causing such errors, we can reduce their undesirable effects by means of coding. The basic idea is simple. To each "word" $(a_1, a_2, \ldots, a_k)$ of k *information digits* which we wish to transmit, we adjoin a further r *parity check digits* and transmit the resulting *code word* of $n = k + r$ digits. We may regard such a code word as a row matrix $\boldsymbol{a} = (a_1, a_2, \ldots, a_n)$, as a vector of a vector space $V_n(F)$ of dimension n over a field F, or as a point of an affine space of n dimensions; and we require to be able to detect errors in one or more of the digits of such a code word when it is received at the destination. The code is then *error detecting*. If further, we can determine from a received word with one or more digits in error the word which was intended by the transmitter, then the code is *error correcting*.

Usually the code words consist of ordered sequences of binary digits, but it is probable that in the future non-binary codes will become more and more frequently used. A code constructed on q digits, say $0, 1, 2, \ldots, q-1$, will be called a *q-ary code*. In the vector space representation of such a code, referred to above, the field F is the finite field (Galois field) of q elements.

In order to be able to explain how error detection and error correction can be effected, we need the concept of the *Hamming distance* $d(\boldsymbol{a}, \boldsymbol{b})$ between two code words

$$\boldsymbol{a} = (a_1, a_2, \ldots, a_n) \quad \text{and} \quad \boldsymbol{b} = (b_1, b_2, \ldots, b_n)$$

of length n. This is defined as the number of places in which the two words differ. It has the properties $d(\boldsymbol{a}, \boldsymbol{a}) = 0$, $d(\boldsymbol{a}, \boldsymbol{b}) = d(\boldsymbol{b}, \boldsymbol{a})$, and $d(\boldsymbol{a}, \boldsymbol{b}) + d(\boldsymbol{b}, \boldsymbol{c}) \geq d(\boldsymbol{a}, \boldsymbol{c})$, as required for a distance function. (See D. W. Hamming [1].)

In the case when the *alphabet* comprises the two digits 0 and 1, the code words may be represented by the points of an n-dimensional cube and the Hamming distance is then equal to the number of edges which must be traversed to get from one point to another.

Whatever the size of the alphabet, we may define the *Hamming sphere*, centre $\boldsymbol{a}$ and radius d, as comprising the set of all points whose Hamming distance from $\boldsymbol{a}$ is equal to d.

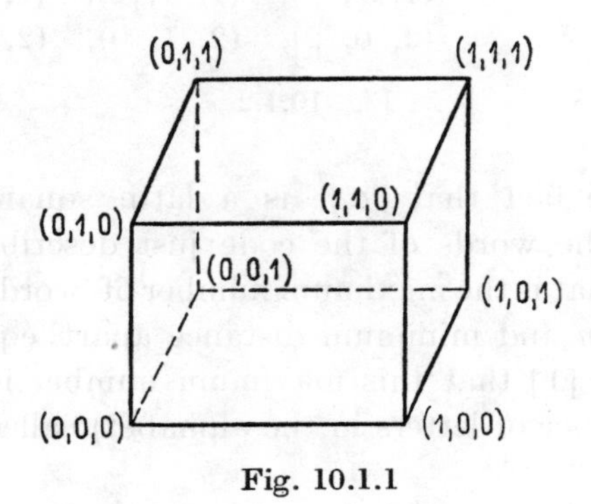

Fig. 10.1.1

EXAMPLE. In Fig. 10.1.1, the sphere of centre (1, 1, 1) and radius 2 comprises the three points (1, 0, 0), (0, 1, 0) and (0, 0, 1).

If a set of code words is such that the minimum distance between any two code words is 2 then the code will be *single error detecting*, since any word which is transmitted with a single error will be received as a meaningless symbol. However, a double error may go undetected.

A set of code words whose minimum distance is 3 will be a *single error correcting* or a *double error detecting* code. If it is assumed that a received word will have at most one error, the received symbol will always represent a point which is closer to the code point representing the intended code word than to any other code point. Consequently, it will be possible to correct the error. Double errors will be detected but may be miscorrected.

More generally, a set of code words whose minimum distance is d will be a $d-1$ error detecting or a $(d-1)/2$ error correcting code.

The purpose of the check digits introd ucedabove is to enable the minimum Hamming distance between the words of a code to be increased from the value which it would have for words comprising message digits only.

It is easy to see that, from a latin square of order q, one can construct a q-ary code with q^2 code words in the following way. Let $\|a_{ij}\|$ be the latin square of order q. Then the q^2 ordered triples (i, j, a_{ij}), $i, j, a_{ij} = 0, 1, 2, \ldots, q-1$ may be regarded as a set of q^2 code words defined over the alphabet $0, 1, 2, \ldots, q-1$.

This correspondence between a latin square and a set of q^2 code words is illustrated by Fig. 10.1.2.

0	1	2	(0, 0, 0),	(0, 1, 1),	(0, 2, 2)
1	2	0	(1, 0, 1),	(1, 1, 2),	(1, 2, 0)
2	0	1	(2, 0, 2),	(2, 1, 0),	(2, 2, 1)

Fig. 10.1.2

By virtue of the fact that $\|a_{ij}\|$ is a latin square, the minimum distance apart of the words of the code just described is equal to 2.

We may ask "What is the maximum number of words in a code whose words have length n and minimum distance apart equal to d?" D. D. Joshi has shown in [1] that this maximum number is at most q^{n-d+1} (where q is the number of letters in the alphabet) called the *Joshibound*. We have

THEOREM 10.1.1. *The maximum number of code words of length n of a q-ary code with minimum Hamming distance between words equal to d is* q^{n-d+1}.

PROOF. Let $\boldsymbol{a}_i = (a_{1i}, a_{2i}, \ldots, a_{ni})$, $i = 1, 2, \ldots, h$, be the words of a code whose minimum distance is d. Let $\alpha_j = (\alpha_{1j}, \alpha_{2j}, \ldots, \alpha_{d-1j}, 0, 0, \ldots, 0)$, $j = 1, 2, \ldots, q^{d-1}$, be vectors whose last $n - d + 1$ entries are zeros. Then, the vectors $\boldsymbol{a}_i + \alpha_j = (a_{1i} + \alpha_{1j}, \ldots, a_{d-1i} + \alpha_{d-1j}, a_{di}, \ldots, a_{ni})$ are necessarily all distinct: for, in the first place, it is immediate that $\boldsymbol{a}_i + \alpha_j \neq \neq \boldsymbol{a}_i + \alpha_k$; and, in the second place, $\boldsymbol{a}_i + \alpha_j = \boldsymbol{a}_l + \alpha_k$ for $i \neq l$ would imply $a_{di} = a_{dl}, a_{d+1i} = a_{d+1l}, \ldots, a_{ni} = a_{nl}$ and so the code words $\boldsymbol{a}_i, \boldsymbol{a}_l$ could differ in at most the first $d - 1$ places, which would imply that their Hamming distance apart was less than the minimum distance d. The number of distinct vectors $\boldsymbol{a}_i + \alpha_j$ is hq^{d-1} and, since this cannot exceed the total number of possible vectors of length n, we have $hq^{d-1} \leq q^n$. Therefore, $h \leq q^{n-d+1}$.

For given values of q, n and d, the Joshibound is not always attainable. Consider, for example, the case when $q = 2$ and $d = 3$. The sphere with centre at a given code point $\boldsymbol{a}_i$ and radius 1 will contain n points of the affine space formed by the 2^n points $(x_1, x_2, \ldots, x_n)$, where each x_i is 0 or 1. Those points which represent code words are at a distance 3 apart and so they are the centres of a set of spheres of radius 1 no two of which have a point in common. The space of 2^n points can comprise at most $2^n/(n + 1)$ such spheres. Consequently, the maximum number of code words is at most $2^n/(n + 1)$. For $n > 3$, this number is less than the Joshibound.

A similar argument applied to the case when the code word alphabet has q letters and $d = 3$ shows that the number of words in a single error detecting code over a q letter alphabet is at most $q^n/\{1 + n(q - 1)\}$. For $n > q + 1$, this number is less than the Joshibound. (In this case, a sphere of radius 1 has $n(q - 1)$ points on its surface and so, with its centre, it occupies $n(q - 1) + 1$ points of the space of q^n points.)

DEFINITION. A single error correcting code having $q^n/\{1 + n\,(q - 1)\}$ code words is called *close-packed*.

For the q-ary code which we constructed above from a latin square of order q, the Joshibound was attained, since $q^{3-2+1} = q^2$. More generally, we have

THEOREM 10.1.2. *When* $n \leq q + 1$ *and* $d = n - 1$ *or when* $n \leq q$ *and* $d = n$, *the Joshibound can always be attained if q is a number for which*

$q - 1$ *mutually orthogonal latin squares exist*. (That is, if q is equal to the order of a finite projective plane.)

PROOF. By hypothesis, there exist $q - 1$ mutually orthogonal latin squares, $L_1, L_2, \ldots, L_{q-1}$ of order q whose entries are the symbols $0, 1, \ldots, q - 1$ of an alphabet of q letters. We form q^2 code words as follows: In the code word $\boldsymbol{a} = (i, j, l_1, l_2, \ldots, l_{q-1})$ of length $n = q + 1$, the entry l_k is equal to the entry in the ith row and jth column of the latin square L_k. Since the latin squares are orthogonal, two such code words have the same entry in at most one place. Thus, $d = q = n - 1$.

If a subset of code words all of which have the same pth entry is taken and this entry is deleted, we get a set of q words of length $n = q$ whose Hamming distance d is still equal to q. Thus, $n = q$, $d = n = q$.

If from the set of q^2 code words of length $n = q + 1$ and minimum Hamming distance $d = q$ constructed above, we remove any one column, we obtain a set of q^2 code words whose length is $n - 1$ and minimum Hamming distance at least $d - 1$. Similarly, if we remove one column

Case when $q = 4$.

$$L_1 = \begin{array}{|cccc} 0 & 1 & 2 & 3 \\ 1 & 0 & 3 & 2 \\ 2 & 3 & 0 & 1 \\ 3 & 2 & 1 & 0 \end{array}$$

$$L_2 = \begin{array}{|cccc} 0 & 1 & 2 & 3 \\ 2 & 3 & 0 & 1 \\ 3 & 2 & 1 & 0 \\ 1 & 0 & 3 & 2 \end{array}$$

$$L_3 = \begin{array}{|cccc} 0 & 1 & 2 & 3 \\ 3 & 2 & 1 & 0 \\ 1 & 0 & 3 & 2 \\ 2 & 3 & 0 & 1 \end{array}$$

The code words are

(0 0 0 0 0)
(0 1 1 1 1)
(0 2 2 2 2)
(0 3 3 3 3)
(1 0 1 2 3)
(1 1 0 3 2)
(1 2 3 0 1)
(1 3 2 1 0)
(2 0 2 3 1)
(2 1 3 2 0)
(2 2 0 1 3)
(2 3 1 0 2)
(3 0 3 1 2)
(3 1 2 0 3)
(3 2 1 3 0)
(3 3 0 2 1)
$[n = 5, d = 4]$

Words with same fourth entry are

(0 1 1 1)
(1 3 2 0)
(2 2 0 3)
(3 0 3 2)
$[n = d = 4]$

Deletion of second entry gives

(0 1 1)
(1 2 0)
(2 0 3)
(3 3 2)
$[n = d = 3]$

Fig. 10.1.3

from the set of q code words of length $n = q$ and minimum Hamming distance $d = q$, we shall obtain a set of q code words whose length is $n - 1$ and minimum Hamming distance also $n - 1$. By repeated deletion of columns, we see that the statement of the theorem is true.

An example is given in Fig. 10.1.3.

In [1], S. W. Golomb and E. C. Posner have expressed the above connections between error correcting codes and sets of mutually orthogonal latin squares also in terms of *rook domains*, a concept which was suggested to them by the game of chess. In order to describe their construction, let us denote the set of all n-tuples from a q-symbol alphabet (which we now take as $1, 2, \ldots, q$) by V_q^n.

Let us define the *rook domain* of any element of V_q^n to be the element itself and all the elements differing from it in exactly one coordinate.

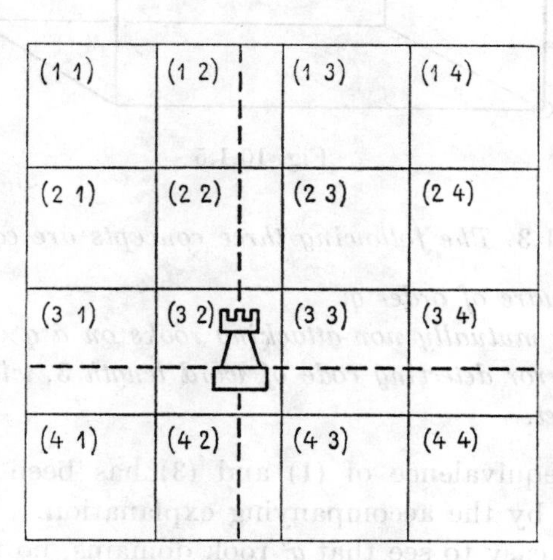

Fig. 10.1.4

For a typical rook domain see Fig. 10.1.4, which shows the case $n = 2$. For the general case, we shall have an n-dimensional cube and the corresponding rook will be able to move along files in each of n directions. Given any cell in the n-dimensional cube representing V_q^n, the number of cells in its rook domain is clearly $1 + n(q - 1)$ since there is the cell itself and $q - 1$ neighbours in each of n directions. The problem of single error correcting codes is thus the problem of rook domain packing in V_q^n: that is, of selecting as many non-overlapping rook domains in V_q^n as possible. Such a code is close packed if it contains $q^n/\{1 + n(q - 1)\}$ words, as we mentioned earlier, and corresponds to the situation in which the rook domains exhaust the space. A necessary condition for this is, of course, that $1 + n(q - 1)$ divides q^n exactly.

The simplest example of a close-packed single error correcting code arises when $q = 2$ and $n = 3$. In this case, $2^3/\{1 + 3(2 - 1)\} = 2$ and so the code contains only two code words, corresponding to the fact that the rook domains of a pair of rooks at opposite corners of the cube shown in Fig. 10.1.5 (say at 111 and 222) exhaust the space.

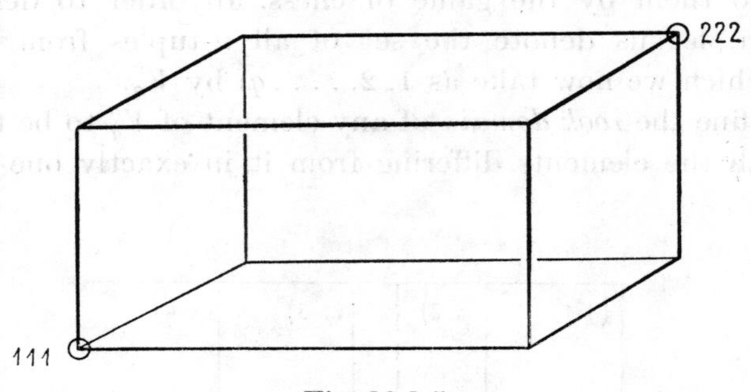

Fig. 10.1.5

Theorem 10.1.3. *The following three concepts are equivalent*:

(1) *a latin square of order* q;

(2) *a set of* q^2 *mutually non-attacking rooks on a* $q \times q \times q$ *board*;

(3) *a single error detecting code of word length* 3, *with* q^2 *words from a* q-*symbol alphabet*.

Proof. The equivalence of (1) and (3) has been demonstrated by Fig. 10.1.2 and by the accompanying explanation.

Further, it is easy to see that q^2 rook domains, no two of which have any cell in common, can be defined by means of the q^2 cells whose co-ordinates are given by the code words obtained from (3). Consequently, (2) and (3) are equivalent.

Theorem 10.1.3 and its generalization, theorem 10.1.4 below, are contained in S. W. Golomb and E. C. Posner [1].

For the statement of theorem 10.1.4, we shall need a further definition.

Definition. A *rook of power* t (or *super rook*) is a rook which can move from any given cell to any new cell which differs from the first in at most t co-ordinates.

THEOREM 10.1.4. *The following three concepts are equivalent:*

(1) *A set of t mutually orthogonal latin squares of order q;*

(2) *A set of q^2 super rooks of power t on a q^{t+2} board* (that is, a $(t+2)$-dimensional cube having q cells along each edge) *such that no two super rooks attack each other;*

(3) *a code of word length $t + 2$ and minimum Hamming distance $t + 1$ having q^2 words constructed from a q-symbol alphabet.*

PROOF. The equivalence of (1) and (3) can be shown by an argument exactly similar to that of theorem 10.1.2 taking t latin squares of order q instead of $q - 1$ such squares; whence $n = t + 2$, $d = t + 1$.

Further, by taking the q^2 cells whose co-ordinates are given by the code words constructed as in (3) as defining cells for q^2 super rooks, it is obvious that we get super rooks which are non-attacking, and so (2) and (3) are equivalent.

An example of the construction is given in Fig. 10.1.6 where the entries in the quadruples are respectively row index, column index, entry in the first square, and entry in the second square. Hence we get a code having sixteen words of length four.

$L_1 =$

1	2	3	4
2	1	4	3
3	4	1	2
4	3	2	1

$L_2 =$

1	2	3	4
3	4	1	2
4	3	2	1
2	1	4	3

The code words are

(1 1 1 1) (3 1 3 4)
(1 2 2 2) (3 2 4 3)
(1 3 3 3) (3 3 1 2)
(1 4 4 4) (3 4 2 1)
(2 1 2 3) (4 1 4 2)
(2 2 1 4) (4 2 3 1)
(2 3 4 1) (4 3 2 4)
(2 4 3 2) (4 4 1 3)

Fig. 10.1.6

By selecting as code words only those which correspond to a common transversal of all the latin squares, one can also obtain from the above construction a code of q words only, of the same length $t + 2$ and on the same q-symbol alphabet, but having minimum Hamming distance $t + 2$.

Theorem 10.1.5. *Let* $\|a_{ij}^{(1)}\|, \|a_{ij}^{(2)}\|, \ldots, \|a_{ij}^{(t)}\|$ *be a set of* t *mutually orthogonal latin squares of order* q *and let* K *be a code whose code words are* $(i, j, a_{ij}^{(1)}, a_{ij}^{(2)}, \ldots, a_{ij}^{(t)})$, *formed as in theorem* 10.1.4. *Then there exists no* q*-ary code* K' *with the same parameters* n, d *as* K *but having more than* q^2 *words.*

Proof. Theorem 10.1.4 ensures that K has q^2 words of length $t + 2$ and with minimum Hamming distance $t + 1$. If K' has the same parameters it contains at most $q^{(t+2)-(t+1)+1} = q^2$ code words by theorem 10.1.1.

In general, it is not known when the Joshibound is attainable: in other words, it is not known for which q, n, d there exists at least one code with q^{n-d+1} code words.

Next, we shall explain a procedure based on orthogonal latin squares which was originally published in G. B. Olderogge [1]. In that paper a technique for synthesizing binary codes based upon orthogonal latin squares is given which is quite different from that described in theorems 10.1.2 and 10.1.4. By means of this technique codes with word length $n^2 + 4n + 1$ and having n^2 information symbols can be constructed from pairs of orthogonal latin squares of order n, n odd. These codes will correct all single and double error patterns, and almost all triple error patterns, and will even detect the majority of patterns of four errors.

The encoding procedure devised by G. B. Olderogge will be demonstrated by an example.

We take the case $n = 5$ and suppose that the primary message of 5^2 binary digits which is to be transmitted is as follows: 0 1 1 0 0, 1 0 1 1 0, 1 1 0 1 1, 0 1 1 0 1, 0 1 0 1 0. In order to obtain the $4.5 + 1 = 21$ additional parity check digits which are to be trans-

	0	0	1	1	0
0	0	1	1	0	0
1	1	0	1	1	0
0	1	1	0	1	1
1	0	1	1	0	1
0	0	1	0	1	0

Fig. 10.1.7

mitted, we first split the primary message up into components of length 5 digits each, as indicated by the commas, and take these as the rows of a 5×5 matrix K, as shown in Fig. 10.1.7. We border the matrix K by parity check digits, one for each row and one for each column. That is, for example, if the sum of the digits of the rth column is i ($i = 0$ or 1), modulo 2, we head that column by i, and similarly for the remaining columns. Likewise, each of the rows of K is preceded by the sum of the digits appearing in that row, taken modulo 2, as shown in Fig. 10.1.7.

Next, we construct an auxiliary 5×5 matrix $M = \| (a_{ij}, b_{ij}) \|$ whose entries are the ordered pairs (a_{ij}, b_{ij}) of elements of a pair of orthogonal latin squares $A = \| a_{ij} \|$ and $B = \| b_{ij} \|$ each of order 5, as in Fig. 10.1.8.

$M =$

11	22	33	44	55
34	45	51	12	23
52	13	24	35	41
25	31	42	53	14
43	54	15	21	32

Fig. 10.1.8

(Such pairs of orthogonal latin squares exist for all orders $n \neq 6$.) By means of M, we construct from K a further 5×5 matrix $L = \| l_{ij} \|$ by defining $l_{ij} = k_{a_{ij}, b_{ij}}$ for all i and j in the range 1 to 5. For example, to decide the entry l_{23}, we look at the ordered pair which appears in the second row and third column of the matrix M. This is (5,1) and so $l_{23} = k_{51}$. That is, the entry of the second row and third column of the matrix L is to be the same as the entry of the fifth row and first column of the matrix K. We border the matrix L by parity check digits in exactly the same way as we did for the matrix K. The result is shown in Fig. 10.1.9.

	1	0	0	1	0
0	0	0	0	0	0
0	1	1	0	1	1
0	0	1	1	0	0
0	1	1	1	1	0
0	1	1	0	1	1

Fig. 10.1.9

To the message to be transmitted we now adjoin in turn the five parity check digits of the rows of K, the five parity check digits of the columns of K, the five parity check digits of the rows of L, the five parity check digits of the columns of L, and finally a parity check digit taken over the whole preceding set of $5^2 + 4.5$ digits, giving a total word length of $5^2 + 4.5 + 1$ digits, or $n^2 + 4n + 1$ digits in the general case. (See also Fig. 10.1.10.)

r_1	r_2	r_3	r_4	r_5	r_6	r_7	r_8	r_9	r_{10}	r_{11}	r_{12}	r_{13}	r_{14}	r_{15}	r_{16}	r_{17}	r_{18}	r_{19}	r_{20}	r_{21}	r_{22}	r_{23}	r_{24}	r_{25}
0	1	1	0	0	1	0	1	1	0	1	1	0	1	1	0	1	1	0	1	0	1	0	1	0

primary message

0 1 0 1 0	0 0 1 1 0	0 0 0 0 0	1 0 0 1 0	0 ↑
row parity check symbols in matrix K	column parity check symbols in matrix K	row parity check symbols in matrix L	column parity check symbols in matrix L	parity check digit taken over all the preceding digits

Fig. 10.1.10

To explain why this code is successful in correcting all double errors we observe firstly that, if the two errors occur in different rows and columns of the matrix K then two of the five parity checks on the rows of K will fail and two on the columns of K will also fail, thus defining four possible cells whose entries are among the two in error. These four cells occur two in each of two rows and two in each of two columns of K. In L, the contents k_{rs} and k_{rt} of two cells of the same row of K necessarily occur in different rows and a similar result holds for columns. Consequently, the particular two cells whose entries are in error can be distinguished by examination of the parity check digits associated with L.

If the two errors in K occur in the same row (column) of K then two of the five parity checks on the columns (rows) of K will alone be in error. To determine which is the row (column) of K which contains the two errors, we again make use of the parity checks on L.

Finally, cases when one or both of the errors occur among the parity checks themselves are easily detected and corrected.

Recently, in [1], D. C. Bossen, M. Y. Hsiao and R. T. Chien have made use of sets of mutually orthogonal latin squares in a construction of a new type of multiple-error correcting code which allows exceptionally rapid decoding.

In section 2.3, we gave some properties of complete latin squares. Such latin squares exist for every even order and also for some odd orders. In our next theorem, we explain a further algorithm for the construction of non-binary error detecting codes which makes use of these squares.

THEOREM 10.1.6. *Let* $A = \| a_{ij} \|$ *denote a complete latin square of order q. Then the set of code words* $(i, j, a_{ij}, a_{i,j+1})$, $i = 0, 1, 2, \ldots, q-1$ *and* $j = 0, 1, 2, \ldots, q-2$ *form a q-ary code having* $q(q-1)$ *code words each of length* 4 *and with Hamming distance* $d = 3$.

PROOF. Since the ordered pairs $(a_{ij}, a_{i,j+1})$, $i = 0, 1, \ldots, q-1$ and $j = 0, 1, \ldots, q-2$ are all different by definition of a complete latin square, two different code words, say $(i, j, a_{ij}, a_{i,j+1})$ and $(i', j', a_{i'j'}, a_{i',j'+1})$ differ from each other in at least one component. Let us suppose that $a_{ij} = a_{i'j'}$ and $a_{i,j+1} \neq a_{i',j'+1}$. Then $i \neq i'$ and $j \neq j'$ since A is latin. On the other hand if $a_{ij} \neq a_{i'j'}$ and $a_{i,j+1} \neq a_{i',j'+1}$ then either $i \neq i'$ or else $j \neq j'$. Consequently, the Hamming distance between any two words of the code is at least 3.

The above construction has the disadvantage that, in general, the number of code words which it gives is less than the Joshibound. (Compare theorem 10.1.2.)

However, for the particular values $q = 2$ and 6 the methods of theorem 10.1.2 and of Golomb and Posner (theorem 10.1.4) fail to work but the construction just given is valid for these cases, as is indicated by Fig. 10.1.11.

In the remaining part of this section we shall mention a number of miscellaneous results concerning connections between coding problems and latin squares. We start with a definition.

DEFINITION. A code whose code words are binary vectors is said to be of *constant weight* if each of its code words contains the same number of ones.

In B. Elspas, R. C. Minnick and R. A. Short [1], the authors proposed to use unipotent symmetric latin squares to design error detecting codes for the protection of constant weight code words against single bit errors

	1	2	3	4	5	6
1	1	6	2	5	3	4
2	6	5	1	4	2	3
3	5	4	6	3	1	2
4	4	3	5	2	6	1
5	3	2	4	1	5	6
6	2	1	3	6	4	5

1116	1262	1325	1453	1534
2165	2251	2314	2442	2523
3154	3246	3363	3431	3512
4143	4235	4352	4426	4561
5132	5224	5341	5415	5556
6121	6213	6336	6464	6545

	1	2
1	1	2
2	2	1

1 1 1 2
2 1 2 1

Fig. 10.1.11

for the case when the code words are of constant weight two. (See also B. Elspas and W. H. Kautz [1].)

A *unipotent symmetric latin square* of order n is defined to be a multiplication table of a commutative unipotent quasigroup. (For the definition of a unipotent quasigroup, see section 2.1; in particular identity (2).) In other words, the matrix of a unipotent symmetric latin square is required to be both symmetric and also to have all the elements of its main left-to-right diagonal equal.

The coding problem can be formulated as follows: Given the $n(n-1)/2$ binary n-bit code words of weight two, it is desired to append check symbols (one of the numbers $1, 2, \ldots, n-1$ written in binary form) to each word in such a way that all the $n-1$ words having a one in a given data position will be assigned different check symbols.

Fig. 10.1.12 makes it clear that this goal can be achieved whenever a unipotent symmetric latin square L of order n exists. When a latin square $L = \|a_{ij}\|$ of this kind exists, the check symbol to be associated with a word which has ones in its ith and jth information places is the element a_{ij} of the latin square L. (It is assumed that the element n appears along the main left-to-right diagonal of L and, in consequence, is not used as a check symbol.)

In this connection, we have the following theorem:

Theorem 10.1.7. *There exist unipotent symmetric latin squares of order n for every even integer n but none exists of any odd order n greater than* 1.

Proof. Let us suppose that $\| a_{ij} \|$ is a unipotent symmetric latin square of odd order. By its definition, all entries of the main left-to-right diagonal of the square are occupied by the same number, say 0. Now, by virtue of the commutativity, every digit i ($i \neq 0$) must occur an even number of times. However the total number of occurrences of any symbol is n, so n must be even.

6	2	3	4	5	1
2	6	4	5	1	3
3	4	6	1	2	5
4	5	1	6	3	2
5	1	2	3	6	4
1	3	5	2	4	6

i	j	a_{ij}	Code words	Check symbols
1	2	2	1 1 0 0 0 0	0 1 0
1	3	3	1 0 1 0 0 0	0 1 1
1	4	4	1 0 0 1 0 0	1 0 0
1	5	5	1 0 0 0 1 0	1 0 1
1	6	1	1 0 0 0 0 1	0 0 1
2	3	4	0 1 1 0 0 0	1 0 0
2	4	5	0 1 0 1 0 0	1 0 1
2	5	1	0 1 0 0 1 0	0 0 1
2	6	3	0 1 0 0 0 1	0 1 1
3	4	1	0 0 1 1 0 0	0 0 1
3	5	2	0 0 1 0 1 0	0 1 0
3	6	5	0 0 1 0 0 1	1 0 1
4	5	3	0 0 0 1 1 0	0 1 1
4	6	2	0 0 0 1 0 1	0 1 0
5	6	4	0 0 0 0 1 1	1 0 0

Fig. 10.1.12

If n is even, it is easy to construct a commutative idempotent quasigroup Q of order $n - 1$. For, by the corollary to theorem 1.4.3 every finite group of odd order $n - 1$ is isotopic to an idempotent quasigroup; and, if the group is commutative, so is the isotope obtained by the construction given in that corollary. Thus, for example, we may take for our group the cyclic group of order $n - 1$ and hence get a commutative idempotent quasigroup Q which is isotopic to it. Then, from the multiplication table of Q one can obtain a symmetric latin square of order n by a prolongation (see section 1.4), taking the main left-to-right diagonal of the multiplication table of Q as transversal for the prolongation.

Theorem 10.1.7 first appeared in B. Elspas, R. C. Minnick and R. A. Short [1], but the proof given there seems to be rather more complicated than that above.

We would like to give an example to show how the construction just described works in practice.

Let us consider the following multiplication table (Fig. 10.1.13) of a commutative idempotent quasigroup of order 5 obtained from the cyclic group C_5 by the construction described in the corollary to theorem 1.4.3 and then standardized. (If $C_5 = \{a/a^5 = e\}$ the digit r in the table denotes a^{r-1}.) By prolongation (using the main diagonal), the symmetric latin square of order 6 given in Fig. 10.1.12 is obtained.

	1	3	5	2	4
1	1	2	3	4	5
3	2	3	4	5	1
5	3	4	5	1	2
2	4	5	1	2	3
4	5	1	2	3	4

Fig. 10.1.13

Our next item concerns a connection between the covering set of a finite abelian group G, latin squares, and error correcting codes.

Let G be a finite abelian group which is the direct product of k cyclic groups each of order n, and let $g_1, g_2, \ldots, g_k$ be generating elements for these cyclic groups, so that a typical element of G can be written in the form $g_1^{\alpha_1} g_2^{\alpha_2} \ldots g_k^{\alpha_k}$.

Let the complex formed by the set theoretical union of the members of these k cyclic groups be denoted by S. That is, S comprises the set of all powers of the generating elements $g_1, g_2, \ldots, g_k$. It follows that S contains $|S| = 1 + k(n - 1)$ distinct elements.

Let H be a subset of G with the property that every element $g \in G$ can be written in the form $g = hs$, where $h \in H$ and $s \in S$. Thus, $G = HS$. H is called a *covering set* for the abelian group G.

A problem posed by O. Taussky and J. Todd in [1] is that of finding the integer $\varrho(k, n)$ which is defined as the smallest number of elements which a subset H of G can contain if it is to serve as a covering set for G.

The problem is not yet solved but we shall list the known values of $\varrho(k, n)$ for cases in which k and n do not exceed 8. (See Fig. 10.1.14, which was first given in R. G. Stanton [1].)

In [1], J. G. Kalbfleisch and P. H. Weiland have shown that, for the case $k = 3$, a covering set may be derived by means of two latin squares

k \ n	2	3	4	5	6	7	8
2	2	3	4	5	6	7	8
3	2	5	8	13	18	25	32
4	4	9	24		72		
5	7	27	4^3				
6	12			5^4			
7	16						
8	32					7^6	

Fig. 10.1.14

(which are of unequal size if n is odd) and the error detecting codes which these squares define. They have given the following example.

Let $n = 2t + 1 = 5$ say. Then $t = 2$ and $t + 1 = 3$. As our two latin squares we take the multiplication tables of the cyclic groups of orders $t = 2$ and $t + 1 = 3$ respectively, as shown in Fig. 10.1.15.

0	1
1	0

2	3	4
3	4	2
4	2	3

Fig. 10.1.15

From these two latin squares we form a t-ary error detecting code of t^2 code words and a $(t + 1)$-ary error detecting code of $(t + 1)^2$ code words respectively, in the manner described on page 350 of this section (and illustrated in Fig. 10.1.2). Each of these codes has a minimum Hamming distance of 2. The code words obtained in the case of our example are as follows: 000, 011, 101, 110, 222, 233, 244, 323, 334, 342, 424, 432, 443, and they define a covering set H for the group $G = \{g_1\} \times \{g_2\} \times \{g_3\}$ whose members are the elements $e = g_1^0 g_2^0 g_3^0$, $g_2 g_3$, $g_1 g_3$, $g_1 g_2$, $g_1^2 g_2^2 g_3^2$, $g_1^2 g_2^3 g_3^3$, $g_1^2 g_2^4 g_3^4$, $g_1^3 g_2^2 g_3^3$, $g_1^3 g_2^3 g_3^4$, $g_1^3 g_2^4 g_3^2$, $g_1^4 g_2^2 g_3^4$, $g_1^4 g_2^3 g_3^2$, $g_1^4 g_2^4 g_3^3$ of G.

We note that this method yields a covering set H which is formed from the union of two error detecting codes each of which has the maximal number of code words.

To understand why the method always determines a covering set, let us observe first that the single error detecting code of s^2 code words of length 3 constructed as described on page 350 from a latin square of order s has the property that each ordered pair of symbols of the code alphabet occurs just once among the set of code words in each pair of positions. Now suppose that $n = t + (t + 1)$, as in the example above, and that the t^2 code words of the t-ary code are constructed on the alphabet $A_1 = \{0, 1, \ldots, t - 1\}$ and that the $(t + 1)^2$ code words of the $(t + 1)$-ary code are constructed on the alphabet $A_2 = \{t, t + 1, \ldots, n - 1\}$. Let $g = g_1^i g_2^j g_3^k$ be an arbitrary element of the group $G = \{g_1\} \times \{g_2\} \times \{g_3\}$, where each of $\{g_1\}$, $\{g_2\}$, $\{g_3\}$ is cyclic of order n. We wish to show that $g = hs$, where $h \in H$, the covering set of the abelian group G, and $s \in S$ is a power of one of g_1, g_2, or g_3. That is, we wish to show that the triad of integers (i, j, k) differs from some one of the $t^2 + (t + 1)^2$ code words which we have constructed in at most one place. However, this is certainly the case since, of the integers i, j, k, at least two belong to the set A_1 or at least two belong to the set A_2. Let us suppose, for example, that i and k both belong to the set A_1. Then there exists exactly one code word constructed on the alphabet A_1 which has i as its first entry and k as its third entry. If this code word is (i, u, k), the group element $h = g_1^i, g_2^u, g_3^k$ belongs to H and $g = g_1^i g_2^j g_3^k = hs$, where $s = g_2^{j-u}$, as required.

Before leaving the subject of covering sets of abelian groups, let us mention that they may be used to solve a well known problem concerning "football pools". k teams are to play football matches on a given date. Each team may win, lose, or draw. Let us denote the teams by $T_1, T_2, \ldots, T_k$. There are 3^k possible "results". What is the least number of football pool coupon lines that it is necessary to fill up to ensure that some one of the lines will differ from the actual result in at most one place? Let $G = \{g_1\} \times \ldots \times \{g_k\}$ be the direct product of k cyclic groups each of order 3. Then the solution to our problem is the size of the minimal covering set for G.

In section 2.2, we introduced the concept of generalized identities and, in particular, the generalized law of associativity. We mentioned that the earliest investigation of the latter was carried out by R. Schauffler (see [2]). We remark here that the same author has shown in another paper [1] that his results can be employed in the design of error detecting and correcting codes and we shall give some details of this application.

Let $(a_1, a_2, \ldots, a_k)$ be a word of k message digits which we wish to transmit, where $a_1, a_2, \ldots, a_k$ are symbols of an alphabet of q distinct symbols. If these q symbols are the integers $0, 1, 2, \ldots, q-1$ and if we adjoin an additional parity check digit $a_{k+1} = a_1 + a_2 + \ldots + a_k$, where addition is modulo q, to our k message digits, the resulting code words $(a_1, a_2, \ldots, a_k, a_{k+1})$ of length $k+1$ each will have minimum Hamming distance 2 and will detect single errors. If we adjoin further check digits $a_{k+2}, a_{k+3}, \ldots$, each defined by means of equations involving some or all of the remaining digits a_i, we may make our code multiple error detecting or multiple error correcting.

Schauffler has pointed out that the equations connecting the code word digits may be chosen so as to involve more than one operation. If Q denotes the code word alphabet of q symbols, we may define a set of quasigroups $(Q, \boldsymbol{i})$ on[(1)] the set Q, exhibiting the multiplication table of each as a bordered latin square. The check digit a_{k+1} may then be determined by an equation of the form

$$a_{k+1} = [\{(a_1 \text{①} a_2) \text{②} a_3\} \text{③} a_4] \text{④} \ldots,$$

where the single operation of addition modulo q previously used has been replaced by a number of different quasigroup operations. The advantage of such a procedure is that the message will be more difficult to decode by an unauthorized receiver. However, an additional complication has been introduced, in that the order in which the various operations are carried out is now significant and has to be indicated by means of bracketing.

Schauffler has proposed to eliminate the bracketing by choosing the quasigroups $(Q, \boldsymbol{i})$ to be members of an associative system of quasigroups and has given some examples of codes so constructed in R. Schauffler [1]. It follows from theorem 2.2.5 that, in such cases, the quasigroups used must all be isotopes of one and the same group. In R. Schauffler [2], the author has pointed out that theorem 2.2.6 also has significance in the construction of error correcting codes of this kind.

The following problem arises in connection with the economical storage of information.

Let us define the *check sum* of two or more k-digit binary code words as the k-digit word obtained by placing a one at the foot of each column which does not consist entirely of zeros as in Fig. 10.1.16. We seek a set, as

(1) The operation $\boldsymbol{i}$ here stands for one of several quasigroup operations ①, ②, ③, ...

large as possible, of k-digit binary code words such that, for a given small positive integer m, every check sum of up to m different code words is distinct from every other sum of m or fewer code words, and so that each such check sum can be decomposed uniquely, apart from the order of the words, into the m code words which were used to form that sum. If such a set can be found, each of the distinct check sums in the set is a k-digit code word which in effect provides m distinct pieces of information represented by the m code words of which it is the check sum.

CODE WORDS	0	1	1	0	0	1
	0	1	0	0	1	0
	1	1	1	0	1	0
CHECK SUM	1	1	1	0	1	1

Fig. 10.1.16

By means of examples, W. H. Kautz and R. C. Singleton have put forward in [1] a suggestion for a construction using orthogonal latin squares which, in the particular cases which they have considered, gives codes with the above properties.

Latin squares and orthogonal systems of such squares have also been applied in connection with the design of computer memory access schemes. Some of the latter applications have been described in R. C. Minnick and J. L. Haynes [1] and some further results on the connection between storage devices and latin squares can be found in T. N. Eingorina and M. J. Eingorin [1].

In [1], E. Knuth has investigated three problems concerning orthogonal pairs of latin squares which arose in connection with the design of telephone exchanges.

Let $(L_{11}, L_{12}), (L_{21}, L_{22}), \ldots, (L_{h1}, L_{h2})$ be a set of h pairs of orthogonal latin squares (L_{i1} being orthogonal to L_{i2} for each separate integer i), all of the same order n, with the two squares of each orthogonal pair written in juxtaposed form so as to form a table of n^2 distinct ordered pairs (u, v) of elements (as in Fig. 10.1.17). Then each table has n rows of ordered pairs and n columns of ordered pairs. We shall call such a row or column a *line*. Thus, there exist all together $2nh$ lines.

In the first problem, it is required to find a set of h tables of the kind described, having h as large as possible and with the property that no two lines have more than one ordered pair (u, v) in common. It is easy to see that there cannot be a solution to the problem if h exceeds $[(n-1)/2]$, where $[x]$ denotes the largest integer not larger than x. For consider, for example, the ordered pair (1,1). This ordered pair occurs in exactly one row and exactly one column of each of the h tables; that is, it occurs in $2h$ lines all together. Each of these lines contain $n-1$ further ordered pairs of elements and the requirement of the problem is that no two of these $2h(n-1)$ ordered pairs of elements should be alike. Since the squares L_{ij} $(i = 1, 2, \ldots, h; j = 1, 2)$ are all latin, none of the ordered pairs of elements in the lines which contain (1,1) can have their first or second component equal to 1. It follows that there are at most $(n-1)^2$ permissible entries. Hence $2h(n-1) \leq$ $\leq (n-1)^2$, from which the result quoted above follows.

An example consisting of three pairs of orthogonal squares of order 7 for which the conditions are satisfied is given in Fig. 10.1.17 and it will be noted that in this example h attains its upper bound. E. Knuth has given a general algorithm by which a solution to the problem with h equal to its upper bound may always be produced when n is a prime number. However, the question whether solutions in which h attains its

11	22	33	44	55	66	77
23	34	45	56	67	71	12
35	46	57	61	72	13	24
47	51	62	73	14	25	26
52	63	74	15	26	37	41
64	75	16	27	31	42	53
76	17	21	32	43	54	65

11	24	37	43	56	62	75
25	31	44	57	63	76	12
32	45	51	64	77	13	26
46	52	65	71	14	27	33
53	66	72	15	21	34	47
67	73	16	22	35	41	54
74	17	23	36	42	55	61

11	26	34	42	57	65	73
27	35	43	51	66	74	12
36	44	52	67	75	13	21
45	53	61	76	14	22	37
54	62	77	15	23	31	46
63	71	16	24	32	47	55
72	17	25	33	41	56	64

Fig. 10.1.17

upper bound exist for other values of n ($n \neq 6$) remains an open one.

Knuth's second and third problems are obtained by relaxing the conditions of the first. In the second problem, the requirement that no two lines chosen from among the h tables have more than one ordered pair of elements in common is replaced by the requirement that no two rows and also no two columns have more than one ordered pair in common. In the third problem, the only requirement is that no two columns have more than one ordered pair in common. It turns out that in each of these cases $n - 1$ is an upper bound for h. Moreover, Knuth has shown that, as in problem one, the bound can be attained for all prime values of n.

A solution to the second problem for the case $n = 5$ is given in Fig. 10.1.18.

If $a_{jk}^{(i)}$ denotes the ordered pair of integers which occurs in the cell of the jth row and kth column of the ith orthogonal pair (L_{i1}, L_{i2}) of latin squares in the set, then Knuth's algorithms are as follows:

For the first problem, $a_{jk}^{(i)} = (j + k - 1, 2i(j + k - 2) - k + 2)$, where $i = 1, 2, \ldots, [(n - 1)/2]$; $j, k = 1, 2, \ldots, n$ and addition is modulo p.

For the second and third problems, $a_{jk}^{(i)} = (j + k - 1,\ i(k - j) + 1)$ where $i = 1, 2, \ldots, n - 1$; $j, k = 1, 2, \ldots, n$ and addition is modulo p as before.

The reader will easily see that Figs 10.1.17 and 10.1.18 respectively have been constructed by means of these algorithms.

11	22	33	44	55	11	23	35	42	54
25	31	42	53	14	24	31	43	55	12
34	45	51	12	23	32	44	51	13	25
43	54	15	21	32	45	52	14	21	33
52	13	24	35	41	53	15	22	34	41
11	24	32	45	53	11	25	34	43	52
23	31	44	52	15	22	31	45	54	13
35	43	51	14	22	33	42	51	15	24
42	55	13	21	34	44	53	12	21	35
54	12	25	33	41	55	14	23	32	41

Fig. 10.1.18

In section 6.3 we mentioned that magic squares have been used in the construction of non-binary error correcting codes. We shall justify this statement by describing a coding procedure devised by S. I. Samoilenko (see [1]) which was originally inspired by the concept of a magic square.

Let $A = ||a_{ij}||$ be a square $n \times n$ matrix each of whose entries is an element of an abelian group $(G, +)$ of order n. We may denote these n elements by $g_1, g_2, \ldots, g_n$. Any collection of n entries of A, no two of which are equal, will be called an *n-pattern* of A. For example, if A were a diagonal latin square (as defined in section 6.1), the n entries of any row, any column, or either main diagonal of A would form such an n-pattern. The *value* $v(N)$ of an n-pattern N is defined to be the sum of the n entries composing it. If A is an $n \times n$ matrix possessing k disjoint n-patterns $N_1, N_2, \ldots, N_k$ such that $v(N_1) = v(N_2) = v(N_3) = \ldots = v(N_k) = h$ say, we shall call A a *generalized magic square* of *order n* and *type k* with *parameter h*. (This definition differs slightly from Samoilenko's original one.)

We note that any latin square of order n whose entries are the elements of an abelian group $(G, +)$ of order n may be regarded as a generalized magic square of order and type n and that the n n-patterns may be selected in many ways. (Moreover, by virtue of a theorem due to L. J. Paige, see page 34 of section 1.4, the parameter of such a generalized magic square will be equal to the identity element e of G unless G has a unique element g^* of order 2, and equal to g^* otherwise.)

Making use of a generalized magic square X of the kind just described, Samoilenko has devised a coding method by which messages of length $n + k$, $0 \leq k \leq n - 1$, may be transmitted as blocks of fixed length n^2 (independent of k) and the value of k may easily be adjusted to suit the level of noise in the transmission channel.

His procedure is as follows:

We choose $n + k$ arbitrary symbols $x_1, x_2, \ldots, x_{n+k}$. We take a latin square X_n of order n whose entries $x_1, x_2, \ldots, x_n$ are the first n of these arbitrary symbols and select within it n disjoint n-patterns which we denote by $N_1, N_2, \ldots, N_n$. From this square we form an $n \times n$ matrix $X_n^{(k)}$ by the rules: (*i*) If $k = 0$, $X_n^{(k)} = X_n$. (*ii*) If $k = 1$, we replace each entry x_i of N_1 by the ordered pair (x_i, x_{n+1}) where x_{n+1} is the $(n + 1)$th member of the set of arbitrary symbols; the resulting square is then $X_n^{(k)}$. (*iii*) If $k = r > 1$, we replace each entry x_i of N_j by the ordered pair (x_i, x_{n+j}) for each value of j $(j = 1, 2, \ldots, r)$ and so obtain $X_n^{(k)}$.

An example illustrating the procedure for the case $n = 3$ and $k = 2$ is exhibited in Fig. 10.1.19 and the two n-patterns used are indicated thus: $N_1 = \{|\underline{x_1}, |\underline{x_2}, |\underline{x_3}\}$, $N_2 = \{\overline{x_1}|, \overline{x_2}|, \overline{x_3}|\}$.

$$X_3 = \begin{array}{|ccc} |\underline{x_2} & |\underline{x_1} & x_3 \\ |\underline{x_3} & x_2 & \overline{x_1}| \\ x_1 & \overline{x_3}| & \overline{x_2}| \end{array} \qquad X_3^{(2)} = \begin{array}{|ccc} (x_2, x_4) & (x_1, x_4) & x_3 \\ (x_3, x_4) & x_2 & (x_1, x_5) \\ x_1 & (x_3, x_5) & (x_2, x_5) \end{array}$$

The group $(G, +)$ is Z_{10}, the group formed by the integers $0, 1, \ldots, 9$ under addition, modulo 10. The message is $(g_1, g_2, g_3, g_4, g_5) = (2, 8, 2, 1, 9)$.

$$X_3(G) = \begin{array}{|ccc} 8 & 2 & 2 \\ 2 & 8 & 2 \\ 2 & 2 & 8 \end{array} \qquad X_3^{(2)}(G) = \begin{array}{|ccc} 9 & 3 & 2 \\ 3 & 8 & 1 \\ 2 & 1 & 7 \end{array}$$

Fig. 10.1.19

Now suppose that the message to be transmitted is $(g_1, g_2, \ldots, g_n, g_{n+1}, \ldots, g_{n+k})$ where $g_1, g_2, \ldots, g_{n+k}$ are elements of an abelian group $(G, +)$ of order equal to the length of the message alphabet. We replace the symbols $x_1, x_2, \ldots, x_n$ in X_n by the group elements $g_1, g_2, \ldots, g_n$, respectively and call the resulting generalized magic square $X_n(G)$. Similarly, we replace the symbols $x_1, x_2, \ldots, x_{n+k}$ in $X_n^{(k)}$ by the group elements $g_1, g_2, \ldots, g_{n+k}$ respectively and at the same time replace the ordered pairs (x_i, x_{n+j}), $1 \leq i \leq n$, $1 \leq j \leq k$ by corresponding group elements $g_i + g_{n+j}$ to obtain an $n \times n$ matrix $X_n^{(k)}(G)$ whose entries are group elements.

We transmit the matrix $X_n^{(k)}(G)$ as a block of n^2 characters.

Let us suppose that the matrix received at the other end of the channel is $\bar{X}_n^{(k)}(G)$ and that $\bar{X}_n(G)$ denotes the corresponding generalized magic square of "best fit".

The message is decoded as follows:

(a) First sum the group elements of each n-pattern of the received matrix $\overline{X}_n^{(k)}(G)$. For the jth n-pattern N_j this sum should be $\sum_{i=1}^{n} g_i + ng_{n+j}$.

Hence, determine best values for the parameter $h = \sum_{i=1}^{n} g_i$ of the generalized magic square $\overline{X}_n(G)$ and for the group elements $g_{n+1}, g_{n+2}, \cdots, g_{n+k}$.

(b) The elements of those n-patterns of the square $\overline{X}_n(G)$ whose values are equal to h are assumed to be correct.

(c) Determine the elements of the remaining n-patterns (some of which elements are presumed to be in error) by solving certain systems of equations whose actual specification is rather too complicated to be given here.

A computer simulation of the above method was described by E. I. Tretakova and E. Rahmatkariev in a lecture given to the Third Symposium on the Use of Redundancy in Information Systems, Leningrad 1968.

Further details of Samoilenko's work will be found in his recently published book, S. I. Samoilenko [2].

Finally, let us mention that the theory of error correcting codes has recently been employed in a new attempt to decide the existence or non-existence of a projective plane of order 10 (see F. J. MacWilliams, N. J. A. Sloane and W. Thompson [1]).

10.2. Latin squares and experimental designs

Since the 1930's, when the idea of doing so was pioneered by R. A. Fisher(1), latin squares and other related "designs" have been much used in the design of statistical experiments whose subsequent interpretation is to be effected with the aid of the procedure known as the "analysis of variance". We have already referred to such statistical applications in several earlier chapters of the present book: notably in sections 2.3, 5.4, and 6.4, where the uses of complete latin squares, of latin cubes and hypercubes, and of Room designs for this purpose were mentioned. In the present section, we shall explain the use of orthogonal sets of latin squares for the same purpose and introduce a number of other related designs. Some of the latter also have relevance to the disproof of the Euler conjecture, which is the subject of the following chapter.

Let us start with an example. It is required to test the effect of five different treatments on yarn. Five looms are available for spinning the

(1) For the earliest references to the use of designs in statistical experiments the reader is referred to R. A. Fisher [4].

yarn and it is believed that both the loom used and the operator employed may affect the final texture of the spun yarn. If time were of no account, it would be desirable ideally to test the effect of every treatment with every operator and on every loom. However, by making use of the statistical technique of *analysis of variance*, a satisfactory test result can be obtained provided that each type of treated yarn is spun once on each loom and also once by each operator. If five operators are employed, the necessary set of twenty-five experiments can be specified by means of the 5×5 latin square illustrated in Fig. 10.2.1. Here, the columns O_1, O_2, O_3, O_4, O_5 specify the operator and the rows L_1, L_2, L_3, L_4, L_5 specify which loom is to be used for the particular experiment. The

	O_1	O_2	O_3	O_4	O_5
L_1	Y_1	Y_4	Y_0	Y_2	Y_3
L_2	Y_3	Y_1	Y_2	Y_4	Y_0
L_3	Y_2	Y_0	Y_1	Y_3	Y_4
L_4	Y_0	Y_3	Y_4	Y_1	Y_2
L_5	Y_4	Y_2	Y_3	Y_0	Y_1

Fig. 10.2.1

entries Y_0, Y_1, Y_2, Y_3, Y_4 in the body of the square indicate which type of treated yarn is to be used. (Usually, one sample of yarn, called the *control*, is left untreated and the suffix 0 is used for this type of treatment.) Thus, in the first experiment, yarn Y_1 is to be spun by operator O_1, on loom L_1. In another experiment, yarn Y_3 is to be spun by operator O_2 on loom L_4. If it were also the case that the efficiency of the loom operators varied with the day of the week (five weekdays), this extra source of variation could be allowed for by using a 5×5 latin square which was orthogonal to the first to specify the day of the week on which each experiment was to be performed. Thus, in the square

1	2	4	3	0
3	4	1	0	2
2	3	0	4	1
0	1	3	2	4
4	0	2	1	3

Fig. 10.2.2

illustrated in Fig. 10.2.2, the integers 0, 1, 2, 3, 4 are used to specify the five days of a week. Then each loom, each operator, and each type of treated yarn are associated with each day of the week just once.

The use of a latin square design for an experiment of the above kind is somewhat restrictive. It might happen, for example, that only four operators were available. This situation could be accommodated by using the same latin square with its last column deleted. It would still be the case that each type of yarn would be spun by each of the four remaining operators but now only four of the five different yarns would be spun on any one of the looms. A development of this observation gives rise to the concept of a *balanced incomplete block design*.

DEFINITION. A *balanced incomplete block design* comprises a set of v *varieties* (or *treatments*) arranged in b *blocks* (or rows) in such a way that (i) each block has the same number k of treatments ($k < v$), no treatment occurring twice in the same block; (ii) each treatment occurs in exactly r ($= \lambda_1$) blocks; (iii) each pair of treatments occur together in exactly λ ($= \lambda_2$) blocks.

Balanced incomplete block designs were first introduced by F. Yates in [4]. An example of such a design is given by the latin square used above when the last column is deleted. This has parameters $b = 5$, $v = 5$, $k = 4$, $r = 4$, $\lambda = 3$.

A design for which property (iii) is satisfied is called *pairwise balanced*. More generally, if each set of t treatments occurs in exactly λ_t blocks, the design is called *t-wise balanced* or, more briefly, a *t-design*. We shall show later that, if s is an integer such that $0 < s < t$, then every t-design is also an s-design. We have the following theorem:

THEOREM 10.2.1. *The parameters b, v, r, k, λ of a (pairwise) balanced incomplete block design* (usually written briefly as a BIBD) *satisfy the relations* $bk = vr$, $\lambda(v - 1) = r(k - 1)$, $r > \lambda$, $b \geq v$.

PROOF. Each side of the first relation is an expression for the total number of treatments to be found in all the blocks. Each side of the second relation is an expression for the total number of treatments (other than v_1) which occur in the various blocks which contain an assigned treatment v_1. Next, if $r > \lambda$ were untrue, then with any given treatment every other treatment would occur in every block, contradicting $k < v$. Finally, to show that $b \geq v$ it is easiest to make use of the *incidence matrix* of the design. Let a BIBD with parameters b, v, r, k, λ be given and let $n_{ij} = 1$ or 0 according as the ith treatment v_i does or does not occur in the jth block. The $v \times b$ matrix so obtained is the incidence matrix of the design. Clearly,

$$\sum_{j=1}^{b} n_{ij}^2 = r \quad \text{and} \quad \sum_{j=1}^{b} n_{ij}\, n_{hj} = \lambda (h \neq i)$$

by definition of r and λ. If possible, let $b < v$ and consider the $v \times v$ matrix

$$N = \begin{bmatrix} n_{11} & n_{12} & \ldots & n_{1b} & 0 & 0 & \ldots & 0 \\ n_{21} & n_{22} & \ldots & n_{2b} & 0 & 0 & \ldots & 0 \\ \cdot & \cdot & \ldots & \cdot & \ldots & \cdot & \ldots & \cdot \\ \cdot & \cdot & \ldots & \cdot & \ldots & \cdot & \ldots & \cdot \\ n_{v1} & n_{v2} & \ldots & n_{vb} & 0 & 0 & \ldots & 0 \end{bmatrix}$$

where the last $v - b$ columns consist of zeros. It follows from the relations above that

$$NN^T = \begin{bmatrix} r & \lambda & \lambda & \ldots & \lambda \\ \lambda & r & \lambda & \ldots & \lambda \\ \cdot & \cdot & \cdot & \ldots & \cdot \\ \lambda & \lambda & \lambda & \ldots & r \end{bmatrix}$$

and so

$$\det(NN^T) = \{r + \lambda(v-1)\} \begin{vmatrix} 1 & 1 & 1 & \ldots & 1 \\ \lambda & r & \lambda & \ldots & \lambda \\ \cdot & \cdot & \cdot & \ldots & \cdot \\ \lambda & \lambda & \lambda & \ldots & r \end{vmatrix}$$

by addition of rows. Then, subtracting the first column from each other one, we get

$$\det(NN^T) = \{r + \lambda(v-1)\}(r-\lambda)^{v-1} = kr(r-\lambda)^{v-1}$$

since $\lambda(v - 1) = r(k - 1)$. But, $\det (NN^T) = (\det N) (\det N^T) = 0$, which implies that $r = \lambda$. This contradiction shows that the supposition $b < v$ is impossible. Therefore, $b \geq v$ as required.

A BIBD for which $b = v$, and consequently $r = k$, is said to be *symmetric* and such a design may also be referred to as a *(v, k, λ)-design*.

Two obvious problems concerning BIBD's are to provide methods of constructing them and to determine for what values of the parameters such designs exist. The second problem is unsolved. As regards the first, many methods have been devised. The simplest means of obtaining such designs is perhaps by use of the finite geometries.

THEOREM 10.2.2. *A BIBD having $(q^{h+1} - 1)/(q - 1)$ varieties arranged in $(q^{h+1} - 1)(q^h - 1) \ldots (q^{h-m+1} - 1)/(q^{m+1} - 1)(q^m - 1) \ldots (q - 1)$ blocks with $(q^{m+1} - 1)/(q - 1)$ varieties in each block, where $q = p^n$, can always be constructed by means of a finite projective space.*

PROOF. A desarguesian projective space $PG(h, p^n)$ of h dimensions defined over the Galois field $GF[p^n]$ has $v = (q^{h+1} - 1)/(q - 1)$ points, where $q = p^n$. We take these as our varieties (treatments) and the sets of points which form the m-dimensional subspaces of the geometry as the blocks, whence

$$b = \frac{(q^{h+1} - 1)(q^h - 1) \ldots (q^{h-m+1} - 1)}{(q^{m+1})(q^m - 1) \ldots (q - 1)}.$$

From the homogeneity of the geometry, it follows that each point occurs in the same number of m-dimensional subspaces and that the same is true of each pair of points. Since $k = (q^{m+1} - 1)/(q - 1)$, it is easy to calculate the values of r and λ.

We note in particular that any finite projective plane, desarguesian or not, gives rise to a symmetric BIBD with parameters

$$b = v = q^2 + q + 1, \quad r = k = q + 1, \quad \lambda = 1,$$

where q is the order of the plane.

A BIBD with $k = 3$ and $\lambda = 1$ is called a *Steiner triple system*. We have already discussed some of the properties of such systems in section 2.2. Here, we content ourselves with remarking that the unique Steiner system with $v = 7$ can be constructed from the projective plane $PG(2,2)$.

Balanced incomplete block designs and their generalizations have been used in the construction of mutually orthogonal latin squares and, in particular, in disproving the Euler conjecture as we shall show in the next chapter. We shall next discuss the two most important of these generalizations: *t-designs* and *partially balanced incomplete block designs.*

Definition. An experimental design comprising v treatments (varieties) arranged in b blocks (subsets) each containing k distinct treatments ($k < v$) and with the property that each set of t treatments is contained in exactly λ_t blocks is a *t-design* or, more accurately, a t-(b, v, k, λ_t) design.

The name t-design seems to have been first used by D. R. Hughes in [3]. The following theorem is from D. R. Hughes [4].

Theorem 10.2.3. *If* $0 < s < t$ *and* $\mathfrak{D}$ *is a* t-(b, v, k, λ_t) *design, then* $\mathfrak{D}$ *is also an* s-(b, v, k, λ_s) *design, where*

$$\lambda_s = \frac{(v-s)(v-s-1)\ldots(v-t+1)}{(k-s)(k-s-1)\ldots(k-t+1)}\lambda_t .$$

Proof. Let Σ be a fixed set of $t - 1$ treatments and suppose that Σ is included as a set in λ_{t-1} blocks of $\mathfrak{D}$. We shall determine the total number of t-sets containing Σ which occur in the design $\mathfrak{D}$.

Each block which contains Σ contains $k - (t - 1)$ other treatments any one of which forms with Σ a t-set containing Σ. Since Σ occurs in λ_{t-1} blocks, there are $\lambda_{t-1}(k - t + 1)$ such t-sets altogether.

On the other hand, since $\mathfrak{D}$ contains a total of $v - (t - 1)$ treatments other than the treatments of Σ and since any one of these treatments forms with Σ a t-set containing Σ and each such t-set occurs in λ_t blocks by hypothesis, we see that there exist $\lambda_t(v - t + 1)$ t-sets all together which contain Σ.

Therefore, $\lambda_{t-1}(k-t+1) = \lambda_t(v-t+1)$, whence $\lambda_{t-1} = \dfrac{v-t+1}{k-t+1}\lambda_t$ and this number is independent of the particular $(t - 1)$-set chosen. Thus, the given t-design is also a $(t - 1)$-design with $\lambda_{t-1} = \dfrac{v-t+1}{k-t+1}\lambda_t$. By $t - s$ repetitions of the argument we easily obtain the result stated in the theorem.

Corollary. *Every t-design is a BIBD with* $r = \lambda_1$, $\lambda = \lambda_2$.

By the corollary to the theorem, it follows immediately that the parameters of a t-design satisfy the relations

$$bk = v\lambda_1, \quad \lambda_2(v-1) = \lambda_1(k-1), \quad \lambda_1 > \lambda_2, \quad b \geq v.$$

We notice that the second and third of these relations also follow as direct consequences of the theorem itself.

An example of a t-design is provided by the $(t + 1) \times t$ latin rectangle which is obtained by deleting the last column of any $(t + 1) \times (t + 1)$ latin square. Here, the rows of the latin rectangle are to be regarded as the blocks and the parameters of the design are $v = t + 1$, $b = t + 1$, $k = t$ and $\lambda_t = 1$.

A partially balanced incomplete block design (PBIBD) may be regarded as the generalization of a BIBD which arises when the parameter λ is no longer kept fixed for every pair of varieties.

DEFINITION. A *partially balanced incomplete block design* comprises a set of v varieties arranged in b blocks of k varieties each $(k < v)$ and having the properties (i) each variety occurs in exactly $r(= \lambda_1)$ blocks; (ii) relative to any specified variety A, the remaining $v - 1$ varieties lie in m disjoint sets $S_1^{(A)}, S_2^{(A)}, \ldots, S_m^{(A)}$ such that the set $S_i^{(A)}$ contains n_i varieties each of which occurs with the specified variety A in μ_i blocks: the numbers n_i and μ_i are the same for each choice of variety A and the n_i varieties in the set $S_i^{(A)}$ are called the *μ_i-associates*[1] of the variety A; (iii) the number of varieties common to the μ_i-associates of one specified variety A and the μ_j-associates of another specified variety B depends solely on the associate relationship between A and B. If A and B occur together in μ_h different blocks, then the number of treatments common to the set $S_i^{(A)}$ of μ_i-associates of A and the set $S_j^{(B)}$ of μ_j associates of B is denoted[2] by p_{ij}^h.

The concept of partially balanced incomplete block design was first defined and studied in R. C. Bose and K. R. Nair [1]. The following result is due to these authors.

THEOREM 10.2.4. *In the notation defined above, the following identities connect the parameters of any PBIBD:*

$$\sum_{i=1}^{m} n_i = v - 1, \quad bk = vr, \quad \sum_{i=1}^{m} n_i \mu_i = r(k-1), \quad p_{ij}^h = p_{ji}^h,$$

$$\sum_{j=1}^{m} p_{ij}^h = n_i \quad \textit{if } h \neq i \textit{ and } = n_i - 1 \textit{ if } h = i,$$

$$\textit{and} \quad n_h p_{ij}^h = n_i p_{hj}^i = n_j p_{hi}^j.$$

(1) For example, if $m = 1$, we have a BIBD with $\mu_1 = \lambda$ $(= \lambda_2)$ and $n_1 = v - 1$. If $m \neq 1$ but all the μ_i are equal to λ we have a BIBD with $\Sigma n_i = v - 1$.

(2) For example, in a BIBD, each variety has $v - 1$ λ-associates and each two varieties are themselves λ-associates. The set $S_1^{(A)}$ comprising the $v - 1$ λ-associates of one variety A and the set $S_1^{(B)}$ comprising those of another variety B have $v - 2$ varieties in common since every variety distinct from A and B is a λ-associate of both. Thus, $p_{11}^1 = v - 2$.

PROOF. It is an immediate consequence of the definitions that

$$\sum_{i=1}^{m} n_i = v - 1 \text{ and that } p_{ij}^h = p_{ji}^h.$$

Also,

$$bk = vr \text{ and } \sum_{i=1}^{m} n_i\mu_i = r(k-1)$$

since both sides of the first identity are expressions for the total number of varieties occurring in all the blocks and both sides of the second identity express the total number of varieties occurring with a specified treatment A in all the blocks. Next, we have

$$\sum_{j=1}^{m} p_{ij}^h = n_i \text{ if } h \neq i \text{ and } = n_i - 1 \text{ if } h = i.$$

For, let A and B be μ_h-associates. There are n_i μ_i-associates of A and each of these is either a $\mu_1, \mu_2, \ldots,$ or μ_m-associate of B. But the number of μ_i-associates of A which are also μ_j-associates of B is p_{ij}^h, equal to the number of varieties common to the sets $S_i^{(A)}$ and $S_j^{(B)}$, so

$$\sum_{j=1}^{m} p_{ij}^h = n_i \text{ if } h \neq i.$$

If A and B are μ_i-associates, there exist $n_i - 1$ μ_i-associates of A distinct from B and each of these is either a $\mu_1, \mu_2, \ldots,$ or μ_m-associate of B, whence we get

$$\sum_{j=1}^{m} p_{ij}^i = n_i - 1.$$

Finally, we show that $n_h p_{ij}^h = n_i p_{hj}^i = n_j p_{hi}^j$. Let a variety A be chosen. A has n_i μ_i-associates $b_1, b_2, \ldots, b_{n_i}$ forming a set $S_i^{(A)}$ and has n_j μ_j-associates $c_1, c_2, \ldots, c_{n_j}$ forming a set $S_j^{(A)}$. Each b_i occurs with A in μ_i different blocks (by definition) and so, by property (*iii*) of a PBIBD, the number of varieties common to the sets $S_j^{(A)}$ and $S_h^{(b_i)}$ is p_{hj}^i. That is, p_{hj}^i of the varieties c_j, $j = 1$ to n_j, are μ_h-associates of b_i. This is true for each choice of b_i, so there exist $n_i p_{hj}^i$ pairs (b_x, c_y), $b_x \in S_i^{(A)}$, $c_y \in S_j^{(A)}$, which are μ_h-associates. Similarly, p_{hi}^j of the b_i, $i = 1$ to n_i, are μ_h-associates of c_j, and so there exist $n_j p_{hi}^j$ pairs (b_x, c_y) which are μ_h-associates. In other words, $n_i p_{hj}^i = n_j p_{hi}^j$. By a similar argument, $n_h p_{ij}^h = n_j p_{ih}^j$. Since $p_{hi}^j = p_{ih}^j$, the identity $n_h p_{ij}^h = n_i p_{hj}^i = n_j p_{hi}^j$ follows.

Perhaps the simplest example of a PBIBD is derived by taking the eight vertices $A_1, A_2, A_3, A_4, A_5, A_6, A_7, A_8$ of a cube as varieties and each set of four vertices which lie in the same face as a block, so that the blocks (see Fig. 10.2.3) are $A_1A_2A_3A_4$, $A_2A_3A_6A_7$, $A_5A_6A_7A_8$, $A_3A_4A_7A_8$, $A_1A_4A_5A_8$, and $A_1A_2A_5A_6$. Then $b = 6$, $v = 8$, $r = 3$, $k = 4$. Two vertices are associates of the first kind (2-associates) if both lie on the same edge. They are associates of the second kind (1-associates) if they lie at the opposite ends of a diagonal of one of the faces. They are associates of the third kind if they lie at the opposite ends of a diagonal of the cube. Thus, we see that $\mu_1 = 2$, $n_1 = 3$; $\mu_2 = 1$, $n_2 = 3$; $\mu_3 = 0$, $n_3 = 1$. For example, with respect to the vertex A_1, the remaining vertices lie in three sets, a set $S_1^{(A_1)} = \{A_2, A_4, A_5\}$ of 2-associates, a set $S_2^{(A_1)} = \{A_3, A_6, A_8\}$ of 1-associates and a set $S_3^{(A_1)} = \{A_7\}$ of

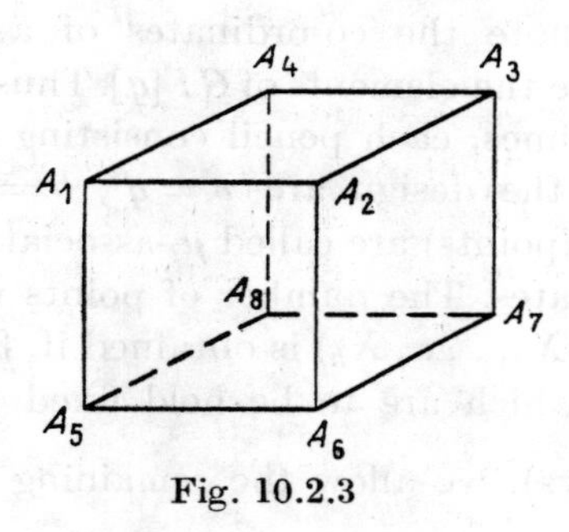

Fig. 10.2.3

0-associates. As regards the parameters p_{jk}^i, we observe that p_{11}^1 denotes the number of vertices which occur on the same edge as one given vertex A_p and on the same edge as another given vertex A_q where A_p and A_q are themselves on the same edge. Thus, $p_{11}^1 = 0$. The parameter p_{12}^1 denotes the number of vertices which occur on the same edge as one given vertex A_p and on the same face diagonal as another given vertex A_q, where A_p and A_q are themselves on the same edge. Hence $p_{12}^1 = 2$. The parameter p_{12}^3 denotes the number of vertices which occur on the same edge as one given vertex A_p and on the same face diagonal as another given vertex A_q, where A_p and A_q are themselves at opposite ends of a cube diagonal. Therefore, $p_{12}^3 = 3$. We may represent the values of the p_{jk}^i by means of three matrices (p_{jk}^i), $i = 1, 2, 3$, as follows:

$$p_{jk}^1 = \begin{pmatrix} 0 & 2 & 0 \\ 2 & 0 & 1 \\ 0 & 1 & 0 \end{pmatrix} \qquad p_{jk}^2 = \begin{pmatrix} 2 & 0 & 1 \\ 0 & 2 & 0 \\ 1 & 0 & 0 \end{pmatrix} \qquad p_{jk}^3 = \begin{pmatrix} 0 & 3 & 0 \\ 3 & 0 & 0 \\ 0 & 0 & 0 \end{pmatrix}$$

The preceding example may be obtained as a special case of the following general method for constructing PBIBD's from finite affine spaces.

Theorem 10.2.5. *If p is any prime, n any integer, and $q = p^n$, a PBIBD with principal parameters $v = q^h$, $b = qh$, and $k = q^{h-1}$ can always be constructed with the aid of a finite desarguesian affine space of dimension h.*

Proof. A desarguesian affine space $AG(h, p^n)$ of h dimensions defined over the Galois field $GF[p^n]$ has $v = q^h$ points, where $q = p^n$. We take these as our varieties and, as blocks, we take a subset of hq of the $(h-1)$-dimensional subspaces (or "primes") of $AG(h, q)$: namely, those with equations $X_i = c_j$ for $i = 1, 2, \ldots, h$ and $j = 0, 1, \ldots, q-1$. Here, $(X_1, X_2, \ldots, X_h)$ denote the co-ordinates of a point of $AG(h, q)$ and $c_0, c_1, \ldots, c_{q-1}$ denote the elements of $GF[q]$. Thus, the blocks comprise h pencils of parallel primes, each pencil consisting of q primes. The principal parameters of the design are $v = q^h$, $b = qh$, $k = q^{h-1}$; whence $r = h$. Two varieties (points) are called μ_i-associates if they have exactly $h - i$ equal co-ordinates. The number of points which are μ_i-associates of a given point $(X_1, X_2, \ldots, X_h)$ is obtained if, for each choice of $h - i$ of the co-ordinates which are to be held fixed (such a choice can be made in $\binom{h}{h-i}$ ways), we allow the remaining i co-ordinates to take all possible values distinct from those of the given point. Thus, for each of $\binom{h}{h-i}$ choices of co-ordinates to be held fixed, we get $(q-1)^i$ μ_i-associates of our given point. Therefore, each point has a total of

$$n_i = \binom{h}{h-i}(q-1)^i = \binom{h}{i}(q-1)^i$$

μ_i-associates. Two points which are μ_i-associates occur together in exactly those primes (blocks) whose equations are satisfied by any one of the $h - i$ equal co-ordinates. Hence, $\mu_i = h - i$.

To find the parameters p^i_{jl} of the above design, we observe that, if two varieties (points), say $\boldsymbol{X} = (X_1, X_2, \ldots, X_h)$ and $\boldsymbol{Y} = (Y_1, Y_2, \ldots, Y_h)$ are μ_i-associates, they have $h - i$ equal co-ordinates and i unequal ones. A variety $\boldsymbol{Z}$ which is a μ_j-associate of $\boldsymbol{X}$ and a μ_l-associate of $\boldsymbol{Y}$ has $h - j$ co-ordinates in common with $\boldsymbol{X}$ and $h - l$ co-ordinates in common with $\boldsymbol{Y}$. Of the $h - j$ former ones, there will be u (any integer from 0 to $h - i$) taken from those common to $\boldsymbol{X}$ and $\boldsymbol{Y}$ (and these will consequently be common to $\boldsymbol{Z}$ and $\boldsymbol{Y}$ as well as to $\boldsymbol{Z}$ and $\boldsymbol{X}$) and $h - j - u$ from the remaining i co-ordinates of $\boldsymbol{X}$. Of the $h - l$ co-ordinates common to $\boldsymbol{Z}$ and $\boldsymbol{Y}$, u will be among those common to $\boldsymbol{X}$ and $\boldsymbol{Y}$ as we have just said, and the remaining $h - l - u$ will be among those co-ordinates not common to $\boldsymbol{X}$ and $\boldsymbol{Y}$ and not common to $\boldsymbol{Z}$ and $\boldsymbol{X}$. (If they were common to $\boldsymbol{Z}$ and $\boldsymbol{X}$ as well as to $\boldsymbol{Z}$ and $\boldsymbol{Y}$, they would be common to $\boldsymbol{X}$ and $\boldsymbol{Y}$, contradicting the hypothesis.) There are $i - (h - j - u)$ such co-ordinates from among which the $h - l - u$ may be chosen. Thus, the number of choices of co-ordinate places in which a variety $\boldsymbol{Z}$ has co-ordinates which are in common with those of $\boldsymbol{X}$ or $\boldsymbol{Y}$ is

$$\binom{h-i}{u}\binom{i}{h-j-u}\binom{i+j+u-h}{h-l-u}.$$

Of the remaining co-ordinate places in $\boldsymbol{Z}$, those $h - i - u$ which are common to $\boldsymbol{X}$ and $\boldsymbol{Y}$ may each take $q - 1$ values (that is, any value distinct from that of the corresponding co-ordinate in $\boldsymbol{X}$ and $\boldsymbol{Y}$), and those $i - (h - j - u) - (h - l - u) = i + j + l - 2h + 2u$ which are not common to $\boldsymbol{X}$ and $\boldsymbol{Y}$ may each take $q - 2$ values (that is, any values distinct from those of $\boldsymbol{X}$ or $\boldsymbol{Y}$ in these co-ordinate places). Therefore, the number of possible choices of the variety $\boldsymbol{Z}$ is

$$p^i{}_l = \sum_{u=0}^{h-i} \binom{h-i}{u}\binom{i}{h-j-u}\binom{i+j+u-h}{h-l-u}(q-1)^{h-i-u}\,(q-2)^{i+j+l-2h+2u}$$

When $h = 3$, $q = 2$, we get a PBIBD in which $v = 8$, $b = 6$, $k = 4$, $r = 3$ and in which the varieties (points) have co-ordinates (0, 0, 0), (1, 0, 0), (0, 1, 0), (0, 0, 1), (0, 1, 1), (1, 0, 1), (1, 1, 0), and (1, 1, 1) and may be represented as the eight vertices of a cube. That is, we have the simple example which we previously considered.

A large number of other methods for constructing PBIBD's are known but, for the most part, their discussion is not very relevant to the subject matter of the present book. However, the following theorem is of considerable interest.

Theorem 10.2.6. *The existence of a PBIBD with an L_h-association scheme (to be defined below) is equivalent to the existence of a set of $h - 2$ mutually orthogonal latin squares.*

Proof. We need the fact, already mentioned in section 5.4, that a set of $h - 2$ mutually orthogonal latin squares of order n defines and is defined by an $h \times n^2$ matrix of n elements such that each ordered pair of elements occurs exactly once as a column in each submatrix of two (not necessarily adjacent) rows. Such a matrix is called an *orthogonal array* of strength 2 and index 1 and we shall give a formal proof of its equivalence to a set of mutually orthogonal latin squares in section 11.1 of the next chapter. (An example of a 4×5^2 array is shown in Fig. 10.2.4.) We construct a PBIBD from such an orthogonal array in the following way. We label the columns of the array by means of n^2 symbols $v_1, v_2, \ldots, v_{n^2}$ which are to be the varieties of the PBIBD. A set of varieties is to be regarded as forming a block of the PBIBD if the columns of the orthogonal array which these varieties represent all have the same element in one particular row. Thus, in the example, the varieties v_1, $v_{10}, v_{14}, v_{18}, v_{22}$ form a block because the columns which they represent all have the same element 0 in the third row. Since each of the elements $0, 1, \ldots, n-1$ occurs n times in each row of the orthogonal array, the total number of blocks is $b = nh$ and each contains $k = n$ varieties. Since there are $v = n^2$ varieties all together, each variety occurs in $r = h$ blocks. Two varieties are called μ_1-associates if they occur together in exactly one block (so $\mu_1 = 1$). They are called μ_2-associates if they do not occur together in any block (so $\mu_2 = 0$). Each variety v_j occurs in just h blocks, these being the blocks corresponding to the set of

v_1	v_2	v_3	v_4	v_5	v_6	v_7	v_8	v_9	v_{10}	v_{11}	v_{12}	v_{13}
0	0	0	0	0	1	1	1	1	1	2	2	2
0	1	2	3	4	0	1	2	3	4	0	1	2
0	1	2	3	4	1	2	3	4	0	2	3	4
0	1	2	3	4	3	4	0	1	2	1	2	3

v_{14}	v_{15}	v_{16}	v_{17}	v_{18}	v_{19}	v_{20}	v_{21}	v_{22}	v_{23}	v_{24}	v_{25}
2	2	3	3	3	3	3	4	4	4	4	4
3	4	0	1	2	3	4	0	1	2	3	4
0	1	3	4	0	1	2	4	0	1	2	3
4	0	4	0	1	2	3	2	3	4	0	1

Fig. 10.2.4

columns which have the same entry in the rth row as does the given column v_j, for $r = 1, 2, \ldots, h$. No two of these blocks can have another variety v_l in common otherwise the same ordered pair of elements of the orthogonal array would occur twice as a column in the particular two-rowed submatrix whose rows defined these two blocks. Thus, no two varieties can occur together in more than one block, so we get only the above-stated two associate classes. Because no column of the orthogonal array can have the same entry as the jth column v_j in more than one row (by virtue of the definition of an orthogonal array) and because every symbol which occurs in v_j occurs in the same row of the array $n-1$ further times, the number of 1-associates of v_j is $(n-1)h$. This result is true for each choice of v_j. Thus, $n_1 = (n-1)h$ and $n_2 = (n^2-1)-(n-1)h = (n-1)(n+1-h)$.

To determine the parameters p^i_{jk}, we first find p^1_{11}. This represents the number of varieties which are 1-associates of each of two varieties v_r, v_s which are themselves 1-associates (that is, occur in the same block). Two varieties are 1-associates if the columns of the orthogonal array which they represent both have the same entry in one particular row. For simplicity of discussion let us take this to be the first row. Then, among the varieties which are 1-associates of these two varieties v_r, v_s are all those which lie in the unique block which contains both the varieties. There are $n-2$ of these and all represent columns with the same first row entry. To find the remaining varieties which are 1-associates of both v_r and v_s, select any two of the remaining $\binom{h-1}{2}$ rows of the orthogonal array and suppose that the rth and sth columns of the two-rowed submatrix which they form are $\binom{r_1}{r_2}$ and $\binom{s_1}{s_2}$. Then those two varieties which have $\binom{r_1}{s_2}$ and $\binom{s_1}{r_2}$ as columns of the two-rowed submatrix, and only those, are 1-associates of both v_r and v_s. Hence, there exist $2\binom{h-1}{2}$ further varieties to be included in the number p^1_{11}, so

$$p^1_{11} = n-2+(h-1)(h-2) = n+h(h-3).$$

Since $p^1_{11}+p^1_{12} = n_1 - 1$, we have

$$p^1_{12} = (n-1)h-1-[n+h(h-3)] = (h-1)(n-h+1).$$

We may similarly compute p^1_{21} and p^1_{22} from the equations $p^1_{21} = p^1_{12}$ and $p^1_{21}+p^1_{22} = n_2$. Then, since $n_1 p^1_{12} = n_2 p^2_{11}$, we easily find p^2_{11}, $p^2_{12} = p^2_{21}$ and p^2_{22}. We get

$$p^1_{jl} = \begin{pmatrix} n+h(h-3) & (h-1)(n-h+1) \\ (h-1)(n-h+1) & (n-h)(n-h+1) \end{pmatrix}$$

and

$$p^2_{jl} = \begin{pmatrix} h(h-1) & h(n-h) \\ h(n-h) & (n-h)^2+h-2 \end{pmatrix}.$$

An association scheme with two associate classes and of the above kind is called an L_h-*scheme*.

If the blocks of a PBIBD with such an association scheme are given, we can construct the corresponding orthogonal array in the following way. Select one block b_{11} and suppose that the varieties which it contains are denoted by $v_1, v_2, \ldots, v_n$. Denote the remaining $h - 1$ blocks which contain the variety v_1 by $b_{12}, b_{13}, \ldots, b_{1h}$. Similarly, for each other variety v_l of the block b_{11} denote the remaining $h - 1$ blocks which contain that variety by $b_{l2}, b_{l3}, \ldots, b_{lh}$ where the order of the suffices is chosen so that, for each choice of c, the blocks $b_{1c}, b_{2c}, \ldots, b_{lc}$ have no variety in common. Such a choice of disjoint blocks must be possible for each integer c, $1 < c \leq h$, otherwise the association scheme could not be L_h: for in the orthogonal array exhibiting the association scheme all the columns representing the varieties of a specified block have the same entry in one row, its "entry row", and the h blocks $b_{11}, b_{s2}, b_{s3}, \ldots, b_{sh}$ containing a specified variety v_s $(0 < s < n)$ each have a different entry row, since each must have one of its entries in the column representing the variety v_s. Evidently, two blocks which have the same entry row cannot have any variety in common. On the other hand, it follows from what we have just said that each of the blocks $b_{l2}, b_{l3}, \ldots, b_{lh}$ must have the same entry row as one of the h blocks $b_{11}, b_{s2}, \ldots, b_{sh}$, so, for each s, there must be a block b_{lc} which has no variety in common with b_{sc}. In fact, from the formula $p_{11}^1 = n + h(h - 3)$, it is easy to deduce that there is only one.(1) With the blocks so labelled, we put the integer i in the jth row of the column representing v_t for each variety v_t that occurs in the block b '. For each

(1) Consider the sets of blocks $b_{s2}, b_{s3}, \ldots, b_{sh}$ and $b_{t2}, b_{t3}, \ldots, b_{th}$. Each block of the first set contains the variety v_s and each block of the second set the variety v_t. There are no other blocks except the block b_{11} which contains either of these varieties. Also, v_s and v_t both occur in the block b_{11} and in no other block do they occur together. The total number of varieties which are 1-associates of both v_s and v_t is $p_{11}^1 = n + h(h - 3)$ and this number includes the $n - 2$ remaining varieties of the block b_{11}. Thus, there are exactly $(h - 1)(h - 2)$ varieties which are 1-associates of both v_s and v_t and which do not occur in b_{11}. Since all such 1-associates of v_s occur among the blocks $b_{s2}, b_{s3}, \ldots, b_{sh}$ and all such 1-associates of v_t occur among the blocks $b_{t2}, b_{t3}, \ldots, b_{th}$, and there are $(h - 1)^2$ pairs b_{sl}, b_{tm}, if follows that just $(h - 1)^2 - (h - 1)(h - 2) = h - 1$ of these pairs have no variety in common: for varieties which are common to the two blocks of a pair are precisely those which are 1-associates of both v_s and v_t. Therefore, since for each choice of l $(2 \leq l \leq h)$ there is one block b_{tm} which has no variety in common with b_{sl}, there cannot be more than one.

$j \neq 1$, the above labelling process labels n blocks, namely the blocks $b_{1j}, b_{2j}, \ldots, b_{nj}$, so all rows of the orthogonal array except the first are thereby filled. There remain $n - 1$ blocks not so far labelled. Because the association scheme is L_h, these blocks have no varieties in common and we label them $b_{21}, b_{31}, \ldots, b_{n1}$ in arbitrary order. Using the same rule as before, we can then fill the remaining places in the first row of the orthogonal array.

In [1], S. S. Shrikhande has shown that if a set of n^2 varieties, $n \geq 2$ but $n \neq 4$, is given with an association scheme having two associate classes and such that $n_1 = 2(n - 1)$, $p_{11}^1 = n - 2$, $p_{11}^2 = 2$ then the association scheme must be L_2 and a PBIBD of $2n$ blocks with n varieties in each can be constructed by putting varieties in the same block if they are 1-associates. In an unpublished thesis to which the present authors have not had access, D. M. Mesner has proved a similar result for a set of n^2 varieties with an association scheme of two associate classes whose parameters n_1, p_{11}^1, p_{11}^2 are those of L_c; namely that in general the association scheme must then be L_c. (See D. M. Mesner [1].) Some further related results are given in D. M. Mesner [2].

S. S. Shrikhande used his result just referred to, to show that the existence of a set of $n - 3$ mutually orthogonal latin squares of order $n > 4$ implies the existence of a complete set of $n - 1$ such squares, as we mentioned in section 9.3. (See also S. S. Shrikhande [2].)

For further information concerning designs and their construction and for an extensive bibliography the reader is referred to the following four recent books: S. Vajda [1], M. Hall [11], P. Dembowski [1] and W. T. Federer and L. N. Balaan [1].

A general account of the importance of latin squares in the field of design of experiments will be found in A. Hedayat and S. S. Shrikhande [1].

The reader's attention is also drawn to the following books on experimental design: R. A. Fisher [4], W. G. Cochran and G. M. Cox [1], M. C. Chakrabarti [1], D. J. Finney [5], and H. B. Mann [4].

In the bibliography of this book, all papers concerned mainly with applications to statistics have been indicated by means of a suffix S. So far as statistical papers which make use of latin squares are concerned, we have tried to make our bibliography as complete as possible but no attempt has been made to refer to all such papers explicitly in the text.

CHAPTER 11

The end of the Euler conjecture

In this chapter, we resume our discussion of construction methods for sets of mutually orthogonal latin squares and, thereby, obtain a number of further results concerning the function $N(n)$ which we defined in chapter 5. In particular, we show the falsity of both the Euler conjecture and also that of MacNeish concerning upper bounds for this function.

11.1. Some general theorems

We showed in section 8.2 that the existence of a set of $h-2$ mutually orthogonal latin squares of order n is equivalent to the existence of an h-net of order n. Also, in section 10.2, we made use of the fact (first mentioned in section 5.4) that a set of mutually orthogonal latin squares is equivalent to a particular type of orthogonal array to show that the existence of a set of $h-2$ mutually orthogonal latin squares is also equivalent to the existence of a PBIBD with two associate classes and parameters $v=n^2$, $b=hn$, $r=h$, $k=n$, $\mu_1=1$, $\mu_2=0$ having a particular type of association scheme, called an L_h-association scheme. Our first task in the present section is to confirm the validity of the last result by formally demonstrating the equivalence between a set of $h-2$ mutually orthogonal latin squares of order n and an orthogonal array of h rows (or constraints), n^2 columns (n levels), strength 2, and index unity. We shall begin with a definition.

DEFINITION. Let $A=\|a_{ij}\|$ and $B=\|b_{ij}\|$ be two $r\times s$ matrices whose entries are the numbers $1, 2, \ldots, n$, where r and s are integers such that $rs=n^2$. The matrices A and B are said to be *n-orthogonal* if, for every ordered pair (a, b) of the integers $1, 2, \ldots, n$, there is (exactly) one pair of indices i, j such that $a_{ij}=a$ and $b_{ij}=b$.

For example, the two $n\times n$ matrices R and C given below are clearly n-orthogonal.

$$R = \begin{pmatrix} 1 & 1 & \dots & 1 \\ 2 & 2 & \dots & 2 \\ . & . & \dots & . \\ . & . & \dots & . \\ n & n & \dots & n \end{pmatrix} \qquad C = \begin{pmatrix} 1 & 2 & \dots & n \\ 1 & 2 & \dots & n \\ . & . & \dots & . \\ . & . & \dots & . \\ 1 & 2 & \dots & n \end{pmatrix}$$

Likewise, the two $1 \times n^2$ matrices

$$R^* = (1\,1 \dots 1\,2\,2 \dots 2\,3\,3 \dots 3 \dots \dots \dots n\,n \dots n)$$

and

$$C^* = (1\,2 \dots n\,1\,2 \dots n\,1\,2 \dots n \dots \dots \dots 1\,2 \dots n)$$

obtained from R, C respectively by writing their rows consecutively are n-orthogonal. Evidently, two n-orthogonal $n \times n$ matrices which are such that each of $1, 2, \dots, n$ occurs exactly once in each row and once in each column are orthogonal latin squares.

THEOREM 11.1.1. *The existence of* $k-2$ *mutually orthogonal latin squares of order* n *implies and is implied by the existence of* k *mutually* n*-orthogonal* $n \times n$ *matrices whose entries are the numbers* $1, 2, \dots, n$.

PROOF. It is clear from the definition that the n-orthogonality of two matrices is unaffected by any permutation of the n^2 cells of all the matrices simultaneously, or by relabelling all of the symbols $1, 2, \dots, n$ simultaneously as, say $1', 2', \dots, n'$, where $1', 2', \dots, n'$ is some permutation of the symbols $1, 2, \dots, n$.

Now suppose that k mutually n-orthogonal matrices $A_1, A_2, \dots, A_k$ are given. Since A_1 and A_2 are n-orthogonal, every ordered pair (i, j), $i = 1, 2, \dots, n$, $j = 1, 2, \dots, n$, occurs just once among the n^2 cells of these matrices. Consequently, by a suitable permutation of these n^2 cells, the two matrices can be transformed simultaneously into the matrices R and C. If this rearrangement of cells is carried out on all the matrices simultaneously, their n-orthogonality is not affected thereby, but the transforms of the matrices $A_3, A_4, \dots, A_k$ are necessarily latin squares: for, since the entries in each row of R are all the same, the entries in each row of any matrix which is n-orthogonal to R must be all different. Likewise, since the entries in each column of C are all the same, the entries in each column of any matrix which is n-orthogonal to C must be all different. Thus, the transforms of the matrices $A_3, A_4, \dots, A_k$ are a set of $k-2$ mutually orthogonal latin squares.

Conversely, suppose that a set $L_1, L_2, \dots, L_{k-2}$ of $k-2$ mutually orthogonal latin squares is given. Since the entries in each row of the square L_i ($i = 1, 2, \ldots$, or $k - 2$) are all different, L_i is an $n \times n$ matrix which is n-orthogonal to the matrix R. Similarly, since the entries in each column of L_i are all different, L_i is n-orthogonal to the matrix C. Thus, $R, C, L_1, L_2, \ldots, L_{k-2}$ are a set of k mutually n-orthogonal $n \times n$ matrices.

Corollary. *The existence of $k - 2$ mutually orthogonal latin squares of order n implies and is implied by the existence of a $k \times n^2$ matrix S whose entries are the numbers $1, 2, \ldots, n$ and such that each two rows of S are n-orthogonal submatrices.*

Proof. The corollary follows by observing that, if the rows of each of a set of k mutually n-orthogonal $n \times n$ matrices are written consecutively so as to form a $1 \times n^2$ row matrix (as, for example, when the matrix R is transformed to R^*), the set of matrices, say $R, C, L_1, L_2, \ldots, L_{k-2}$, becomes equivalent to a $k \times n^2$ matrix of the kind described, with R^* and C^* as its first two rows.

A $k \times n^2$ matrix S of the type described in the corollary is called an *orthogonal array* of k *constraints*, n *levels*, *strength* 2, and *index* 1, as already mentioned in section 5.4. In the present chapter, we shall denote such an orthogonal array by the symbol $OA(n, k)$.

Theorem 11.1.2. *If there exists both an $OA(n_1, s)$ and an $OA(n_2, s)$ then an $OA(n_1 n_2, s)$ can be constructed.*

Proof. We shall give two proofs of this important result.

For the first, let S_1 denote the $OA(n_1, s)$ and S_2 denote the $OA(n_2, s)$, and suppose that the elements of S_1 are called $x_1, x_2, \ldots, x_{n_1}$ and that the elements of S_2 are called $y_1, y_2, \ldots, y_{n_2}$. Let $a_{1p}, a_{2p}, \ldots, a_{sp}$ denote the elements of the pth column of S_1, where the a_{ip} are a rearrangement of the elements $x_1, x_2, \ldots, x_{n_1}$, and let $b_{1q}, b_{2q}, \ldots, b_{sq}$ denote the elements of the qth column of S_2, where the b_{iq} are a rearrangement of the elements $y_1, y_2, \ldots, y_{n_2}$. Then the $s \times n_1^2 n_2^2$ matrix S^* whose $[p + n_1^2(q - 1)]$th column, $p = 1, 2, \ldots, n_1^2$, $q = 1, 2, \ldots, n_2^2$, contains the entries $(a_{1p}, b_{1q}), (a_{2p}, b_{2q}), \ldots, (a_{sp}, b_{sq})$ is an $OA(n_1 n_2, s)$ whose $n_1 n_2$ distinct elements are the ordered pairs $z_{ij} = (x_i, y_j)$, $i = 1, 2, \ldots, n_1$, $j = 1, 2, \ldots, n_2$. For, consider the kth and lth rows of S^* and let z_{tu} and z_{vw} be any two of the $n_1 n_2$ distinct elements of S^*. Since the kth and lth rows of S_1 are n_1-orthogonal matrices, there exists one column, say the pth, in which the elements a_{kp}, a_{lp} of these rows are respectively equal to x_t and x_v. Since the kth and lth rows of S_2 are n_2-orthogonal matrices, there exists one column, say the qth, in which the elements b_{kq}, b_{lq} of these rows are respectively equal to y_u and y_w. Then the elements (a_{kp}, b_{kq}) and (a_{lp}, b_{lq}) which occur in the $[p+n_1^2(q-1)]$th column of the kth and lth rows of S^* are respectively equal to z_{tu} and z_{w}. Consequently, the kth and lth rows of S^* are n_1n_2-orthogonal. This is true for arbitrary choices of k and l.

Our second proof is intrinsically the same as the first, but provides an actual construction for an $OA\ (n_1n_2, s)$ whose entries are the integers $1, 2, \ldots, n_1n_2$ from an $OA\ (n_1, s)$ with entries $1, 2, \ldots, n_1$ and an $OA\ (n_2, s)$ with entries $1, 2, \ldots, n_2$. Let $OA\ (n_1, s)$ be the matrix $A = ||a_{ij}||$, $i = 1, 2, \ldots, s$, $j = 1, 2, \ldots, n_1^2$, and let $OA\ (n_2, s)$ be the matrix $B = ||b_{ij}||$, $i = 1, 2, \ldots, s$, $j = 1, 2, \ldots, n_2^2$. Form a new matrix $C = ||c_{ir}||$, where, for $r = p + n_1^2(q-1)$ with $1 \leq p \leq n_1^2$, $1 \leq q \leq n_2^2$, $c_{ir} = a_{ip} + n_1(b_{iq} - 1)$. We shall show that C is an $OA\ (n_1n_2, s)$.

In the first place, since $1 \leq a_{ip} \leq n_1$ and $O \leq b_{iq} - 1 \leq n_2 - 1$, each c_{ir} is one of the integers $1, 2, \ldots, n_1n_2$. Secondly, we easily show that for any choice of k and l, the kth and lth rows of C are n_1n_2-orthogonal matrices. Let u and v be any two integers between 1 and n_1n_2. Then $u = u_1 + n_1(u_2 - 1)$ and $v = v_1 + n_1(v_2 - 1)$ for suitable integers u_1, v_1 between 1 and n_1 and u_2, v_2 between 1 and n_2. Since the kth and lth rows of A are n_1-orthogonal matrices, there exists one column, say the pth, whose entries a_{kp} and a_{lp} are respectively equal to u_1 and v_1. Similarly, for a suitable integer q, we have $b_{kq} = u_2$ and $b_{lq} = v_2$. But then, $c_{kr} = u$ and $c_{lr} = v$, where $r = p + n_1^2(q-1)$, and so the kth and lth rows of C are n_1n_2-orthogonal. This completes the proof.

Corollary. *If there exists a set of t mutually orthogonal latin squares of order n_1 and a set of t mutually orthogonal latin squares of order n_2, then there exists a set of t mutually orthogonal latin squares of order n_1n_2.*

Proof. The corollary is an immediate consequence of the fact that the existence of a set t mutually orthogonal latin squares of order n implies and is implied by the existence of an $OA\ (n, t+2)$.

As an example, we shall outline the construction of a pair of orthogonal latin squares of order 15 (or equivalently, an OA (15, 4)) from the pairs of orthogonal latin squares of orders 3 and 5 given in Fig. 11.1.1, using the method given in our second proof above.

We first form sets of four 3-orthogonal 3×3 matrices and 5-orthogonal 5×5 matrices by appending the matrices R and C of appropriate size

$$\begin{array}{|ccc}\hline 1 & 2 & 3 \\ 2 & 3 & 1 \\ 3 & 1 & 2 \end{array} \qquad \begin{array}{|ccc}\hline 1 & 2 & 3 \\ 3 & 1 & 2 \\ 2 & 3 & 1 \end{array}$$

$$\begin{array}{|ccccc}\hline 1 & 2 & 3 & 4 & 5 \\ 4 & 3 & 1 & 5 & 2 \\ 5 & 1 & 4 & 2 & 3 \\ 2 & 4 & 5 & 3 & 1 \\ 3 & 5 & 2 & 1 & 4 \end{array} \qquad \begin{array}{|ccccc}\hline 1 & 2 & 3 & 4 & 5 \\ 3 & 5 & 2 & 1 & 4 \\ 2 & 4 & 5 & 3 & 1 \\ 5 & 1 & 4 & 2 & 3 \\ 4 & 3 & 1 & 5 & 2 \end{array}$$

Fig. 11.1.1

to the squares given. Then, writing the rows of each matrix in a single row, we get an OA (3,4) and an OA (5,4) as in Fig. 11.1.2.

$$\begin{array}{ccccccccc} 1 & 1 & 1 & 2 & 2 & 2 & 3 & 3 & 3 \\ 1 & 2 & 3 & 1 & 2 & 3 & 1 & 2 & 3 \\ 1 & 2 & 3 & 2 & 3 & 1 & 3 & 1 & 2 \\ 1 & 2 & 3 & 3 & 1 & 2 & 2 & 3 & 1 \end{array}$$

$$\begin{array}{ccccccccccccccccccccccccc} 1&1&1&1&1&2&2&2&2&2&3&3&3&3&3&4&4&4&4&4&5&5&5&5&5 \\ 1&2&3&4&5&1&2&3&4&5&1&2&3&4&5&1&2&3&4&5&1&2&3&4&5 \\ 1&2&3&4&5&4&3&1&5&2&5&1&4&2&3&2&4&5&3&1&3&5&2&1&4 \\ 1&2&3&4&5&3&5&2&1&4&2&4&5&3&1&5&1&4&2&3&4&3&1&5&2 \end{array}$$

Fig. 11.1.2

From these, we derive an OA (15,4) of 255 columns, and, of these columns, we exhibit only the first fifty-four in Fig. 11.1.3.

We reorder these 225 columns in such a way that the first two rows of the matrix represent the rows, written consecutively, of two 15×15 matrices R and C, and the remaining two rows then represent, in order, the rows of the two desired orthogonal latin squares of order 15. Hence the latter can be written down.

The result stated in theorem 11.1.2 has probably been more used than any other in constructions of sets of mutually orthogonal latin squares. In particular, it leads at once to a property of the function $N(n)$ which was first pointed out by H. F. MacNeish in [2]: namely,

MacNeish's theorem. *If* $n = p_1^{\alpha_1} p_2^{\alpha_2} \ldots p_r^{\alpha_r}$ *is the factorization of the integer* n *into powers of the distinct primes* $p_1, p_2, \ldots, p_r$ *then* $N(n) \geq$ $\geq \min (p_i^{\alpha_i} - 1)$.

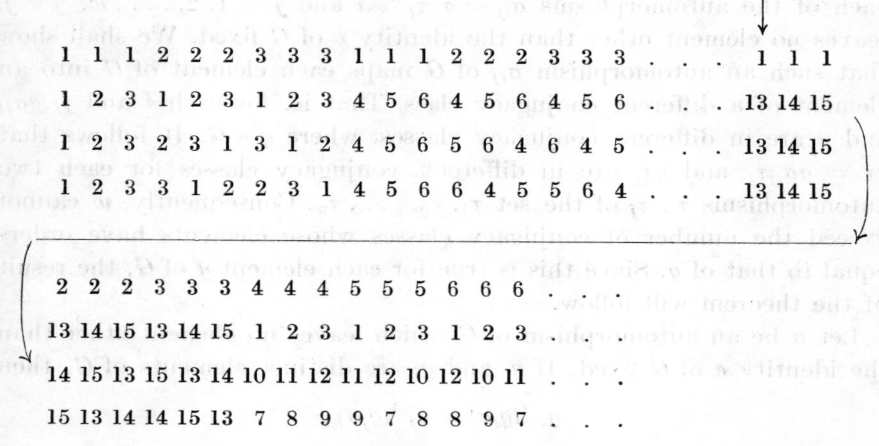

Fig. 11.1.3

The theorem asserts that there exist at least $t = \min(p_i^{\alpha_i} - 1)$ mutually orthogonal latin squares of order n. Since we know by theorem 5.2.3 that there are at least $p_i^{\alpha_i} - 1$ mutually orthogonal latin squares of order $p_i^{\alpha_i}$ for each i, $i = 1, 2, \ldots, r$, the result follows by repeated application of the corollary to theorem 11.1.2.

MacNeish's theorem gives a lower bound for $N(n)$. MacNeish himself conjectured that this was also an upper bound as, for example, is certainly the case when n itself is a prime power. Further evidence for this conjecture seemed to have been provided when, in [1], H. B. Mann proved:

Theorem 11.1.3. *Let L be a latin square based on a group G* (that is, L represents the multiplication table of the group G), *and let c_q be the number of conjugacy classes of elements with the property that all the elements of each class are elements of order q, in the given group G. Let $s = \min c_q$ when q ranges through the factors of the order of G. Then not more than s mutually orthogonal latin squares containing L can be constructed from G by the automorphism method.*

Proof. The automorphism method was described in section 7.2. The number of mutually orthogonal squares which can be constructed by it is equal to the largest number w of automorphisms $\tau_1, \tau_2, \ldots, \tau_w$ of G which can be found such that, for each pair τ_i, τ_j of distinct automorphisms of the set, $u\tau_i \neq u\tau_j$ for any element u of G other than the identity. This requires that $u\tau_i\tau_j^{-1} \neq u$ for $u \neq e$; in other words, that each of the automorphisms $\sigma_{ij} = \tau_i\tau_j^{-1}$, i and $j = 1, 2, \ldots, w$, $i \neq j$, leaves no element other than the identity e of G fixed. We shall show that such an automorphism σ_{ij} of G maps each element of G into an element of a different conjugacy class. That is, for each i and j, $g\sigma_{ij}$ and g are in different conjugacy classes, where $g \in G$. It follows that $g\tau_i = g\sigma_{ij}\tau_j$ and $g\tau_j$ are in different conjugacy classes for each two automorphisms τ_i, τ_j of the set $\tau_1, \tau_2, \ldots, \tau_w$. Consequently, w cannot exceed the number of conjugacy classes whose elements have orders equal to that of g. Since this is true for each element g of G, the result of the theorem will follow.

Let σ be an automorphism of G which leaves no element other than the identity e of G fixed. If g_i and g_j are distinct elements of G, then

$$g_i^{-1}(g_i\sigma) \neq g_j^{-1}(g_j\sigma),$$

since equality would imply

$$(g_i\sigma)\,(g_j\sigma)^{-1} = g_ig_j^{-1}$$

and so $(g_ig_j^{-1})\sigma = g_ig_j^{-1}$, whence $g_ig_j^{-1} = e$ and g_i, g_j would not be distinct. It follows that, as g_i ranges through the elements of G, so does $g_i^{-1}(g_i\sigma)$. Now let a be an element of G of order q. Then the element $a\sigma$ is likewise of order q. Suppose, if possible, that a and $a\sigma$ are in the same conjugacy class so that $a\sigma = h^{-1}ah$ for some element h in G. We can represent h in the form $g^{-1}(g\sigma)$ for some g in G. Then $gh = g\sigma$ and so

$$(gag^{-1})\sigma = g\sigma.a\sigma.(g\sigma)^{-1} = gh.h^{-1}ah.h^{-1}g^{-1} = gag^{-1}$$

implying $gag^{-1} = e$ and $a = e$. Thus, a and $a\sigma$ must be in different conjugacy classes, as claimed above. This completes the proof.

Corollary. *If $n = p_1^{\alpha_1}, p_2^{\alpha_2} \ldots p_r^{\alpha_r}$, where $p_1, p_2, \ldots, p_r$ are distinct primes, then not more than $t = \min\,(p_i^{\alpha_i} - 1)$ mutually orthogonal latin squares of order n can be constructed from any group G by the automorphism method.*

Proof. Since the Sylow p_i-subgroups of G are all conjugate (for fixed i), each conjugacy class of elements whose orders are powers of p_i has a representative element in each Sylow p_i-subgroup. Since the number of elements in a Sylow p_i-subgroup is $p_i^{\alpha_i}$, c_q cannot exceed $p_i^{\alpha_i} - 1$ when q is equal to p_i. (At least one element of G has order p_i, by Cauchy's theorem.) Hence, $s = \min c_q$ cannot exceed $\min\,(p_i^{\alpha_i} - 1)$.

The MacNeish conjecture was not disproved until 1959 when, in [1], E. T. Parker showed (as a counter example) that there exists a set of at least 4 mutually orthogonal latin squares of order 21. To this end, Parker proved the following theorem:

THEOREM 11.1.4. *If there exists a balanced incomplete block design with* $\lambda = 1$ *and* k *equal to the order of a projective plane, then there exists a set of* $k - 2$ *mutually orthogonal latin squares of order* v.

PROOF. Since k is the order of a projective plane π say, we can introduce co-ordinates into π in the manner described at the beginning of section 8.3, first labelling the lines through two particular points A and B of π (excepting the line AB itself) by means of a set Σ of k symbols and thence representing each of the $k(k-1)$ lines of π which do not pass through A or B by means of a permutation $\begin{pmatrix} \ldots x_i \ldots \\ \ldots y_i \ldots \end{pmatrix}$ of the k symbols indicating the points (x_i, y_i) of the plane $\pi \setminus AB$ through which that particular line passes. Let us call this set of permutations S. Since one (and only one) line $\mathfrak{L}$ of π passes through each two distinct points (x_i, y_i) and (x_j, y_j) of $\pi \setminus AB$ and since $\mathfrak{L}$ does not pass through A except when $x_i = x_j$ and does not pass through B except when $y_i = y_j$, the set of permutations S is doubly transitive on the set of symbols Σ. (In fact, as in section 8.3, we can always assign the symbols Σ to the lines of the pencils with vertices A and B in such a way that S contains the identity permutation and also contains a subset $\bar{S}$ of exactly k permutations of which none but the identity leaves any symbol fixed.) Moreover, if P_1 and P_2 are distinct permutations of the set S, then the permutation $P_1P_2^{-1}$ fixes at most one of the symbols of Σ. For, suppose that $P_1P_2^{-1}$ fixed both of the symbols x_i and x_j. Then it would follow that P_1 and P_2 both transformed x_i into the same symbol y_i and both transformed x_j into the same symbol y_j. In that case, P_1 and P_2 would both represent lines through the two points (x_i, y_i) and (x_j, y_j) of the plane $\pi \setminus AB$, contradicting the fact that there is only one such line. Thus, we are assured that, when k is equal to the order of a projective plane, a doubly transitive set S of $k(k-1)$ permutations on k symbols and with the property that, for any distinct permutations P_1 and P_2 of S, $P_1P_2^{-1}$ fixes at most one symbol, always exists.

Now fix an ordering of the k symbols (selected from the total of v symbols, which we may take as $1, 2, \ldots, v$) in each block of the given balanced incomplete block design. Retaining these digits, permute their positions in the block by all permutations of the doubly transitive set S

obtained above, thereby generating $k(k-1)$ ordered k-tuples from each block. To the set of ordered k-tuples so found (to be written as columns, and of which there will be $bk(k-1) = vr(k-1)$ altogether) adjoin the additional k-tuples $(1, 1, \ldots, 1)$, $(2, 2, \ldots, 2)$, $\ldots$, $(v, v, \ldots, v)$. This gives a $k \times v^2$ matrix[1] whose entries are the integers $1, 2, \ldots, v$ and with the property that each two of its rows are v-orthogonal $1 \times v^2$ submatrices; in other words, an orthogonal array $OA(v, k)$. The v-orthogonality follows from the facts that, in the original block design, each pair of distinct symbols occur together in exactly one block, and that the permutations of S are a doubly transitive set. Using the corollary to theorem 11.1.1, we can deduce the existence of $k-2$ mutually orthogonal latin squares of order v.

By a specialization of the above theorem and by making use of the fact that every Galois plane possesses a cyclic collineation (to be proved below), we get:

THEOREM 11.1.5. *If m is a Mersenne prime greater than 3 or if $m+1$ is a Fermat prime greater than 3, then there exists a set of m mutually orthogonal latin squares of order m^2+m+1.*

PROOF. For the proof (which is due to E. T. Parker, see [1]) we shall need to use the following result.

Every Galois plane (finite desarguesian projective plane) *π of order $q = p^r$ has a collineation σ of order $N = q^2+q+1$ which is cyclic on its points.*

To show this, we observe firstly that the points of π are represented by homogeneous co-ordinate triples (x_0, x_1, x_2) where $x_i \in GF[p^r]$ and $(\lambda x_0, \lambda x_1, \lambda x_2)$ represents the same point as (x_0, x_1, x_2) for all non-zero values of λ taken from $GF[p^r] \equiv F$ say. The q^3-1 distinct triples (x_0, x_1, x_2) form a vector space of dimension 3 over F and may be separated into $N = q^2+q+1$ equivalence classes of triples, such that all the members of each equivalence class represent the same point of π.

Consider now the Galois field $GF[p^{3r}] \equiv K$ say. This may be regarded as a cubic extension of F by a root of an irreducible cubic equation with coefficients in F. Let ω be a primitive root of K so that $\omega^{q^3-1} = 1$, $\omega^h \neq 1$ for $0 < h < q^3-1$, and the q^3-1 distinct powers of ω give the q^3-1 non-zero elements of K. If $\omega^h \in F$, then it satisfies the equa-

[1] Since $r(k-1) = \lambda(v-1)$, we have $vr(k-1)+v = v(v-1)+v = v^2$ when $\lambda = 1$.

tion $x^{q-1} - 1 = 0$ and so $h \equiv 0 \pmod{N}$. Thus, $\omega \notin F$ and so $K = F(\omega)$ whence each element ω^x of K is expressible in the form

$$\omega^x = x_0 + x_1\omega + x_2\omega^2, \quad x_i \in F.$$

Moreover, there are as many distinct powers of ω as there are distinct vector triples (x_0, x_1, x_2). By this means, the vectors (x_0, x_1, x_2) may be represented by elements ω^x of K. Two vectors give the same point of π (belong to the same equivalence class of vectors) if and only if the elements ω^x and ω^y representing them are related by $x \equiv y \pmod{N}$: for then $\omega^{y-x} = \lambda$ say, belongs to F and we have

$$\omega^y = \lambda\omega^x = \lambda x_0 + \lambda x_1\omega + \lambda x_2\omega^2.$$

It now follows that the mapping $\sigma\colon \omega^x \to \omega^{x+1}$ effects a permutation of the points of π in a cycle of length N. Moreover, σ is obviously a collineation since the points $\omega^x + \lambda\omega^y$ of the line joining the points ω^x and ω^y are mapped onto the points $\omega^{x+1} + \lambda\omega^{y+1}$ of the line joining the images of ω^x and ω^y under σ. Precisely, we have

$$\omega^x + \lambda\omega^y = (x_0 + \lambda y_0) + (x_1 + \lambda y_1)\omega + (x_2 + \lambda y_2)\omega^2$$

whence

$$\sigma\colon (\omega^x + \lambda\omega^y) \to (\omega^x + \lambda\omega^y)\,\omega = \omega^{x+1} + \lambda\omega^{y+1}.$$

The proof of the existence of a cyclic collineation is now complete. This property of Galois planes was first noted and proved by J. Singer(1) in [1].

Next, we return to the proof of theorem 11.1.5. The hypothesis of the theorem implies that both m and $m+1$ are prime powers. Since a projective plane π of order m is a balanced incomplete block design with $\lambda = 1$, $v = m^2 + m + 1$, and $k = m + 1$, the hypothesis on m implies that such a design exists and satisfies the hypotheses of theorem 11.1.4. The lines of π are the blocks of the design. Let σ be the collineation of order $v = m^2 + m + 1$ which is cyclic on the points of π and whose existence we proved above. Let the labelling of the points P_i $(i = 1, 2, \ldots, v)$ of π be so chosen that $P_i^\sigma = P_{i+1}$ for $i = 1, 2, \ldots, v-1$ and $P_v^\sigma = P_1$. Also, suppose that, with this labelling, $P_{x_1}, P_{x_2}, \ldots, P_{x_k}$ are the points of some line of π. Then, because σ maps lines into lines, the remaining lines of π contain the sets of points $P_{x_1+t}, P_{x_2+t}, \ldots, P_{x_k+t}$

(1) For an interesting historical comment connected with this result see also J. Singer [3].

for $t = 1, 2, \ldots, v-1$ where suffices are taken modulo v. Hence, the blocks of the design may be taken as the k-ads

$$\begin{array}{cccc} x_1 & x_1+1 & \ldots & x_1+(v-1) \\ x_2 & x_2+1 & \ldots & x_2+(v-1) \\ \cdot & \cdot & \cdots & \cdot \\ \cdot & \cdot & \cdots & \cdot \\ x_k & x_k+1 & \ldots & x_k+(v-1) \end{array}$$

As in the proof of theorem 11.1.4, each of these blocks generates a set of $k(k-1)$ blocks when its elements are permuted by the permutations of a doubly transitive set S, and a further v blocks of the form

$$\begin{array}{cccc} 1 & 2 & \ldots & v \\ 1 & 2 & \ldots & v \\ \cdot & \cdot & \cdots & \cdot \\ \cdot & \cdot & \cdots & \cdot \\ 1 & 2 & \ldots & v \end{array}$$

are then adjoined so as to produce finally a $k \times v^2$ matrix which is an $OA\ (v, k)$. It is clear from the foregoing that if this matrix has $i, j, a_1, a_2, \ldots, a_{k-2}$ as a column, then it has $i+t, j+t, a_1+t, a_2+t, \ldots, a_{k-2}+t$ for $t = 1, 2, \ldots$ or $v-1$ as another, where all addition is modulo v. Thus, the associated latin squares $L_1, L_2, \ldots, L_{k-1}$ each have the property that every diagonal contains every one of the symbols $1, 2, \ldots, v$ exactly once: for, when a_l is the symbol in the (i, j)th cell of the lth latin square, $a_l + t$ is the symbol in the $(i+t, j+t)$th cell, $t = 1, 2, \ldots, v-1$. Under these circumstances, the further latin square

$$L^* = \begin{array}{|ccccc} \hline 1 & 2 & 3 & \ldots & v \\ v & 1 & 2 & \ldots & v-1 \\ v-1 & v & 1 & \ldots & v-2 \\ \cdot & \cdot & \cdot & \ldots & \cdot \\ \cdot & \cdot & \cdot & \ldots & \cdot \\ 2 & 3 & 4 & \ldots & 1 \end{array}$$

will be orthogonal to each of $L_1, L_2, \ldots, L_{k-2}$ and so we shall have a set of $k-1=m$ mutually orthogonal latin squares of order $v = m^2 + m + 1$ all together.

By taking $m+1$ equal to the Fermat prime 2^2+1, Parker deduced from his theorem 11.1.5 that there exist sets of at least 4 mutually orthogonal latin squares of order 21. Since $21 = 3\times 7$, the MacNeish conjecture would have given an upper bound of 2 for the number of such squares. The conjecture was thus disproved.

11.2. The end of the Euler conjecture

Next, it was the turn of the Euler conjecture to suffer a similar fate. Only a few months after the above work of Parker's was completed, R. C. Bose and S. S. Shrikhande (see [1]) made a somewhat similar use of statistical designs to construct a pair of orthogonal latin squares of order 22 and thus disproved Euler's conjecture. We shall explain their method below. (We remark here that so long as the MacNeish conjecture remained unchallenged, it lent support to the belief that the Euler conjecture was true, since, for $n = 4m+2$, $\min (p_i^{\alpha_i} - 1) = 1$.) No sooner had Bose and Shrikhande found a counter-example to Euler's conjecture than E. T. Parker derived a further construction in [2] by means of which, among others, he was able to obtain a pair of orthogonal latin squares of order 10. His result was as follows:

Theorem 11.2.1. *If q is a prime power which is congruent to* 3 *modulo* 4, *then there exists a pair of orthogonal latin squares of order* $(3q-1)/2$.

Proof. The proof consists in constructing an orthogonal array $OA\,[(3q-1)/2, 4]$. Let ω be a fixed primitive element (generator of the multiplicative cyclic group) of the Galois field $GF[q]$, x be a variable element of $GF[q]$, and $Y_1, Y_2, \ldots, Y_{\frac{1}{2}(q-1)}$ be $(q-1)/2$ symbols not belonging to $GF[q]$. With these elements we form the following column quadruples:

$$\begin{array}{cccc}
Y_i & \omega^{2i}(\omega+1)+x & \omega^{2i}+x & x \\
x & Y_i & \omega^{2i}(\omega+1)+x & \omega^{2i}+x \\
\omega^{2i}+x & x & Y_i & \omega^{2i}(\omega+1)+x \\
\omega^{2i}(\omega+1)+x & \omega^{2i}+x & x & Y_i
\end{array}$$

Here x varies through the q elements of $GF[q]$ and i takes the values $1, 2, \ldots, \frac{1}{2}(q-1)$. Hence, we get a total of $4q[\frac{1}{2}(q-1)] = 2q(q-1)$ columns. The four columns shown are obtained by cyclic rearrangements of the first. It is easily seen, therefore, that as x varies with i fixed, a particular symbol Y_i occurs among the columns once and only once with each element of $GF[q]$ in each ordered pair of row positions. No two of the symbols Y_i (like or distinct) occur in any one of the columns, and no element of $GF[q]$ is repeated in a column. Since each of the expressions x, $\omega^{2i} + x$, $\omega^{2i}(\omega + 1) + x$ ranges through the whole set of elements of $GF[q]$ as x varies and since $(\omega^{2i} + x) - x = \omega^{2i}$ is always a non-zero square element of $GF[q]$ while

$$[\omega^{2i}(\omega + 1) + x] - (\omega^{2i} + x) = \omega^{2i+1}$$

is always a non-square element of $GF[q]$, each pair of distinct elements of $GF[q]$ occurs once in each adjacent pair of row positions once in either order. (Here the fourth and first rows are regarded as being adjacent.) Since q is congruent to 3 modulo 4, -1 is a non-square in $GF[q]$: for suppose on the contrary that $-1 = x^2$, with x in $GF[q]$. Since each element x of $GF[q]$ satisfies the equation $x^{q-1} - 1 = 0$, we should have $(-1)^{\frac{1}{2}(q-1)} = 1$. But $q = 4m + 3$ for some integer m and so this would give $1 = (-1)^{2m+1} = -1$, a contradiction. From the fact that -1 is a non-square in $GF[q]$ it follows that

$$[\omega^{2i}(\omega + 1) + x] - x \quad \text{and} \quad x - [\omega(\omega^{2i} + 1) + x]$$

are one a square element and one a non-square element of $GF[q]$.

It follows that each ordered pair of distinct elements of $GF[q]$ occurs among the set of columns once in the first and third row positions and once in the second and fourth row positions.

Let us now adjoin the set of q columns $\begin{pmatrix} x \\ x \\ x \\ x \end{pmatrix}$, where x again varies through the q elements of $GF[q]$. These additional columns provide for one occurrence of each element x with itself in each pair of row positions.

Next from a pair of orthogonal latin squares of order $(q-1)/2$ (such a pair always exists because q congruent to 3 modulo 4 implies that $(q-1)/2$ is an odd number) we form an orthogonal array $OA[(q-1)/2, 4]$ whose elements are the symbols $Y_1, Y_2, \ldots, Y_{\frac{1}{2}(q-1)}$ and thence adjoin a further $[(q-1)/2]^2$ columns to our set. Among this latter set of column quadruples, each ordered pair of the symbols $Y_1, Y_2, \ldots, Y_{\frac{1}{2}(q-1)}$ occurs exactly once in each ordered pair of row positions. Since the total number of columns is

$$2q(q-1) + q + [\tfrac{1}{2}(q-1)]^2 = [\tfrac{1}{2}(3q-1)]^2$$

and since the total number of distinct symbols occurring in them is

$$q + \tfrac{1}{2}(q-1) = \tfrac{1}{2}(3q-1)$$

it follows from the above analysis that they form an $OA\ [(3q-1)/2, 4]$. By virtue of the corollary to theorem 11.1.1, this proves the present theorem.

As an illustration, we shall demonstrate Parker's construction of a pair of orthogonal latin squares of order 10.

We take $q = 7$ and denote the elements of $GF[7]$ by 0, 1, 2, 3, 4, 5, 6, taken modulo 7 with $\omega = 3$ as primitive element. Then $\omega^2 = 2$, $\omega^3 = 6$, $\omega^4 = 4$, $\omega^5 = 5$, $\omega^6 = 1$. As additional symbols, we take $Y_1 = 8$, $Y_2 = 9$, and $Y_3 = 7$. Taking $i = 1$, we get the columns

```
8 1 2 0   8 2 3 1   8 3 4 2   8 4 5 3
0 8 1 2   1 8 2 3   2 8 3 4   3 8 4 5
2 0 8 1   3 1 8 2   4 2 8 3   5 3 8 4
1 2 0 8   2 3 1 8   3 4 2 8   4 5 3 8

     8 5 6 4   8 6 0 5   8 0 1 6
     4 8 5 6   5 8 6 0   6 8 0 1
     6 4 8 5   0 5 8 6   1 6 8 0
     5 6 4 8   6 0 5 8   0 1 6 8
```

Similarly, taking $i = 2$ and 3, we get two further sets of 28 columns. To these 84 columns, we adjoin the 7 columns

```
0 1 2 3 4 5 6 7
0 1 2 3 4 5 6 7
0 1 2 3 4 5 6 7
0 1 2 3 4 5 6 7
```

and also the 9 columns of an OA (3,4), say

```
7 7 7 8 8 8 9 9 9
7 8 9 7 8 9 7 8 9
7 8 9 8 9 7 9 7 8
7 8 9 9 7 8 8 9 7
```

which corresponds to the two orthogonal squares

$$L_1 = \begin{array}{|ccc} 7 & 8 & 9 \\ 8 & 9 & 7 \\ 9 & 7 & 8 \end{array} \qquad \text{and} \qquad L_2 = \begin{array}{|ccc} 7 & 8 & 9 \\ 9 & 7 & 8 \\ 8 & 9 & 7 \end{array}$$

Thence we get an orthogonal array OA (10,4) which takes the form

$$\begin{array}{l}
0\,0\,0\,0\,0\,0\,0\,0\,0\,0\,1\,1\,1\,1\,1\,1\,1\,1\,1\,1\,2\,2\,2 \ldots\ldots\ 9\,9 \\
0\,1\,2\,3\,4\,5\,6\,7\,8\,9\,0\,1\,2\,3\,4\,5\,6\,7\,8\,9\,0\,1\,2 \ldots\ldots\ 8\,9 \\
0\,4\,1\,7\,2\,9\,8\,3\,6\,5\,8\,1\,5\,2\,7\,3\,9\,4\,0\,6\,9\,8\,2 \ldots\ldots\ 7\,8 \\
0\,7\,8\,6\,9\,3\,5\,4\,1\,2\,6\,1\,7\,8\,0\,9\,4\,5\,2\,3\,5\,0\,2 \ldots\ldots\ 9\,7
\end{array}$$

and yields the two orthogonal latin squares exhibited in Fig. 11.2.1

$$\begin{array}{|ccccccc ccc}
0 & 4 & 1 & 7 & 2 & 9 & 8 & 3 & 6 & 5 \\
8 & 1 & 5 & 2 & 7 & 3 & 9 & 4 & 0 & 6 \\
9 & 8 & 2 & 6 & 3 & 7 & 4 & 5 & 1 & 0 \\
5 & 9 & 8 & 3 & 0 & 4 & 7 & 6 & 2 & 1 \\
7 & 6 & 9 & 8 & 4 & 1 & 5 & 0 & 3 & 2 \\
6 & 7 & 0 & 9 & 8 & 5 & 2 & 1 & 4 & 3 \\
3 & 0 & 7 & 1 & 9 & 8 & 6 & 2 & 5 & 4 \\
\\
1 & 2 & 3 & 4 & 5 & 6 & 0 & 7 & 8 & 9 \\
2 & 3 & 4 & 5 & 6 & 0 & 1 & 8 & 9 & 7 \\
4 & 5 & 6 & 0 & 1 & 2 & 3 & 9 & 7 & 8
\end{array}
\qquad
\begin{array}{|ccccccc ccc}
0 & 7 & 8 & 6 & 9 & 3 & 5 & 4 & 1 & 2 \\
6 & 1 & 7 & 8 & 0 & 9 & 4 & 5 & 2 & 3 \\
5 & 0 & 2 & 7 & 8 & 1 & 9 & 6 & 3 & 4 \\
9 & 6 & 1 & 3 & 7 & 8 & 2 & 0 & 4 & 5 \\
3 & 9 & 0 & 2 & 4 & 7 & 8 & 1 & 5 & 6 \\
8 & 4 & 9 & 1 & 3 & 5 & 7 & 2 & 6 & 0 \\
7 & 8 & 5 & 9 & 2 & 4 & 6 & 3 & 0 & 1 \\
\\
4 & 5 & 6 & 0 & 1 & 2 & 3 & 7 & 8 & 9 \\
1 & 2 & 3 & 4 & 5 & 6 & 0 & 9 & 7 & 8 \\
2 & 3 & 4 & 5 & 6 & 0 & 1 & 8 & 9 & 7
\end{array}$$

Fig. 11.2.1

The method used by R. C. Bose and S. S. Shrikhande to construct a pair of orthogonal latin squares of order 22 which we referred to above, employs a generalization of the concept of a balanced incomplete block design, called by its authors a *pairwise balanced block design of index unity and type* $(v;\ k_1, k_2, \ldots, k_m)$. Such a design comprises a set of v treatments (elements) arranged in $b = \sum_{i=1}^{m} b_i$ blocks such that there are b_1 blocks each of which contains k_1 treatments, b_2 blocks each of which contains k_2 treatments, . . . , b_m blocks each of which contains k_m treatments ($k_i \leq v$ for $i = 1, 2, \ldots, m$), and in which each pair of the v distinct treatments occur together in exactly one of the b blocks. The latter condition implies that

$$v(v-1) = \sum_{i=1}^{m} b_i k_i (k_i - 1)$$

since each side of this equality is an expression for the total number of (ordered) pairs of distinct treatments which occur among the blocks of the design. Bose and Shrikhande's main theorem[1] is as follows:

THEOREM 11.2.2. *Suppose that there exists a pairwise balanced block design* $\mathfrak{D}$ *of index unity and type* $(v; k_1, k_2, \ldots, k_m)$ *and that, for each* k_i, *a set of* $q_i - 1$ *mutually orthogonal latin squares of that order exists. Then, if*

$$q = \min(q_1, q_2, \ldots, q_m),$$

it is possible to construct a set of $q - 2$ *mutually orthogonal latin squares of order* v.

PROOF. Let the treatments of the design $\mathfrak{D}$ be denoted by $t_1, t_2, \ldots t_v$ and let those blocks of the design (written as columns) each of which contains k_i treatments be denoted by $\gamma_{i1}, \gamma_{i2}, \ldots, \gamma_{ib_i}$. Since there are at least $q - 1$ mutually orthogonal latin squares of order k_i, an orthogonal array $OA(k_i, q+1)$ on the symbols $1, 2, \ldots, k_i$ can be constructed. Moreover, if the set of mutually orthogonal latin squares are put into standard form, it is easy to see (from the method of construction given in theorem 11.1.1) that the orthogonal array will have every entry in its jth column except the first[2] equal to j for $j = 1, 2, \ldots, k_i$. Let these first k_i columns be deleted and also let the first row of the entire array be deleted. Then the remaining matrix array A_{k_i} of q rows and $k_i^2 - k_i$ columns has the property that every ordered pair $\begin{pmatrix} i \\ j \end{pmatrix}$, $i \neq j$, of symbols from the set $1, 2, \ldots, k_i$ occurs exactly once as a column in any two-rowed submatrix of the array. In A_{k_i}, let each occurrence

[1] We shall give a later generalization of this theorem in the next section.
[2] The first row of the orthogonal array is the $i \times k_i^2$ matrix

$$R^* = (1\ 1\ \ldots\ 1\ 2\ 2\ \ldots\ 2\ \ldots\ \ldots\ \ldots\ k_i\ k_i\ \ldots\ k_i).$$

of the symbol j be replaced by that one of the treatments t_r $(1 \leq r \leq v)$ which occurs in the jth place of the block γ_{ip} $(1 \leq p \leq b_i)$ of the design $\mathfrak{D}$. Since $q \leq q_i \leq k_i$, this is always possible. Denote the $q \times (k_i^2 - k_i)$ matrix so obtained by $A_{k_i}(\gamma_{ip})$. Then, from the property of the matrix A_{k_i} that every ordered pair $\begin{pmatrix} i \\ j \end{pmatrix}$, $i \neq j$, of symbols from the set $1, 2, \ldots, k_i$ occurs exactly once as a column in any two-rowed submatrix and from the property of the design $\mathfrak{D}$ that each pair of treatments occurs in exactly one of the blocks, it is clear that the $q \times \sum_{i=1}^{m} b_i k_i(k_i - 1)$ matrix $[A_{k_1}(\gamma_{11}), A_{k_1}(\gamma_{12}), \ldots, A_{k_1}(\gamma_{1b_1}), A_{k_2}(\gamma_{21}), A_{k_2}(\gamma_{22}), \ldots, A_{k_2}(\gamma_{2b_2}), \ldots, \ldots, \ldots, A_{k_m}(\gamma_{m1}), A_{k_m}(\gamma_{m2}), \ldots, A_{k_m}(\gamma_{mb_m})]$ has the property that every ordered pair $\begin{pmatrix} t_r \\ t_s \end{pmatrix}$, $r \neq s$, of treatments from the set $t_1, t_2, \ldots, t_v$ occurs exactly once as a column in each of its two-rowed submatrices. If we now adjoin the $q \times v$ matrix A in which every entry of the rth column is the treatment t_r, $r = 1, 2, \ldots, v$, it is evident that we shall obtain a

$$q \times [v + \sum_{i=1}^{m} b_i k_i(k_i - 1)] = q \times v^2$$

matrix which is an orthogonal array $OA\ (v, q)$. So, as in theorem 11.1.1, $q - 2$ mutually orthogonal latin squares of order v can be constructed.

Bose and Shrikhande made use of their theorem by first observing that the three dimensional finite projective space of order two gives rise to a balanced incomplete block design with parameters $v = 15$, $b = 35$, $r = 7$, $k = 3$, and $\lambda = 1$. Here the varieties are the points and the blocks are the lines. They knew (see R. C. Bose, S. S. Shrikhande, and K. N. Bhattacharya [1]) that this design is resolvable into seven sets of five blocks such that each set of five blocks contains every treatment exactly once. Thence, by adding a new treatment θ_i to each block of the ith set of five blocks $(i = 1, 2, \ldots, 7)$ and then adjoining a single new block consisting of the seven new treatments $\theta_1, \theta_2, \ldots, \theta_7$ they were able to construct the pairwise balanced block design of index unity and type (22; 4, 7) which we exhibit in Fig. 11.2.2. Since there exist three mutually orthogonal latin squares of order four and six of order seven, the construction of the theorem yields an orthogonal array $OA\ (22, 4)$ and hence a pair of orthogonal latin squares of order 22. We shall not give the complete construction but, as an illustration of the method, we shall construct the matrix array $A_4(\gamma_{41})$. (See Fig. 11.2.3.)

Design $\mathfrak{D}$ *of type (22; 4, 7)*

γ_{41}	γ_{42}	γ_{43}	γ_{44}	γ_{45}	γ_{46}	γ_{47}	γ_{48}	γ_{49}	$\gamma_{4(10)}$
θ_1	θ_1	θ_1	θ_1	θ_1	θ_2	θ_2	θ_2	θ_2	θ_2
t_1	t_3	t_6	t_5	t_7	t_2	t_4	t_7	t_6	t_1
t_2	t_8	t_9	t_{11}	t_{14}	t_3	t_9	t_{10}	t_{12}	t_8
t_4	t_{12}	t_{10}	t_{13}	t_{15}	t_5	t_{13}	t_{11}	t_{14}	t_{15}

$\gamma_{4(11)}$	$\gamma_{4(12)}$	$\gamma_{4(13)}$	$\gamma_{4(14)}$	$\gamma_{4(15)}$	$\gamma_{4(16)}$	$\gamma_{4(17)}$	$\gamma_{4(18)}$	$\gamma_{4(19)}$	$\gamma_{4(20)}$
θ_3	θ_3	θ_3	θ_3	θ_3	θ_4	θ_4	θ_4	θ_4	θ_4
t_3	t_5	t_1	t_7	t_2	t_4	t_6	t_2	t_1	t_3
t_4	t_{10}	t_{11}	t_{13}	t_9	t_5	t_{11}	t_{12}	t_{14}	t_{10}
t_6	t_{14}	t_{12}	t_8	t_{15}	t_7	t_8	t_{13}	t_9	t_{15}

$\gamma_{4(21)}$	$\gamma_{4(22)}$	$\gamma_{4(23)}$	$\gamma_{4(24)}$	$\gamma_{4(25)}$	$\gamma_{4(26)}$	$\gamma_{4(27)}$	$\gamma_{4(28)}$	$\gamma_{4(29)}$	$\gamma_{4(30)}$
θ_5	θ_5	θ_5	θ_5	θ_5	θ_6	θ_6	θ_6	θ_6	θ_6
t_5	t_7	t_3	t_2	t_4	t_6	t_1	t_4	t_3	t_5
t_6	t_{12}	t_{13}	t_8	t_{11}	t_7	t_{13}	t_{14}	t_9	t_{12}
t_1	t_9	t_{14}	t_{10}	t_{15}	t_2	t_{10}	t_8	t_{11}	t_{15}

$\gamma_{4(31)}$	$\gamma_{4(32)}$	$\gamma_{4(33)}$	$\gamma_{4(34)}$	$\gamma_{4(35)}$	γ_{71}
θ_7	θ_7	θ_7	θ_7	θ_7	θ_1
t_7	t_2	t_5	t_4	t_6	θ_2
t_1	t_{14}	t_8	t_{10}	t_{13}	θ_3
t_3	t_{11}	t_9	t_{12}	t_{15}	θ_4
					θ_5
					θ_6
					θ_7

Fig. 11.2.2

Latin squares of order 4 and matrix array $A_4(\gamma_{41})$.

$$L_1 = \begin{array}{|cccc} 1 & 2 & 3 & 4 \\ 2 & 1 & 4 & 3 \\ 3 & 4 & 1 & 2 \\ 4 & 3 & 2 & 1 \end{array} \qquad L_2 = \begin{array}{|cccc} 1 & 2 & 3 & 4 \\ 3 & 4 & 1 & 2 \\ 4 & 3 & 2 & 1 \\ 2 & 1 & 4 & 3 \end{array} \qquad L_3 = \begin{array}{|cccc} 1 & 2 & 3 & 4 \\ 4 & 3 & 2 & 1 \\ 2 & 1 & 4 & 3 \\ 3 & 4 & 1 & 2 \end{array}$$

$$OA\,(4,5) = \left\{\begin{array}{cccc|cccccccccccc} 1 & 1 & 1 & 1 & 2 & 2 & 2 & 2 & 3 & 3 & 3 & 3 & 4 & 4 & 4 & 4 \\ 1 & 2 & 3 & 4 & 1 & 2 & 3 & 4 & 1 & 2 & 3 & 4 & 1 & 2 & 3 & 4 \\ 1 & 2 & 3 & 4 & 2 & 1 & 4 & 3 & 3 & 4 & 1 & 2 & 4 & 3 & 2 & 1 \\ 1 & 2 & 3 & 4 & 3 & 4 & 1 & 2 & 4 & 3 & 2 & 1 & 2 & 1 & 4 & 3 \\ 1 & 2 & 3 & 4 & 4 & 3 & 2 & 1 & 2 & 1 & 4 & 3 & 3 & 4 & 1 & 2 \end{array}\right\}$$

$$A_4(\gamma_{41}) = \begin{array}{|cccccccccccc} \theta_1 & t_1 & t_2 & t_4 & \theta_1 & t_1 & t_2 & t_4 & \theta_1 & t_1 & t_2 & t_4 \\ t_1 & \theta_1 & t_4 & t_2 & t_2 & t_4 & \theta_1 & t_1 & t_4 & t_2 & t_1 & \theta_1 \\ t_2 & t_4 & \theta_1 & t_1 & t_4 & t_2 & t_1 & \theta_1 & t_1 & \theta_1 & t_4 & t_2 \\ t_4 & t_2 & t_1 & \theta_1 & t_1 & \theta_1 & t_4 & t_2 & t_2 & t_4 & \theta_1 & t_1 \end{array}$$

Fig. 11.2.3

11.3. Some lower bounds for $N(n)$

Once counter-examples to the Euler conjecture had been constructed by Bose and Shrikhande[1] for $n=22$ and by Parker for $n = 10$, these three authors got together and, using a generalization of theorem 11.2.2 above as their principal tool, were able to show that the number $N(n)$ of mutually orthogonal latin squares of order n is greater than or equal to 2 for all $n \neq 6$. For the original method, the reader is referred to R. C. Bose, S. S. Shrikhande, and E. T. Parker [1]. Subsequently, R. Guérin proved by a similar type of argument that $N(n) \geq 4$ for all $n \geq 53$ (see R. Guérin [5]). Very recently, again using methods of a very similar kind, H. Hanani has shown that $N(n) \geq 5$ for all $n \geq 63$ and, among other results, has reaffirmed that $N(n) \geq 3$ for all $n \geq 52$, a result implicit in the paper of Guérin (see H. Hanani [1]). More recently

(1) In R. C. Bose and S. S. Shrikhande [2], the original counter-example for $n = 22$ was generalized to give pairs of orthogonal latin squares of all orders n of the form $36m + 22$.

still, R. M. Wilson has shown that $N(n) \geq 6$ for all $n > 90$. Again the method used is very similar to that of R. Guérin (see R. M. Wilson [1]).

In this section, we shall first prove Bose, Shrikhande and Parker's main theorem and then use it to derive a number of lower bounds for the integers $N(km)$, $N(km + x)$, $N(km - y)$ and $N(km + x - y)$. We shall make use of these in the following section to give a proof, which substantially follows the methods of R. Guérin, that the Euler conjecture is false for all $n > 6$.

Let BIB$(v; k_1, k_2, \ldots, k_m)$ denote a pairwise balanced block design of index unity having v elements arranged into blocks in such a way that there are b_1 blocks of k_1 elements each, b_2 blocks of k_2 elements each, $\ldots$, b_m blocks of k_m elements each. The set of b_i blocks comprising all those containing k_i elements is called *the* ith *equiblock component* of the design. A union of equiblock components such that no two of the blocks contained in it have an element in common is called a *clear set* of equiblock components. Note that, if a clear set contains blocks of k_i elements then it necessarily contains <u>all</u> the blocks of the design which have this number of elements. That is, we wish to emphasize that a clear set is a union of complete equiblock components.

Example. In the BIB(12; 3, 4) whose blocks are given below, the blocks which contain three elements form a clear set.

$$\begin{array}{llll}
a_2 a_3 b_1 c_1 & a_1 c_1 d_2 d_3 & b_2 b_3 c_1 d_1 & a_1 b_1 d_1 \\
a_3 a_1 b_2 c_2 & a_2 c_2 d_3 d_1 & b_3 b_1 c_2 d_2 & a_2 b_2 d_2 \\
a_1 a_2 b_3 c_3 & a_3 c_3 d_1 d_2 & b_1 b_2 c_3 d_3 & a_3 b_3 d_3 \\
 & & & c_1 c_2 c_3
\end{array}$$

If we think of the elements of a BIB $(v; k_1, k_2, \ldots, k_m)$ as points and its blocks as lines, a point being incident with a line when the element (point) belongs to the block (line), we may, following R. Guérin, regard the block design as a *system of lines* satisfying the axiom that each two points are incident with exactly one line. Blocks with no element in common are represented as parallel lines and a set of parallel lines is a *partial bundle*. (R. Guérin reserves the word *bundle* for a set of parallel lines such that there is one line of the set incident with every point of the system.) A partial bundle N_S is called a *special nucleus*(1) if no line

(1) In [5], R. Guérin uses the term *nucleus* both for a special nucleus and also later in his paper, for an arbitrary partial bundle as well.

not in N_S is incident with the same number of points as any line of N_S. Thus, the concept of a special nucleus is the same as that of a clear set.

The generalization of theorem 11.2.2 which becomes our main tool is as follows:

Theorem 11.3.1. *If there exists a pairwise balanced block design* $\mathfrak{D} = \text{BIB}(v; k_1, k_2, \ldots, k_m)$ *of index unity such that the union of the first, second, . . . , rth equiblock components is a clear set, then*

$$N(v) \geq \min [N(k_1), N(k_2), \ldots, N(k_r), N(k_{r+1}) - 1, \ldots, N(k_m) - 1] .$$

Proof. Let $q - 2 = \min [N(k_1), N(k_2), \ldots, N(k_r), N(k_{r+1}) - 1, \ldots, N(k_m) - 1]$. We deal first with the equiblock components of the block design $\mathfrak{D}$ which do not form part of the clear set. We make use of these in much the same way as in theorem 11.2.2.

Since there exist at least $q - 1$ mutually orthogonal latin squares of order k_i, $i = r + 1, r + 2, \ldots, m$, we can construct an orthogonal array $OA(k_i, q + 1)$ on the symbols $1, 2, \ldots, k_i$. Also, as in theorem 11.2.2, it is possible to form this array in such a way that every entry in its jth column except the first is equal to j for $j = 1, 2, \ldots, k_i$. We delete the entire first row of this $OA(k_i, q + 1)$ and also its first k_i columns and hence form a matrix array A_{k_i} of q rows and $k_i^2 - k_i$ columns with the property that every ordered pair $\binom{i}{j}$, $i \neq j$, of symbols from the set $1, 2, \ldots, k_i$ occurs exactly once as a column in any two-rowed submatrix of the array. If $\gamma_{i1}, \gamma_{i2}, \ldots \gamma_{ib_i}$ are the blocks of the ith equiblock component of the given block design, we form from A_{k_i} a $q \times b_i(k_i^2 - k_i)$ matrix array $B_{k_i} = [A_{k_i}(\gamma_{i1}), A_{k_i}(\gamma_{i2}), \ldots, A_{k_i}(\gamma_{ib_i})]$ where the submatrix $A_{k_i}(\gamma_{ip})$ is obtained by replacing each occurrence of the symbol j in A_{k_i} by that one of the treatments t_r $(1 \leq r \leq v)$ which occurs in the jth place of the block γ_{ip} $(1 \leq p \leq b_i)$ of the design $\mathfrak{D}$.

Next we deal with the blocks of $\mathfrak{D}$ which belong to the clear set. By our definition of q, there exist at least $q - 2$ mutually orthogonal latin squares of order k_i, $i = 1, 2, \ldots, r$, and so we can construct an orthogonal array $OA(k_i, q) = A_{k_i}$, where A_{k_i} is a matrix array of q rows and k_i^2 columns with the property that every ordered pair $\binom{i}{j}$ of symbols, i and j not necessarily distinct, from the set $1, 2, \ldots, k_i$ occurs exactly once as a column in any two-rowed submatrix, of the array. We now form a matrix $B_{k_i} = [A_{k_i}(\gamma_{i1}), A_{k_i}(\gamma_{i2}), \ldots A_{k_i}(\gamma_{ib_i})]$ of q rows and $b_i k_i^2$ columns where the submatrix $A_{k_i}(\gamma_{ip})$ is derived from the matrix A_{k_i} in exactly the same way as was described above, but each γ_{ip} is now a block belonging to the clear set of $\mathfrak{D}$. Since each treatment which occurs in the blocks of the clear set of $\mathfrak{D}$ occurs in only one such block, each ordered pair $\begin{pmatrix} t_i \\ t_j \end{pmatrix}$ occurs only once in any two-rowed submatrix of B_{k_i} if $1 \leq i \leq r$ and where t_j is a treatment occurring in the clear set of $\mathfrak{D}$.

Consider now the matrix $B = [B_1, B_2, \ldots, B_m]$. This has q rows and $\sum_{i=1}^{r} b_i k_i^2 + \sum_{i=r+1}^{m} b_i(k_i^2 - k_i)$ columns and, in any two-rowed submatrix of it, each ordered pair $\begin{pmatrix} t_j \\ t_l \end{pmatrix}$ of distinct treatments of $\mathfrak{D}$ occurs exactly once and each ordered pair $\begin{pmatrix} t_j \\ t_l \end{pmatrix}$, where t_j is a treatment occurring among the blocks of the clear set, occurs exactly once. There exist $v - \sum_{i=1}^{r} b_i k^i$ treatments not in the clear set. If we adjoin to B a further submatrix of q rows and $v - \sum_{i=1}^{r} b_i k_i$ columns each column of which consists of a single treatment repeated q times and such that the treatments occurring in these various columns are those not occurring in the blocks of the clear set of $\mathfrak{D}$, we shall get an orthogonal array $OA(q, v)$ since the final matrix will have $\sum_{i=1}^{r} b_i k_i^2 + \sum_{i=r+1}^{m} b_i(k_i^2 - k_i) + v - \sum_{i=1}^{r} b_i k_i = v^2$ columns and each two of its rows will be v-orthogonal. It follows that at least $q - 2$ mutually orthogonal latin squares exist, as stated in the theorem.

As an example, the projective plane of order 4 is a system of lines satisfying the axiom that each two points are incident with exactly one line. This axiom is unaffected if a subset of the points is deleted. When we delete three points not all on one line we shall consequently still be left with a system of lines in the Guérin sense: that is, we shall have a pairwise balanced block design BIB(18; 5, 4, 3) of index unity. It turns out that the blocks of three points form a clear set, as illustrated in Fig. 11.3.1. Consequently, by theorem 11.3.1,

$$N(18) \geq \min [N(3), N(4) - 1, N(5) - 1] = \min [2, 2, 3] = 2.$$

In other words, there exist at least two mutually orthogonal latin squares of order 18, which provides a further counter-example to Euler's conjecture.

Projective plane of order 4.		BIB(18; 5, 4, 3)
$a_2\ a_3\ b_1\ c_1\ d_1$	$a_1 = (1\ 0\ 0)$	$a_2\ a_3\ b_1\ c_1\ d_1$
$a_3\ a_1\ b_2\ c_2\ d_2$	$a_2 = (0\ 1\ 0)$	$a_3\ a_1\ b_2\ c_2\ d_2$
$a_1\ a_2\ b_3\ c_3\ d_3$	$a_3 = (0\ 0\ 1)$	$a_1\ a_2\ b_3\ c_3\ d_3$
$a_1\ b_1\ e_1\ f_1\ u$	$b_1 = (0\ 1\ 1)$	$b_1\ c_2\ d_3\ e_2\ e_3$
$a_2\ b_2\ e_2\ f_2\ u$	$b_2 = (1\ 0\ 1)$	$b_2\ c_3\ d_1\ e_3\ e_1$
$a_3\ b_3\ e_3\ f_3\ u$	$b_3 = (1\ 1\ 0)$	$b_3\ c_1\ d_2\ e_1\ e_2$
$a_1\ c_1\ e_3\ f_2\ v$	$c_1 = (0\ 1\ \alpha)$	$b_1\ c_3\ d_2\ f_2\ f_3$
$a_2\ c_2\ e_1\ f_3\ v$	$c_2 = (\alpha\ 0\ 1)$	$b_2\ c_1\ d_3\ f_3\ f_1$
$a_3\ c_3\ e_2\ f_1\ v$	$c_3 = (1\ \alpha\ 0)$	$b_3\ c_2\ d_1\ f_1\ f_2$
$a_1\ d_1\ e_2\ f_3\ w$	$d_1 = (0\ \alpha\ 1)$	$a_1\ b_1\ e_1\ f_1$
$a_2\ d_2\ e_3\ f_1\ w$	$d_2 = (1\ 0\ \alpha)$	$a_2\ b_2\ e_2\ f_2$
$a_3\ d_3\ e_1\ f_2\ w$	$d_3 = (\alpha\ 1\ 0)$	$a_3\ b_3\ e_3\ f_3$
$b_1\ c_2\ d_3\ e_2\ e_3$	$e_1 = (\alpha\ 1\ 1)$	$a_1\ c_1\ e_3\ f_2$
$b_2\ c_3\ d_1\ e_3\ e_1$	$e_2 = (1\ \alpha\ 1)$	$a_2\ c_2\ e_1\ f_3$
$b_3\ c_1\ d_2\ e_1\ e_2$	$e_3 = (1\ 1\ \alpha)$	$a_3\ c_3\ e_2\ f_1$
$b_1\ c_3\ d_2\ f_2\ f_3$	$f_1 = (\alpha^2\ 1\ 1)$	$a_1\ d_1\ e_2\ f_3$
$b_2\ c_1\ d_3\ f_3\ f_1$	$f_2 = (1\ \alpha^2\ 1)$	$a_2\ d_2\ e_3\ f_1$
$b_3\ c_2\ d_1\ f_1\ f_2$	$f_3 = (1\ 1\ \alpha^2)$	$a_3\ d_3\ e_1\ f_2$
$b_1\ b_2\ b_3\ v\ w$	$u = (1\ 1\ 1)$	$b_1\ b_2\ b_3$
$c_1\ c_2\ c_3\ w\ u$	$v = (1\ \alpha\ \alpha^2)$	$c_1\ c_2\ c_3$
$d_1\ d_2\ d_3\ u\ v$	$w = (1\ \alpha^2\ \alpha)$	$d_1\ d_2\ d_3$

where $\alpha^3 = 1$, $\alpha \neq 1$.

Fig. 11.3.1

In [5], R. Guérin has described a constructive method for obtaining directly the $N(v)$ mutually orthogonal latin squares whose existence is proved in theorem 11.3.1. We may state his result as follows:

THEOREM 11.3.2. *Let a set of elements (points)* $E = \{x, y, \ldots\}$ *arranged into blocks (lines)* $\mathfrak{L}_1, \mathfrak{L}_2, \ldots, \mathfrak{L}_b$ *be given such that each pair of elements occur in exactly one block and such that the blocks* $\mathfrak{L}_1, \mathfrak{L}_2, \ldots, \mathfrak{L}_j$, $j < b$, *form a clear set (special nucleus), N say. Suppose further that*

(*i*) *there exists a system of at least q mutually orthogonal quasigroups* $(\mathfrak{L}_i, \varphi_{ih})$, $h = 1$ *to q, for each* i, $i = 1, 2, \ldots, j$;

(*ii*) *there exists a system of at least q mutually orthogonal idempotent quasigroups* $(\mathfrak{L}_i, \varphi_{ih})$, $h = 1$ *to q, for each* i, $i = j + 1, j + 2, \ldots, b$.

Let us define on the set E, the q operations θ_h, $h = 1$ *to q, by the rules* (*a*) *for all* $x, y \in \mathfrak{L}_i$, $1 \leq i \leq j$, $x\theta_h y = x\varphi_{ih} y$,

(*b*) *for all* $x, y \in \mathfrak{L}_i$, $j + 1 \leq i \leq b$, $x\theta_h y = x\varphi_{ih} y$ *if* $x \neq y$,

(*c*) *for all* $x \notin N$, $x\theta_h x = x$.

(Since a particular element x occurs in only one block of the special nucleus, $x\theta_h x$ is uniquely defined by definition (*a*) for such an element x.)

Then, the operations θ_h, $h = 1$ *to* q, *define* q *mutually orthogonal quasigroups on the set* E.

PROOF. We shall write $x\theta_h E$, $E\theta_h x$ to denote respectively the sets $\{x\theta_h x_1, x\theta_h x_2, \ldots, x\theta_h x_s\}$, $\{x_1\theta_h x, x_2\theta_h x, \ldots, x_s\theta_h x\}$, where $E = \{x_1, x_2, \ldots \ldots, x_s\}$. Then, in order that (E, θ_h) be a quasigroup, it is necessary and sufficient that, for all $x \in E$, $x\theta_h E = E = E\theta_h x$.

If x is any element of E, let X denote the subset of indices of the set $1, 2, \ldots, b$ such that $x \in \mathfrak{L}_i$ if and only if $i \in X$. Then the family of subsets of points $\{\mathfrak{L}_i \setminus \{x\} / i \in X\}$ form a partition of the set $E \setminus \{x\}$ in virtue of the fact that x occurs with any other element y of E in exactly one block $\mathfrak{L}_i$, $i \in X$.

We separate our proof that $x\theta_h E = E$ into two parts according as x does or does not belong to the special nucleus N.

Case (*A*) *One block* $\mathfrak{L}_\sigma$ *of the nucleus contains* x.

In this case,

$$E = \mathfrak{L}_\sigma \bigcup_{t \in X \setminus \{\sigma\}} (\mathfrak{L}_t \setminus \{x\}).$$

Then, by definitions (*a*) and (*b*),

$$x\theta_h E = (x\varphi_{\sigma h}\mathfrak{L}_\sigma) \bigcup_{t \in X \setminus \{\sigma\}} [x\varphi_{th}(\mathfrak{L}_t \setminus \{x\})].$$

The quasigroups $(\mathfrak{L}_t, \varphi_{th})$ for $t \in X \setminus \{\sigma\}$ are all idempotent whence, for all $t \in X \setminus \{\sigma\}$,

$$x\varphi_{th}(\mathfrak{L}_t \setminus \{x\}) = \mathfrak{L}_t \setminus \{x\}.$$

Also, since $(\mathfrak{L}_\sigma, \varphi_{\sigma h})$ is a quasigroup for each h, $x\varphi_{\sigma h}\mathfrak{L}_\sigma = \mathfrak{L}_\sigma$. It follows that $x\theta_h E = E$.

Case (*B*) *No block of the nucleus contains* x.

In this case,

$$E = \{x\} \bigcup_{t \in X} (\mathfrak{L}_t \setminus \{x\}).$$

By definitions (*b*) and (*c*),

$$x\theta_h E = \{x\} \bigcup_{t \in X} [x\varphi_{th}(\mathfrak{L}_t \setminus \{x\})]$$

and so, since the quasigroups $(\mathfrak{L}_t, \varphi_{th})$ for $t \in X$ are all idempotent, we again have $x\theta_h E = E$.

An exactly similar argument shows that $E\theta_h x = E$ in all cases, whence it follows that (E, θ_h) is a quasigroup for each h.

To show that two quasigroups (E, θ_h) and (E, θ_k), $h \neq k$, $h, k \in 1, 2, \ldots, q$ are orthogonal, it is necessary and sufficient to show that $w\theta_h x = y\theta_h z$ and $w\theta_k x = y\theta_k z$ imply $w = y$ and $x = z$.

Let $\mathfrak{L}_r$ and $\mathfrak{L}_s$ be respectively the blocks which contain the pairs of elements w, x and y, z. In virtue of the definitions (*a*), (*b*), and (*c*), the orthogonality conditions just stated are meaningful only when the blocks $\mathfrak{L}_r$ and $\mathfrak{L}_s$ have an element v in common. (Note that two distinct blocks cannot have more than one element in common.) Since $\mathfrak{L}_r$ and $\mathfrak{L}_s$ have an element v in common, these blocks cannot both belong to the nucleus, so suppose that $\mathfrak{L}_s \notin N$. In that case, the quasigroups $(\mathfrak{L}_s, \varphi_{sh})$ and $(\mathfrak{L}_s, \varphi_{sk})$ are idempotent as well as orthogonal and so $y\varphi_{sh}z = v$ and $y\varphi_{sk}z = v$ imply that $y = z = v$, since only the elements of the leading diagonal can be the same in the multiplication tables of two idempotent orthogonal quasigroups. But then, since $v \in \mathfrak{L}_r$, $v\varphi_{rh}v = v$ and $v\varphi_{rk}v = v$ if the operation φ_{rh} is to correspond to the operation θ_h. Also, $w\varphi_{rh}x = v$ and $w\varphi_{rk}x = v$. Since the quasigroups $(\mathfrak{L}_r, \varphi_{rh})$ and $(\mathfrak{L}_r, \varphi_{rk})$ are orthogonal, we get $w = v$ and $x = v$. Thus $w = v = y$ and $x = v = z$. The alternative possibility is that $\mathfrak{L}_r = \mathfrak{L}_s$ and w, x, y, z all belong to the same block. In that case, the original relations read $w\varphi_{sh}x = y\varphi_{sh}z$ and $w\varphi_{sk}x = y\varphi_{sk}z$. Since the quasigroups $(\mathfrak{L}_s, \varphi_{sh})$ and $(\mathfrak{L}_s, \varphi_{sk})$ are orthogonal, we again have $w = y$ and $x = z$. We conclude that each pair of quasigroups (E, θ_h) and (E, θ_k) are orthogonal. This completes the proof of the theorem.

Since, see theorem 5.3.4, the existence of $q + 1$ mutually orthogonal quasigroups implies the existence of q orthogonal and idempotent quasigroups, the conditions (*i*) and (*ii*) of the theorem are satisfied when $q = \min [N(k_1), N(k_2), \ldots, N(k_r), N(k_{r+1}) - 1, \ldots, N(k_m) - 1]$ where $k_1, k_2, \ldots, k_r$ are the number of elements in the blocks belonging to the special nucleus and $k_{r+1}, \ldots, k_m$ are the numbers of elements in the remaining blocks, whence $N(v) \geq \min [N(k_1), N(k_2), \ldots, N(k_r), N(k_{r+1}) - 1, \ldots, N(k_m) - 1]$ as before, where v is the cardinal of the set E.

A careful examination of the above proof and of the proof of theorem 11.3.1 shows that these are unaffected if the clear set (special nucleus) is replaced by an arbitrary partial bundle (nucleus). If the blocks of the nucleus have numbers of elements $k_1, k_2, \ldots, k_r, k_{r+1}, \ldots, k_s$ and the remaining blocks have numbers of elements $k_{r+1}, \ldots, k_s, k_{s+1}, \ldots, k_m$, we then get

$$N(v) \geq \min [N(k_1), \ldots, N(k_r), N(k_{r+1}), \ldots, N(k_s), N(k_{r+1}) - 1, \ldots, N(k_s) - 1, \ldots, N(k_m) - 1]$$

which is no improvement on our previous result. Nevertheless, it will enable some of our later formulae to be expressed in a more general form. As an illustration, we give an example of the construction of a quasigroup of order 11 by the method of theorem 11.3.2 above using a BIB(11; 4, 3, 2) and a nucleus of blocks of three and two elements which is not a special nucleus.

The BIB(11; 4, 3, 2) is the following and its first three blocks form a nucleus (or partial bundle).

$$\left.\begin{matrix} v & w & \\ b_1 & c_1 & d_1 \\ b_2 & c_2 & d_2 \end{matrix}\right\} \text{nucleus} \qquad \begin{matrix} b_3 & c_3 & d_3 \\ b_1 & c_2 & d_3 \\ b_2 & c_3 & d_1 \\ b_3 & c_1 & d_2 \end{matrix} \qquad \begin{matrix} b_1 & c_3 & d_2 & v \\ b_2 & c_1 & d_3 & v \\ b_3 & c_2 & d_1 & v \end{matrix} \qquad \begin{matrix} b_1 & b_2 & b_3 & w \\ c_1 & c_2 & c_3 & w \\ d_1 & d_2 & d_3 & w \end{matrix}$$

For the construction of the quasigroup of order 11, the idempotent quasigroups

	0	1	2
0	0	2	1
1	2	1	0
2	1	0	2

and

	0	1	2	3
0	0	2	3	1
1	3	1	0	2
2	1	3	2	0
3	2	0	1	3

are used for the blocks not belonging to the nucleus. Also products arising from the elements which are contained in the nucleus are shown enclosed in squares. For example, the quasigroup

	b_1	c_1	d_1
b_1	b_1	c_1	d_1
c_1	c_1	d_1	b_1
d_1	d_1	b_1	c_1

is used for the second block of the nucleus, while the first block not belonging to the nucleus gives rise to the idempotent quasigroup

	b_3	c_3	d_3
b_3	b_3	d_3	c_3
c_3	d_3	c_3	b_3
d_3	c_3	b_3	d_3

We get finally the quasigroup of order 11 shown in Fig. 11.3.2.

	b_1	b_2	b_3	c_1	c_2	c_3	d_1	d_2	d_3	v	w
b_1	$\boxed{b_1}$	b_3	w	$\boxed{c_1}$	d_3	d_2	$\boxed{d_1}$	v	c_2	c_3	b_2
b_2	w	$\boxed{b_2}$	b_1	d_3	$\boxed{c_2}$	d_1	c_3	$\boxed{d_2}$	v	c_1	b_3
b_3	b_2	w	b_3	d_2	d_1	d_3	v	c_1	c_3	c_2	b_1
c_1	$\boxed{c_1}$	v	d_2	$\boxed{d_1}$	c_3	w	$\boxed{b_1}$	b_3	b_2	d_3	c_2
c_2	d_3	$\boxed{c_2}$	v	w	$\boxed{d_2}$	c_1	b_3	$\boxed{b_2}$	b_1	d_1	c_3
c_3	v	d_1	d_3	c_2	w	c_3	b_2	b_1	b_3	d_2	c_1
d_1	$\boxed{d_1}$	c_3	c_2	$\boxed{b_1}$	v	b_2	$\boxed{c_1}$	d_3	w	b_3	d_2
d_2	c_3	$\boxed{d_1}$	c_1	b_3	$\boxed{b_2}$	v	w	$\boxed{c_2}$	d_1	b_1	d_3
d_3	c_2	c_1	c_3	v	b_1	b_3	d_2	w	d_3	b_2	d_1
v	d_2	d_3	d_1	b_2	b_3	b_1	c_2	c_3	c_1	$\boxed{v}$	$\boxed{w}$
w	b_3	b_1	b_2	c_3	c_1	c_2	d_3	d_1	d_2	$\boxed{w}$	$\boxed{v}$

Fig. 11.3.2

Our next step is to give some general methods for constructing pairwise balanced block designs of index unity from geometrical nets.

Let $k \leq N(m) + 1$ ($\leq m$). Then, as we showed in section 8.2, a geometric net of degree $k + 1$ and order m exists. Such a net has $k + 1$ parallel classes of lines and each pair of lines belonging to distinct parallel classes have exactly one point in common. If we adjoin one additional point to all the lines of a parallel class, we shall then have a system satisfying the axiom that each two lines have exactly one point in common. Let the additional points be called $Q_0, Q_1, \ldots, Q_k$ and let the lines through Q_0 be denoted by $l_1, l_2, \ldots, l_m$ as in Fig. 11.3.3. If we interpret the lines as treatments and each set of concurrent lines (pencil of lines) as a block, we shall have a pairwise balanced block design of index unity since each two lines belong to exactly one pencil. If some lines are deleted or if a line l_∞ incident with the points $Q_0, Q_1, \ldots, Q_k$ and no others is adjoined, this property will be unaffected, so we shall still have a block design of the same type.

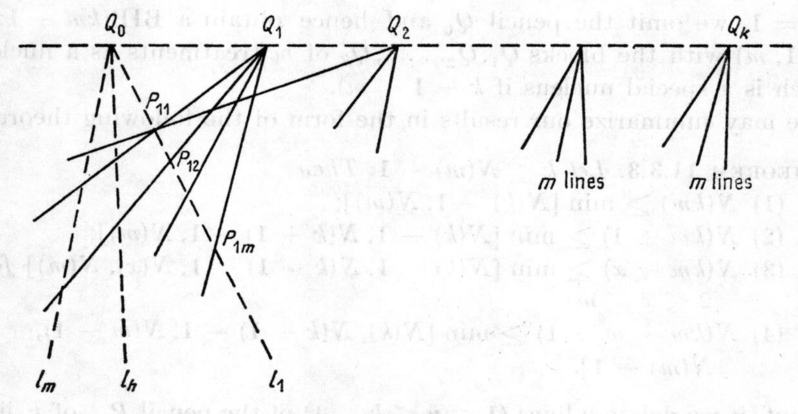

Fig. 11.3.3

First let us note that deletion of the pencil with vertex Q_0 and all its lines leaves us with a BIB(km; k, m) in which the blocks $Q_1, Q_2, \ldots, Q_k$ of m treatments together form a nucleus (partial bundle) which is a special nucleus (clear set) if $k < m$.

Let us denote the points of the line l_h through Q_0 by $P_{h1}, P_{h2}, \ldots, P_{hm}$, for $h = 1, 2, \ldots, m$. From the BIB(km; k, m) just constructed, we can form other BIB's by deletion of further lines or by restoration of some of the lines previously deleted.

In the first place, if we restore the point Q_0 and the lines $l_1, l_2, \ldots, l_x$ ($x \leq m$) through it, the pencil with vertex Q_0 will then have x lines, the pencils P_{hi} ($1 \leq h \leq x$) will each have $k+1$ lines, the pencils P_{hi} ($x < h \leq m$) will each have k lines, and the pencils Q_i ($1 \leq i \leq k$) will each have m lines. We shall thus obtain a design $\mathfrak{D}^*$ which is a BIB($km + x$; k, $k+1$, x, m). The blocks $Q_1, Q_2, \ldots, Q_k$ of m treatments form a nucleus which is a special nucleus if $k + 1 < m$ and $x < m$. Indeed, the blocks $Q_0, Q_1, Q_2, \ldots, Q_k$ of x and m treatments also form a nucleus which is a special nucleus if $k + 1 < m$ and $x \neq k$, $k + 1$. Hence, making use of theorem 11.3.1, we may deduce that

$$N(km + x) \geq \min\left[N(k) - 1, N(k+1) - 1, N(x), N(m)\right]$$

in any event since, if $x = k$, $k + 1$, or m, or if $k + 1 = m$, the value of the right-hand side is not affected. Let us observe also that if $x < m$ the blocks $Q_0, P_{m1}, P_{m2}, \ldots, P_{mm}$ form a second nucleus which is a special nucleus if $x = m - 1$ and $k < m$, $k \neq m - 2$. Hence we have the further result

$$N(km + m - 1) \geq \min [N(k), N(k + 1) - 1, N(m - 1), N(m) - 1].$$

If $x = 1$, we omit the pencil Q_0 and hence obtain a BIB$(km + 1; k, k + 1, m)$ with the blocks $Q_1, Q_2, \ldots, Q_k$ of m treatments as a nucleus (which is a special nucleus if $k + 1 < m$).

We may summarize our results in the form of the following theorem:

Theorem 11.3.3. *Let* $k \leq N(m) + 1$. *Then*

(1) $N(km) \geq \min [N(k) - 1, N(m)]$;

(2) $N(km + 1) \geq \min [N(k) - 1, N(k + 1) - 1, N(m)]$;

(3) $N(km + x) \geq \min [N(k) - 1, N(k + 1) - 1, N(x), N(m)]$ *for* $2 \leq x < m$;

(4) $N(km + m - 1) \geq \min [N(k), N(k + 1) - 1, N(m - 1), N(m) - 1]$.

Next, if we delete y lines $(1 \leq y \leq k - 1)$ of the pencil P_{m1} of k lines from the design $\mathfrak{D}^*$ constructed above, we shall obtain a BIB$(km + x - y;\ k - y, k - 1, k, k + 1, x, m - 1, m)$. For, after the deletion, the pencil P_{m1} will have $k - y$ lines, the pencils P_{mi} $(i \neq 1)$ will each have k lines, the pencils P_{hi} $(x < h < m)$ will have either k or $k - 1$ lines and the pencils P_{hi} $(1 \leq h \leq x)$ either $k + 1$ or k lines since no two of the deleted lines through P_{m1} can have another point in common, the pencils $Q_1, Q_2, \ldots, Q_y$ will have $m - 1$ lines each, and the pencils $Q_{y+1}, Q_{y+2}, \ldots, Q_k$ will each have m lines. The blocks $Q_0, Q_1, \ldots, Q_k$ of $x, m - 1$, and m treatments will form a nucleus, which is special if $k < m - 2$ and $x \neq k - y, k - 1, k, k + 1$. The blocks $Q_0, Q_1, Q_2, \ldots, Q_y$ and P_{m1} of $x, m - 1, k - y$ treatments respectively will also form a nucleus provided $x \neq m$ (special, if $x \neq k - 1, k, k + 1$) since the lines common to the pencils P_{m1} and Q_i $(i = 1, 2, \ldots, y)$ have been deleted.

If, as an alternative to the above, we delete y lines $(1 \leq y \leq m - 1)$ of the pencil Q_1 of m lines from the design $\mathfrak{D}^*$, we shall obtain a BIB $(km + x - y;\ k - 1, k, k + 1, x, m - y, m)$ wherein the blocks of x, $m - y$, and m treatments form a nucleus (special, if $k < m - 1$ and $x, m - y \neq k - 1, k, k + 1$).

Using theorem 11.3.1, we may deduce the following results from the existence of the above designs:

Theorem 11.3.4. *Let* $k \leq N(m) + 1$, *and* $1 \leq y \leq k - 1$. *Then*

(1) $N(km + x - y) \geq \min [N(k - y) - 1, N(k - 1) - 1, N(k) - 1, N(k + 1) - 1, N(x), N(m - 1), N(m)]$ *for* $0 < x \leq m$;

(2) $N(km + x - y) \geq \min [N(k - y), N(k - 1) - 1, N(k) - 1, N(k + 1) - 1, N(x), N(m - 1), N(m) - 1]$ *for* $0 < x < m$; *where in both* (1) *and* (2) *we interpret* $N(1)$ *as infinite.*

Theorem 11.3.5. *Let* $k \leq N(m) + 1$ *and* $1 \leq y \leq m - 1$. *Then* $N(km + x - y) \geq \min [N(k - 1) - 1, N(k) - 1, N(k + 1) - 1, N(x), N(m - y), N(m)]$ *for* $0 < x \leq m$; *where we interpret* $N(1)$ *as infinite.*

Finally, we derive formulae for $N(km - y)$, corresponding to the case when $x = 0$ in theorems 11.3.4 and 11.3.5. We consider again the BIB $(km; k, m)$ obtained from our original $(k + 1)$-net of order m by introducing the additional points $Q_0, Q_1, Q_2, \ldots, Q_k$ and then deleting the pencil with vertex Q_0 and all its lines. From this design we delete either y lines of the pencil P_{m1} $(1 \leq y \leq k - 1)$ or alternatively y lines of the pencil Q_1 $(1 \leq y \leq m - 1)$. In either case, after the deletions all the pencils P_{hi} (except P_{m1} in the case of the first alternative) will have either k or $k - 1$ lines remaining. Hence, arguing in a similar way to that before, we get

Theorem 11.3.6. *Let* $k \leq N(m) + 1$. *Then*

(1) $N(km - y) \geq \min [N(k - y) - 1, N(k - 1) - 1, N(k) - 1, N(m - 1), N(m)]$ *for* $1 \leq y \leq k$;

(2) $N(km - y) \geq \min [N(k - y), N(k - 1) - 1, N(k) - 1, N(m - 1), N(m) - 1]$ *for* $1 \leq y \leq k - 1$;

(3) $N(km - y) \geq \min [N(k - 1) - 1, N(k) - 1, N(m - y), N(m)]$ *for* $1 \leq y \leq m - 1$; *where, in each of* (1), (2), *and* (3), *we interpret* $N(1)$ *as infinite.*

Remark 1. In theorems 11.3.4, 11.3.5, and 11.3.6, the block of the design which has only one element when y has its maximum value or x has its minimum value is omitted. For this reason, $N(1)$ is interpreted as infinite in all the formulae.

Remark 2. Observe that, from the corollary to theorem 11.1.2, we may deduce at once that $N(km) \geq \min [N(k), N(m)]$ for every pair of positive integers k, m. This result is much stronger than that of theorem 11.3.3 (1), so we shall state it as an additional theorem:

Theorem 11.3.7. $N(km) \geq \min [N(k), N(m)]$ *for every pair of positive integers* k, m.

11.4. Falsity of the Euler conjecture for all $n > 6$

We begin by proving the following important result.

Theorem 11.4.1. *If m is an integer such that $N(m) \geq 7$ and $m \geq 16$, then $N(v) \geq 4$ for every integer v lying in the interval $7m + 13 \leq v \leq 8m + 13$.*

Proof. Since $N(m) + 1 \geq 8$, we may take $k = 8$ in the formulae of theorems 11.3.3, 11.3.4, 11.3.5, and 11.3.6.

From theorem 11.3.7, we have

$$N(8m) \geq \min [N(8), N(m)] = 7,$$

since $N(m) \geq 7$ and $N(8) = 7$. (By contrast, theorem 11.3.3(1) gives $N(8m) \geq 6$.)

From theorem 11.3.3 (2), we have

$$N(8m + 1) \geq \min [N(8) - 1, N(9) - 1, N(m)] = 6.$$

From theorem 11.3.3 (3), we have

$$N(8m + x) \geq \min [N(8) - 1, N(9) - 1, N(x), N(m)].$$

Therefore, for $x = 5, 7, 8, 9, 11, 12, 13$, we have $N(8m + x) \geq 4$.

Together, these results show that, under the conditions of the theorem, $N(8m + x) \geq 4$ for all x in the range $0 \leq x \leq 13$ except the values $x = 2, 3, 4, 6, 10$.

From theorem 11.3.5, we have

$$N(8m + x - y) \geq \min [5, N(x), N(m - y)]$$

for all choices of x in the range $0 < x \leq m$, y in the range $1 \leq y \leq m - 1$, and $m \geq 16$. Here we have used the fact that $\min [N(7) - 1, N(8) - 1, N(9) - 1, N(m)] = \min [5, 6, 7, 7] = 5$.

In particular,

$$N(8m + z) = N(8m + h - \overline{h - z}) \geq \min [5, N(h), N(m - h + z)],$$

where $0 < h \leq m$ $(m \geq 16)$ and $1 \leq h - z \leq m - 1$ $(m - 1 \geq 15)$.

For $z = 2, 4, 6, 10$ and m odd, or $z = 3$ and m even, choose $h = 12$ and $h = 16$. Since $N(12) \geq 5$ (see section 7.1) and $N(16) = 15$, we can deduce that

$$N(8m + z) \geq \min [5, N(m - 12 + z)] \text{ and}$$
$$N(8m + z) \geq \min [5, N(m - 16 + z)]$$

for $z = 2, 3, 4, 6, 10$. If m is odd and $z = 2, 4, 6$ or 10, or if m is even and $z = 3$, $m - 12 + z$ and $m - 16 + z$ are both odd and, since their difference is 4, not both are divisible by 3. Therefore, one of them has 5 as smallest possible prime divisor. But if the integer u has 5 as smallest prime divisor, $N(u) \geq 4$ by MacNeish's theorem (see section 11.1). Therefore, $N(8m + z) \geq 4$ for $z = 2, 4, 6, 10$ and m odd or for $z = 3$ and m even.

For $z = 2, 4, 6, 10$ and m even, or $z = 3$ and m odd, choose $h = 11$ and $h = 13$. Since $N(11) = 10$ and $N(13) = 12$, we can deduce that

$$N(8m + z) \geq \min [5, N(m - 11 + z)] \text{ and}$$
$$N(8m + z) \geq \min [5, N(m - 13 + z)]$$

for $z = 2, 3, 4, 6, 10$. If m is even and $z = 2, 4, 6$, or 10, or if m is odd and $z = 3$, $m - 11 + z$ and $m - 13 + z$ are both odd, and, since their difference is 2, not both are divisible by 3. Therefore, as before, one of them has 5 as its smallest possible prime divisor and so at least one of $N(m - 11 + z)$ and $N(m - 13 + z)$ is greater than or equal to 4, whence $N(8m + z) \geq 4$ for $z = 2, 4, 6, 10$ and m even, or $z = 3$ and m odd.

In summary, therefore, we have proved that $N(8m + x) \geq 4$ for all x in the range $0 \leq x \leq 13$. That is, $N(v) \geq 4$ for all v in the range $8m \leq v \leq 8m + 13$.

To complete the proof of the theorem we have to show that $N(8m + z) \geq 4$ for all values of z in the range $13 - m \leq z < 0$ since then $7m + 13 \leq 8m + z < 8m$, implying that $N(v) \geq 4$ for all v in the range $7m + 13 \leq v < 8m$.

Let us write $u = -z$ so that $0 < u \leq m - 13$. As above, we have

$$N(8m - u) = N(8m + h - \overline{h + u}) \geq \min [5, N(h), N(m - h - u)],$$

where $0 < h \leq m$ $(m \geq 16)$ and $1 \leq h + u \leq m - 1$ $(m - 1 \geq 15)$. For $h = 8, 9, 11$ or 12 and $0 < u \leq m - 13$, the inequalities are satisfied, whence, since $N(8), N(9), N(11)$ and $N(13)$ are equal to 7, 8, 10 and 12 respectively, we have

$$N(8m - u) \geq \min [5, N(m - h - u)]$$

for $h = 8, 9, 11$ and 12. When m and u have the same parity (both odd or both even), $m - 9 - u$ and $m - 11 - u$ are both odd and, since their difference is 2, not both are divisible by 3. Hence, as above, at least one of the integers $N(m - 9 - u)$ and $N(m - 11 - u)$ is greater than or equal to 4. When m and u have opposite parity, $m - 8 - u$ and

$m - 12 - u$ are both odd and, since their difference is 4, we again have that not both are divisible by 3. So, at least one of the integers $N(m - - 8 - u)$ and $N(m - 12 - u)$ is greater than or equal to 4. Therefore, in all cases $N(8m - u) \geq 4$ for $0 < u \leq m - 13$. This completes the proof of our theorem.

We remark next that, if m_1 and m_2 are two integers such that $m_1 < m_2$ then the intervals I_{m_1} $(7m_1 + 13 \leq v \leq 8m_1 + 13)$ and I_{m_2} $(7m_2 + + 13 \leq v \leq 8m_2 + 13)$ adjoin each other or overlap if and only if $8m_1 + 13 \geq 7m_2 + 13$. That is, if $m_2 - m_1 \leq m_2/8$ and, a fortiori, if $m_2 - m_1 \leq m_1/8$.

Hence, we can deduce from theorem 11.4.1 that, if m is an integer such that $N(m) \geq 7$ and $m \geq 16$, a sufficient condition in order that the interval $7m + 13 \leq v \leq 8m + 13$ in which $N(v) \geq 4$ can be extended to the right is that there exist an integer m' such that $N(m') \geq 7$ and $m < m' \leq 9m/8$. So, if we can find a sequence of integers $m_1, m_2, \ldots, m_i, \ldots$ satisfying the conditions (a) $N(m_i) \geq 7$ and (b) $m_i < m_{i+1} \leq \leq 9m_i/8$ for $i = 1, 2, \ldots$, it will follow from theorem 11.4.1 above that $N(v) \geq 4$ for all $v \geq 7m_1 + 13$ (provided that $m_1 \geq 16$). As a first step in this direction we have

Theorem 11.4.2. *If m is an integer such that $N(m) \geq 7$ and $m \geq 2^9$, then there exists an integer m' such that $N(m') \geq 7$ and $m < m' \leq 9m/8$.*

Proof. Given the positive integer m, let q be an integer such that $2^q \leq m < 2^{q+1}$. Then $2^{q-3} \leq m/8 < 2^{q-2}$. It follows that

$$2^q \leq m < 9m/8 < 9.2^{q-2}. \quad \ldots\ldots\ldots\ldots\ldots (1)$$

Let us consider the successive integer multiples of 2^{q-6} which lie in the interval $(m, 9m/8)$. They can be written

$$m < 2^{q-6}a < 2^{q-6}(a+1) < \ldots < 2^{q-6}(a+j) < 9\,m/8. \quad \ldots (2)$$

In order to prove the theorem, it suffices to show that, for all integers $q \geq 9$, there exists at least one integer b, $b \in \{a, a+1, \ldots, a+j\}$ such that $N(b) \geq 7$. For, in that case, the integer $m' = 2^{q-6}\,b$ will satisfy $m < m' < 9m/8$ and also we shall have $N(m') \geq \min\,[N(2^{q-6}), N(b)]$, by theorem 11.3.7, $= \min\,[2^{q-6} - 1, N(b)] \geq 7$, since $N(2^{q-6}) = = 2^{q-6} - 1 \geq 7$ for $q \geq 9$.

Since

$$m/2^{q-6} < a < a+1 < \ldots < a+j < (1+\tfrac{1}{8})m/2^{q-6} =$$

$$= (m/2^{q-6}) + (m/2^{q-3})$$

by equation (2) and $m/2^{q-3} \geq 8$ when equation (1) holds, there exist at least eight constructive integers $a, a+1, \ldots, a+7$, which satisfy the equation (2). Since $2^q \leq m < 2^{q-6} a$ implies $a > 64$ and $2^{q-6}(a+7) < < 9m/8 < 9 . 2^{q-2}$ implies $a < 137$, the integer a lies always between these limits when equation (2) holds with $2^q \leq m < 2^{q+1}$. Hence, to complete the proof of the theorem, we have only to check that, for all integers a in the range $64 < a < 137$, there exists an integer b belonging to the set $\{a, a+1, \ldots, a+7\}$ for which $N(b) \geq 7$. A list of integers which are primes or powers of primes is given in the table below (Fig. 11.4.1) and

TABLE OF PRIMES AND PRIME POWERS

$m_i =$	$d_i = m_{i+1} - m_i =$	$m_i =$	$d_i = m_{i+1} - m_i =$	$m_i =$	$d_i = m_{i+1} - m_i =$	$m_i =$	$d_i = m_{i+1} - m_i =$
13	3	101	2	229	4	367	6
$16 = 2^4$	1	103	4	233	6	373	6
17	2	107	2	239	2	379	4
19	4	109	4	241	2	383	6
23	2	113	8	$243 = 3^5$	8	389	8
$25 = 5^2$	2	$121 = 11^2$	4	251	5	397	4
$27 = 3^3$	2	$125 = 5^3$	2	$256 = 2^3$	1	401	8
29	2	127	1	257	6	409	10
31	1	$128 = 2^7$	3	263	6	419	2
$32 = 2^5$	5	131	6	269	2	421	10
37	4	137	2	271	6	431	2
41	2	139	10	277	4	433	6
43	4	149	2	281	2	439	4
47	2	151	6	283	6	443	6
$49 = 7^2$	4	157	6	$289 = 17^2$	4	449	8
53	6	163	4	293	14	457	4
59	2	167	2	307	4	461	2
61	3	$169 = 13^2$	4	311	2	463	4
$64 = 2^6$	3	173	6	313	4	467	2
67	4	179	2	317	14	479	8
71	2	181	10	331	6	487	4
73	6	191	2	337	6	491	8
79	2	193	4	$343 = 7^3$	4	499	4
$81 = 3^4$	2	197	2	347	2	503	6
83	6	199	12	349	4	509	3
89	8	211	12	353	6	$512 = 2^9$	9
97	4	223	4	359	2	521	2
		227	2	$361 = 19^2$	6	523	—

Fig. 11.4.1

we see, by inspection, that no two consecutive members of it which lie between 64 and 137 differ by more than 8, so our theorem holds.

With the aid of theorems 11.4.1 and 11.4.2, we can now prove

Theorem 11.4.3. *For all integers* $v \geq 125$, $N(v) \geq 4$.

Proof. We first prove that $N(v) \geq 4$ for all integers $v \geq 3660$. We observe from the table of primes and prime powers given in Fig. 11.4.1 below that the smallest prime number which exceeds 2^9 is 521. Taking $m = 521$, the hypotheses of theorems 11.4.1 and 11.4.2 are both satisfied and so we may conclude firstly that $N(v) \geq 4$ for all integers v in the range $7.521 + 13 \leq v \leq 8.521 + 13$ and secondly that there exists a positive integer $m' > m$ such that $N(m') \geq 7$ and $7m' + 13 \leq 8m + 13$. The hypotheses of theorems 11.4.1 and 11.4.2 are both satisfied by the integer m' and, by successive repetitions of the argument, we deduce that $N(v) \geq 4$ for all integers $v \geq 7.521 + 13 = 3660$.

Next, we show that for all integers v in the range $482 \leq v \leq 4109$, $N(v) \geq 4$. For this, it suffices to examine again our table giving all primes and powers of primes, say $m_1, m_2, \ldots, m_i, \ldots$, which lie in the interval $67 \leq m_i < 521$. We easily verify that, in the interval $113 \leq m_i < 521$, the maximum distance $d_i = m_{i+1} - m_i$ between successive entries of the table is $d_i = 14 = [113/8]$ where [ ] denotes "integer part". Also, in the interval $67 \leq m_i < 113$, the maximum distance $d_i = 8 = [67/8]$. But, since 67 is a prime, $N(67) = 66 > 7$ and because $d_i = m_{i+1} - m_i \leq [m_i/8]$ for each pair of consecutive prime powers m_i and m_{i+1} in the interval $67 \leq m_i < 521$ (since $[113/8] \leq [m_i/8]$ for every prime power m_i in the first interval and $[67/8] \leq [m_i/8]$ for every prime power m_i in the second interval), each pair of consecutive intervals I_{m_i} $(7m_i + 13 \leq v \leq 8m_i + 13)$ and $I_{m_{i+1}}(7m_{i+1} + 13 \leq v \leq 8m_{i+1} + 13)$ either adjoin each other or overlap, whence $N(v) \geq 4$ for all v in the interval $7.67 + 13 \leq v \leq 8.512 + 13$ by theorem 11.4.1. That is, $N(v) \geq 4$ for $482 \leq v \leq 4109$. Note that 512 is the largest prime power which is less than 521.

Again from our table of primes and prime powers given in Fig. 11.4.1, we may verify that, in the interval $16 \leq m_i < 67$, $d_i = m_{i+1} - m_i \leq m_{i+1}/8$ for all pairs of consecutive entries m_i and m_{i+1} except $m_i = 19$, $m_{i+1} = 23$ and $m_i = 32$, $m_{i+1} = 37$. This, as shown earlier, is sufficient to prove that the pairs of consecutive intervals I_{m_i} and $I_{m_{i+1}}$ overlap with the exception of the pairs of intervals $I_{19}(7.19 + 13 \leq v \leq 8.19 + 13)$, $I_{23}(7.23 + 13 \leq v \leq 8.23 + 13)$ and $I_{32}(7.32 + 13 \leq v \leq$

$\leq 8.32 + 13)$, $I_{37}(7.37 + 13 \leq v \leq 8.37 + 13)$. Hence, using theorem 11.4.1, we have $N(v) \geq 4$ for all integers v in the intervals $125 = 7.16 + + 13 \leq v \leq 8.19 + 13 = 165$, $174 = 7.23 + 13 \leq v \leq 8.32 + 13 = = 269$ and $272 = 7.37 + 13 \leq v \leq 8.64 + 13 = 525$.

It remains to show that $N(v) \geq 4$ for $v = 166, 167, 168, 169, 170, 171, 172, 173$, and for $v = 270, 271$. The integers 167, 169, 173, 271 are either primes or prime powers so $N(v) > 4$ for each of these values of v.

$N(166) = N(8.19 + 16 - 2) \geq \min [N(7) - 1, N(8) - 1, N(9) - 1, N(16), N(17), N(19)] = 5$ by theorem 11.3.5.

$N(168) = N(8.19 + 16) \geq \min [N(8) - 1, N(9) - 1, N(16), N(19)] = = 6$ by theorem 11.3.3(3).

$N(170) = N(8.23 - 14) \geq \min [N(7) - 1, N(8) - 1, N(9), N(23)] = = 5$ by theorem 11.3.6(3).

$N(171) = N(9.19) \geq \min [N(9), N(19)] = 8$ by theorem 11.3.7.

$N(172) = N(8.23 - 12) \geq \min [N(7) - 1, N(8) - 1, N(11), N(23)] = = 5$ by theorem 11.3.6(3).

$N(270) = N(8.32 + 17 - 3) \geq \min [N(7) - 1, N(8) - 1, N(9) - 1, N(17), N(29), N(32)] = 5$ by theorem 11.3.5.

We conclude that the theorem is true.

Theorem 11.4.4. *For all integers v in the range* $53 \leq v < 125$, $N(v) \geq 4$, $N(52) \geq 3$ *and, for all integers v in the range* $7 \leq v < 52$, $N(v) \geq 2$.

Proof. If v is odd and has 5 as its smallest prime divisor, then $N(v) \geq 4$ by MacNeish's theorem. For the first part of the theorem, it remains to examine all other integers which lie between 51 and 125 and are not prime powers. We make use of theorems 11.3.3, 11.3.4, 11.3.5, 11.3.6, 11.3.7 and draw up a table of the results obtained. (See Fig. 11.4.2.)

Since $N(v) \geq 4$ in all cases (excepting that $N(52) \geq 3$), the first part of the theorem is true.

For the second part, we note first that if v is odd (not a prime) or a multiple of 4 then $N(v) \geq 2$ by theorem 11.3.7. Consequently, only integers v of the form $4n + 2$ need be considered. We have shown earlier in the present chapter that each of $N(10)$, $N(18)$, and $N(22)$ is greater than or equal to two. By theorem 11.3.7, $N(30) \geq \min [N(3), N(10)] = 2$, $N(42) \geq \min [N(3), N(14)]$ and $N(50) \geq \min [N(5), N(10)] \geq 2$. Hence, our proof will be complete if we can show that there exist at least two mutually orthogonal latin squares of each of the orders 14, 26, 34, 38 and 46. We find it necessary to treat each of these cases separately.

$v =$	$N(v) \geq$	Theorem	$v =$	$N(v) \geq$	Theorem
$52 = 4.13$	3	11.3.7	$92 = 8.11+7-3$	5	11.3.5
$54 = 7.7+5$	4	11.3.3(3)	$93 = 8.11+5$	4	11.3.3(3)
$56 = 7.8$	6	11.3.7	$94 = 8.11+8-2$	5	11.3.5
$57 = 8.8-7$	5	11.3.6(3)	$96 = 8.12$	7	11.3.7
$58 = 8.8+1-7$	5	11.3.5	$98 = 8.13+1-7$	5	11.3.4(2)
$60 = 5.12$	4	11.3.7	$99 = 9.11$	8	11.3.7
$62 = 8.8+1-3$	4	11.3.5	$100 = 8.13-4$	5	11.3.6(3)
$63 = 7.9$	6	11.3.7	$102 = 8.13-2$	5	11.3.6(3)
$66 = 8.8+5-3$	4	11.3.5	$104 = 8.13$	7	11.3.7
$68 = 8.8+5-1$	4	11.3.5	$105 = 8.13+1$	6	11.3.3(2)
$69 = 8.8+5$	4	11.3.3(3)	$106 = 8.13+8-6$	5	11.3.5
$70 = 8.8+7-1$	5	11.3.5	$108 = 9.12$	8	11.3.7
$72 = 8.9$	7	11.3.7	$110 = 8.13+7-1$	5	11.3.5
$74 = 8.9+5-3$	4	11.3.4(2)	$111 = 8.13+7$	6	11.3.3(3)
$75 = 8.9+7-4$	4	11.3.5	$112 = 7.16$	6	11.3.7
$76 = 8.9+8-4$	4	11.3.5	$114 = 8.13+11-1$	5	11.3.5
$78 = 8.9+7-1$	5	11.3.5	$116 = 8.13+12$	5	11.3.3(3)
$80 = 5.16$	4	11.3.7	$117 = 9.13$	8	11.3.7
$82 = 7.11+5$	4	11.3.3(3)	$118 = 8.16+1-11$	4	11.3.5
$84 = 7.12$	6	11.3.7	$120 = 8.16-8$	5	11.3.6(3)
$86 = 8.11-2$	5	11.3.6(3)	$122 = 8.16+1-7$	5	11.3.5
$87 = 8.11+1-2$	5	11.3.5	$123 = 8.16-5$	5	11.3.6(3)
$88 = 8.11$	7	11.3.7	$124 = 8.16-4$	5	11.3.6(3)
$90 = 8.11+5-3$	4	11.3.5			

Fig. 11.4.2

For $v = 14$, consider the 4×16 matrix A_i defined by

$$A_i = \left(\begin{array}{cccc|cccc} i & x_1 & x_2 & x_3 & 1+i & i & i & i \\ 1+i & i & i & i & 4+i & 4+i & 6+i & 9+i \\ 4+i & 4+i & 6+i & 9+i & 6+i & 1+i & 2+i & 8+i \\ 6+i & 1+i & 2+i & 8+i & i & x_1 & x_2 & x_3 \end{array}\right.$$

$$\left.\begin{array}{cccc|cccc} 4+i & 4+i & 6+i & 9+i & 6+i & 1+i & 2+i & 8+i \\ 6+i & 1+i & 2+i & 8+i & i & x_1 & x_2 & x_3 \\ i & x_1 & x_2 & x_3 & 1+i & i & i & i \\ 1+i & i & i & i & 4+i & 4+i & 6+i & 9+i \end{array}\right)$$

where the second, third, and fourth sets of four columns are obtained by cyclically permuting the rows of the first set of four columns.

If the integers $0, 1, 2, 3, 4, 5, 6, 7, 8, 9, 10, i$ are regarded as residue classes taken modulo 11, we see that among the columns $\begin{pmatrix} j \\ k \end{pmatrix}$ of each two-rowed submatrix of A_i, every non-zero residue occurs exactly once among the ten differences $j - k$ not involving the elements x_1, x_2, x_3. It is easy to deduce from this that every two-rowed submatrix of the 4×176 matrix $A = (A_0, A_1, A_2, \ldots, A_{10})$ contains every ordered pair $\begin{pmatrix} j \\ k \end{pmatrix}$, $j \neq k$, just once as a column and that it also contains each pair $\begin{pmatrix} x_l \\ i \end{pmatrix}$ and each pair $\begin{pmatrix} i \\ x_l \end{pmatrix}$. If we adjoin to this matrix the $OA(3, 4)$ which is defined by the two mutually orthogonal latin squares of order 3 defined on the elements x_1, x_2, x_3, and also adjoin a 4×11 matrix whose ith column contains the residue class i in every place, $i = 0, 1, \ldots, 10$, the resulting 4×196 matrix will be an $OA(14, 4)$ from which two mutually orthogonal latin squares of order 14 can be constructed.

For $v = 26$, we consider the 4×28 matrix A_i defined in a similar way to that given above for $v = 14$ by cyclically permuting the rows of the array

$$\begin{array}{ccccccc}
i & i & i & i & x_1 & x_2 & x_3 \\
3+i & 6+i & 2+i & 1+i & i & i & i \\
8+i & 20+i & 12+i & 16+i & 20+i & 17+i & 8+i \\
12+i & 16+i & 7+i & 2+i & 19+i & 6+i & 21+i
\end{array}$$

This time we regard the various integers as residue classes modulo 23 and we form the matrix $A = (A_0, A_1, A_2, \ldots, A_{22})$. We adjoin to this the same $OA(3, 4)$ as before and also a 4×23 matrix whose ith column contains the residue class i in every place, $i = 0, 1, \ldots, 22$. The resulting matrix has four rows and $(28 \times 23) + 9 + 23 = 676 = (26)^2$ columns and is an $OA(26, 4)$, whence $N(26) \geq 2$.

For $v = 38$, we first construct a balanced incomplete block design (see section 10.2) with parameters $v = 41$, $b = 82$, $k = 5$, $r = 10$, and $\lambda = 1$, as follows. For our elements (treatments) we take the residue classes modulo 41 and as blocks we take the following

$$A_i = (i, i + 1, i + 4, i + 11, i + 29), \qquad i = 0, 1, 2, \ldots, 40,$$

$$B_i = (i + 1, i + 10, i + 16, i + 18, i + 27), \qquad i = 0, 1, 2, \ldots, 40.$$

We observe that every non-zero residue class d arises as a difference $d \equiv x_i - x_j \pmod{41}$ where x_i and x_j both belong either to the set $\{0, 1, 4, 11, 29\}$ or to the set $\{1, 10, 16, 18, 27\}$. Since there exist just

40 such differences, it follows that each pair of elements occur together in just one of the 82 blocks which we have just defined. Since every 2-design is a 1-design by section 10.2, it follows that the design is a BIBD and, since $bk = vr$, we have $r = 10$.

Let us delete three elements which do not occur in the same block, say the elements i, j, and k. The design so produced still has the property that each two of its elements occur together in exactly one block, so it is a pairwise balanced block design of index unity, which we may denote by BIB(38; 5, 4, 3). The three blocks which each contain three elements form a clear set: for suppose two of them contained the element l, say those from which the pairs j, k and k, i have been deleted. In that case, k and l would both occur in both blocks of the original BIBD, a contradiction. From theorem 11.3.1, it now follows that $N(38) \geq \min [N(5) - 1, N(4) - 1, N(3)] = 2$.

Finally, to show that $N(34) \geq 2$ and $N(46) \geq 2$, we take $m = 11$ and $m = 15$ in theorem 11.4.5 which follows below. This completes the proof of the present theorem.

Theorem 11.4.5. *If m is a positive integer such that $N(m) \geq 2$, then $N(3m + 1) \geq 2$.*

Proof. Consider the $4 \times 4m$ matrix A_i defined by

$$A_i = \left(\begin{array}{cccc|cccc|cccc|cccc} i & i & \dots & i & i+1 & i+2 & \dots & i+m & i+2m & i+2m-1 & \dots & i+m+1 & x_1 & x_2 & \dots & x_m \\ i+1 & i+2 & \dots & i+m & i & i & \dots & i & x_1 & x_2 & \dots & x_m & i+2m & i+2m-1 & \dots & i+m+1 \\ i+2m & i+2m-1 & \dots & i+m+1 & x_1 & x_2 & \dots & x_m & i & i & \dots & i & i+1 & i+2 & \dots & i+m \\ x_1 & x_2 & \dots & x_m & i+2m & i+2m-1 & \dots & i+m+1 & i+1 & i+2 & \dots & i+m & i & i & \dots & i \end{array}\right)$$

If the integers $0, 1, 2, \dots, m$ and i are regarded as residue classes taken modulo $2m + 1$, we see that among the columns $\binom{j}{k}$ of each two-rowed submatrix of A_i, every non-zero residue occurs exactly once among the $2m$ differences $j - k$ not involving the elements $x_1, x_2, \dots, x_m$. It is easy to deduce from this that every two-rowed submatrix of the $4 \times 4m(2m+1)$ matrix $A = (A_0, A_1, \dots, A_{2m})$ contains every ordered pair $\binom{j}{k}$, $j \neq k$, just once as a column and that it also contains each pair $\begin{pmatrix} x_l \\ i \end{pmatrix}$ and each pair $\begin{pmatrix} i \\ x_l \end{pmatrix}$. If we adjoin to this matrix the $OA(m, 4)$ which can be constructed from the pair of orthogonal latin squares on the symbols $x_1, x_2, \ldots, x_m$ whose existence is postulated in the statement of the theorem, and also adjoin a $4 \times (2m + 1)$ matrix whose ith column contains the residue class i in every place, $i = 0, 1, \ldots, 2m$, we shall obtain a matrix of four rows and $4m(2m + 1) + m^2 + 2m + 1 = 9m^2 + 6m + 1 = (3m + 1)^2$ columns which is an $OA(3m + 1, 4)$. Hence, $N(3m + 1) \geq 2$.

Since, by MacNeish's theorem, there exist pairs of orthogonal latin squares of every odd order, we may take $m = 4s + 3$ in the above theorem and obtain

THEOREM 11.4.6. *For every positive integer s*, $N(12s + 10) \geq 2$.

This result was used by R. C. Bose, S. S. Shrikhande, and E. T. Parker (see [1]) in their original proof that the Euler conjecture is false for all $n = 6$. We note that theorem 11.4.6 shows at once the existence of pairs of orthogonal latin squares of orders 10, 22, 34, and 46, and that theorem 11.4.5 describes a means of constructing such pairs. In particular, we obtain the pair of squares of order 10 exhibited in Fig. 11.4.3.

On combining the results of theorems 11.4.3 and 11.4.4, we see that $N(v) \geq 4$ for all $v > 52$, that $N(52) \geq 3$, and that $N(v) \geq 2$ for all $v > 6$. The latter result was originally obtained by R. C. Bose, S. S. Shrikhande, and E. T. Parker in [1], as already mentioned, and the former by R. Guérin in [5]. Recently, H. Hanani has defined the integer v_r as being the smallest positive integer with the property that $N(v) \geq r$ for every $v > v_r$. Thus, $v_2 = 6$ and $v_3 \leq 51$. Using a method which requires considerably more enumeration of individual cases than that given above, Hanani has reproved the Guérin result that $v_3 \leq 51$ and has also obtained the further results $v_5 \leq 62$ and $v_{29} \leq 34, 115, 553$ (see H. Hanani [1]). More recently still yet another variant of a proof that $v_2 = 6$ has appeared in R. M. Wilson's paper [1], and it has been shown in the same paper that $v_6 \leq 90$ and $v_3 \leq 46$, the latter being an improvement on R. Guérin's value.

Another relevant result is that obtained by W. H. Mills in [1]. Mills has proved that $N(v) \geq 3$ for all values of $v \equiv 0$ or $1 \bmod 4$. For $v > 46$, this result is already covered by the result $v_3 \leq 46$ proved by R. M. Wilson. However, in regard to values of v less than 46, it is worth noting that Mills' result is an improvement on that given by MacNeish's theorem for the three values $v = 21$, 24 and 33, although $N(33) \geq 3$ had already been proved much earlier in R. C. Bose, S. S. Shrikhande and E. T. Parker [1]. Some further known results are $N(35) \geq 4$, $N(40) \geq 4$ and $N(45) \geq$ ≥ 4. (See F. de Sousa [1] and section 12.5.) These latter results are also a consequence of theorem 11.3.7.

0	6	5	4	x_3	x_2	x_1	1	2	3
x_1	1	0	6	5	x_3	x_2	2	3	4
x_2	x_1	2	1	0	6	x_3	3	4	5
x_3	x_2	x_1	3	2	1	0	4	5	6
1	x_3	x_2	x_1	4	3	2	5	6	0
3	2	x_3	x_2	x_1	5	4	6	0	1
5	4	3	x_3	x_2	x_1	6	0	1	2
2	3	4	5	6	0	1	x_1	x_2	x_3
4	5	6	0	1	2	3	x_2	x_3	x_1
6	0	1	2	3	4	5	x_3	x_1	x_2

0	x_1	x_2	x_3	1	3	5	2	4	6
6	1	x_1	x_2	x_3	2	4	3	5	0
5	0	2	x_1	x_2	x_3	3	4	6	1
4	6	1	3	x_1	x_2	x_3	5	0	2
x_3	5	0	2	4	x_1	x_2	6	1	3
x_2	x_3	6	1	3	5	x_1	0	2	4
x_1	x_2	x_3	0	2	4	6	1	3	5
1	2	3	4	5	6	0	x_1	x_2	x_3
2	3	4	5	6	0	1	x_3	x_1	x_2
3	4	5	6	0	1	2	x_2	x_3	x_1

Fig. 11.4.3

CHAPTER 12

Further constructions of orthogonal latin squares and miscellaneous results

In this chapter, we begin by discussing some further methods for constructing pairs of orthogonal latin squares: in particular, we describe the methods of A. Sade and K. Yamamoto, both of which allow the construction of counter-examples to the Euler conjecture. Next we give some sufficient conditions that a given latin square (or set of squares) be not extendible to a larger set of orthogonal latin squares. We end the chapter with a number of miscellaneous results. Especially, we discuss the problem of constructing latin squares which are orthogonal to their own transposes.

12.1. The direct product and singular direct product of quasigroups

MacNeish's theorem (chapter 11) may be translated into the language of quasigroups by means of the concept of *direct product*. We have the following theorem:

THEOREM 12.1.1. *Let* $(G, \cdot)$ *and* $(G, \odot)$, $(H, \times)$ *and* $(H, \otimes)$ *be pairs of orthogonal quasigroups* (section 5.3) *and let us define the operations* $(\circ)$ *and* $(\circledcirc)$ *on the cartesian product* $F = G \times H$ *by the statements* $(x_1, y_1) \circ (x_2, y_2) = (x_1 . x_2, y_1 \times y_2)$ *and* $(x_1, y_1) \circledcirc (x_2, y_2) = (x_1 \odot x_2, y_1 \otimes y_2)$ *where* $x_1, x_2 \in G$; $y_1, y_2 \in H$. *Then the groupoids* $(F, \circ)$ *and* $(F, \circledcirc)$ *are quasigroups and are orthogonal.*

PROOF. Let us show first that the groupoid $(F, \circ)$ is a quasigroup. The proof that $(F, \circledcirc)$ is also a quasigroup is similar. Note firstly that the equation $(x_1, y_1) \circ (x_2, y_2) = (x_1, y_1) \circ (x_3, y_3)$ can be written $(x_1 . x_2, y_1 \times y_2) = (x_1 . x_3, y_1 \times y_3)$ and is equivalent to the equations $x_1 . x_2 = x_1 . x_3$ and $y_1 \times y_2 = y_1 \times y_3$, which imply $x_2 = x_3$ and $y_2 = y_3$, since $(G, \cdot)$ and $(H, \times)$ are quasigroups. Therefore, $(x_2, y_2) = (x_3, y_3)$

from which we conclude that the equation $(x_1, y_1) \circ (x_2, y_2) = (X, Y)$ is uniquely soluble for (x_2, y_2). It is equally easy to show that it is uniquely soluble for (x_1, y_1), and so $(F, \circ)$ is a quasigroup.

To show the orthogonality of $(F, \circ)$ and $(F, \circledcirc)$ it is necessary and sufficient to show that $(x_1, y_1) \circ (x_2, y_2) = (x_3, y_3) \circ (x_4, y_4)$ and $(x_1, y_1) \circledcirc (x_2, y_2) = (x_3, y_3) \circledcirc (x_4, y_4)$ together imply $(x_1, y_1) = (x_3, y_3)$ and $(x_2, y_2) = (x_4, y_4)$. The equations can be written $(x_1 \cdot x_2, y_1 \times y_2) = (x_3 \cdot x_4, y_3 \times y_4)$ and $(x_1 \odot x_2, y_1 \otimes y_2) = (x_3 \odot x_4, y_3 \otimes y_4)$. That is, $x_1 \cdot x_2 = x_3 \cdot x_4$, $x_1 \odot x_2 = x_3 \odot x_4$, $y_1 \times y_2 = y_3 \times y_4$, $y_1 \otimes y_2 = y_3 \otimes y_4$. The first two of the latter equations imply $x_1 = x_3$ and $x_2 = x_4$ by the orthogonality of the quasigroups $(G, \cdot)$ and $(G, \odot)$ and the last two similarly imply $y_1 = y_3$ and $y_2 = y_4$. The result now follows.

If G has n_1 elements and H has n_2 elements, then F has $n_1 n_2$ elements. If there exist $N(n_1)$ mutually orthogonal latin squares of order n_1, then there exist $N(n_1)$ mutually orthogonal quasigroups of that order. Hence, by forming the direct products of $\min \{N(n_1), N(n_2)\}$ pairs of quasigroups of orders n_1 and n_2, we can obtain an equal number of mutually orthogonal quasigroups of order $n_1 n_2$. Thus, $N(n) \geq \min \{N(n_1), N(n_2)\}$, as we obtained in section 11.1 by an argument in terms of orthogonal arrays. In particular, if $n = \prod_{i=1}^{r} p_i^{\alpha_i}$ where the p_i are distinct primes, we have $N(n) \geq \min (p_i^{\alpha_i} - 1)$, which is MacNeish's theorem.

By use of a generalized form of the above direct product method which he calls "produit direct singulier", A. Sade has shown how a latin square of order $m + lk$ may be constructed from given latin squares L_g, L_r, and L_h; where L_g is a latin square of order $m + l$ which contains a subsquare of order m and L_r, L_h have orders l, k respectively. Taking advantage of the fact that the direct product of two (or more) pairs of orthogonal quasigroups is again a pair of orthogonal quasigroups, as proved above, he has thence been able to obtain sets of orthogonal latin squares of order $m + lk$ in cases when the values of the integers m, l, k are suitable. In particular, the method provides counter-examples to the Euler conjecture. For the details, see A. Sade [15]. We shall give only the two main theorems.

Let $(G, \cdot)$ be a quasigroup of order $n = m + l$ which contains a subquasigroup $(Q, \cdot)$ of order m, and let $(R, \times)$ be a quasigroup of order l defined on the set $R = G \setminus Q$. Let $(H, +)$ be an idempotent quasigroup of order k. The set $T = Q \cup (R \times H)$ is then of order $m + lk$ and we define an operation $(\circ)$ on it in such a way that the structure $(T, \circ)$ is a quasigroup, called by A. Sade the *singular direct product* of the four given quasigroups.

The operation ($\circ$) is defined by the following five statements:

(*i*) for all $q_1, q_2 \in Q$, $\qquad q_1 \circ q_2 = q_1 q_2$;

(*ii*) for all $q \in Q$, $r \in R$, $h \in H$, $\qquad q \circ (r, h) = (qr, h)$ and

$$(r, h) \circ q = (rq, h);$$

(*iii*) for $r_1, r_2 \in R$, $h_1, h_2 \in H$, $h_1 \neq h_2$, $\qquad (r_1, h_1) \circ (r_2, h_2) = (r_1 \times r_2, h_1 + h_2)$;

(*iv*) for $r_1, r_2 \in R$ with $r_1 r_2 \in R$,
and for $h \in H$, $\qquad (r_1, h) \circ (r_2, h) = (r_1 r_2, h)$;

(*v*) for $r_1, r_2 \in R$ with $r_1 r_2 \in Q$,
and for $h \in H$, $\qquad (r_1, h) \circ (r_2, h) = r_1 r_2$.

By way of explanation of these definitions, let us remark firstly that, whereas $r_1 \times r_2 \in R$ for all choices of $r_1, r_2 \in R$, we cannot make the same statement about the products $r_1 r_2$. We have $r_1 r_2 \in G$ but, in general, some of the products will be in Q. However, as may clearly be seen from Fig. 12.1.1, we always have $qr \in R$ and $rq \in R$ whatever the choices of $q \in Q$ and $r \in R$: for, since Q is a quasigroup and since each element of Q is to occur exactly once in each row and each column of the multiplication table of G, no element of Q can occur in either of the subsquares A or B shown in Fig. 12.1.1. Again from Fig. 12.1.1, we may easily see

	q_1 q_2 . . . q_m	r_1 r_2 . . . r_l
q_1 q_2 . . . q_m	Q	A
r_1 r_2 . . . r_l	B	C

Fig. 12.1.1

that the ordered pairs (qr_i, h) and (rr_i, h) for values of r such that $rr_i \in R$ just cover the set $R \times \{h\}$ as q and r vary through the sets Q and R respectively with $r_i \in R$ kept fixed. The same is true of the ordered pairs $(r_i q, h)$ and $(r_i r, h)$.

THEOREM 12.1.2. *The groupoid $(T, \circ)$ defined as above is a quasigroup.*

PROOF. To show that $(T, \circ)$ is a quasigroup, it is necessary and sufficient to show that, for each $t \in T$, $t \circ T = T = T \circ t$.

For each $q \in Q$, we have $qR = R$, as is evident from Fig. 12.1.1. So from (ii), $q \circ (R \times H) = qR \times H = R \times H$. Also, from (i), $q \circ Q = qQ = Q$. Thus, for all $q \in Q$, we have $q \circ T = T$, (*relation A*).

For each element $(r, h) \in R \times H$, we have $r \times R = R$ because $(R, \times)$ is a quasigroup; and $h + (H \setminus \{h\}) = H \setminus \{h\}$ because $(H, +)$ is a quasigroup and $h + h = h$ since it is idempotent. Therefore, by virtue of (iii), $(r, h) \circ [R \times (H \setminus \{h\})] = (r \times R) \times [h + (H \setminus \{h\})] = R \times (H \setminus \{h\})$, (*relation B*).

Observe next that $rg_i \in Q$ with $g_i \in G$ implies $g_i \in R$, since $rq_i \in R$ for every $q_i \in Q$. Using this, we may define $R_r \subset R$ by the statement that $r_i \in R_r$ if and only if $rr_i \in Q$. Then, by virtue of (v), we have $(r, h) \circ (r_i, h) = rr_i$ for every $r_i \in R_r$. It follows that $(r, h) \circ (R_r \times \{h\}) = Q$, (*relation C*).

Now $rG = G$. So, since $R \setminus R_r$ comprises all $r_i \in R$ such that $rr_i \in R$ and since $rq_i \in R$ for every $q_i \in Q$, we have $r(R \setminus R_r) = R \setminus rQ$. Therefore, by (iv), $(r, h) \circ [(R \setminus R_r) \times \{h\}] = r(R \setminus R_r) \times \{h\} = (R \setminus rQ) \times \{h\}$, (*relation D*).

By (ii), we have $(r, h) \circ Q = rQ \times \{h\}$, (*relation E*).

The union of the right-hand sides of relations B, C, D, E is T. Also, the union of the right-hand members of the left sides of relations B, C, D, E is again T, whence $(r, h) \circ T = T$. This, together with relation A, gives $t \circ T = T$ for every $t \in T$. In a similar fashion, we can prove that $T \circ t = T$ for every $t \in T$. Hence T is a quasigroup, as required.

It is now easy to prove (as in A. Sade [15]):

THEOREM 12.1.3. *Let $(G, \cdot)$ and $(G, \odot)$ be orthogonal quasigroups defined on the set G of $n = m + l$ elements and suppose that these quasigroups contain subquasigroups $(Q, \cdot)$ and $(Q, \odot)$ respectively of order m which are themselves orthogonal. Let $(R, \times)$ and $(R, \otimes)$ be orthogonal quasigroups of order l defined on the set $R = G \setminus Q$, and let $(H, +)$ and $(H, \oplus)$ be orthogonal idempotent quasigroups of order k. Then, if $T = Q \cup (R \times H)$, the singular direct product quasigroups $(T, \circ)$ and $(T, \circledcirc)$ are themselves orthogonal.*

PROOF. We have to show that, for any elements $t_1, t_2, t_3, t_4 \in T$, the relations $t_1 \circ t_2 = t_3 \circ t_4$ and $t_1 \circledcirc t_2 = t_3 \circledcirc t_4$ together imply $t_1 = t_3$ and $t_2 = t_4$. When $t_1 \circ t_2$ is in $R \times H$, there are four cases to consider, two in which t_1 and t_2 are both in $R \times H$ and two in which one or the other is in Q. In each case, $t_3 \circ t_4$ must also be in $R \times H$ and so we have (apparently) sixteen cases of this kind. When $t_1 \circ t_2$ is in Q so must $t_3 \circ t_4$ be also, and this, by (*i*) and (*v*), gives three further (apparent) possibilities according as all four of t_1, t_2, t_3, t_4 are in Q, all four are in $R \times H$, or one pair, say t_1 and t_2, are in $R \times H$ and the other pairs are in Q.

We find that some of these cases are impossible and that the others imply the required conclusion that $t_1 = t_2$ and $t_3 = t_4$. We shall illustrate the method by discussing two cases of the first kind and one of the second.

First case. $t_i = (r_i, h_i) \in R \times H$, $i = 1, 2, 3, 4$, and $h_1 \neq h_2$, $h_3 \neq h_4$. We have $(r_1, h_1) \circ (r_2, h_2) = (r_3, h_3) \circ (r_4, h_4)$ and $(r_1, h_1) \circledcirc (r_2, h_2) = (r_3, h_3) \circledcirc (r_4, h_4)$. Thence, $(r_1 \times r_2, h_1 + h_2) = (r_3 \times r_4, h_3 + h_4)$ and $(r_1 \otimes r_2, h_1 \oplus h_2) = (r_3 \otimes r_4, h_1 \oplus h_4)$, implying that $r_1 \otimes r_2 = r_3 \otimes r_4$, $r_1 \otimes r_2 = r_1 \otimes r_4$, and $h_1 + h_2 = h_1 + h_4$, $h_1 \oplus h_2 = h_1 \oplus h_4$. The orthogonality of the pairs of quasigroups $(R, \times)$, $(R, \otimes)$ and $(H, +)$, $(H, \oplus)$ then gives $r_1 = r_3$, $r_2 = r_4$ and $h_1 = h_3$, $h_2 = h_4$, whence $t_1 = t_3$ and $t_2 = t_4$.

Second case. $t_i = (r_i, h_i) \in R \times H$, $i = 1, 2, 3, 4$, and $h_1 = h_2$, $h_3 \neq h_4$. We have $(r_1, h_1) \circ (r_2, h_1) = (r_3, h_3) \circ (r_4, h_4)$, which implies that $(r_1 r_2, h_1) = (r_3 \times r_4, h_3 + h_4)$ if $r_1 r_2 \in R$ and is impossible otherwise. In the first case, $r_1 r_2 = r_3 \times r_4$, $h_1 = h_3 + h_4$ and, similarly, we have $r_1 \odot r_2 = r_3 \otimes r_4$, $h_1 = h_3 \oplus h_4$. Then $h_3 + h_4 = h_1 + h_1$ and $h_3 \oplus h_4 = h_1 \oplus h_1$ because the quasigroups $(H, +)$ and $(H, \oplus)$ are idempotent. But then, because these quasigroups are orthogonal, $h_1 = h_3$, and $h_1 = h_4$ in contradiction to the hypothesis that $h_3 \neq h_4$.

Third case. $t_i = (r_i, h_i) \in R \times H$, $i = 1, 2$, and $t_i = q_i \in Q$, $i = 3, 4$. We have $(r_1, h_1) \circ (r_2, h_2) = q_3 \circ q_4$ which implies that $r_1 r_2 = q_3 q_4$ if $r_1 r_2 \in Q$ and is impossible otherwise. Similarly, we have $r_1 \odot r_2 = q_3 \odot q_4$. Since the quasigroups $(G, \cdot)$ and $(G, \odot)$ are orthogonal, these relations imply $r_1 = q_3$ and $r_2 = q_4$ in contradiction to the fact that $Q \cap R$ is the null set. Hence this case is impossible.

By similar analysis of all the other cases, we are led to conclude that the theorem is true.

From the preceding two theorems, we easily deduce the following one, which was originally given by A. Sade in [15].

Theorem 12.1.4. *If r mutually orthogonal latin squares of order $n = m + l$, each containing a latin subsquare of order m such that the r latin subsquares are also mutually orthogonal, are given and if there exist s mutually orthogonal latin squares of order l and t mutually orthogonal latin squares of order k, then there exist at least $u = \min(r, s, t-1)$ mutually orthogonal latin squares of order $m + lk$.*

Proof. Each of the r latin squares may be used to define the multiplication table of a quasigroup $(G, \cdot)$ of order n which contains a subquasigroup $(Q, \cdot)$ of order m, and each of the s latin squares of order l may be used to specify the multiplication table of a quasigroup $(R, \times)$ defined on the set $R = G \setminus Q$. Also, by theorem 5.3.4, the t mutually orthogonal latin squares of order k give rise to $t-1$ idempotent quasigroups of order k, all defined on the same subset H. Thence, using theorems 12.1.2 and 12.1.3, we can construct u mutually orthogonal quasigroups (and latin squares) of order $m + lk$ on the set $T = Q \cup \cup (R \times H)$.

Corollary. $N(1 + lk) \geq \min[N(1 + l), N(l), N(k) - 1]$.

Proof. To obtain the corollary, it is only necessary to put $m = 1$ in the theorem.

As an example, we have $N(22) = N(1 + 3.7) \geq \min[N(4), N(3), N(7) - 1] = \min(3, 2, 5) = 2$, so there exist at least two orthogonal latin squares of order 22. (Compare section 11.2.)

By the *conjoint* or transpose $(Q, *)$ of a quasigroup $(Q, \cdot)$ is meant the quasigroup which is derived from $(Q, \cdot)$ by the rule $x * y = z$ if and only if $y \cdot x = z$. A. Sade has called a quasigroup *anti-abelian* if it is orthogonal to its conjoint[(1)]: that is, if $xy = zt$ and $yx = tz$ $(x * y = z * t)$ imply $x = z$ and $y = t$. In such a quasigroup, $xy = yx$ is impossible for distinct elements x, y since then the orthogonality condition would give $x = y$. If two quasigroups are both anti-abelian, it is trivial to verify that their direct product is also anti-abelian.

Sade has shown that

Theorem 12.1.5. *If $a, b, a + b$ and $a - b$ are integers prime to h in the ring Z_h of residue classes modulo h, then the law of composition*

(1) It seems worthwhile to point out here that, under certain circumstances, a quasigroup may be isotopic to its conjoint. See theorem 4.2.2 for some cases in which it occurs. Isotopies between a quasigroup and its conjoint have been studied in detail in A. Sade [20].

$x.y = ax + by + c$ *defines an anti-abelian quasigroup* $(Z_h, \cdot)$ *on the set* Z_h.

Proof. The proof is trivial. The conditions $xy = zt$ and $yx = tz$ become $ax + by \equiv az + bt \pmod h$ and $ay + bx \equiv at + bz \pmod h$. Therefore, $a(x - z) \equiv b(t - y) \pmod h$ and $b(x - z) \equiv a(t - y)$. Hence, $ab(x - z) \equiv \equiv b^2(t - y) \equiv a^2(t - y)$. In other words, $(a - b)(a + b)(t - y) \equiv 0 \pmod h$, and so $y = t$ and $x = z$.

Since a, b, $a + b$ and $a - b$ are all to be prime to h, so also must be the integers $2a$, $3a$, $2b$, and $3b$. Consequently, h has to be prime to 6 for these conditions to hold. For all such integers h anti-abelian quasigroups, and consequently pairs of orthogonal latin squares of order h, actually exist. Sade takes as an example, $a = 2$ and $b = -1$, from which we see that the operation $x.y = 2x - y$ gives an anti-abelian (and idempotent) quasigroup whenever h is an integer prime to 6.

The above results and also the following one will all be found in A. Sade [15].

Theorem 12.1.6. *In any Galois field* $GF[p^n]$, *every quasigroup defined on the elements of the field by a law of composition of the form* $x.y = ax + by + c$, *where* a *and* b *are non-zero elements of the field such that* $a^2 - b^2 \neq 0$, *is anti-abelian. Moreover, such quasigroups exist in all finite fields except* $GF[2]$ *and* $GF[3]$.

Proof. Exactly as in theorem 12.1.5, the conditions $xy = zt$ and $yx = tz$ together imply $(a^2 - b^2)(t - y) = 0$ if a and b are non-zero. Then, provided that $a^2 - b^2 \neq 0$, we have $y = t$ and $x = z$, as required for an anti-abelian quasigroup. In any field except $GF[2]$ and $GF[3]$ non-zero elements a and b with distinct squares exist. For example, we can take $a = 1$ and b any element of the field distinct from $0, 1, -1$.

Recently, C. C. Lindner has made use of the singular direct product to obtain a number of interesting results, and he has also generalized the construction in several ways. In his earliest paper [3] on the subject, Lindner has proved that the singular direct product preserves the Stein identity $x(xy) = yx$ (identity (11) of section 2.1). As was pointed out in section 5.3, quasigroups which satisfy the Stein identity (called Stein quasigroups) are orthogonal to their own transposes (that is, they are anti-abelian) and, using this fact, Lindner has constructed several new classes of quasigroups having this property. (We give further details in section 12.4.) In a subsequent paper [12], he has investigated the general question of what kinds of quasigroup identity are preserved by the singular direct product.

In two further papers [4] and [11], Lindner has generalized the singular direct product in two different ways. Later in [13] and [17], he has proved some results similar in nature to those of [12], but applicable to a modification of the first of these generalizations of the singular direct product.

Lindner's first generalization of the singular direct product is as follows:

First generalized singular direct product.

Let $(G, \cdot)$ be a quasigroup of order $n = m + l$ which contains a subquasigroup $(Q, \cdot)$ of order m and let R denote the set $G \setminus Q$. Let $(H, +)$ be an idempotent quasigroup of order k. For each ordered pair of distinct elements h_1, h_2 of H let $\varphi_{h_1h_2}$ be a binary operation defined on the set R such that $(R, \varphi_{h_1h_2})$ is a quasigroup. We now define a binary operation $(\circ)$ on the set $T = Q \cup (R \times H)$ in such a way that $(T, \circ)$ is a quasigroup. This quasigroup $(T, \circ)$ of order $m + lk$ is the first generalized singular direct product of the $k^2 - k$ quasigroups $(R, \varphi_{h_1 h_2})$ and the quasigroups $(G, \cdot)$, $(Q, \cdot)$ and $(H, +)$. We have:

(*i*) for all $q_1, q_2 \in Q$, $\quad q_1 \circ q_2 = q_1 q_2$;

(*ii*) for all $q \in Q$, $r \in R$, $h \in H$, $\quad q \circ (r, h) = (qr, h)$ and $(r, h) \circ q = (rq, h)$;

(*iii*) for $r_1, r_2 \in R, h_1, h_2 \in H, h_1 \neq h_2$, $\quad (r_1 h_1) \circ (r_2 h_2) = (r_1 \varphi_{h_1h_2} r_2, h_1 + h_2)$;

(*iv*) for $r_1, r_2 \in R$ with $r_1 r_2 \in R$, and for $h \in H$, $\quad (r_1, h) \circ (r_2, h) = (r_1 r_2, h)$;

(*v*) for $r_1, r_2 \in R$ with $r_1 r_2 \in Q$, and for $h \in H$, $\quad (r_1, h) \circ (r_2, h) = r_1 r_2$.

We note that if all the $k^2 - k$ binary operations $\varphi_{h_i h_j}$ coincide, the above generalized singular direct product reduces to the singular direct product of A. Sade as previously defined.

Lindner's second generalization of the singular direct product is as follows:

Second generalized singular direct product.

Let G be a set of order $n = m + l$ which contains a subset Q of order m and let $(R, \times)$ be a quasigroup of order l defined on the set $R = G \setminus Q$.

Let $(H,+)$ be an idempotent quasigroup of order k. For each element h_i of H let φ_{h_i} be a binary operation defined on the set G such that (G, φ_{h_i}) is a quasigroup. Further suppose that on the subset Q of G the effect of all the operations φ_{h_i} is the same: say $q_1 \varphi_{h_i} q_2 = q_1 q_2$ for all $h_i \in H$, and that $(Q, \cdot)$ is a subquasigroup of each one of the k quasigroups (G, φ_{h_i}). We define a binary operation on the set $T = Q \cup (R \times H)$ by the following rules:

(*i*) for all $q_1, q_2 \in Q$, $\quad q_1 \circ q_2 = q_1 q_2$;

(*ii*) for all $q \in R$, $r \in R$, $h \in H$, $\quad q \circ (r, h) = (q \varphi_h r, h)$ and $(r, h) \circ q = (r \varphi_h q, h)$;

(*iii*) for $r_1, r_2 \in R$, $h_1, h_2 \in H, h_1 \neq h_2$, $\quad (r_1, h_1) \circ (r_2, h_2) = (r_1 \times r_2, h_1 + h_2)$;

(*iv*) for $r_1, r_2 \in R$ with $r_1 r_2 \in R$, and for $h \in H$, $\quad (r_1, h) \circ (r_2, h) = (r_1 \varphi_h r_2, h)$;

(*v*) for $r_1, r_2 \in R$ with $r_1 r_2 \in Q$ and for $h \in H$, $\quad (r_1, h) \circ (r_2, h) = r_1 \varphi_h r_2$.

$(T, \circ)$ is then a quasigroup called the second generalized singular direct product of the k quasigroups (G, φ_h) and the quasigroups $(Q, \cdot)$, $(R, \times)$ and $(H, +)$.

We note that if all the k binary operations φ_{h_i} coincide on the whole of the set G (instead of only on the subset Q), the above generalized singular direct product reduces to the singular direct product of A. Sade as previously defined.

In [4], C. C. Lindner has used the first generalized singular direct product to construct quasigroups which are orthogonal to their transposes. Further details are given in section 12.4. In [11], he has used the second generalized singular direct product to construct Steiner quasigroups[1] and also to generalize E. H. Moore's construction for Steiner triple systems.

In [5], Lindner has used a third generalization of the singular direct product which combines the two generalizations just defined to construct a large number of latin squares each of which is orthogonal to a given latin square L and no two of which are isomorphic. In particular, he has proved:

[1] The concept of a Steiner quasigroup is defined in section 2.2.

Theorem 12.1.7. *If there are s mutually orthogonal quasigroups of order v, t mutually orthogonal quasigroups of order q containing t mutually orthogonal subquasigroups of order p, and r mutually orthogonal quasigroups of order $q - p$, then there is a quasigroup of order $v(q - p) + p$ having at least $(s - 2)(t - 1)^v(r - 1)^{v^2-v}$ orthogonal complements no two of which are isomorphic.*

As an example, Lindner points out that, since $17 = 4(5 - 1) + 1$ and there are three mutually orthogonal latin squares (quasigroups) of order 4 and four mutually orthogonal latin squares (quasigroups) of order 5 containing four mutually orthogonal subquasigroups of order 1, there is a latin square of order 17 having at least 331, 776 isomorphically distinct orthogonal mates.

In a combined paper (C. C. Lindner and N. S. Mendelsohn [1]), Lindner and Mendelsohn have used the first generalized singular direct product to construct pairs of perpendicular Steiner quasigroups. Two commutative quasigroups $(Q, \cdot)$ and $(Q, \odot)$ of order n are said to be *perpendicular* if there exist exactly $n(n + 1)/2$ distinct ordered pairs $(x \cdot y, x \odot y)$ as x, y vary in Q. This generalizes the concept of orthogonal quasigroups defined in section 5.3. Commutative quasigroups cannot be orthogonal since the maximum possible number of distinct ordered pairs $(x \cdot y, x \odot y)$, is $n(n + 1)/2$ instead of n^2, as required for orthogonality. We shall mention some further results concerning perpendicular commutative quasigroups in section 12.5.

The second generalized singular direct product has been employed by C. C. Lindner and T. H. Straley in their joint paper [1].

12.2. K. Yamamoto's construction

In [8], K. Yamamoto has introduced a new kind of direct product construction for obtaining a set of t mutually orthogonal latin squares of order $m + n$ from suitable sets of $t + 1$ mutually orthogonal latin squares of order m and t mutually orthogonal latin squares of order n, where $tn \leq m$. The squares of orders m and n have to satisfy certain conditions, which we shall explain. The method has been generalized and also re-presented in a more easily understood form by R. Guérin in [5], and this author has christened it "Yamamoto's method". Quite recently, the method has been rediscovered by A. Hedayat and E. Seiden (see [1], [2] and [4]) but these authors write as if unaware of Yamamoto's and Guérin's earlier work.

For clarity and simplicity of explanation, we shall first discuss only the case $t = 2$.

Let H_1, H_2, and H_3 be three mutually orthogonal latin squares of order $m \geq 2n$ defined on a set H of m elements. We require $n \geq 3$ as a necessary condition so that we can have a pair K_1 and K_2 of mutually orthogonal latin squares of order n. We take the latter to be defined on a set K of n elements such that $H \cap K$ is empty. Since $n \geq 3$, $m \geq 7$ and we also require m to be an integer for which triads of mutually orthogonal latin squares exist. We use the square H_3 to determine $2n$ disjoint transversals which are common to the squares H_1 and H_2. From H_1 and K_1 we form a latin square L_1 of order $m + n$ by a process very similar to the method of prolongation described in section 1.4. We replace the elements of the cells of a set of n disjoint transversals of H_1 by the elements of K_1, all the elements of the cells of any one transversal being replaced by the same element of K_1 and the transversals in question being from the set of $2n$ which are common to H_1 and H_2. The elements displaced from the ith transversal become the first m elements of the $(m + i)$th column of the enlarged square L_1 and are written into that column in the order in which they appear in the rows of H_1. These same elements become the first m elements of the $(m + i)$th row of L_1 and are written into that row in the order in which they appear in the columns of H_1. The square L_1 is completed by inserting the complete square K_1 into its bottom right hand corner. A square L_2, also of order $m + n$, is formed from the squares H_2 and K_2 in an exactly similar way but this time using the second set of n transversals which are common to H_1 and H_2.

In general, the squares L_1 and L_2 so formed will not be orthogonal, as is clear from the example given in Fig. 12.2.1 where, for instance, the ordered pair (5, 6) appears both in the cell of the second row and fifth column and in the cell of the tenth row and seventh column when L_1 and L_2 are juxtaposed. However, it is immediately evident from the mode of construction that, as the cells of the subsquare formed by the first m rows and first m columns of the juxtaposed pair are traversed, all ordered pairs of each of the sets $H \times K$ and $K \times H$ will be included exactly once among the set of ordered pairs obtained and that, as the cells of the subsquare formed by the last n rows and n columns are traversed, all ordered pairs of the set $K \times K$ also will be obtained exactly once. We need therefore to determine sufficient conditions on the structures of the squares H_1, H_2 and H_3 in order that the remaining ordered pairs be all different and hence exactly cover the set $H \times H$.

$$H_1 = \left|\begin{array}{ccccccc}
0 & 1_1 & 2_2 & 3_3 & 4 & 5 & 6 \\
1_3 & 2 & 3 & 4 & 5 & 6_1 & 0_2 \\
2 & 3 & 4_1 & 5_2 & 6_3 & 0 & 1 \\
3_2 & 4_3 & 5 & 6 & 0 & 1 & 2_1 \\
4 & 5 & 6 & 0_1 & 1_2 & 2_3 & 3 \\
5_1 & 6_2 & 0_3 & 1 & 2 & 3 & 4 \\
6 & 0 & 1 & 2 & 3_1 & 4_2 & 5_3
\end{array}\right. \qquad
H_2 = \left|\begin{array}{ccccccc}
0 & 1 & 2 & 3 & 4_4 & 5_5 & 6_6 \\
2 & 3_4 & 4_5 & 5_6 & 6 & 0 & 1 \\
4_6 & 5 & 6 & 0 & 1 & 2_4 & 3_5 \\
6 & 0 & 1_4 & 2_5 & 3_6 & 4 & 5 \\
1_5 & 2_6 & 3 & 4 & 5 & 6 & 0_4 \\
3 & 4 & 5 & 6_4 & 0_5 & 1_6 & 2 \\
5_4 & 6_5 & 0_6 & 1 & 2 & 3 & 4
\end{array}\right.$$

$$H_3 = \left|\begin{array}{ccccccc}
0 & 1 & 2 & 3 & 4 & 5 & 6 \\
3 & 4 & 5 & 6 & 0 & 1 & 2 \\
6 & 0 & 1 & 2 & 3 & 4 & 5 \\
2 & 3 & 4 & 5 & 6 & 0 & 1 \\
5 & 6 & 0 & 1 & 2 & 3 & 4 \\
1 & 2 & 3 & 4 & 5 & 6 & 0 \\
4 & 5 & 6 & 0 & 1 & 2 & 3
\end{array}\right.$$

$$K_1 = \left|\begin{array}{ccc}
7 & 8 & 9 \\
8 & 9 & 7 \\
9 & 7 & 8
\end{array}\right. \qquad
K_2 = \left|\begin{array}{ccc}
7 & 8 & 9 \\
9 & 7 & 8 \\
8 & 9 & 7
\end{array}\right.$$

$$L_1 = \left|\begin{array}{cccccccccc}
0 & 7 & 8 & 9 & 4 & 5 & 6 & 1 & 2 & 3 \\
9 & 2 & 3 & 4 & 5 & 7 & 8 & 6 & 0 & 1 \\
2 & 3 & 7 & 8 & 9 & 0 & 1 & 4 & 5 & 6 \\
8 & 9 & 5 & 6 & 0 & 1 & 7 & 2 & 3 & 4 \\
4 & 5 & 6 & 7 & 8 & 9 & 3 & 0 & 1 & 2 \\
7 & 8 & 9 & 1 & 2 & 3 & 4 & 5 & 6 & 0 \\
6 & 0 & 1 & 2 & 7 & 8 & 9 & 3 & 4 & 5 \\
5 & 1 & 4 & 0 & 3 & 6 & 2 & 7 & 8 & 9 \\
3 & 6 & 2 & 5 & 1 & 4 & 0 & 8 & 9 & 7 \\
1 & 4 & 0 & 3 & 6 & 2 & 5 & 9 & 7 & 8
\end{array}\right.$$

$$L_2 = \left|\begin{array}{cccccccccc}
0 & 1 & 2 & 3 & 7 & 8 & 9 & 4 & 5 & 6 \\
2 & 7 & 8 & 9 & 6 & 0 & 1 & 3 & 4 & 5 \\
9 & 5 & 6 & 0 & 1 & 7 & 8 & 2 & 3 & 4 \\
6 & 0 & 7 & 8 & 9 & 4 & 5 & 1 & 2 & 3 \\
8 & 9 & 3 & 4 & 5 & 6 & 7 & 0 & 1 & 2 \\
3 & 4 & 5 & 7 & 8 & 9 & 2 & 6 & 0 & 1 \\
7 & 8 & 9 & 1 & 2 & 3 & 4 & 5 & 6 & 0 \\
5 & 3 & 1 & 6 & 4 & 2 & 0 & 7 & 8 & 9 \\
1 & 6 & 4 & 2 & 0 & 5 & 3 & 9 & 7 & 8 \\
4 & 2 & 0 & 5 & 3 & 1 & 6 & 8 & 9 & 7
\end{array}\right.$$

Fig. 12.2.1

Let us first investigate the structure of a single one of the squares L_1 or L_2 in more detail. The latin squares H_1 and K_1 by means of which L_1 is constructed may be regarded as arising from the multiplication tables of two quasigroups $(H, \cdot)$ and $(K, \times)$, where $H = \{h_1, h_2, \ldots, h_m\}$ and $K = \{k_1, k_2, \ldots, k_n\}$. The latin square H_3 arises from the multiplication table of a third quasigroup (H, σ) which is orthogonal to $(H, \cdot)$ and is used to define n disjoint transversals $T_1, T_2, \ldots, T_n$ of the latin square H_1 by the definition $T_l = \{(i, j) : h_i \sigma h_j = h_l^*\}$ where $l = 1, 2, \ldots, n$ and $h_1^*, h_2^*, \ldots, h_n^*$ are n chosen elements of the set H. If L denotes the union of the sets H and K, the latin square L_1 can be regarded as the multiplication table of the quasigroup $(L, \circ)$ where the operation $(\circ)$ is defined by the five statements:

(*i*) for all $h_i, h_j \in H$ such that $(i, j) \notin \bigcup_{l=1}^{n} T_l$, $\quad h_i \circ h_j = h_i . h_j$,

(*ii*) for all $h_i, h_j \in H$ such that $(i, j) \in T_l$, $\quad h_i \circ h_j = k_l$, $(l = 1, 2, \ldots, n)$

(*iii*) for all $k_i, k_j \in K$, $\quad k_i \circ k_j = k_i \times k_j$,

(*iv*)′ for $h_i \in H$ and $k_l \in K$, let j be defined by $(i, j) \in T_l$, then $\quad h_i \circ k_l = h_i . h_j$,

(*v*)′ for $k_l \in K$ and $h_j \in H$, let i be defined by $(i, j) \in T_l$, then $\quad k_l \circ h_j = h_i . h_j$.

We notice at this point that an additional degree of freedom may be introduced by observing that the "latinness" of the square L_1 is unaffected if the leading m-tuples of any two of the last n columns are interchanged or if, independently, the leading m-tuples of any two of the last n rows are interchanged. Either kind of interchange corresponds to a re-numbering of the transversals $T_1, T_2, \ldots, T_n$. If such rearrangements are effected, the statements (*iv*)′ and (*v*)′ become modified to (*iv*) and (*v*) below:

(*iv*) for $h_i \in H$ and $k_l \in K$ let j be defined by $(i, j) \in T_{l\theta}$, then $\quad h_i \circ k_l = h_i . h_j$,

(*v*) for $k_l \in K$ and $h_j \in K$ let i be defined by $(i, j) \in T_{l\varphi}$, then $\quad k_l \circ h_j = h_i . h_j$.

Here θ and φ are permutation mappings of the set of integers $1, 2, \ldots, n$.

If the latin squares H_2 and K_2 arise from the multiplication tables of two quasigroups $(H, \odot)$ and $(K, \otimes)$ respectively orthogonal to $(H, \cdot)$ and $(K, \times)$, the latin square L_2 can be regarded as the multiplication table of a quasigroup $(L, \circledcirc)$, where the operation $(\circledcirc)$ is defined in a similar manner to the operation $(\circ)$ above. However, in the construction of $(L, \circledcirc)$ we use a set of transversals $T'_l = \{(i, j) : h_i \sigma h_j = h_l^{\otimes}\}$, $l = 1, 2, \ldots, n$, where the subset $\{h_1^{\otimes}, h_2^{\otimes}, \ldots, h_n^{\otimes}\}$ of H has no members in common with the subset $\{h_1^*, h_2^*, \ldots, h_n^*\}$. Also, when writing the statements corresponding to (iv) and (v) above, we replace the permutations θ, φ of the integers $1, 2, \ldots, n$ by two different permutations θ', φ'.

The quasigroups $(L, \circ)$ and $(L, \circledcirc)$ and hence the latin squares L_1 and L_2 will be orthogonal provided that, as the ordered pair (x, y) of $L \times L$ traverses the set $L \times L$, so also does the ordered pair $(x \circ y, x \circledcirc y)$. From statement (iii), we see that the ordered pair $(x \circ y, x \circledcirc y)$ traverses the set $K \times K$ and, from statement (ii), together with the fact that the subsets $\{h_1^*, h_2^*, \ldots, h_n^*\}$ and $\{h_1^{\otimes}, h_2^{\otimes}, \ldots, h_n^{\otimes}\}$ are disjoint, we see that it also traverses the sets $H \times K$ and $K \times H$. From statements (i), (iv) and (v), we conclude that $(L, \circ)$ and $(L, \circledcirc)$ are orthogonal quasigroups provided that $A = B_1 \cup B_2$, where

$$A = \{(h_i . h_j, h_i \odot h_j) : (i, j) \in (\bigcup_{l=1}^{n} T_l) \cup (\bigcup_{l=1}^{n} T'_l)\},$$

$$B_1 = \{(h_i . h_j, h_i \odot h_{j'}) : (i, j, i, j') \in \bigcup_{l=1}^{n} (T_{l\theta} \times T'_{l\theta'})\},$$

$$B_2 = \{(h_i . h_j, h_{i'} \odot h_j) : (i, j, i', j) \in \bigcup_{i=1}^{n} (T_{l\varphi} \times T'_{l\varphi'})\}.$$

Now let $R = (H, +, \cdot)$ be an associative and commutative ring defined on the set H, and let multiplication in this ring be indicated by juxtaposition of elements. Let $\alpha_1, \beta_1, \alpha_2, \beta_2$ be four arbitrarily chosen elements of this ring and define the three quasigroups $H_1 = (H, \cdot)$, $H_2 = (H, \odot)$, and $H_3 = (H, \sigma)$ by the following equations: $x \sigma y = x + y$, $x . y = \alpha_1 x + \beta_1 y$, $x \odot y = \alpha_2 x + \beta_2 y$. Then H_1 and H_2 will be quasigroups provided that $\alpha_1, \beta_1, \alpha_2, \beta_2$ are all regular elements of the ring R. That is, if each of the mappings $x \to \alpha_1 x$, $x \to \alpha_2 x$, $x \to \beta_1 x$, $x \to \beta_2 x$ effects a permutation of the non-zero elements of R. H_1 and H_3 will be orthogonal quasigroups if the equations $x + y = u$ and $\alpha_1 x + \beta_1 y = v$ are uniquely soluble for x and y for each given pair u, v in H. This requires $(\alpha_1 - \beta_1)x = v - \beta_1 u$ to be uniquely soluble for x for each element

$v - \beta_1 u$ in H. Since v takes all values in H, so does $v - \beta_1 u$. It is necessary and sufficient therefore that $\alpha_1 - \beta_1$ be a regular element of R. Similarly, H_2 and H_3 will be orthogonal quasigroups if and only if $\alpha_2 - \beta_2$ is a regular element of R. H_1 and H_2 will be orthogonal quasigroups provided that the equations $\alpha_1 x + \beta_1 y = u$ and $\alpha_2 x + \beta_2 y = v$ are uniquely soluble for x and y for each given pair u, v in H. This requires that $(\alpha_1\beta_2 - \alpha_2\beta_1)x = \beta_2 u - \beta_1 v$ and thence it is easy to see that the only extra condition required is that $\alpha_1\beta_2 - \alpha_2\beta_1$ be a regular element of R. Assuming that these requirements are satisfied, we may define T_l by $T_l = \{(i,j) : h_i + h_j = h_l^*\}$ and T_l' by $T_l' = \{(i,j): h_i + h_j = h_l^{\otimes}\}$.

We may summarize the above discussion into the form of a theorem:

THEOREM 12.2.1. *Let R be an associative and commutative ring defined on a set $H = \{h_1, h_2, \ldots, h_m\}$ of m distinct elements. Let (H, σ), $(H, \cdot)$ and $(H, \odot)$ be three quasigroups defined on the set H by the equations $x\sigma y = x + y$, $x \cdot y = \alpha_1 x + \beta_1 y$ and $x \odot y = \alpha_2 x + \beta_2 y$ for all x, y in H, where $\alpha_1, \beta_1, \alpha_2, \beta_2, \alpha_1 - \beta_1, \alpha_2 - \beta_2$ and $\alpha_1\beta_2 - \alpha_2\beta_1$ are all regular elements of the ring R. Let $(K, \times)$ and $K, \otimes)$ be a pair of orthogonal quasigroups defined on a set $K = \{k_1, k_2, \ldots, k_n\}$ of n distinct elements, where $2n \leq m$. Let $T_l = \{(i,j) : h_i + h_j = h_l^*\}$ and $T'_l = \{(i,j) : h_i + h_j = h_l^{\otimes}\}$, where $l = 1, 2, \ldots, n$, and the two sets $\{h_1^*, h_2^*, \ldots, h_l^*\}$ and $\{h_1^{\otimes}, h_2^{\otimes}, \ldots, h_l^{\otimes}\}$ each of which contains n fixed elements of H, are disjoint. Then the quasigroups $(L, \circ)$ and $(L, ⊚)$ of order $m + n$ defined on the set $L = H \cup K$ by the relations (i) to (v) and (i)′ to (v)′ given below are orthogonal quasigroups provided that $A = B_1 \cup B_2$, where*

$$A = \{(\alpha_1 x + \beta_1 y, \alpha_2 x + \beta_2 y) : x, y \in H, \quad x + y \in h_1^*, h_2^*, \ldots, h_n^*, h_1^{\otimes}, h_2^{\otimes}, \ldots, h_n^{\otimes}\},$$

$$B_1 = \{\alpha_1 x + \beta_1 y, \ \alpha_2 x + \beta_2 z) : x, y, z \in H, \ x + y = h_{r\theta}^*, \ x + z = h_{r\theta'}^{\otimes}, \ r = 1, 2, \ldots, n\},$$

$$B_2 = \{(\alpha_1 x + \beta_1 y, \alpha_2 z + \beta_2 y) : x, y, z \in H, x + y = h_{r\varphi}^*, \ z + y = h_{r\varphi'}^{\otimes}, \ r = 1, 2, \ldots, n\};$$

and $\theta, \theta', \varphi, \varphi'$ are permutations of the set $1, 2, \ldots, n$. We have the relations:

(i) *for all $h_i, h_j \in H$ such that $(i,j) \notin \bigcup_{l=1}^{n} T_l$,* $\quad h_i \circ h_j = h \,. h_j$,

(i)′ *for all $h_i, h_j \in H$ such that $(i,j) \notin \bigcup_{l=1}^{n} T_l'$,* $\quad h_i ⊚ h_j = h_i \odot h_j$,

(ii) *for all $h_i, h_j \in H$ such that $(i,j) \in T_l$,* $\quad h_i \circ h_j = k_l, (l = 1, 2, \ldots, n)$,

$(ii)'$ *for all* $h_i, h_j \in H$ *such that* $(i, j) \in T'_l$, $h_i \circledcirc h_j = k_l$, $(l = 1, 2, \ldots, n)$,

$(iii), (iii)'$ *for all* $k_i, k_j \in K$, $k_i \circ k_j = k_i \times k_j$ *and* $k_i \circledcirc k_j = k_i \otimes k_j$,

$(iv), (iv)'$ for $h_i \in H$ *and* $k_l \in K$, *let* j_1, j_2 *be defined by* $(i, j_1) \in T_{l\theta}$, $(i, j_2) \in T'_{l\theta'}$, *then* $h_i \circ k_l = h_i . h_{j_1}$, *and* $h_i \circledcirc k_l = h_i \odot h_{j_2}$,

$(v), (v)'$ *for* $k_l \in K$ *and* $h_j \in H$, *let* i_1, i_2 *be defined by* $(i_1, j) \in T_{l\varphi}$, $(i_2, j) \in T'_{l\varphi'}$, *then* $k_l \circ h_j = h_{i_1} . h_j$ *and* $k_l \circledcirc h_j = h_{i_2} \odot h_j$.

It will be clear to the reader how this theorem can be generalized to deal with the case of t mutually orthogonal quasigroups defined on the set $L = H \cup K$, as stated at the beginning of this section.

K. Yamamoto showed that, with the aid of the above theorem and using suitably chosen direct products of finite fields as his ring R, pairs of orthogonal latin squares of all orders $t \equiv 2$ modulo 4 $(t > 6)$ could be constructed.

R. Guérin later investigated in detail under what circumstances a ring R and elements $\alpha_1, \beta_1, \alpha_2, \beta_2$ would satisfy all the conditions of theorem 12.2.1. As a by-product of his researches, he was able to show that, as t tends to infinity, the number of non-equivalent pairs of orthogonal latin squares of order t also tends to infinity. For full details of these investigations and some further results, the reader is referred to chapter 6 of R. Guérin [5]. See also J. R. Barra and R. Guérin [1] and [2] and R. Guérin [1], [2] and [3].

In [1] and [4] A. Hedayat and E. Seiden have given some constructions of Yamamoto type for obtaining pairs of orthogonal quasigroups (latin squares) which are much simpler than those which Yamamoto himself gave. Their first theorem is as follows:

THEOREM 12.2.2. *Let* $m = p^s$, $m \neq 13$, *where* p *is an odd prime and* s *any positive integer. Then, if* $n = (m - 1)/2$, *there exists a pair of orthogonal latin squares of order* $m + n$.

PROOF. We take the Galois field $GF[p^s]$ as the ring R of theorem 12.2.1 and, in this field, we choose $\alpha_1 = \alpha \neq \pm 1, \alpha_2 = \alpha^{-1}, \beta_1 = \beta_2 = 1$. The necessary conditions that the quasigroups (H, σ), $(H, \cdot)$ and $(H, \odot)$ be orthogonal are fulfilled by these choices of $\alpha_1, \beta_1, \alpha_2, \beta_2$. Also, since $n \neq 6$, there exists a pair of orthogonal quasigroups $(K, \times)$ and $(K, \otimes)$ of order n. We take an arbitrary fixed element λ of $GF[p^s]$. Then the elements of $GF[p^s]$ distinct from $\frac{1}{2}\lambda$ can be separated into n pairs h_r^* and $h_r^{\otimes} = \lambda - h_r^*$, $r = 1, 2, \ldots, n$, which we use to define the transversals T_l and T'_l, $l = 1, 2, \ldots, n$. It follows that, if each of the permutations θ, θ', φ and φ' is taken to be the identity permutation, the necessary condition $A = B_1 \cup B_2$ in order that the quasigroups $(L, \circ)$ and $(L, \circledcirc)$ be orthogonal is fulfilled, as we shall show.

We have $A = \{(\alpha x + y, \alpha^{-1}x + y) : x + y = \mu \in GF[p^s],\ \mu \neq \frac{1}{2}\lambda\}$. Also,

$$B_1 = \{(\alpha x + y, \alpha^{-1}x + z) : x + y = h_r^*, x + z = h_r^\otimes = \lambda - h_r^*, \quad r = 1, 2, \ldots, n\}$$

and

$$B_2 = \{(\alpha x + y, \alpha^{-1}z + y) : x + y = h_r^*, z + y = h_r^\otimes = \lambda - h_r^*, \quad r = 1, 2, \ldots, n\}.$$

We consider the set B_1 first and show that there exist elements x_1, $y_1 \in GF[p^s]$ such that $\alpha x + y = \alpha x_1 + y_1$ and $\alpha^{-1}x + z = \alpha^{-1}x_1 + y_1$. By subtraction, and using the fact that $y - z = h_r^* - h_r^\otimes$, we get $(\alpha - \alpha^{-1})x + h_r^* - h_r^\otimes = (\alpha - \alpha^{-1})x_1$. Hence we find that $x_1 = x + (h_r^* - h_r^\otimes)/(\alpha - \alpha^{-1})$ and $y_1 = y - \alpha(h_r^* - h_r^\otimes)/(\alpha - \alpha^{-1})$. We deduce that the set B_1 contains all the ordered pairs $(\alpha x_1 + y_1, \alpha^{-1}x_1 + y_1)$ where

$$x_1 + y_1 = \frac{h_r^* + \alpha h_r^\otimes}{1 + \alpha} = \frac{\alpha}{1 + \alpha}\lambda - \frac{1 - \alpha}{1 + \alpha}h_r^*, \qquad r = 1, 2, \ldots, n.$$

Next we consider the set B_2 and show that there exist elements $x_2, y_2 \in GF[p^s]$ such that $\alpha x + y = \alpha x_2 + y_2$ and $\alpha^{-1}z + y = \alpha^{-1}x_2 + y_2$. Using the fact that $x - z = h_r^* - h_r^\otimes$, we find this time that $x_2 = x - \alpha(h_r^* - h_r^\otimes)/(\alpha^{-1} - \alpha)$ and $y_2 = y + (h_r^* - h_r^\otimes)/(\alpha^{-1} - \alpha)$ and we deduce that the set B_2 contains all the ordered pairs $(\alpha x_2 + y_2, \alpha^{-1}x_2 + y_2)$ where

$$x_2 + y_2 = \frac{\alpha h_r^* + h_r^\otimes}{1 + \alpha} = \frac{\alpha}{1 + \alpha}\lambda - \frac{1 - \alpha}{1 + \alpha}h_r^\otimes, \qquad r = 1, 2, \ldots, n.$$

Since the h_r^*'s and $h_r^\otimes$'s $(r = 1, 2, \ldots, n)$ are all different, it follows immediately that $A = B_1 \cup B_2$ as required.

As an illustration of the construction we take the case when $m = 7$ and $n = 3$. In Fig. 12.2.2 a construction of a pair of orthogonal latin squares of order $10 = 7 + 3$ is exhibited corresponding to the values $\alpha = 2$, $\alpha^{-1} = 4$, $\lambda = 0$. With $\lambda = 0$, we have $h_1^* = 1$, $h_2^* = 2$, $h_3^* = 3$, $h_1^\otimes = 6$, $h_2^\otimes = 5$, $h_3^\otimes = 4$ and, in Fig. 12.2.2, the corresponding transversals are indicated by suffices.

By methods of construction very similar to that used in theorem 12.2.2, Hedayat and Seiden have also shown how to obtain pairs

$$H_1 = \left|\begin{array}{ccccccc} 0 & 1_1 & 2_2 & 3_3 & 4 & 5 & 6 \\ 2_1 & 3_2 & 4_3 & 5 & 6 & 0 & 1 \\ 4_2 & 5_3 & 6 & 0 & 1 & 2 & 3_1 \\ 6_3 & 0 & 1 & 2 & 3 & 4_1 & 5_2 \\ 1 & 2 & 3 & 4 & 5_1 & 6_2 & 0_3 \\ 3 & 4 & 5 & 6_1 & 0_2 & 1_3 & 2 \\ 5 & 6 & 0_1 & 1_2 & 2_3 & 3 & 4 \end{array}\right. \qquad H_2 = \left|\begin{array}{ccccccc} 0 & 1 & 2 & 3 & 4_4 & 5_5 & 6_6 \\ 4 & 5 & 6 & 0_4 & 1_5 & 2_6 & 3 \\ 1 & 2 & 3_4 & 4_5 & 5_6 & 6 & 0 \\ 5 & 6_4 & 0_5 & 1_6 & 2 & 3 & 4 \\ 2_4 & 3_5 & 4_6 & 5 & 6 & 0 & 1 \\ 6_5 & 0_6 & 1 & 2 & 3 & 4 & 5_4 \\ 3_6 & 4 & 5 & 6 & 0 & 1_4 & 2_5 \end{array}\right.$$

$$H_3 = \left|\begin{array}{ccccccc} 0 & 1 & 2 & 3 & 4 & 5 & 6 \\ 1 & 2 & 3 & 4 & 5 & 6 & 0 \\ 2 & 3 & 4 & 5 & 6 & 0 & 1 \\ 3 & 4 & 5 & 6 & 0 & 1 & 2 \\ 4 & 5 & 6 & 0 & 1 & 2 & 3 \\ 5 & 6 & 0 & 1 & 2 & 3 & 4 \\ 6 & 0 & 1 & 2 & 3 & 4 & 5 \end{array}\right.$$

$$K_1 = \left|\begin{array}{ccc} 7 & 8 & 9 \\ 8 & 9 & 7 \\ 9 & 7 & 8 \end{array}\right. \qquad K_2 = \left|\begin{array}{ccc} 7 & 8 & 9 \\ 9 & 7 & 8 \\ 8 & 9 & 7 \end{array}\right.$$

$$L_1 = \left|\begin{array}{cccccccccc} 0 & 7 & 8 & 9 & 4 & 5 & 6 & 1 & 2 & 3 \\ 7 & 8 & 9 & 5 & 6 & 0 & 1 & 2 & 3 & 4 \\ 8 & 9 & 6 & 0 & 1 & 2 & 7 & 3 & 4 & 5 \\ 9 & 0 & 1 & 2 & 3 & 7 & 8 & 4 & 5 & 6 \\ 1 & 2 & 3 & 4 & 7 & 8 & 9 & 5 & 6 & 0 \\ 3 & 4 & 5 & 7 & 8 & 9 & 2 & 6 & 0 & 1 \\ 5 & 6 & 7 & 8 & 9 & 3 & 4 & 0 & 1 & 2 \\ 2 & 1 & 0 & 6 & 5 & 4 & 3 & 7 & 8 & 9 \\ 4 & 3 & 2 & 1 & 0 & 6 & 5 & 8 & 9 & 7 \\ 6 & 5 & 4 & 3 & 2 & 1 & 0 & 9 & 7 & 8 \end{array}\right.$$

$$L_2 = \left|\begin{array}{cccccccccc} 0 & 1 & 2 & 3 & 7 & 8 & 9 & 6 & 5 & 4 \\ 4 & 5 & 6 & 7 & 8 & 9 & 3 & 2 & 1 & 0 \\ 1 & 2 & 7 & 8 & 9 & 6 & 0 & 5 & 4 & 3 \\ 5 & 7 & 8 & 9 & 2 & 3 & 4 & 1 & 0 & 6 \\ 7 & 8 & 9 & 5 & 6 & 0 & 1 & 4 & 3 & 2 \\ 8 & 9 & 1 & 2 & 3 & 4 & 7 & 0 & 6 & 5 \\ 9 & 4 & 5 & 6 & 0 & 7 & 8 & 3 & 2 & 1 \\ 3 & 0 & 4 & 1 & 5 & 2 & 6 & 7 & 8 & 9 \\ 6 & 3 & 0 & 4 & 1 & 5 & 2 & 9 & 7 & 8 \\ 2 & 6 & 3 & 0 & 4 & 1 & 5 & 8 & 9 & 7 \end{array}\right.$$

Fig. 12.2.2

of orthogonal latin squares of order $2^\alpha + 2^{\alpha-1}$ for all positive integers $\alpha \geq 3$ and of order $p^\alpha + 3$ when α is any positive integer and p is a prime of one the forms $3m + 1$, $8m + 1$, $8m + 3$, $24m + 11$, $60m + 23$, or $60m + 47$. These authors have also written a second paper (A. Hedayat and E. Seiden [2]) in which they have made a more general analysis of the Yamamoto method. However, the investigation carried out by R. Guérin and which we mentioned earlier is, in the opinion of the present authors, more far reaching.

Some further refinements of the results of Hedayat and Seiden will be found in F. Ruiz and E. Seiden [1].

A completely different method for the construction of pairs and triples of mutually orthogonal latin squares of order $v = p + q$, where p is prime and q is an integer divisor of $p - 1$, has been described in Shin-Chi-Chi [1]. The author has obtained a triple of mutually orthogonal latin squares of order 46 by means of his construction method and has suggested that it may be possible to use the method to produce a triple of mutually orthogonal atin squares of order ten (taking $p = 7$ and $q = 3$).

12.3. Conditions that a given latin square (or set of squares) have no orthogonal mate

In contrast to the various constructive procedures for the formation of sets of mutually orthogonal latin squares given in the preceding chapters and sections, several conditions that a given latin square (or set of squares) be not extendible to a larger set of orthogonal latin squares are known. In particular L. Euler (in [2]), E. Maillet (see [2]), H. B. Mann (see [3]) and E. T. Parker (see [6] and [9]) have all given such conditions.

The results of Euler and Maillet concern latin squares of a particular kind which the latter author has called *q-step type*.

DEFINITION. A latin square L of order $n = mq$ is said to be of *q-step type* if it can be represented by a matrix C of the form shown in Fig. 12.3.1 where, for each fixed choice of k, the $A_{ij}^{(k)}$ are latin subsquares of L all of which contain the same q elements.

For example, the latin square shown in Fig. 12.3.2 is of 2-step type while both the squares shown is Fig. 12.3.3 are of 3-step type.

As early as 1779, L. Euler was able to prove that a cyclic latin square of even order cannot have any transversals and so has no orthogonal

$$
C = \begin{Vmatrix}
A_{00}^{(0)} & A_{01}^{(1)} & A_{02}^{(2)} & \cdots & \cdots & A_{0\,m-2}^{(m-2)} & A_{0\,m-1}^{(m-1)} \\
A_{10}^{(1)} & A_{11}^{(2)} & A_{12}^{(3)} & \cdots & \cdots & A_{1\,m-2}^{(m-1)} & A_{1\,m-1}^{(0)} \\
A_{20}^{(2)} & A_{21}^{(3)} & A_{22}^{(4)} & \cdots & \cdots & A_{2\,m-2}^{(0)} & A_{2\,m-1}^{(1)} \\
\cdot & \cdot & \cdot & \cdots & \cdots & \cdot & \cdot \\
\cdot & \cdot & \cdot & \cdots & \cdots & \cdot & \cdot \\
A_{m-1\,0}^{(m-1)} & A_{m-1\,1}^{(0)} & A_{m-1\,2}^{(i)} & \cdots & \cdots & A_{m-1\,m-2}^{(m-3)} & A_{m-1\,m-1}^{(m-2)}
\end{Vmatrix}
$$

Fig. 12.3.1

0	1	2	3	4	5
1	0	3	2	5	4
2	3	5	4	1	0
3	2	4	5	0	1
4	5	0	1	2	3
5	4	1	0	3	2

Fig. 12.3.2

0	1	2	3	4	5
1	2	0	4	5	3
2	0	1	5	3	4
3	4	5	0	1	2
4	5	3	1	2	0
5	3	4	2	0	1

0	1	2	3	4	5
1	2	0	5	3	4
2	0	1	4	5	3
3	5	4	0	2	1
4	3	5	2	1	0
5	4	3	1	0	2

Fig. 12.3.3

mate.(1) In the same paper [2], he showed that the same is true for latin squares of order 6 or 12 which are of 3-step type and that it is also true for all latin squares of order $2q$ and of q-step type when q is odd.

In 1894 in his paper [2], E. Maillet noted that a cyclic latin square is a square of 1-step type and that all the above mentioned results of Euler are special cases of the following general theorem.

(1) In a recent paper [1], published in 1970, A. Hedayat and W. T. Federer have claimed this as a new result.

THEOREM 12.3.1. *A latin square L of order $n = mq$ and of q-step type has no transversals if m is even and q is odd.*

PROOF. Our proof substantially follows that of Maillet. Since L is of q-step type we may represent it by means of the matrix C given in Fig. 12.1.1, where each $A_{ij}^{(k)}$ represents a $q \times q$ latin subsquare of L.

If we take the elements of L to be the integers $0, 1, 2, \ldots, n-1$, there will be no loss of generality in supposing that (by a change of labelling if necessary) all the subsquares $A_{ij}^{(k)}$ which correspond to the same fixed value of k contain the same q elements $kq + b$, where k has this fixed value and $0 \leq b \leq q-1$: for every one of the integers $0, 1, 2, \ldots, n-1$ has a unique representation in the form $aq + b$ with $0 \leq a \leq m-1$ and $0 \leq b \leq q-1$. Let us note for use later that, for each latin subsquare $A_{ij}^{(k)}$ of the matrix C, k is determined by the relation $k \equiv i + j \bmod m$

Let us suppose that the theorem is false and that τ is a transversal of L. Then τ will contain just q cells belonging to the set of latin subsquares $A_{00}, A_{10}, \ldots, A_{m-1\,0}$ which form the first column of the matrix C. This is because the first column of matrix C represents the first q columns of L and each of these columns contains exactly one cell of τ. Let the entries in these q cells be the integers $a_{00}q + b_{00}, a_{10}q + b_{10}, \ldots, a_{q-1\,0}q + b_{q-1\,0}$ and suppose that the cells belong to the latin subsquares $A_{c_{00}0}, A_{c_{10}0}, \ldots, A_{c_{q-1\,0}0}$ respectively, the integers $c_{00}, c_{10}, \ldots, c_{q-1\,0}$ being not necessarily all distinct.

The transversal τ will also contain just q cells belonging to the latin subsquares of the jth column of the matrix C for the same reason as before. (This is true for each fixed j in the range $0 \leq j \leq m-1$.) Let the entries in these q cells be the integers $a_{0j}q + b_{0j}, a_{1j}q + b_{1j}, \ldots, a_{q-1\,j}q + b_{q-1\,j}$ and suppose that the cells belong to the latin subsquares $A_{c_{0j}j}, A_{c_{1j}j}, \ldots, A_{c_{q-1j}j}$ respectively, the integers $c_{0j}, c_{1j}, \ldots, c_{q-1\,j}$ being again not necessarily all distinct. Since $a_{ij}q + b_{ij}$ is an element of $A_{c_{ij}j}$ we have $a_{ij} \equiv c_{ij} + j \bmod m$ in consequence of the special choice of labelling which we established at the beginning.

The entries in the n cells of the transversal τ are equal in some order to the n integers $0, 1, \ldots, n-1$. When these n integers are expressed in the form $a_{ij}q + b_{ij}$, there exist just q of them which have the same fixed value of a_{ij} in the range $0 \leq a_{ij} \leq m-1$ but have different values of b_{ij}, since b_{ij} varies through the integers $0, 1, \ldots, q-1$. Hence, $\sum_{\tau} a_{ij} = q(0 + 1 + \ldots + \overline{m-1}) = q\dfrac{m(m-1)}{2}$ where the summation is over the entries in all the cells of τ.

Also $\sum_\tau a_{ij} \equiv \sum_\tau (c_{ij}+j) \bmod m$. The right-hand side of this congruence is equal to $\sum_\tau c_{ij} + q(0+1+\ldots+\overline{m-1})$ since j varies between 0 and $m-1$ and takes each of these values for exactly q of the cells of τ: namely it takes the value j for all the cells occurring in the q columns of L which are included in the latin subsquares of the jth column of the matrix C. Further, since τ has exactly one cell in each row of L, it has exactly q cells belonging to the set of latin subsquares $A_{h0}, A_{h1}, \ldots, A_{hm-1}$ which form the hth row of the matrix C. For each of these cells, $c_{ij} = h$. Hence, as h varies between 0 and $m-1$, c_{ij} takes each of these values q times, and so $\sum_\tau c_{ij} = q(0+1+\ldots+\overline{m-1})$ where the summation is over the entries in all the cells of τ as before. Hence, $\sum_\tau (c_{ij}+j) = q\dfrac{m(m-1)}{2} + q\dfrac{m(m-1)}{2} = qm(m-1)$. It follows that $\sum_\tau a_{ij} \equiv 0 \bmod m$.

But, we have shown already that $\sum_\tau a_{ij} = q\dfrac{m(m-1)}{2}$. These conditions are not consistent unless $q(m-1)$ is divisible by 2, so no transversal exists if q is odd and m is even.

The results of H. B. Mann and E. T. Parker are slightly different in kind from those of L. Euler and E. Maillet. As we have seen, the squares discussed by the latter are ones which have no transversals at all. However, the results of Mann and Parker concern single latin squares or sets of latin squares which, though not necessarily lacking transversals all together, nevertheless have no orthogonal mates.

Mann's criterion is concerned only with single squares and is as follows:

Theorem 12.3.2. (*a*) *Let L be a latin square of order $4n+2$ whose entries are the symbols* $1, 2, \ldots, 4n+2$. *Then if L contains a $(2n+1)\times(2n+1)$ submatrix A such that less than $n+1$ of its cells contain elements distinct from the symbols* $1, 2, \ldots, 2n+1$, *L has no orthogonal mate.*

(*b*) *Let L be a latin square of order $4n+1$ whose entries are the symbols* $1, 2, \ldots, 4n+1$. *Then, if L contains a $2n\times 2n$ submatrix A such that less than $n/2$ of its cells contain elements distinct from the symbols* $1, 2, \ldots, 2n$, *L has no orthogonal mate.*

Proof. (*a*) We may suppose that the rows and columns of L have been rearranged so that the submatrix A occurs in the first $2n+1$ rows and columns. Let $L = \left(\frac{A \mid B}{C \mid D}\right)$, where each of A, B, C, D is a $(2n+1) \times (2n+1)$ submatrix.

Suppose that the symbol x occurs r times in the submatrix A. In that case, since x must appear exactly once in each of the first $2n+1$ rows of L, it must appear $(2n+1) - r$ times in the submatrix B. But then, since x must appear exactly once in each of the last $2n+1$ columns of L, it must appear r times in the submatrix D. Thus, each symbol of L appears as many times in the submatrix D as it does in the submatrix A and an even number of times among the cells of A and D combined.

Let k $(< n+1)$ be the number of cells of A which contain entries different from the symbols $1, 2, \ldots, 2n+1$. From what we have just said, it follows that there must be just k cells of D also which contain entries distinct from the symbols $1, 2, \ldots, 2n+1$. (Each "foreign" symbol which occurs in A occurs in D an equal number of times.) Since each of the symbols $1, 2, \ldots, 2n+1$ occurs in L altogether $4n+2$ times and since A and D combined contain $(2n+1) \times (4n+2)$ cells, there exist exactly $2k$ cells not in A or D whose entries are among the subset of symbols $1, 2, \ldots, 2n+1$. Now let us suppose that L has an orthogonal mate L^* and consequently has $4n+2$ disjoint transversals. Then the preceding remarks imply that at least $4n+2-2k$ transversals have the $2n+1$ of their cells which contain the symbols $1, 2, \ldots, 2n+1$ included among the cells of A or D. Also, since only $2k$ of the cells of A and D combined contain symbols distinct from $1, 2, \ldots, 2n+1$, at most $2k$ transversals of L have cells containing symbols other than $1, 2, \ldots, 2n+1$ in A or D. Therefore, not less than $(4n+2-2k) - 2k$ transversals have the $2n+1$ of their cells which contain the symbols $1, 2, \ldots, 2n+1$ included among the cells of A and D and have no other cells in A or D. For $k < n+1$, this number of transversals is at least two. The cells of L^* which correspond to the cells of such a transversal of L all contain the same symbol x. So, in L^*, x occurs $2n+1$ times (that is, an odd number of times) among the cells of A^* and D^* combined. (We suppose that L^* has been partitioned in the same way as L.) But, as shown above for L, this is impossible. This contradiction shows that no orthogonal mate L^* can exist for L if $k < n+1$.

(b) As before, we may suppose that the rows and columns of L have been rearranged so that the submatrix A occurs in the first $2n$ rows and columns. Let $L = \left(\frac{A}{C}\middle|\frac{B}{D}\right)$ where A, B, C, D are submatrices of sizes $2n \times 2n$, $2n \times (2n+1)$, $(2n+1) \times 2n$, $(2n+1) \times (2n+1)$ respectively.

Suppose that the symbol x occurs r times in the submatrix A. In that case, by an argument similar to that of (a), it must occur $2n - r$ times in the submatrix B and $(2n+1) - (2n-r) = r+1$ times in the submatrix D. That is, any symbol occurs an odd number of times among the cells of A and D combined. In particular, a symbol x that does not occur in the submatrix A at all occurs exactly once in the submatrix D.

Let k ($< n/2$) be the number of cells of A which contain entries distinct from the symbols $1, 2, \ldots, 2n$. If these entries are all the same, equal to the symbol x say, the symbol x occurs $k+1$ times in D and the $(4n+1) - 2n - 1$ symbols of L which do not occur at all in A, each occur just once in D, so there exist $(k+1) + 2n$ cells of D which contain symbols distinct from $1, 2, \ldots, 2n$ and $2k + 1 + 2n$ cells of A and D combined which contain such symbols. If, on the other hand, the k cells of A which contain entries distinct from $1, 2, \ldots, 2n$ all contain different symbols, say the symbols $x_1, x_2, \ldots, x_k$, then each of these symbols occurs twice in D and D also has $(4n+1) - 2n - k$ further cells which contain symbols distinct from $1, 2, \ldots, 2n$, equal to the number of symbols of L which do not occur at all in A. Thus, in this case, there exist $k + 2k + [(4n+1) - 2n - k] = 2n + 2k + 1$ cells of A and D combined which contain entries distinct from $1, 2, \ldots, 2n$. If the symbols in the k cells of A under discussion are some different and some the same, we shall still get the number $2n + 2k + 1$ of cells of A and D combined which contain symbols distinct from $1, 2, \ldots, 2n$.

Since each of the symbols $1, 2, \ldots, 2n$ occurs $4n+1$ times in L and since A and D have $(2n)^2 + (2n+1)^2$ cells all together of which at most $2n + 2k + 1$ contain symbols distinct from this subset, there exist at most $(4n+1)2n - [(2n)^2 + (2n+1)^2 - (2n + 2k + 1)] = 2k$ cells not in A or D whose entries are among the subset of symbols $1, 2, \ldots, 2n$. If L had an orthogonal mate L^*, it would have $4n+1$ disjoint transversals. Of these, at least $(4n+1) - 2k$ would have the $2n$ of their cells which contain the symbols $1, 2, \ldots, 2n$ included among the cells of A and D. Also, at most $2n + 2k + 1$ transversals could have cells containing symbols other than $1, 2, \ldots, 2n$ included among the cells of A or D, and so at least $(4n+1) - 2k - (2n + 2k + 1) = 2n - 4k$ of the transversals would have the $2n$ of their cells which contain the symbols $1, 2, \ldots, 2n$ included among the cells of A and D and have no other cells in A or D. For $k < n/2$, this number of transversals is at least two

$[= 2n - 4(\frac{1}{2}n - \frac{1}{2})]$. The cells of L^* which correspond to the cells of such a transversal of L all contain the same symbol x. So, in L^*, x occurs $2n$ times (that is, an even number of times) among the cells of A^* and D^*, which (compare part (a)) gives us a contradiction.

M. Hall [11] contains an interesting alternative proof of the above theorem, using orthogonal arrays.

It has been shown by R.T. Ostrowski and K. D. van Duren (see [1]) that Mann's result (a) is "best possible". With the aid of a computer, these authors have constructed a pair of orthogonal latin squares L and L^* of order 10 such that one square L of the pair contains a 5×5 submatrix A with only 3 entries distinct from the symbols 1, 2, 3, 4, 5, thus showing that for a square L with $n = 2$ and $k = n + 1 = 3$, an orthogonal mate can exist. The squares are as exhibited in Figure 12.3.4.

In J. W. Brown [2], an extension of theorem 12.3.2 (a) to the special case of a triple of mutually orthogonal latin squares of order ten has been

$$L = \left|\begin{array}{ccccc|ccccc}
1 & 2 & 3 & 4 & 5 & 6 & 7 & 8 & 9 & 0 \\
4 & 5 & 1 & 2 & 3 & 8 & 0 & 9 & 7 & 6 \\
5 & 4 & 2 & 3 & 1 & 0 & 8 & 7 & 6 & 9 \\
2 & 3 & 5 & 1 & 8 & 9 & 6 & 4 & 0 & 7 \\
3 & 1 & 4 & 8 & 6 & 7 & 9 & 0 & 5 & 2 \\
\hline
6 & 8 & 7 & 0 & 9 & 4 & 5 & 2 & 3 & 1 \\
9 & 0 & 8 & 6 & 7 & 2 & 3 & 1 & 4 & 5 \\
7 & 6 & 0 & 9 & 2 & 5 & 4 & 3 & 1 & 8 \\
0 & 9 & 6 & 7 & 4 & 1 & 2 & 5 & 8 & 3 \\
8 & 7 & 9 & 5 & 0 & 3 & 1 & 6 & 2 & 4
\end{array}\right.$$

$$L^* = \left|\begin{array}{ccccc|ccccc}
1 & 2 & 0 & 3 & 4 & 9 & 5 & 7 & 6 & 8 \\
7 & 8 & 9 & 0 & 6 & 3 & 4 & 2 & 1 & 5 \\
0 & 4 & 8 & 5 & 7 & 6 & 9 & 3 & 2 & 1 \\
4 & 9 & 3 & 6 & 5 & 8 & 0 & 1 & 7 & 2 \\
2 & 5 & 6 & 1 & 8 & 4 & 7 & 0 & 9 & 3 \\
\hline
3 & 6 & 7 & 2 & 0 & 5 & 1 & 9 & 8 & 4 \\
5 & 1 & 2 & 4 & 9 & 7 & 3 & 8 & 0 & 6 \\
6 & 7 & 5 & 9 & 1 & 2 & 8 & 4 & 3 & 0 \\
9 & 3 & 1 & 8 & 2 & 0 & 6 & 5 & 4 & 7 \\
8 & 0 & 4 & 7 & 3 & 1 & 2 & 6 & 5 & 9
\end{array}\right.$$

Fig. 12.3.4

given. If such a triple existed (an unsolved question), it would be equivalent to an orthogonal array $OA(10,5)$. Brown's result is that, given an $OA(10,5)$, at least one of the ten $OA(10,3)$'s obtainable from it by selecting some three of its five rows would represent a latin square containing a subsquare of order five in which some five of the ten distinct entries occurred at least nine and most twelve times.

It follows as a corollary to Mann's theorem that if a latin square L is of order $4m + 2$ and contains a latin subsquare of order $2m + 1$ or if L is of order $4m + 1$ and contains a latin subsquare of order $2m$ then L has no orthogonal mate. The second of our next two theorems, both due to E. T. Parker (see [6]), provides a generalization of these statements and includes them as special cases when $t = 1$.

Theorem 12.3.3. *If a set of t mutually orthogonal latin squares of order n has a set of t mutually orthogonal latin subsquares of order r $(r < n)$, then $n \geq (t + 1)r$.*

Proof. Let $L_1, L_2, \ldots, L_t$ be the squares with $L_i = \left(\begin{array}{c|c} A_i & B_i \\ \hline C_i & D_i \end{array}\right)$ where A_i, B_i, C_i, D_i are submatrices of sizes $(n - r)\times(n - r)$, $(n - r)\times r$, $r\times(n - r)$, and $r\times r$ respectively. We suppose the rows and columns of the squares to have been arranged in such a way that $D_1, D_2, \ldots, D_t$ are the mutually orthogonal subsquares of order r. Let $x_1, x_2, \ldots, x_r$ be the symbols which occur in these subsquares. Since each of these symbols occurs exactly once in each row of D_i, none of them occurs in C_i, and since each of them occurs exactly once in each column of D_i, none of them occurs in B_i. However, each x_u occurs all together n times in the square L_i and so each must occur $n - r$ times in the submatrix A_i and these occurrences must be once in each row of A_i because A_i is latin. Thus, the unordered set $x_1, x_2, \ldots, x_r$ must occupy r of the $n - r$ places in each row of the submatrix A_i. But because the latin squares L_i and L_j (for any $j \neq i$) are orthogonal, and because the subsquares D_i and D_j are also orthogonal, no cell of A_i can contain any symbol x_u for more than one value of i: for suppose A_i contained x_u in a certain cell and A_j $(j \neq i)$ contained x_v in the corresponding cell, then the ordered pair (x_u, x_v) would appear twice when L_j was superimposed on L_i, once in a cell of (A_i, A_j) and once in a cell of (D_i, D_j). Hence, the unordered set $x_1, x_2, \ldots, x_r$ must occupy a different set of r cells in, say, the kth row $(1 \leq k \leq n - r)$ of each of the submatrices $A_1, A_2, \ldots, A_t$. Consequently, we must have $tr \leq n - r$. That is, $n \geq (t + 1)r$.

We note, in particular, that a single latin square cannot contain a latin subsquare of order greater than $n/2$ and that a pair of orthogonal latin squares of order n cannot contain orthogonal subsquares of order greater than $n/3$. The latter fact is of interest in connection with Parker's construction (described in theorem 11.2.1) of a pair of orthogonal squares of order 10, since these are constructed with the aid of orthogonal subsquares of order 3: that is, of the maximum possible size.

THEOREM 12.3.4. *If a set of t mutually orthogonal latin squares of order n has a set of t mutually orthogonal subsquares of order r, with $r < n$, and if a latin square of order n orthogonal to all t squares exists, then*

$$r \leq \frac{n-r}{t+1} + \left[\frac{r^2}{n}\right],$$

where $[m]$ denotes the integer part of m.

(If r^2/n is an integer, this condition reduces merely to the condition on r that t mutually orthogonal squares of order n having t mutually orthogonal subsquares of order r can exist: namely, $n \geq (t+1)r$).

PROOF. We shall define a *t-transversal* of a set of t mutually orthogonal latin squares of order n as a set of n cells, common to all the squares and taken one from each row and one from each column, whose entries in each separate square form a transversal of that square. Thus, a set of t mutually orthogonal squares of order n can be extended to a set of $(t+1)$ such squares if and only if the n^2 cells of the given squares form n disjoint t-transversals. We shall use the same notation as for theorem 12.3.3, denoting the symbols which occur in the latin subsquares D_i $(i = 1, 2, \ldots, n)$ by $x_1, x_2, \ldots, x_r$ as before. For brevity let us write X to denote the set $x_1, x_2, \ldots, x_r$.

Let T be an arbitrary t-transversal of the squares $L_1, L_2, \ldots, L_t$ and let d be the number of cells of T which occur in the submatrix D_i $(i = 1, 2, \ldots, \text{or } t)$. Since T contains one cell in each row and column of L_i, it must contain $r - d$ cells in each of the submatrices B_i and C_i and consequently $(n - r) - (r - d) = n - 2r + d$ cells in the submatrix A_i. If we count the symbols of each square of the set separately, there are td symbols of the subset X which occur among the cells of T which lie in the subsquares D_i and consequently $tr - td$ which occur elsewhere among the cells of T. No symbol of the subset X appears in B_i or C_i for any value of i. Moreover, no cell of A_i can contain a symbol of the subset X for more than one value of i, as shown in theorem 12.3.3 above, and so A_i

must include at least $tr - td$ cells of T. That is, $n - 2r + d \geq tr - td$ or, equivalently, $d \geq r + \dfrac{r-n}{t+1}$. The n disjoint t-transversals together account for the r^2 cells of D_i and so, for at least one t-transversal T, $d \leq \left[\dfrac{r^2}{n}\right]$, because d is an integer. Hence $\left[\dfrac{r^2}{n}\right] \geq r - \dfrac{n-r}{t+1}$, from which the result follows.

On taking $t = 1$, $n = 4m + 2$, and $r = 2m + 1$ in the theorem, we see that the necessary condition for the existence of a square orthogonal to a given square L of order $4m + 2$ which contains a latin subsquare of order $2m + 1$ is that $2m + 1 \leq \frac{1}{2}(2m + 1) + [\frac{1}{2}(2m + 1)]$. That is, the condition is $2m + 1 \leq m + \frac{1}{2} + m$. Since $1 \nleq \frac{1}{2}$, the condition cannot be satisfied and so no orthogonal mate for L exists. Similarly, on putting $t = 1, n = 4m + 1$, and $r = 2m$, the condition becomes

$$2m \leq \frac{2m+1}{2} + \left[\frac{4m^2}{4m+1}\right].$$

or $2m \leq m + \frac{1}{2} + m - 1$, since

$$\frac{4m^2}{4m+1} = (m-1) + \frac{3m+1}{4m+1}.$$

This condition too cannot be satisfied. Thus, we see that the corollary to Mann's theorem can be deduced as a special case of Parker's theorem above, as we stated.

The following is an interesting deduction from theorems 12.3.3 and 12.3.4. *If a set of $r - 1$ mutually orthogonal latin squares of order n has a set of $r - 1$ mutually orthogonal latin subsquares of order r ($r < n$), then $n \geq r^2$. If $n > r^2$, a necessary condition that there exists a further latin square of order n orthogonal to all those given is that $n \geq r^2 + r$.* (See also E. T. Parker [7].)

This result may be interpreted geometrically. The $r - 1$ mutually orthogonal latin squares of order n represent an $(r + 1)$-net of order n ($n > r$) and the $r - 1$ subsquares of order r represent a projective subplane of order r embedded in this net. The statement is that, under such circumstances, the order n of the net is not less than r^2 and, if $n > r^2$, the net cannot be extended to an $(r + 2)$-net unless n is at least $r^2 + r$. For comparison, we state a well-known result of R. H. Bruck (see lemma 3.1. of [4]). *If a finite projective plane of order n* (or $(n + 1)$-net) *contains a projective subplane of order r, with $r < n$, then $n \geq r^2$; if $n > r^2$ then $n \geq r^2 + r$.*

In a very recent paper [9], E. T. Parker has called a latin square of order *n pathological* if it cannot form one member of a complete set of mutually orthogonal latin squares of that order. He has given the following sufficient condition for a latin square of even order $2t$ to be pathological.

THEOREM 12.3.5. *If a latin square of order* $2t$ *includes a triangular array of* $2t - 2$ *rows, columns and symbols of the form shown in* Fig. 12.3.5 (in common with a cyclic latin square of order $2t$) *then the latin square is pathological.*

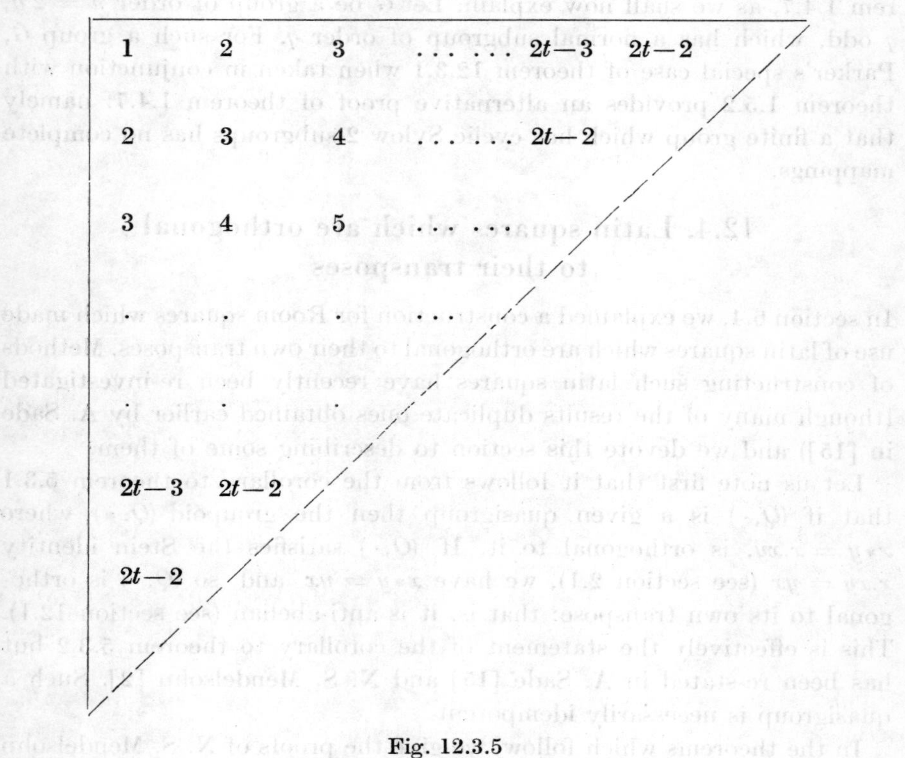

Fig. 12.3.5

As Parker has remarked, further theorems of this type would be of considerable interest.

In the same paper [9], E. T. Parker has also re-discovered a special case of E. Maillet's theorem which he states as follows:

Assume that a latin square L of even order $2^r q$, *q an odd integer, has its rows, columns and symbols each independently partitioned into* 2^r *sets of* *q members each in such a way that each $q \times q$ minor determined by the row and column partitions contains symbols of only one subset of the symbol partition. Suppose further that the latin square of order 2^r which is determined by the q-to-one mappings of the 2^r subsets of rows, columns and symbols respectively, defined by the partitioning, is the multiplication table of a cyclic group. Then L has no transversal.*

The reader will notice that this is the special case of theorem 12.3.1 which we get when we put $m = 2^r$. However, as Parker has pointed out, this special case is of some interest because of its connections with theorem 1.4.7, as we shall now explain. Let G be a group of order $n = 2^r q$, q odd, which has a normal subgroup of order q. For such a group G, Parker's special case of theorem 12.3.1 when taken in conjunction with theorem 1.5.2 provides an alternative proof of theorem 1.4.7: namely that a finite group which has cyclic Sylow 2-subgroups has no complete mappings.

12.4. Latin squares which are orthogonal to their transposes

In section 6.4, we explained a construction for Room squares which made use of latin squares which are orthogonal to their own transposes. Methods of constructing such latin squares have recently been re-investigated (though many of the results duplicate ones obtained earlier by A. Sade in [15]) and we devote this section to describing some of them.

Let us note first that it follows from the corollary to theorem 5.3.1 that if $(Q, \cdot)$ is a given quasigroup then the groupoid $(Q, *)$, where $x * y = x.xy$, is orthogonal to it. If $(Q, \cdot)$ satisfies the Stein identity $x.xy = yx$ (see section 2.1), we have $x * y = yx$ and so $(Q, \cdot)$ is orthogonal to its own transpose: that is, it is anti-abelian (see section 12.1). This is effectively the statement of the corollary to theorem 5.3.2 but has been re-stated in A. Sade [15] and N. S. Mendelsohn [2]. Such a quasigroup is necessarily idempotent.

In the theorems which follow, we give the proofs of N. S. Mendelsohn but at the same time pointing out that almost all the results actually date from the work of A. Sade published 9 years earlier. We begin with a re-statement of the corollary to theorem 5.3.2 (first proved by S. K. Stein in [2]).

THEOREM 12.4.1. *Let $(Q, \cdot)$ be a quasigroup which satisfies the Stein identity $yx = x.xy$. Then the multiplication table of Q defines a latin square L which is orthogonal to its transpose.*

PROOF. Let the rows and columns of the multiplication table of $(Q, \cdot)$ be labelled by the elements of Q. Then the element at the intersection of the ath row and bth column of L is ab and the corresponding element of L^T is ba. If L and L^T were not orthogonal, we would have equality between two ordered pairs (ab, ba) and (cd, dc). But then $a(cd) = a(ab) = ba = dc = c(cd)$ whence $a = c$ and $b = d$.

The above proof is taken from N. S. Mendelsohn [2] and our proof of the next theorem is from the same paper.

THEOREM 12.4.2. *Let $n = 4^k m$ where $k \geq 0$ and m is an odd integer such that every prime divisor p of m for which 5 is a quadratic non-residue occurs an even number of times in the decomposition of m into prime factors. Then there exists a latin square L of order n which is orthogonal to its transpose.*

PROOF. We make use of the fact, proved in section 12.1, that the direct products of pairs of orthogonal quasigroups are again orthogonal quasigroups to obtain L from squares of smaller order.

First suppose that $n = 4$. We form a quasigroup $(Q_4, \cdot)$ of order 4 by taking the elements of the Galois field $GF[2^2]$ as elements of Q_4 and defining the quasigroup operation by $x.y = \lambda x + (1 + \lambda)y$ where λ is such that $\lambda^2 = \lambda + 1$ generates the multiplicative group of $GF[2^2]$ and $(+)$ denotes addition in this field.

Next suppose that $n = 5$. We form a quasigroup $(Q_5, \cdot)$ of order 5 by taking the elements of the Galois field $GF[5]$ as elements of Q_5 and defining the quasigroup operation by $x.y = 4x + 2y$ where $(+)$ denotes addition in the field $GF[5]$.

If $n = p$, p prime, where 5 is a quadratic residue modulo p (that is, the equation $\mu^2 = 5$ is soluble for μ modulo p), we form a quasigroup $(Q_p, \cdot)$ of order p by taking the elements of the Galois field $GF[p]$ as elements of Q_p and defining the quasigroup operation by $x.y = \frac{1}{2}(3 - \mu)x + \frac{1}{2}(\mu - 1)y$ where $(+)$ again denotes addition in the field.

Each of the quasigroups $(Q_4, \cdot)$, $(Q_5, \cdot)$ and $(Q_p, \cdot)$ so constructed is a quasigroup which satisfies the Stein identity and so is orthogonal to its own transpose. It follows from theorem 12.1.1, or by direct verification that the direct product of any number of such quasigroups is again a quasigroup which satisfies the Stein identity and so gives rise to a latin square L which is orthogonal to its own transpose. This proves the theorem except for cases in which the integer m has prime divisors for which 5 is a quadratic non-residue.

Suppose that p is a prime for which 5 is a quadratic non-residue, so that the equation $z^2 - 5 = 0$ is irreducible over $GF[p]$. We adjoin a root μ of this equation to form a field $GF[p^2]$ in which the equation has a root and then define an operation $(\cdot)$ on the elements of $GF[p^2]$ by $x.y = \frac{1}{2}(3 - \mu)x + \frac{1}{2}(\mu - 1)y$.

In this way we get a quasigroup $(Q_{p^2}, \cdot)$ of order p^2 which satisfies the Stein identity.

It now follows easily that the statement of the theorem is true.

In [3], N. S. Mendelsohn has devised an alternative construction of latin squares which are orthogonal to their transposes. This yields a slight improvement of theorem 12.4.2 as follows:

THEOREM 12.4.3. *Let n be a positive integer such that $n \not\equiv 2$ mod 4, $n \not\equiv 3$ mod 9, and $n \not\equiv 6$ mod 9. Then there exists a latin square L of order n which is orthogonal to its transpose.*

PROOF. Let F be any finite field other then $GF[2]$ or $GF[3]$, and let λ be an element of F such that $\lambda \neq 0$, $\lambda \neq 1$, $2\lambda \neq 1$. Then it is easy to check that the operation $(\cdot)$ given by $x.y = \lambda x + (1 - \lambda)y$ defines a quasigroup on the elements of F which is orthogonal to its own transpose. For if not, we should have $(a.b, b.a)$ and $(c.d, d.c)$ equal for some choices of the elements a, b, c, d. But then $\lambda a + (1 - \lambda)b = \lambda c + (1 - \lambda)d$ and $\lambda b + (1 - \lambda)a = \lambda d + (1 - \lambda)c$, implying that $a + b = c + d$. If so, the first equation gives $(1 - 2\lambda)b = (1 - 2\lambda)d$. Hence $b = d$ and $a = c$.

By taking the direct product of a suitable number of such quasigroups, we can get a latin square which is orthogonal to its transpose for any order n whose decomposition into primes does not contain either of the primes 2 or 3 to the first power only; and this is the statement of the theorem.

Let us note that the construction of theorem 12.4.3 is merely a special case of that of theorem 12.1.6 which latter was taken from A. Sade [15] and that this fact when combined with Sade's statement on page 93 of the same paper that the direct product of two anti-abelian quasigroups is again anti-abelian leads us to the realization that theorem 12.4.3 is merely a re-discovery of an old result. Again, it is clear from the remarks made by A. Sade on page 94 of [15] that the latter author was aware of the construction used in theorem 12.4.2. Using his theorem 12.1.6, he noted that a quasigroup satisfying the Stein identity $x.xy = yx$ could easily be constructed from the elements of the Galois field $GF[h]$ provided that it was possible to find elements a and b of the field such that $ax + b(ax + by) = ay + bx$ for all elements x and y in the field. This requires that $a + ba = b$ and $b^2 = a$. That is, $b^2 + b \cdot b^2 = b$; whence $b + b^2 = 1$ or $(b + \frac{1}{2})^2 = \frac{5}{4}$. If 5 is a quadratic residue of h so that an element $\mu \in GF[h]$ exists satisfying the equation $\mu^2 = 5$, then $b = \frac{1}{2}(\mu - 1)$ and $a = b^2 = 1 - b = (3 - \mu)$ as in theorem 12.4.2. Hence follows Sade's statement that quasigroups satisfying the Stein identity can easily be constructed from the Galois field of h elements provided that $h > 5$ and that 5 is a quadratic residue of h.

As an illustration of theorem 12.4.3, we take the example given in N. S. Mendelsohn [3]. Suppose $n = 8$. The elements of $GF[2^3]$ are $x_0 = 0$, $x_1 = 1$, $x_2 = \lambda$, $x_3 = 1 + \lambda$, $x_4 = \lambda^2$, $x_5 = \lambda + \lambda^2$, $x_6 = 1 + \lambda^2$. We define our quasigroup operation by $x_i \cdot x_j = \lambda x_i + (1 - \lambda)x_j$ and hence get the multiplication table shown in Fig. 12.4.1, where we have replaced each element x_i by its suffix i. The unbordered multiplication table is a latin square which is orthogonal to its transpose.

	0	1	2	3	4	5	6	7
0	0	3	5	6	7	1	4	2
1	2	1	4	7	6	3	5	0
2	4	7	2	1	3	6	0	5
3	5	6	0	3	1	7	2	4
4	3	0	6	5	4	2	7	1
5	7	4	1	2	0	5	3	6
6	1	2	7	4	5	0	6	3
7	6	5	3	0	2	4	1	7

Fig. 12.4.1

In K. Byleen and D. W. Crowe [1], a construction which is equivalent to that of theorem 12.4.3 above has been given and the authors have used the latin squares which they obtain by means of it to construct Room designs, as we mentioned in section 6.4. As neither the paper of N. S. Mendelsohn nor that of K. Byleen and D. W. Crowe has been published at the time of writing, it is difficult to decide the official priority of discovery, but it is known to the present authors that the two papers were written quite independently and that neither author was aware of Sade's earlier work on the same subject.

From a given finite field $GF[q]$, Byleen and Crowe have constructed a $q \times q$ matrix $A_q(d)$ whose entries are ordered pairs of the elements of

$GF[q]$ defined as follows. We suppose that the rows and columns of the matrix $A_q(d)$ are labelled by the elements of $GF[q]$ and then the entry of the xth row and yth column of $A_q(d)$ is the ordered pair $[d(y - x) + x, d(x - y) + y]$, where d is a fixed element of $GF[q]$, $d \neq 0$, $d \neq 1$, $2d \neq 1$. If $x = y$, the entry of the xth row and yth column is the ordered pair (∞, x), where ∞ is an additional symbol. Thus, $A_q(d)$ is equivalent to the matrix obtained by juxtaposing L and L^T, where L is a latin square of order q which is orthogonal to its own transpose, and then replacing all the elements of the main left-to-right diagonal of L by the symbol ∞.

Byleen and Crowe have denoted the broken left-to-right diagonal formed by the cells $a_{i\ i+k}$ of $A_q(d)$, where $i \in GF[q]$ and k is fixed in $GF[q]$, by Δ_k. Now let $q = 1 + 2^s m$ where $m \neq 1$ is odd, and let λ be a generator of the multiplicative group of $GF[q]$. Byleen and Crowe have proved that if the diagonals Δ_{λ^r} are all deleted, where r is a member of the set $\{r/r = 2^s h + t$ for some h and for $1 \leq t \leq 2^{s-1}\}$, then the remaining part of $A_q(d)$ will be a Room square.

In R. C. Mullin and E. Németh [5], a construction for latin squares which are orthogonal to their transposes has been given which makes use of a particular type of Room square, thus reversing the point of view of Byleen and Crowe.

We saw in theorem 12.4.2 that the direct product of two quasigroups which satisfy the Stein identity is again a quasigroup which satisfies this identity. In [3], C. C. Lindner has proved that this identity is also preserved by the singular direct product of A. Sade which we described in section 12.1. This fact clearly provides a further means of constructing latin squares of composite order which are orthogonal to their own transposes from similar squares of smaller order. More recently, as already mentioned in section 12.1, Lindner has made use of his first generalized singular direct product of quasigroups to obtain a recursive construction for anti-abelian quasigroups. Precisely he has proved that if $(G, \cdot)$ of order n and $(H, +)$ of order k are anti-abelian quasigroups and if $(G, \cdot)$ contains a subquasigroup $(Q, \cdot)$ of order m then, provided that $n - m \neq 6$, there exists an anti-abelian quasigroup of order $m + (n - m)k$. By this means, he has added to the results of A. Sade and N. S. Mendelsohn by showing that there are an infinite number of anti-abelian quasigroups of order n whenever $n \equiv 2 \pmod 4$, $n \equiv 3 \pmod 9$ or $n \equiv 6 \pmod 9$. (See C. C. Lindner [4].) In particular, there exist anti-abelian quasigroups of order ten, a result which has also been obtained by A. Hedayat. He has given an example of such a quasigroup in [2].

Since the above remarks were written, A. Sade has announced a new construction by which to obtain an anti-abelian quasigroup from an abelian group, using specified automorphisms of the latter. (See A. Sade [31].) Details have not yet been published.

Added in proof. The question of existence of anti-abelian quasigroups has now been settled in the affirmative for all orders n except 2, 3 and 6 by some recent work of R. K. Brayton, D. Coppersmith and A. J. Hoffman. Using quite different methods from those of A. J. Hoffman and his co-authors, D. J. Crampin and A. J. W. Hilton have independently shown the existence of anti-abelian quasigroups for all but 217 values of n. (For the details, see R. K. Brayton, D. Coppersmith and A. J. Hoffman [1] and D. J. Crampin and A. J. W. Hilton [1] and [2].

For a generalization to three dimensions, see T. Evans [3].

12.5. Miscellaneous results

In the last section of this chapter we discuss a number of miscellaneous results.

Firstly, we mention that, from the statistical point of view, the non-existence of a pair of orthogonal latin squares of order 6 is very inconvenient and a way of partially overcoming the difficulty is to use instead of a latin square a generalized form of square in which each element appears the same number of times in each row and each column, but where this number of times may be more than once. The idea of using such generalized latin squares was first conceived by D. J. Finney (see for example, D. J. Finney [1], [2] and [4]) and has recently been developed much more fully by A. Hedayat and E. Seiden in [3]. These authors have called such a generalized latin square a *frequency square* or, more briefly, an *F-square*. Precisely, they have made the following definition:

Definition. Let $A = \|a_{ij}\|$ be an $n \times n$ matrix and let $\Sigma = (c_1, c_2, \ldots, c_m)$, $m \leq n$, be the set of distinct elements of A. Suppose further that, for each i, $i = 1, 2, \ldots, m$, the element c_i appears precisely λ_i times ($\lambda_i \geq 1$) in each row and column of A. Then A is called an *F-square* of order n on the set Σ and with *frequency vector* $(\lambda_1, \lambda_2, \ldots, \lambda_m)$.

Briefly, we say that the matrix A is an $F(n; \lambda_1, \lambda_2, \ldots, \lambda_m)$ square. We may abbreviate the notation further by allowing powers of the λ_i to stand for repeated occurrences of the same symbol. Thus, an $F(n; \lambda_1, \lambda_1, \lambda_3, \lambda_4, \lambda_4, \lambda_6)$ square may be denoted by $F(n; \lambda_1^2, \lambda_3, \lambda_4^2, \lambda_6)$.

As illustration of the definition and notation, we exhibit in Fig. 12.5.1 examples of an $F(5; 2, 1^3)$ square and an $F(6; 2^3)$ square on the set $\Sigma = \{1, 2, 3\}$.

1	4	3	2	1
2	1	1	4	3
1	1	4	3	2
3	2	1	1	4
4	3	2	1	1

1	2	3	2	1	3
2	3	1	1	3	2
3	1	2	3	2	1
3	2	1	1	2	3
1	3	2	2	3	1
2	1	3	3	1	2

Fig. 12.5.1

Hedayat and Seiden have also defined orthogonality of F-squares as follows:

Definition. The two F-squares $F_1(n; \lambda_1, \lambda_2, \ldots, \lambda_l)$ defined on the set $\Sigma_1 = \{a_1, a_2, \ldots, a_l\}$ and $F_2(n; \mu_1, \mu_2, \ldots, \mu_m)$ defined on the set $\Sigma_2 = \{b_1, b_2, \ldots, b_m\}$ are *orthogonal* if the ordered pair (a_i, b_j) appears $\lambda_i \mu_j$ times when the square F_1 is superimposed on the square F_2.

They have given the following example of a set of three mutually orthogonal F-squares.

$F_1(5; 2^2, 1)$

1	2	3	1	2
2	1	2	3	1
1	2	1	2	3
3	1	2	1	2
2	3	1	2	1

$F_2(5; 1^2, 3)$

1	2	3	3	3
3	3	1	2	3
2	3	3	3	1
3	1	2	3	3
3	3	3	1	2

$F_3(5; 1^3, 2)$

1	2	3	4	4
3	4	4	1	2
4	1	2	3	4
2	3	4	4	1
4	4	1	2	3

With the aid of these concepts and a further generalization of the definition of orthogonality to give a stronger relation between F-squares which they call *orthogonal richness*, Hedayat and Seiden have obtained certain new criteria for a given latin square to have an orthogonal mate and they have also suggested some applications of their ideas to the provision of new experimental designs for statistical experiments. It is, for example, possible to construct pairs of orthogonal F-squares of order 6.

Another generalization of the latin square whose use for statistical work has been proposed is the so-called *modified latin square*, itself a generalization of a design called a *semi-latin square* which had been introduced earlier by F. Yates (see [2] and [3]).

DEFINITION. A *modified latin square* has b rows and b sections arranged into a $b \times b$ square. Each section has a cells (or "plots") and ab different elements (or "treatments") are distributed among these cells in such a way that every element occurs once in each row and once in each section. When $a = 1$ we get an ordinary $b \times b$ latin square, and when $a = 2$ we get a semi-latin square.

The concept was introduced by B. Rojas and R. F. White (see [1]). We content ourselves here with giving an example for the case when $a = 4$ and $b = 3$.

	Section (1)				Section (2)				Section (3)			
Row (1)	2	6	3	5	1	11	10	8	4	9	7	12
Row (2)	11	7	12	8	2	4	5	9	3	1	10	6
Row (3)	1	9	4	10	3	7	12	6	2	8	11	5

In 1898, that is to say very much earlier than either of the above generalizations of the latin square concept, a way of generalizing the concept of a pair of mutually orthogonal latin squares of somewhat similar kind was investigated by E. Barbette in [2]. He proposed the problem of arranging m^2 officers of m different ranks and chosen from m different regiments in a square $m \times m$ array in such a way that the officers in each row should be one from each regiment but that, in any particular row, only h ranks should be represented while the officers in each column should be one of every rank but, in any particular column, only h of the regiments would be represented, where h is a divisor of m. In the case $h = m$, the problem becomes that of finding a pair of orthogonal latin squares of order m.

Barbette asked for solutions for each of the cases $h = m/2^r$, $r = 0, 1, 2, \ldots, k$, where $m = 2^k(2p+1)$ calling these the first, second, ..., $(k+1)$th problems. However apart from the case $h = m$ (the *first* problem), he exhibited solutions for the case $h = 2$ only, under the heading of *second* problem. As he gave no explanation of the discrepancy, the present authors are not able to understand fully what he intended. (Similar lack of understanding of the precise problem to be solved seems to have been experienced by B. Heffter, who wrote a later note [1] in which he proposed an erroneous general solution to what he believed to be Barbette's problem. In this connection, see also H. Dellanoy and E. Barbette [1].)

We give Barbette's solutions for the cases $m = 6$ and $m = 8$ in Fig. 12.5.2. (In each case $h = 2$.) Barbette pointed out that each of the arrays which he gave could be used to construct a magic square by replacing the ordered pairs x_y in the various cells of the array by numbers of the form $(x - 1)m + y$. The reader may like to note the similarity between this and the construction used in theorem 6.2.2.

$$
\begin{array}{|cccccc}
\hline
1_1 & 6_5 & 6_4 & 1_3 & 6_2 & 1_6 \\
5_6 & 2_2 & 5_3 & 5_4 & 2_5 & 2_1 \\
4_6 & 3_5 & 3_3 & 3_4 & 4_2 & 4_1 \\
3_1 & 4_5 & 4_3 & 4_4 & 3_2 & 3_6 \\
2_6 & 5_2 & 2_4 & 2_3 & 5_5 & 5_1 \\
6_1 & 1_2 & 1_4 & 6_3 & 1_2 & 6_6
\end{array}
\qquad\qquad
\begin{array}{|cccccccc}
\hline
1_1 & 8_7 & 8_6 & 1_4 & 1_5 & 8_3 & 8_2 & 1_8 \\
7_8 & 2_2 & 2_3 & 7_5 & 7_4 & 2_6 & 2_7 & 7_1 \\
6_8 & 3_2 & 3_3 & 6_5 & 6_4 & 3_6 & 3_7 & 6_1 \\
4_1 & 5_7 & 5_6 & 4_4 & 4_5 & 5_3 & 5_2 & 4_8 \\
5_1 & 4_7 & 4_6 & 5_4 & 5_5 & 4_3 & 4_2 & 5_8 \\
3_8 & 6_2 & 6_3 & 3_5 & 3_4 & 6_6 & 6_7 & 3_1 \\
2_8 & 7_2 & 7_3 & 2_5 & 2_4 & 7_6 & 7_7 & 2_1 \\
8_1 & 1_7 & 1_6 & 8_4 & 8_5 & 1_3 & 1_2 & 8_8
\end{array}
$$

Fig. 12.5.2

Another problem which concerns a modification of the concept of a pair of mutually orthogonal latin squares and to which a number of authors have recently turned their attention is that of constructing perpendicular commutative quasigroups, as we mentioned at the end of section 12.1. If the order of a commutative quasigroup is odd, it is necessarily diagonal, as shown in section 1.4, and, in the constructions which have been considered up to the present time, the latin squares representing the multiplication tables of perpendicular commutative quasigroups have always been so arranged that the elements of their main left-to-right diagonals are in natural order: that is, so that the squares are idempotent.

In K. B. Gross, R. C. Mullin and W. D. Wallis [1], the maximum number $\nu(r)$ of idempotent symmetric latin squares of order r which can occur in a pairwise perpendicular set has been investigated. The procedure used by these authors makes use of a connection with Room designs via the fact that every Room design defines and is defined by a pair of Room quasigroups (as shown in section 6.4) and that a pair of Room quasigroups is a pair of perpendicular idempotent symmetric latin squares.

Steiner quasigroups (defined in section 2.2) also have multiplication tables which are idempotent symmetric latin squares. Consequently, orthogonal Steiner triple systems and their associated perpendicular Steiner quasigroups provide examples of pairs of perpendicular idempotent symmetric latin squares. Perpendicular commutative quasigroups have been investigated from this point of view in C. D. O'Shaughnessy [1], N. S. Mendelsohn [4], C. C. Lindner and N. S. Mendelsohn [1], R. C. Mullin and E. Németh [2] and F. Rado [1]. (See also section 5.4.)

Let us end by making brief reference to a variety of papers. In D. A. Norton and S. K. Stein [1], idempotent latin squares of given order n have been discussed and an integer N introduced whose value is dependent on the disposition of the off-diagonal elements of the square (that is, those elements not on the main left-to-right diagonal). It has been proved that the relation $N \equiv n(n-1)/2$ modulo 2 always holds whatever the explicit value of N.

In [5], N. S. Mendelsohn has proved that if n is a prime power, $n = p^r$, then there exist sets of $n-2$ pairwise orthogonal idempotent latin squares, this being the maximum possible number that can exist for any n. He has given an explicit method for constructing such sets.

In U. Colombo [1], an expression for the total number of non-equivalent sets of s mutually orthogonal latin squares of order $p = p_1^{\alpha_1}p_2^{\alpha_2} \cdots p_r^{\alpha_r}$ has been obtained under the conditions that (i) the sets of squares all have one square (the *starting square*) in common, (ii) that the p_i are primes and $s \leq \min (p_i^{\alpha_i} - 1)$ and (iii) that the elements of the squares are derived from the r Galois fields $GF[p_i^{\alpha_i}]$, $i = 1, 2, \ldots, r$.

C. C. Lindner has shown in [10] how to construct quasigroups having a large number of orthogonal mates no two of which are isomorphic. In particular, he has obtained a quasigroup of order 22 which has 512 distinct orthogonal mates. His construction makes use of the singular direct product (see section 12.1).

In an interesting paper, [3], written by A. Hedayat and W. T. Federer, an investigation is made of the conditions under which a set of r mutually orthogonal latin squares of order n can be embedded in a set of t such squares, $1 \leq r < t \leq n - 1$. One of the main results obtained is the following. Let $n = p_1^{\alpha_1}p_2^{\alpha_2} \ldots p_s^{\alpha_s}$ be the decomposition of n into the product of powers of primes and let

$$W(n) = [(n-1)!]^3 \prod_{i=1}^{s} p_i \Big/ \prod_{i=1}^{s} (p_i - 1).$$

Then, if n is even, there exist at least $W(n)$ latin squares of order n having no orthogonal mate. If n is odd, there exist at least $W(n)$ latin squares of order n which do have an orthogonal mate. As an example of the kind of deduction which the authors are able to make we mention that, if L is a latin square of order 5, then either L has no orthogonal mate or else L can be embedded in a set of four mutually orthogonal 5×5 latin squares.

In P. Kesava Menon [1], an elegant direct method for constructing pairs of orthogonal latin squares of order $3h + 1$, h an integer, is given. The squares obtained are the same as those obtained by another method in R. C. Bose, S. S. Shrikhande and E. T. Parker [1].

In F. de Sousa [1], a correspondence has been established between the existence of sets of mutually orthogonal latin squares and the existence of certain block designs. P. Hall's theorem on subset representatives has then been employed to yield a new criterion for the existence of t mutually orthogonal latin squares of order n. In particular, the author has been able to deduce the existence of at least four mutually orthogonal latin squares of each of the orders 35, 40 and 45. (For some more general results on this subject, see section 11.4.)

Let the rows of an $n \times n$ latin square L be represented as permutations $P_0 = I, P_1, P_2, \ldots, P_{n-1}$ of its first row. Then these n permutations generate a group L_G which is a subgroup of the symmetric group. We shall call such a group L_G a *latin square group*. In J. J. Carroll, G. A. Fisher, A. M. Odlyzko and J. A. Sloane [1], the following questions have been raised: (*i*) Which groups contain a latin square group as a subgroup? (*ii*) How can the latin square groups be characterized? The questions have relevance in the theory of sequential machines.

Added In Proof. The following papers treat certain aspects of the subject matter of this book from novel points of view and it seems worthwhile to bring them to the attention of the reader: B. Balakrishnan 1], D. A. Klarn er and M. L. J. Hautus [1], J. Q. Longyear [1].

CHAPTER 13

The application of computers to the solution of latin square problems

Since the advent of fast digital computers, a number of authors have realized that these could provide a useful extra tool in the investigation of combinatorial problems in addition to their everyday use for solving problems of a numerical type. There are two principal ways in which they can be so used. The first is in generating sets of "basic" combinatorial patterns for use in connection with a larger combinatorial problem. The second is in carrying out complete constructions of, or searches for, combinatorial structures of a particular kind where the nature of the structure is such that constructions or searches by hand would not be feasible. The necessity of using a computer in these cases could be due either to the large size of the structure or to the length of the search required.

However, many combinatorial problems (indeed, probably the majority) are of such a nature that, while they can readily be tackled by hand when only a small number n of elements are involved (usually n is less than or equal to 10) increase very rapidly in magnitude when n becomes larger. Such problems may still remain within the capacity of present day computers when n is increased to $n + 2$ or $n + 3$ say, but usually become quite unmanageable if n is increased further. In practice, for most computational problems values of n greater than 20 are unmanageable. For example any problem which involves finding all permutations of n elements which have certain specified properties is of this nature. This aspect of the situation is particularly relevant to problems involving latin squares.

For investigations involving latin squares, the computing procedure which is most commonly used is that which has come to be known as the *backtrack* method, a name originally given to it by D. H. Lehmer in [1]. This method(1) has been very well described in general terms by a number of authors. See, for example, S. W. Golomb and L. D. Baumert [1], M. Hall and D. E. Knuth [1], D. H. Lehmer [1], and R. J. Walker [1]. The method is often used in conjunction with other techniques, such as temporary identification of symbols, which are designed to reduce the amount of construction or searching which has to be carried out. We shall give some examples of such techniques in the following pages. (For further information about these techniques, the reader may consult C. B. Tomkins [1] and M. B. Wells [2].) We shall first consider problems which concern single latin squares and then investigations which involve orthogonal sets.

13.1. Single latin squares

The simplest problem which can be dealt with by exhaustive trial is that of constructing "all" latin squares of a given order n or a "random" latin square of specified order. In each case, it is first necessary to define precisely what is to be meant by the term "all" or the term "random" as the case may be. (The reader is referred to R. T. Ostrowski and K. D. Van Duren [1] for a description of one computer programme for generating "random" 10×10 latin squares.)

If one requires literally to determine all latin squares, then the simplest method which comes to mind is first to fill in the first row with the digits 0 to $n - 1$ (or 1 to n) and then to ask the computer to fill each cell of the second row in turn, trying successively the digits 0 to $n - 1$ until a digit is found which does not violate the conditions that the square be latin. The same is repeated for the cells of the third row and so on. If at any stage, a cell cannot be filled, the computer is required to backtrack to the previous cell and try again with the digit in that cell increased by one.

When a square has been completed, it is printed out and the computer programme is then re-entered at the last cell of the next to last row with the digit of that cell increased by one. The programme will then automatically backtrack as far as is necessary to enable the computer to generate the "next" possible latin square. By this procedure, the

(1) In a note to the authors, D. E. Knuth has suggested that the backtrack method can be traced back much further into history, perhaps to L. Euler and the Königsberg Bridges problem. For example, it is essentially the method for finding a route in a graph which passes just once in each direction along every edge.

squares will be pointed out in lexicographical order. This is a classic example of a primitive use of the backtrack method.

A more practical and efficient way of generating latin squares is the following. We first write n 1's into an $n \times n$ square in such a way that the n cells occupied by these digits occur one in each row and one in each column. Next, we write n 2's into n of the remaining $n^2 - n$ cells of the square, again choosing the cells to be such that there is one of them in each row and one in each column of the latin square. We repeat the procedure with n 3's, then n 4's, and so on, until either a complete latin square is obtained or the procedure stops because there is no suitable place into which one of the digits can be written. When the latter occurs, it is generally sufficient to change the position of the immediately preceding entry filled or to backtrack a few steps and enter the particular digit which is being placed in a new sequence of cells selected so as to occur one in each row and one in each column. It is rarely necessary to make any changes in the placing of earlier digits.

The computer programme for such a procedure is easy to write. The following flow-chart (Fig. 13.1.1) illustrates the essential part of the programme, that is, the method by which the successive elements are written in. Essentially, the procedure is the following. If one of the elements has not yet been written into all of the n rows, one can make correspond to each of the rows which do not contain this element certain *degrees of freedom*, expressed in terms of the number of cells in that row into which the element in question can be written in. As one can observe by taking a glance at the flow-chart, at the beginning of the procedure for entering a particular element the degree of freedom is prescribed by the programme. If out of n elements $n - k$ have been already written in, this initial degree of freedom is obviously k. Then, the computer is programmed to search for the row with the minimum degree of freedom. If several rows have the same degree of freedom, it simply selects the first one. The element in question is then written into any one of the available places in that row chosen at random. After the new element has been written in, the number of degrees of freedom for each row has to be recalculated accordingly. If the degree of freedom of a row becomes 0; that is, no more entries are available, it is usually sufficient to change the positions of the previous element. Practical experience has shown that with this procedure, the second trial proves to be successful in almost every case (see W. von Lockemann [1]).

The above procedure has been programmed and run on a Siemens 2002 computer. The programme contains 300 instructions and the run-

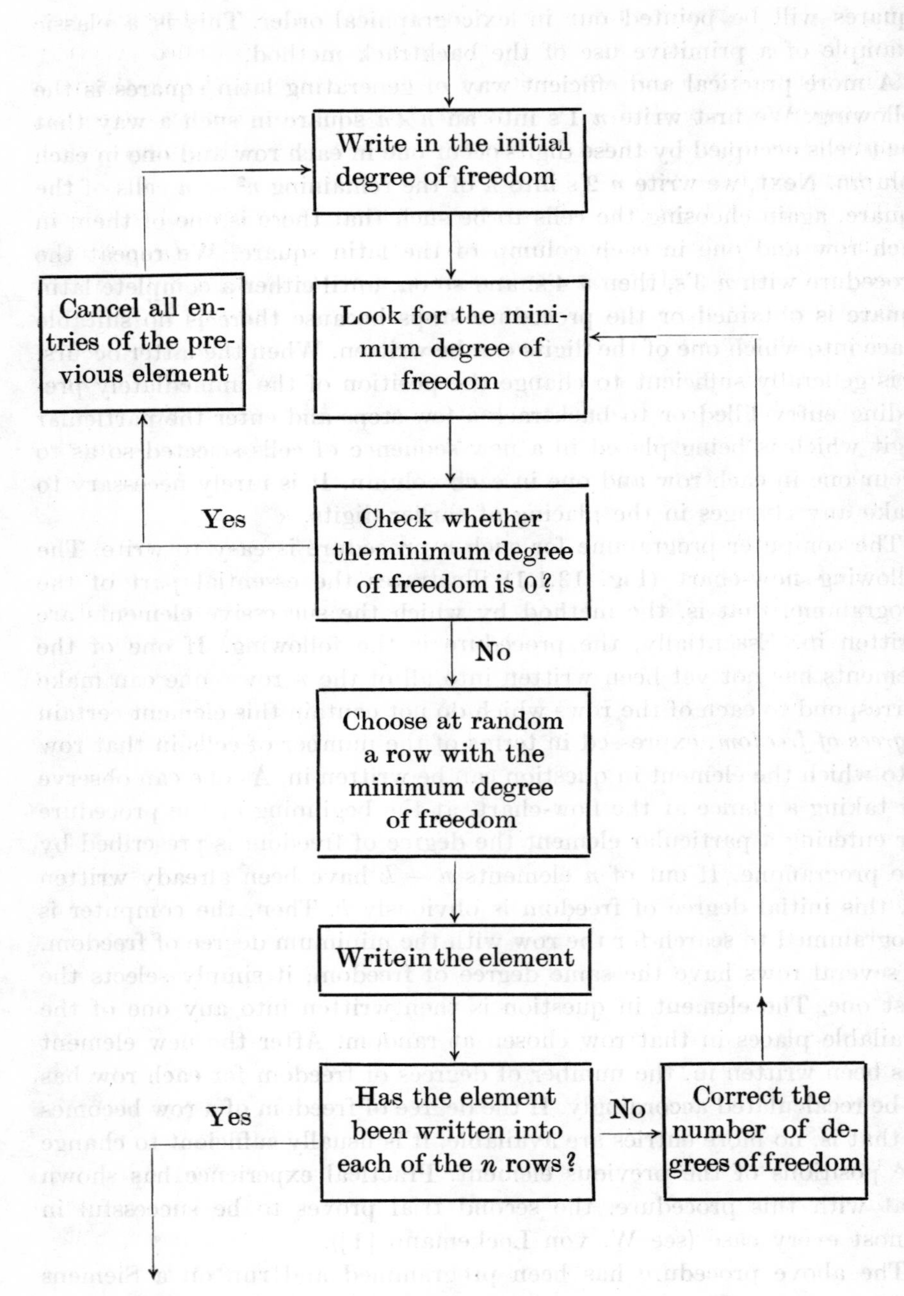
Write in the initial degree of freedom
Look for the minimum degree of freedom
Cancel all entries of the previous element
Yes
Check whether the minimum degree of freedom is 0?
No
Choose at random a row with the minimum degree of freedom
Write in the element
Yes
Has the element been written into each of the n rows?
No
Correct the number of degrees of freedom

Fig. 13.1.1

ning of the programme takes only a few minutes. It is of interest to note that the element last written in has seldom to be cancelled. Usually, cancellation of the previous element is necessary only if two thirds of the square has already been completed. This fact provides an interesting empirical result relevant to the theory of incomplete latin squares which we discussed in section 3.3.

The reader of chapter 4 will have realized that the problem of enumerating all latin squares of an assigned order n is not yet solved. In particular cases, such as that of enumerating all latin squares of order 8, computers have been used as an aid as, for example, by J. W. Brown in [1].

This author used an IBM 7094 computer for his investigation and, in order to effect some economy in the computation, he devised a number of criteria for determining whether two given latin rectangles of equal size are transformable one into the other by permuting rows, permuting columns, and interchanging elements. (In J. W. Brown [1], two latin rectangles related in this way are called *isomorphic*.)

Also, as already mentioned in chapter 4, B. F. Bryant and H. Schneider used a computer as an aid in their solution of the problem of determining all loops of order 6. Their method involved an enumeration of the non-isomorphic principal loop-isotopes of the quasigroups of order 6 and, for this purpose, they made use of the class representatives given in R. A. Fisher and F. Yates [1] for the 22 isotopy classes of quasigroups of order 6. They showed that there exist all together 109 isomorphically distinct loops of order 6. (For the details, see B. F. Bryant and H. Schneider [1].)

In G. B. Belyavskaya [6], a computer algorithm for generating class representatives of all latin squares up to order 32 inclusive has been devised. In addition, Belyavskaya has developed a number of other useful computer algorithms relating to latin square. These include an algorithm for determining whether a given latin square has a transversal and also a procedure for counting the number of transversals possessed by a given latin square. Further, Belyavskaya has written a computer algorithm for obtaining a prolongation of a given latin square; provided, of course, that the given square has a transversal. (For an explanation of the concept of a prolongation, see section 1.4.)

The reader will remember that in section 2.3 we investigated complete latin squares and showed their close connection with sequenceable groups. An enumeration of sequenceable groups of low order has been attempted by one of the present authors in collaboration with É. Török, and, once again, a computer was used. The computer employed was an ICT 1905 and full details of the computational method which was used has been given in É. Török [1].

The results show that, among the non-abelian groups of order 12 or less, only the dihedral groups D_5 and D_6 are sequenceable, and we give some details below of the computational results obtained. (The dihedral group D_m is the group generated by two elements a and b which satisfy the generating relations $a^m = b^2 = e$ and $ab = ba^{-1}$, where e denotes the identity element of the group.)

A product $g_0, g_1, \ldots, g_{k-1}$ of k distinct group elements is said to be a *partial sequencing* of length k if $g_0 = e$ and the partial products g_0, $g_0g_1, g_0g_1g_2, \ldots, g_0g_1 \ldots g_{k-1}$ are all distinct. A group G of order n is sequenceable if and only if it possesses one or more sequencings of length n.

RESULTS.

Dihedral group D_3 of order 6

Number of partial sequencings of length 3 obtained = 18
of length 4 = 12
of length 5 = 0

The group is not sequenceable. (Total number of products tested = 30)

Dihedral group D_4 of order 8

Number of partial sequencings of length 4 obtained = 24
of length 5 = 152
of length 6 = 272
of length 7 = 0

The group is not sequenceable. (Total number of products tested = 448)

Quaternion group Q_3 of order 8

Number of partial sequencings of length 4 obtained = 72
of length 5 = 216
of length 6 = 48
of length 7 = 0

The group is not sequenceable. (Total number of products tested = 336)

Dihedral group D_5 of order 10

Number of partial sequencings of length 5 obtained = 280
of length 6 = 1920
of length 7 = 3920
of length 8 = 2240
of length 9 = 320

The group is sequenceable. (Total number of products tested = 8680)

The first sequencing of length 10 obtained is

$e \quad a \quad a^2 \quad b \quad ba \quad ba^4 \quad ba^2 \quad ba^3 \quad a^3 \quad a^4$

Partial products:

$e \quad a \quad a^3 \quad ba^2 \quad a^4 \quad b \quad a^2 \quad ba \quad ba^4 \quad ba^3$

Dihedral group D_6 of order 12

Number of partial sequencings			
of length	6	obtained =	936
of length	7	=	17,520
of length	8	=	71,580
of length	9	=	108,840
of length	10	=	57,312
of length	11	=	3,072

The group is sequenceable. (Total number of products tested = = 259,260.)

The first sequencing of length 12 obtained is

$e \quad a \quad a^2 \quad b \quad a^3 \quad ba^2 \quad ba \quad ba^4 \quad ba^3 \quad a^4 \quad a^5 \quad ba^5$

Partial products:

$e \quad a \quad a^3 \quad ba^3 \quad b \quad a^2 \quad ba^5 \quad a^5 \quad ba^4 \quad ba^2 \quad ba \quad a^4$

Group $G = \{a, b / a^3 = b^4 = e,\ ab = ba^{-1}\}$ *of order* 12

Number of partial sequencings			
of length	6	obtained =	1,152
of length	7	=	14,832
of length	8	=	64,560
of length	9	=	85,824
of length	10	=	39,792
of length	11	=	0

The group is not sequenceable. (Total number of products tested = = 206,160.)

Alternating group A_4 of order 12

Number of partial sequencings			
of length	6	obtained =	1,032
of length	7	=	16,224
of length	8	=	63,480
of length	9	=	91,248
of length	10	=	41,472
of length	11	=	0

The group is not sequenceable. (Total number of products tested = = 213,456.)

The amount of computer running time required for each of the three non-abelian groups of order 12 was 45 to 55 minutes. For groups of orders greater than 12, the running time would have been much longer than this and so, for these groups, the programme was not run to completion but was stopped either after a pre-set time or as soon as a sequencing (of length equal to the order of the group) was obtained.

By this means, the following results were obtained:

For the dihedral group D_7 of order 14, the following elements form a sequencing,

$e \quad a \quad a^2 \quad a^3 \quad b \quad a^4 \quad a^6 \quad a^5 \quad ba^6 \quad ba^3 \quad ba^4 \quad ba \quad ba^5 \quad ba^2$

Partial products:

$e \quad a \quad a^3 \quad a^6 \quad ba \quad ba^5 \quad ba^4 \quad ba^2 \quad a^4 \quad ba^6 \quad a^5 \quad ba^3 \quad a^2 \quad b$

For the dihedral group D_8 of order 16, the following elements form a sequencing,

$e \quad a \quad a^2 \quad a^3 \quad b \quad a^4 \quad a^5 \quad a^6 \quad ba^5 \quad ba^4 \quad a^7 \quad ba^6 \quad ba^3 \quad ba \quad ba^2 \quad ba^7$

Partial products:

$e \quad a \quad a^3 \quad a^6 \quad ba^2 \quad ba^6 \quad ba^3 \quad ba \quad a^4 \quad b \quad ba^7 \quad a^7 \quad ba^4 \quad a^5 \quad ba^5 \quad a^2$

For the non-abelian group of order 21 generated by two elements a and b which satisfy the generating relations $a^3 = b^7 = e$ and $a^{-1}ba = b^2$, the following lists of elements form sequencings,

$e, b, b^2, b^3, b^5, a, b^4, ab, b^6, ab^3, a^2b, a^2b^2, ab^6, a^2, ab^2, a^2b^4, ab^5, a^2b^6, ab^4, a^2b^3, a^2b^5$

$e, b, b^2, b^3, b^5, a, b^4, ab, b^6, a^2b^2, ab^6, a^2, ab^2, ab^3, a^2b, a^2b^4, ab^4, a^2b^3, ab^5, a^2b^5, a^2b^6$

$e, b, b^2, b^3, b^5, a, b^4, ab, b^6, a^2b^2, ab^6, a^2b^6, ab^5, a^2b^5, ab^3, a^2, ab^4, ab^2, a^2b, a^2b^3, a^2b^4$

$e, b, b^2, b^3, b^5, a, b^4, ab, b^6, a^2b^2, ab^6, a^2b^6, a^2, a^2b^4, a^2b^3, ab^4, a^2b^4, ab^3, a^2b^5, ab^2, ab^5$

$e, b, b^2, b^3, b^5, a, b^4, ab, b^6, a^2b^4, ab^4, ab^3, a^2b, a^2, ab^2, a^2b^2, ab^6, a^2b^5, ab^5, a^2b^3, a^2b^6$

$e, b, b^2, b^3, b^5, a, b^4, ab, b^6, a^2b^5, ab^2, a^2b^2, ab^4, a^2b^3, a^2, a^2b^4, a^2b, ab^6, a^2b^6, ab^3, ab^5$

$e, b, b^2, b^3, b^5, a, b^4, ab, b^6, a^2b^5, ab^2, a^2b^2, ab^4, a^2b^6, ab^6, a^2b, a^2, a^2b^4, a^2b^3, ab^3, ab^5$

$e, b, b^2, b^3, b^5, a, b^4, ab, b^6, a^2b^5, ab^3, a^2b^6, ab^6, a^2b, a^2, a^2b^4, a^2b^3, ab^4, a^2b^2, ab^2, ab^5$

$e, b, b^2, b^3, b^5, a, b^4, ab, b^6, a^2b^5, ab^3, a^2b^3, a^2, a^2b^4, a^2b, ab^6, a^2b^6, ab^4, a^2b^2, ab^2, ab^5$

$e, b, b^2, b^3, b^5, a, b^4, ab, b^6, a^2b^6, ab^4, ab^2, a^2b, a^2, ab^6, a^2b^3, ab^3, a^2b^5, ab^5, a^2b^2, a^2b^4$

$e, b, b^2, b^3, b^5, a, b^4, ab, b^6, a^2b^6, ab^4, ab^2, a^2b, a^2b^2, ab^5, a^2b^5, ab^3, a^2b^3, ab^6, a^2, a^2b^4$

$e, b, b^2, b^3, b^5, a, b^4, ab, b^6, a^2b^6, ab^6, a^2b^2, ab^2, ab^4, a^2b, a^2b^5, ab^3, a^2b^3, ab^5, a^2b^4, a^2$

$e, b, b^2, b^3, b^5, a, b^4, ab, b^6, a^2b^6, a^2, a^2b^4, a^2b, ab^6, a^2b^3, ab^4, a^2b^2, ab^2, a^2b^5, ab^3, ab^5$

$e, b, b^2, b^3, b^5, a, b^4, ab, ab^4, a^2b^6, a^2b, ab^6, a^2b^3, ab^5, b^6, a^2b^2, ab^2, a^2b^5, ab^3, a^2b^4, a^2$

$e, b, b^2, b^3, b^5, a, b^4, ab, a^2, ab^2, a^2b, ab^6, ab^5, a^2b^6, a^2b^4, ab^4, b^6, a^2b^5, ab^3, a^2b^3, a^2b^2$

N. S. Mendelsohn has obtained some new complete latin squares by the following simple method. For a given group G, he has formed a partial sequencing $g_0g_1g_2 \cdots g_{k-1}$ of the group elements by hand trial up to the point at which the partial sequencing has reached a length, say length k, after which no group element suitable for continuing the product is available. He has then fed his partial sequencing into a computer equipped with a (fairly primitive) backtracking programme by means of which such partial sequencings may be modified as necessary (by backtracking) and thence completed by successive trials. This method has been eminently successful in producing new sequencings of the non-abelian group of order 21. (For further details, see N. S. Mendelsohn [1].)

We turn next to problems concerning the construction of orthogonal latin squares.

13.2. Orthogonal latin squares

Computers have been used as an aid in essentially two kinds of orthogonal latin square problems. In the first place, they have been used in searching for sets of mutually orthogonal latin squares of an assigned order n which shall contain as many squares as possible. In the second place, they have been used in attempts to construct new non-desarguesian finite projective planes and hence complete sets of mutually orthogonal latin squares which would be non-equivalent to any previously known set.

The earliest search of the first kind known to the authors was that of L. J. Paige and C. B. Tomkins, (see [1]) a description of which was published(1) in 1960. These authors attempted to find a counter-example to the Euler conjecture by finding a pair of orthogonal latin squares of order 10, ten being the smallest integer greater than two and six which is an odd multiple of 2.

The first method they contemplated was to take a randomly generated 10×10 latin square and to use their SWAC computer to search directly for an orthogonal mate, where, without loss of generality the first rows of the two squares could be assumed to be the same. They estimated from trial runs that such a complete search would take of the order of $4 \cdot 8 \times 10^{11}$ machine hours for each initially selected square and they pointed out the difficulties of devising a method which would reject equivalent pairs of orthogonal squares and hence eliminate duplication of trials. They called two pairs *equivalent* if by rearranging rows, rearranging columns, or renaming the symbols of both squares of one pair simultaneously it would become either formally identical to the second pair or would differ from the second pair only in having its two squares oppositely ordered. (In the latter case, if L_1 and L_2 were the squares of one orthogonal pair, L_2 and L_1 would be those of the other.)

They then discussed a method by which only pairs of squares having a combinatorial pattern of such a kind that orthogonality would be possible would be constructed. This method involved identifying some of the symbols used in each latin square and is said to have been originally proposed by E. Seiden and W. Munro. Let us suppose that the symbols used in a given 10×10 latin square are the digits 0 to 9, and let each of the digits 0 to 4 be replaced by α and each of the digits 5 to 9 be

(1) However, see S. Cairns [1] for a comment on an attempted computer investigation carried out in the early 1950's.

replaced by β. In the resulting square, the symbols α and β each occur five times in each row and five times in each column. If there exists a second latin square orthogonal to the first which is similarly treated and if the two squares are then placed in juxtaposition, we shall get a 10×10 square matrix each element of which is one of the ordered pairs of symbols (or *digraphs*) $\alpha\alpha$, $\alpha\beta$, $\beta\alpha$, or $\beta\beta$, and with the following properties. Each of the four digraphs will occur all together 25 times in the matrix and in each row of the matrix there will be five digraphs whose first member is α and five whose first member is β, likewise five digraphs whose second member is α and five whose second member is β. The same statements will be true for the columns of the matrix. Such a square matrix was called *admissible* by Paige and Tomkins. They obtained further conditions which would have to be satisfied by a matrix constructed from two orthogonal latin squares in this way. The essentials of their computational procedure were first to compute an "admissible" 10×10 matrix and then to reverse the process just described by replacing the first member of each digraph by one of the digits 0 to 9 so as to get a latin square (α being replaced by one of the digits 0 to 4 and β by one of the digits 5 to 9) and subsequently to use the computer again to assign digits to all the cells of the partially specified orthogonal mate. That is, to enter one of the digits 0 to 4 in each α-cell and one of the digits 5 to 9 in each β-cell until one of the orthogonality conditions was violated. When a violation occurred, the programme backtracked in the usual manner to an earlier cell.

Unfortunately neither this nor the previous method was successful in producing an orthogonal pair of 10×10 latin squares despite the fact that more than 100 hours of computing time was used and, shortly afterwards, indeed even before the publication of Paige and Tomkins' paper, the Euler conjecture was disproved by theoretical means. (See chapter 11.)

Almost immediately after the first counter-example to the Euler conjecture was obtained, E. T. Parker found a pair of orthogonal latin squares of order 10 by a construction which made use of statistical designs, and this sequence of events was regarded by many as a triumph for mathematical theory over computer search. However, not long afterwards, E. T. Parker himself initiated a new computer search for latin squares of order 10 which have orthogonal mates using a faster computer and a more sophisticated computer programme. The essential refinement was to generate and store all the transversals of a given 10×10 latin square and to search for disjoint sets of ten transversals (equivalent to finding an orthogonal mate) rather than to attempt to fill the cells of a partial orthogonal mate individually. E. T. Parker has given a full description of his method and results in [8]. He discovered that 10×10 latin squares with orthogonal mates are not, in fact, particularly scarce and he also showed that there exist such squares with a large number of alternative orthogonal mates. His most striking result concerns the square displayed in Fig. 13.2.1 which has 5504 transversals and an estimated one million alternative orthogonal mates (that is, sets of 10 disjoint transversals). However, Parker was able to show by a partly theoretical argument that no two of these alternative orthogonal mates are themselves orthogonal and so, much to his own disappointment, he was not able to obtain a triad of mutually orthogonal 10×10 latin squares. The existence or non-existence of such triads remains an open question.

5	1	7	3	4	0	6	2	8	9
1	2	3	4	5	6	7	8	9	0
7	3	4	5	6	2	8	9	0	1
3	4	5	6	7	8	9	0	1	2
4	5	6	7	8	9	0	1	2	3
0	6	2	8	9	5	1	7	3	4
6	7	8	9	0	1	2	3	4	5
2	8	9	0	1	7	3	4	5	6
8	9	0	1	2	3	4	5	6	7
9	0	1	2	3	4	5	6	7	8

Fig. 13.2.1

Quite recently (1966), one of the present authors has carried out a partial search by computer for a triple of 10×10 mutually orthogonal latin squares of a special kind, the possible existence of such a triple having been suggested to him by his diagonal method of construction described in section 7.5. The search was for 9×9 arrays having the properties specified in the following theorem 13.2.1. A proof of the theorem and further details are given in A. D. Keedwell [4]. No complete arrays were obtained in the 90 minutes of machine time allocated to the search but only about a millioneth part of the search could be carried out in the time available. An ICT "Atlas" computer was used and the programme was coded in Atlas Basic Language.

Theorem 13.2.1. *If an array of the kind shown in Fig.* 13.2.2 *can be set up in which each* a_{ij} *is one of the integers* $0, 1, \ldots, 9$ *and the following* *properties (i), (ii), and (iii) hold, then there exists a triad of mutually orthogonal latin squares of order* 10. *The required properties are: (i) each of the integers* 0, 1, . . . , 9 *occurs exactly once in each row and column (brackets being disregarded for this purpose); (ii) if* $a_{1,j_1} = a_{2,j_2} = \ldots = a_{9,j_9} = r$ *then (a) the integers* $r + 1$, a_{1,j_1+1}, a_{2,j_2+1}, . . . a_{9,j_9+1} *are all different and (b) the integers* $r + 2$, a_{1,j_1+2}, a_{2,j_2+2}, . . . , a_{9,j_9+2} *are all different* (all addition being modulo 9), $r = 0, 1, \ldots, 8$; *(iii) if* $a_{1,j_1} = a_{2,j_2} = \ldots = a_{9,j_9} = 9$ *then (a) the integers* 9, a_{1,j_1+1}, a_{2,j_2+1} . . . a_{9,j_9+1} *are all different, and (b) the integers* 9, a_{1,j_1+2}, a_{2,j_2+2}, . . . , a_{9,j_9+2} *are all different.*

$$\begin{array}{llllllllll} (9)(0 & 1 & 2 & 3 & 4 & 5 & 6 & 7 & 8) \\ (8)(a_{10} & a_{11} & a_{12} & a_{13} & a_{14} & a_{15} & a_{16} & a_{17} & a_{18}) \\ \cdot\ \ \cdot & \cdot & \cdot & \cdot & \cdot & \cdot & \cdot & \cdot & \cdot \\ (0)(a_{90} & a_{91} & a_{92} & a_{93} & a_{94} & a_{95} & a_{96} & a_{97} & a_{98}) \end{array}$$

Fig. 13.2.2

The disproof of the Euler conjecture had been preceded by the disproof by means of a counter-example of the MacNeish conjecture concerning an upper bound for the number of latin squares of assigned order n in a mutually orthogonal set. (For the details, see section 11.1.) This led a number of authors to seek further counter-examples and, as we stated in chapter 7, two separate groups of mathematicians each obtained sets of five mutually orthogonal latin squares of order 12 at about the same time (1960) but by quite different methods.

The first group, consisting of D. M. Johnson, A. L. Dulmage, and N. S. Mendelsohn, used the method of orthomorphisms which we described in chapter 7. They defined two orthomorphisms to be *orthogonal* if the latin squares constructed by means of them are orthogonal, and they used a machine algorithm to compute maximal sets of mutually orthogonal orthomorphisms for all abelian groups up to and including order 12. They found, in particular, that the cyclic groups of orders 8, 10, and 12 have no non-identity orthomorphisms, that the cyclic group of order 9 has no triad of mutually orthogonal non-identity orthomorphisms, but that the group $G = C_6 \times C_2$ of order 12 has maximal sets of four mutually orthogonal non-identity orthomorphisms. Such a set, together with the identity orthomorphism, yields a set of five mutually orthogonal latin squares of order 12. Further details and a listing of the four mutually orthogonal non-identity orthomorphisms of the group $C_6 \times C_2$ will be found in D. M. Johnson, A. L. Dulmage and N. S. Mendelsohn [2].

Later, the orthomorphisms of all groups up to order 14 inclusive were computed by L. Chang, S. Tai and K. Hsiang. For the details, see L. Chang and S. Tai [1] and also L. Chang, K. Hsiang and S. Tai [1]. The results obtained confirmed those of Johnson, Dulmage and Mendelsohn.

The second group consisted of R. C. Bose, I. M. Chakravarti, and D. E. Knuth. These authors represented the group $G = C_{2t} \times C_2$ of order $4t$ as a module, the elements being vectors (a_i, b_i) where $a_i = 0$ or 1 and b_i is an integer in the range $0 \leq b_i \leq 2t - 1$ and the group operation being vector addition $(a_i, b_i) + (a_j, b_j) = (a_k, b_k)$ where $a_k = a_i + a_j$ modulo 2 and $b_k = b_i + b_j$ modulo $2t$. As we mentioned briefly in section 7.1, a set of orthogonal mappings $M_1 \equiv I, M_2, \ldots, M_l$ is a set of one-to-one mappings $g \to gM_h$ of G onto itself such that, for $h \neq k$, the equation $xM_h - xM_k = g$, with g a given element of G, is uniquely soluble for x in G. A set of l such orthogonal mappings gives rise to a set of l mutually orthogonal latin squares of order $4t$ and based on the group G. That is, one of the squares represents the addition table of G and the others are obtained by rearrangements of the rows (or columns) of the first. Bose, Chakravarti, and Knuth set out to construct a $(4t - 1) \times 4t$ matrix V whose first row would comprise the elements $g_1 = (0, 0)$, $g_2 = (0, 1), \ldots, g_{2t} = (0, 2t - 1)$, $g_{2t+1} = (1, 0)$, $g_{2t+2} = (1, 1), \ldots, g_{4t} = (1, 2t - 1)$ of G in the order just written and whose ith row would consist of the elements $g_j M_i$, $j = 1, 2, \ldots, 4t$. They considered what properties such a matrix would need to have if $M_1 \equiv I, M_2, \ldots, M_{4t-1}$ were to be a set of orthogonal mappings and noted that the matrix V^* formed by the first members of each of the ordered pairs occurring as elements of V would be a matrix of 0's and 1's such that the submatrix formed by any two of its rows would contain each of the ordered pairs $\begin{pmatrix}0\\0\end{pmatrix}, \begin{pmatrix}0\\1\end{pmatrix}, \begin{pmatrix}1\\0\end{pmatrix}, \begin{pmatrix}1\\1\end{pmatrix}$ equally often as columns. They constructed such a matrix V^* of 0's and 1's with the aid of the incidence matrix of a balanced incomplete block design with $v = b = 4t - 1$, $r = k = 2t - 1$, and $\lambda = t - 1$. (For an explanation of this notation, see section 10.2.) For the particular case $t = 3$, they then enlisted the aid of a computer in an attempt to fill in the second entries of the elements of V row by row, so as to define as many mutually orthogonal mappings as possible. In this way, they were able to obtain two distinct sets of five mutually orthogonal mappings but no set of six such mappings. These two sets of mappings gave them corresponding sets of five mutually orthogonal latin squares of order 12. (For a diagram showing one of these sets of five squares and for further details see R. C. Bose, I. M. Chakravarti and D. E. Knuth [1].)

Bose, Chakravarti, and Knuth also proposed to consider the case $t = 5$ but this proved to be too large a problem for the available computer. However, a number of additional and supplementary results were obtained of which details are given in R. C. Bose, I. M. Chakravarti, and D. E. Knuth [2].

We proved in section 7.5 that, for every order n for which a property D-neofield exists, a pair of orthogonal latin squares can be obtained. As already remarked in that section, one of the present authors wrote a computer programme for enumerating all the D-neofields of an assigned order and found that the number of non-isomorphic D-neofields of order n appeared to increase with n. He obtained a complete list of isomorphically distinct D-neofields for all orders n up to and including 17 and these are listed in A. D. Keedwell [5].

We turn now to the construction of complete sets of mutually orthogonal latin squares. In this connection, computer aided searches have led to a number of useful and significant results. A survey article on the subject which the reader may find interesting has been written by M. Hall under the title "Numerical Analysis of finite geometries" (see M. Hall [10]).

We recall from section 5.2 that a complete set of mutually orthogonal latin squares of order n is equivalent to a finite projective plane of that order and, from section 8.4, that non-equivalent complete sets of squares correspond to finite projective planes whose co-ordinatizing planar ternary rings are non-isomorphic. However, non-isomorphic planar ternary rings may co-ordinatize the same projective plane. For example, the unique translation plane of order 9 can be co-ordinatized by four isomorphically distinct right quasifields (Veblen–Wedderburn systems) as is shown in M. Hall [1]. Exceptionally, all co-ordinatizing systems for a Desarguesian plane are isomorphic and are fields.

We mentioned in section 5.2 that, up to isomorphism, only one finite projective plane of each of the orders 2, 3, 4, 5, 7 and 8 exists, which in each case is desarguesian, and that there is no plane of order 6. Thus, there exists up to equivalence only one complete set of mutually orthogonal latin squares of each of the orders 2, 3, 4, 5, 7 and 8. For the values 2, 3, 4, 5 and 7, this conclusion was rearched entirely by theoretical means; but, for the case $n = 8$, a computer assisted search was used to obtain the result. The method employed is of particular interest because it depended on making use of the list of latin squares of order 7 which are class representatives of the 147 different main classes of such squares and which had earlier been obtained by H. W. Norton and A. Sade. (Sade noted an omission in Norton's list of 146 squares, see section 4.3.)

Let A, B, C be the vertices of a proper triangle in a projective plane of order 8. Let AB be taken as the line at infinity l_∞, AC as the line $x = 0$ and BC as the line $y = 0$ (see section 8.3). Label the seven remaining lines through A as $x = 1, x = 2, \ldots, x = 7$ in any order and also the seven remaining lines through B as $y = 1, y = 2, \ldots, y = 7$ in any order. A point P, not on l_∞, will then lie on a unique line $x = x_p$ and a unique line $y = y_p$ and will be assigned co-ordinates (x_p, y_p).

The seven lines through the point $C(0, 0)$ which are distinct from CA and CB will intersect each of the lines $x = i$ once and each of the lines $y = i$ once, $i = 1, 2, \ldots, 7$. If the line L_k through C meets the line $x = i$ at the point (i, a_i) for $i = 1, 2, \ldots, 7$, we associate L_k with the permutation $\begin{pmatrix} 0 & 1 & 2 & \ldots & 7 \\ 0 & a_{k_1} & a_{k_2} & \ldots & a_{k_7} \end{pmatrix}$. From the incidence properties of the projective plane it follows that the 7×7 matrix $\boldsymbol{A} = \|a_{k_i}\|$ is a latin square. For, since a line L_k through C meets each of $y = 1, y = 2, \ldots, y = 7$ once, each row of $\boldsymbol{A}$ contains each of the integers $1, 2, \ldots, 7$ once. Since a line $x = i$ intersects each of $L_1, L_2, \ldots, L_7$ in a different point, the ith column contains each of $1, 2, \ldots, 7$ once. The three constraints row, column, and digit correspond respectively to a line through C, a line $x = i$, and a line $y = j$.

Permutations of rows, columns, or digits, and interchanges of the roles of the three constraints correspond to re-coordinatizing the projective plane in various ways and do not affect its structure. Norton's list of 7×7 latin squares contains just one representative of each main class, where two squares are in the same main class exactly when one can be obtained from the other by permutations of rows, columns and digits, and interchange of the three constraints. Thus, distinct projective planes will have their pencil of lines through C represented by distinct 7×7 latin squares from Norton's list. Further, a consideration of the number of quadrangles of a plane of order 8 which can have ABC as diagonal point triangle shows that a latin square which represents the pencil of lines through C can have at most 12 intercalates. (For the definition of this concept, see section 4.3.) Consequently, 47 squares of Norton's list can be excluded immediately.

Hence, it is necessary to find in how many of the one hundred remaining possible cases sufficient lines not through A, B, or C can be adjoined to the structure to give a complete projective plane. The method for so doing was partly theoretical and partly computer aided. Full details are given in M. Hall, J. D. Swift and R. J. Walker [1]. Only one of the one hundred possible 7×7 squares permitted completion of the structure to a projective plane and the unique plane obtained was the desarguesian one.

At the time when the above search was carried out (1956), four projective planes of order 9 were known. In addition to the desarguesian plane, there was a self dual non-desarguesian plane which had been originally discovered by O. Veblen and J. H. Wedderburn (see [1]) and whose representational set of 8 mutually orthogonal latin squares are exhibited in section 8.5, the translation plane of order 9 which we described in section 8.4, and the dual of the latter. For all these planes, the additive loop of an appropriately chosen planar ternary ring is the elementary abelian group of order 9. M. Hall and J. D. Swift, this time in collaboration with R. Killgrove, decided to carry out a computer assisted search for further planes with the same additive loop. Part of the work was done by hand and the extent of the remaining search was reduced by theoretical arguments involving the properties of the automorphism group of the plane. The results of the search were given in the form of a list of permutations defining the lines of the plane. The latin square representation was not used. No further planes were found. Full details of the theoretical and computing methods employed will be found in M. Hall, J. D. Swift, and R. Killgrove [1]. Shortly afterwards, R. B. Killgrove (see [1]) carried out a similar search for planes whose additive loop was the cyclic group, using as far as possible the same computer routines and the same type of theoretical argument as before. He found that no plane of this kind exists. Thus, the combined effect of these searches was to prove that any plane of order 9 which may exist, other than the four already known, cannot be co-ordinatized by a ternary ring with an abelian group as its additive loop. Up to the present time, no further planes of order 9 have been found but neither has any proof of their non-existence been given. Consequently, the number of non-equivalent sets of eight mutually orthogonal latin squares of order 9 remains unknown.

A further investigation on the subject of projective planes of order 9 carried out by E. T. Parker and R. B. Killgrove and reported by them in [1] gave the additional result that "If L is a latin square in a complete set S of mutually orthogonal latin squares of order 9, where S includes a square based on the elementary abelian group of order 9, then any complete set of mutually orthogonal latin squares which includes L as one member corresponds to one of the four known projective planes of order 9".

As regards planes of order 10, no proof of their non-existence has been found[1] but, on the other hand, as we explained earlier in the present section, not even as many as three mutually orthogonal squares of order 10 have been constructed despite the fact that it is known that very many orthogonal pairs exist.

The question of the existence of complete sets of mutually orthogonal latin squares of orders 12 or 15 is also an open one. However, it is known that no plane of order 14 exists (see section 5.2) and that there are unique planes of orders 11 and 13, both desarguesian. The next order to be investigated was 16, and again a computer was used.

In 1960, E. Kleinfeld decided to try to enumerate all right quasifields of order 16 and hence to find how many isomorphically distinct translation planes exist of that order. His method involved representing each quasifield as a left vector space over its nucleus. The *nucleus* N of a right quasifield Q comprises that subset of its elements which has the properties that, for each n in N and each two elements q_1 and q_2 of Q the relations $n(q_1q_2) = (nq_1)q_2$ and $n(q_1 + q_2) = nq_1 + nq_2$ hold. If Q is finite, N is a field. It is quite easy to show that if Q has order 16 then the additive group of Q must be the abelian group $C_2 \times C_2 \times C_2 \times C_2$ and N must be one of the Galois fields $GF[2]$, $GF[4]$, or $GF[16]$. If N is the field $GF[16]$ so is Q, so this case need not be considered further.

For the case when N is the field $GF[4]$, Kleinfeld was able to determine the first column and first three rows of the multiplication table of Q (with zero element omitted) by theoretical arguments. He was able to show further that the rest of the multiplication table of Q, and hence Q itself, was then determined uniquely by its fourth row. A SWAC computer was used as an aid to choosing this fourth row in such a way that the resulting multiplication table would in fact be that of a right quasifield. It turned out that there are 75 distinct right quasifields of order 16 and having $GF[4]$ as nucleus. However, these 75 quasifields determine only two distinct non-desarguesian projective planes. One of these planes is determined by 25 of the quasifields and the other by the remaining 50, among which are five division rings.

For the case when N is the field $GF[2]$, the task of determining all right quasifields appeared to be a very formidable one and so Kleinfeld

[1] For some recent new results concerning the question of existence of projective planes of orders 10 and 12, see F. T. MacWilliams, N. J. A. Sloane and W. Thompson [1], A. Bruen and J. C. Fisher [1] and also L. D. Baumert and M. Hall [1].

confined himself to determining all division rings with $GF[2]$ as nucleus. This time he was able to proceed by hand computation alone. He showed that there exist 18 division rings of this kind and that they all co-ordinatize the same non-desarguesian plane which is distinct from the two previously obtained.

Some further details of the theoretical and computational methods used in the above investigation of translation planes of order 16 are given in E. Kleinfeld [1].

Shortly after this investigation of Kleinfeld's was completed, R. J. Walker and, independently, D. E. Knuth decided to attempt the enumeration of all isomorphically distinct division rings of 32 elements. The method used by the first author has been published in R. J. Walker [2] and is somewhat similar to that which had been used by Kleinfeld for determining the division rings of order 16 with $GF[2]$ as nucleus. However, this time it would have been impossible to complete the task without the aid of a computer. It required about 100 hours of computer time to determine the complete set of non-isomorphic systems and there turned out to be 2502 of these. However, these 2502 division rings yielded only five distinct non-desarguesian translation planes.

It will be clear from the various investigations just described that even when a combination of subtle theoretical argument and computer searching is used, a complete determination of all non-equivalent complete sets of mutually orthogonal latin squares of order n is at present quite out of the question for values of n much in excess of 8.

Let us end this account by mentioning that one valuable byproduct of the investigation last described was that, by examination of the tables of results obtained by R. J. Walker, D. E. Knuth was able to deduce a general method for constructing division rings of all orders 2^{2k+1} with $k > 1$. Subsequently, by generalization of his construction, he obtained a complete class of division rings of orders 2^n, $n \geq 5$ and not equal to a power of 2, and hence was able to construct non-desarguesian planes of each of these orders. He has described in more detail how he achieved this result in D. E. Knuth [2]. The interested reader will find some further results on the subject of division ring planes of orders 2^n in D. E. Knuth [1]. The latter paper also discusses division ring planes in general, not confining its attention only to those of order 2^n.

Problems

Chapter 1

1.1 What is the maximum number of associative triples which a quasigroup may have but still not be a group? (page 20)

1.2 Find necessary and sufficient conditions on a loop G in order that every loop isotopic to G be isomorphic to G. (page 25)

1.3 Do there exist quasigroups of odd order which have no complete mappings? (page 32)

1.4 Do there exist any finite groups which are not P-groups? (page 35)

1.5 Do there exist non-soluble groups whose Sylow 2-subgroups are non-cyclic and such that, despite this, the groups have no complete mappings? (pages 36, 37)

1.6 Is the necessary condition given in theorem 1.4.5 for the existence of a complete mapping also sufficient? (page 38)

1.7 Is it true that, for all sufficiently large n, there exist quasigroups of order n which contain no proper subquasigroups? (page 44)

1.8 Given an arbitrary integer n, does there always exist a latin square of order n which contains a latin subsquare of every order m such that $m \leq n/2$? (page 55)

1.9 If n is any positive integer and $n = n_1 + n_2 + \cdots + n_k$ any fixed partition of n, is it possible to find a quasiqroup Q_n of order n which contains subquasiqroups $Q_{n_1}, Q_{n_2}, \ldots, Q_{n_k}$ of orders $n_1, n_2, \ldots, n_k$ respectively whose set theoretical union is Q_n? (page 56)

1.10 The same question as in problem 1.9 but this time Q_n is required to satisfy the condition for every partition of n. (page 56).

Chapter 2

2.1 Do there exist row complete latin squares which cannot be made column complete by permutation of their rows? (page 83)

2.2 For which integers n do there exist sets of $n - 1$ latin squares in which each ordered triad of distinct elements occurs just once in a row of some member of the set? (page 84)

2.3 For which odd integers n do complete latin squares exist? (page 89)
2.4 What are the necessary and sufficient conditions that a non-abelian group be sequenceable? (page 89)
2.5 Do all complete latin squares satisfy the quadrangle criterion? (page 89)
2.6 Is it true that every finite non-abelian group of odd order on two generators is sequenceable? (page 89)

Chapter 3

3.1 If n is a given odd integer, what is the maximum value of m_n such that a complete latin rectangle of m_n rows and n columns exists? (page 98)
3.2 For which values of k_n is it true that an arbitrary complete latin rectangle of k_n rows and n columns ($k_n < n$) can be completed to a complete latin square of order n? (page 98)
3.3 What is the largest number of cells of an arbitrarily chosen latin rectangle of given size which can be contained in a single partial transversal? (page 99)
3.4 Is it true that every latin square of order n has a partial transversal containing at least $n-1$ cells? (page 103)
[Note that the answer is in the affirmative for latin squares which are Cayley tables of abelian groups, see L. J. Paige [1].]
3.5 If k elements are deleted at random from the Cayley table of a finite abelian group G of order n, what is the greatest value which k may have in order that the remaining part of the table still determines G up to isomorphism? (page 106)
3.6 What is the maximum number of squares which a set of latin squares satisfying the quadrangle criterion and all of the same order n can contain if each pair of squares in the set are to differ from each other in at most m places? (page 111)
3.7 Is it possible to place $n-1$ elements chosen from the set $\{0, 1, 2, \ldots, n-1\}$ in an $n \times n$ matrix in such a way that no two elements of any row or column of the matrix are equal and so that, despite this condition, it is <u>not</u> possible to complete the matrix to an $n \times n$ latin square? (page 115)
3.8 How many elements of a latin square of even order n and which satisfies the quadrangle criterion can be located arbitrarily subject only to the condition that no row or column shall contain any element more than once? (page 115)

3.9 Can a pair of $n \times n$ incomplete latin squares which are orthogonal (insofar as the condition for orthogonality applies to the incomplete squares) be respectively embedded in a pair of $t \times t$ orthogonal latin squares; and, if so, what is the smallest value of t for each value of n? (page 116)

3.10 Can every Brandt groupoid B be embedded in a quasigroup Q which is group isotopic? Alternatively, can every Brandt groupoid be embedded in a group? (pages 119, 120)

Chapter 4

4.1 Is it true that no two distinct sets of n disjoint transversals of a latin square of odd order have any transversal in common? (page 149)

4.2 What is the widest class of latin squares of order n with the property that their number of transversals is a multiple of n? (page 149)

4.3 Find formulae for the numbers $R(5, n)$ and $V(5, n)$ of $5 \times n$ reduced and very reduced latin rectangles. (page 152)

Chapter 5

5.1 Do there exist complete sets of mutually orthogonal latin squares of non-prime power order? (page 160)

5.2 Is it true that there do not exist sets of $n - 1$ mutually orthogonal latin squares based on a cyclic group unless n is prime? (page 170)

5.3 If m denotes the maximum number of mutually orthogonal $n \times n$ latin squares none of which satisfies the quadrangle criterion, what is the value of m? (page 170)

5.4 Do there exist triads of mutually orthogonal latin squares of order 10? What is the value of $N(10)$? (page 173)

5.5 For a given order n which is not a prime power, what is the maximum number of latin squares in a set with the property that all the squares of the set have a transversal in common? (page 173)

5.6 Can a complete classification of Stein quasigroups be given? (page 177)

5.7 What is the maximum number of m-dimensional permutation cubes of order n in a variational set? (page 186)

5.8 What is the maximum number k of constraints possible in an orthogonal array of n levels, strength t and index λ? (page 191)

5.9 Is it true that orthogonal Steiner triple systems exist for all orders congruent to 1 modulo 6? (page 192)

Chapter 6

6.1 Do there exist orthogonal pairs of diagonal latin squares of every order n distinct from 2, 3, and 6? (page 212)

6.2 For what orders n do pandiagonal magic squares whose n^2 entries are consecutive integers exist? (page 214)

6.3 For what orders n do addition-multiplication magic squares exist? (page 215)

6.4 For which orders do patterned Room squares not exist? (page 225)

6.5 How many non-equivalent Room designs of order $2n$ exist? (page 228)

6.6 How many non-isomorphic Room designs of order $2n$ exist? (page 228)

Chapter 7

7.1 What is the maximum number of latin squares of order 12 in a pairwise orthogonal set? (page 233)

7.2 Do any triads of mutually orthogonal latin squares of order 10 constructed by the method described in A. D. Keedwell [4] exist? (pages 235, 478)

7.3 Do there exist property D neofields of all finite orders r except 6? (page 248)

7.4 Can it be proved that both commutative and non-commutative D-neofields exist for all $r > 14$ and that the number of isomorphically distinct D-neofields of assigned order r increases with r? (page 248)

7.5 Do there exist planar D-neofields which are not fields? (page 248)

Chapter 8

8.1 If p is prime, is it true that all sets of $p - 1$ mutually orthogonal latin squares are equivalent? That is, are all projective planes of prime order isomorphic and consequently Galois planes? (page 276)

8.2 How many non-equivalent complete sets of mutually orthogonal latin squares of order 9 exist? (page 280)

8.3 For what orders n do digraph complete sets of latin squares which are not mutually orthogonal exist? That is, for what orders n do finite projective planes which contain K-configurations exist? (page 295)

8.4 For what orders n and for which values of k do k-nets which contain K-configurations exist? (page 295)

8.5 Which types of finite and infinite projective planes contain K-configurations? (page 295)

8.6 Do all projective planes with characteristic p have p-power order? (page 296)

8.7 Do there exist non-desarguesian projective planes with characteristic p? (page 296)

Chapter 9

9.1 For which odd integers n do there exist decompositions of the complete undirected graph G_n on n vertices into n nearly linear factors with the property that the union of every two of them is a Hamiltonian path of G_n? (page 306)

9.2 Determine the number of isomorphically distinct symmetric latin squares of order $2k - 1$ for arbitrary $k > 4$. (page 306)

9.3 For what values of n does a P-quasigroup exist which defines a partition of the complete undirected graph G_n into a single closed path? (page 307)

9.4 What are the distinguishing features of a P-quasigroup of the type mentioned in problem 9.3? (page 307)

9.5 What criteria must a colour graph satisfy if it is to be the Cayley colour graph of a quasigroup? (page 315)

Chapter 10

10.1 For which values of q, n and d can the Joshibound be attained? (page 356)

10.2 What is the least number $\varrho(k, n)$ of elements which a covering set of an abelian group G which is the direct sum of k cyclic groups each of order n can contain? (page 362)

10.3 Find a set, as large as possible, of k-digit binary code words such that, for a given small positive integer m, every check sum of up to m different code words is distinct from every other sum of m or fewer code words, and so that each such check sum can be decomposed uniquely, apart from the order of the words, into the m code words which were used to form that sum. (page 366)

10.4 Can E. Knuth's upper bounds be attained for non-prime values of n? (pages 367, 368)

10.5 For what values of the parameters b, v, r, k, λ do balanced incomplete block designs exist? (page 375)

Chapter 11

11.1 What are the exact values of v_n for $n = 3, 4, 5, \ldots$? (page 425)

11.2 What are the exact values of $N(n)$ for $n \geq 15$, n not a prime power? (page 426)

Chapter 12

12.1 For what values of m and h are E. Barbette's problems soluble? Is it always possible to find solutions which consist of superimposing a row latin square on its own transpose as in Fig. 12.5.2? (page 463)

12.2 Which groups contain a latin square group as a subgroup? (page 466)

12.3 How can the latin square groups be characterized? (page 466)

Chapter 13

13.1 Do projective planes of orders 12 or 15 exist? (page 484)

13.2 How many right quasifields of order 16 with $GF[2]$ as nucleus exist? How many isomorphically distinct projective planes of order 16 do these give rise to? (page 484)

Bibliography

Notes on the Bibliography

(1) The italicized numbers which follow each bibliographic item indicate the pages on which that item is referred to in the text.
(2) A suffix *S* following the number designating an item indicates that the item in question is principally concerned with applications to Statistical designs.
(3) All papers concerned specifically with latin squares or their applications are marked with an asterisk and, for such papers, we have attempted to make the bibliography exhaustive. (See also the Preface.)

ACZÉL J.
[1] Kvázicsoportok — hálózatok — nomogramok. Mat. Lapok 15(1964), 114—162. Hungarian [Quasigroups — nets — nomograms.] MR30(1965) # 3171. *68, 80, 251, 266, 267, 269*
[2] Quasigroups, nets and nomograms. Advances in Math. 1 (1965), 383—450. MR 33 (1967) # 1395. *68, 80, 251, 266, 267, 269*
[3] Conditions for a loop to be a group and for a groupoid to be a semigroup. Amer. Math. Monthly 76(1969), 547—549. Not reviewed in MR. *19*

ACZÉL J., PICKERT G. and RADÓ F.
[1] Nomogramme, Gewebe und Quasigruppen. Mathematica (Cluj) (2) 25(1960), 5—24. MR 23(1962) # A1679.

AKAR A.
*[1] Deuxième réponse à la question numéro 261. Interméd. Math. 2(1895), 79—80. *158*

ALBERT A. A.
[1] Quasigroups I. Trans. Amer. Math. Soc. 54(1943), 507—519. MR 5(1944), page 229. *23, 26, 27, 80*
[2] Quasigroups II. Trans. Amer. Math. Soc. 55(1944), 401—419. MR 6(1945), page 42. *23, 80, 145*

ALIMENA B. S.
*[1]$_S$ A method of determining unbiassed distribution in the Latin square. Psychometrika 27(1962), 315—317. MR 26 (1963) # 865. *84*

ARCHBOLD J. W.

[1] A combinatorial problem of T. G. Room. Mathematika 7(1960), 50—55. MR 23(1962) # A2334. *221*

ARCHBOLD J. W. and JOHNSON N. L.

[1] A construction for Room's squares and an application in experimental design. Ann. Math. Statist. 29(1958), 219—225. MR 21(1960) # 950. *220*

[ARGUNOV B. I.] Аргунов Б. И.

[1] Конфигурационные постулаты и их алгебраические эквиваленты. Мат. Сборник (Н. С.) 26(68)(1950), 425—456. [Configurational postulates and their algebraic equivalents. Mat. Sb. (N. S.) 26(68)(1950), 425—456.] MR 12(1951), page 525. *263, 265, 266*

ARKIN J.

*[1] The first solution of the classical Eulerian magic cube problem of order ten. Fibonacci Quarterly. 11(1973), 174—178. *185, 186*

*[2] A solution to the classical problem of finding systems of three mutually orthogonal numbers in a cube formed by three superposed $10 \times 10 \times 10$ cubes. To appear.

ARKIN J. and HOGGATT V. E. Jnr.

*[1] The Arkin-Hoggatt game and the solution of a classical problem. J. Recreat. Math. To appear.

ARMSTRONG E. J.

*[1] A note on latin squares. Math. Gaz. 39(1955), 215—217. Not reviewed in MR. *194*

BAER R.

[1] Nets and groups I. Trans. Amer. Math. Soc. 46(1939), 110—141. MR 1(1940), page 6. *23, 251, 260*

[2] Nets and groups II. Trans. Amer. Math. Soc. 47(1940), 435—439. MR 2(1941), page 4. *23, 251*

BALAAN L. N. and FEDERER W. T.

[1] Bibliography on experimental design pre 1968. Oliver & Boyd, Edinburgh, 1972. *385*

BALAKRISHNAN R.

*[1] On the algebra of magic matrices. Math. Student 34 (1966), 201—206. *466*

BALL W. W. R.

*[1] Mathematical recreations and essays. Macmillan, New York, Eleventh Ed. 1947, revised by H. S. M. Coxeter. MR 8(1947), page 440. See also MR 22(1961) # 3649. *80, 194, 211, 212, 213, 214*

BAMMEL S. E. and ROTHSTEIN J.

*[1] The number of 9×9 latin squares. J. Discrete Math. To appear. *144*

Barbette E.
*[1] Question numéro 767. Interméd. Math. 3 (1896), 54; 12(1905), 101. *212*
*[2] Sur le problème d'Euler dit des 36 officiers. Généralisation. Interméd. Math. 5(1898), 83—85. *158, 463*

Barbette E. and Dellanoy H.
*[1] Problème des 36 officiers. Interméd. Math. 5(1898), 252. *158, 464*

Barra J. R.
*[1] A propos d'un théorème de R. C. Bose. C. R. Acad. Sci. Paris 256(1963), 5502—5504. MR 27(1964) #1383. *178*
*[2] Carrés Latins et euleriens. Rev. Inst. Internat. Statist. 33(1965), 16—23. MR 34(1967) #1214.

Barra J. R. and Guérin R.
*[1] Extension des carrés gréco-latins cycliques. Publ. Inst. Statist. Univ. Paris 12 (1963), 67—82. MR 28(1964) #1138. *16, 442*
*[2] Utilisation pratique de la méthode de Yamamoto pour la construction systématique de carrés gréco-latins. Publ. Inst. Statist. Univ. Paris 12(1963), 131—136. MR 28 (1964) #1139. *442*

Bateman P. T.
[1] A remark on infinite groups. Amer. Math. Monthly 57 (1950), 623—624. MR 12(1951), page 670. *39*

Baumert L. D. and Golomb S. W.
[1] Backtrack programming. J. Assoc. Comput. Mach. 12 (1965), 516—524. MR 33(1967) #3783. *468*

Baumert L. D. and Hall M.
[1] Non-existence of certain planes of order 10 and 12. J. Combinatorial Theory, Ser. A, 14(1973), 273—280. *484*

Baumgartner L.
[1] Gruppentheorie. De Gruyter, Berlin, 1921. *45*

Beineke L. W.
[1] Decomposition of complete graphs into forests. Publ. Math. Inst. Hung. Acad. Sci. 9(1964), 589—593. MR 32 (1966) #4031. *82*

[Belousov V. D.] Белоусов В. Д.
[1] Ассоциативные системы квазигрупп. Успехи Мат. Наук 13(1958), 243. [Associative systems of quasigroups. Uspehi Mat. Nauk. 13(1958), 243.] Not reviewed in MR. *78*

[2] Ассоциативные в целом системы квазигрупп. Мат. Сборник (Н. С.) 55(97)(1961), 221—236. [Globally associative systems of quasigroups. Mat. Sb. (N. S.) 55 (97)(1961), 221—236.] MR 27(1964) #2578. *78*

[3] Замкнутые системы взаимно ортогональных квазигрупп. IV. Всесоюзное совешание по общей алгебре (Киев, 16-22 Мая 1962 г.) Успехи Мат. Наук 17(1962), №. 6(108), 202—203. [Closed systems of mutually orthogonal quasigroups. Usp. Nat. Nauk. 17(1962), No. 6(108), 202—203.] Not reviewed in MR. *177*

[4] Системы квазигрупп с обобщенными тождествами. Успехи Мат. Наук 20(1965), №. 1(121), 75—146. [Systems of quasigroups with generalized identities Usp. Mat. Nauk. 20 (1965), No. 1(121), 75—146.] Translated as Russian Math. Surveys 20(1965), 73—143. MR 30(1965) # 3934. *57, 65, 66, 76, 80, 175, 176*

[5] Уравновешенные тождества в квазигруппах. Мат. Сборник (Н. С.) 70(112) (1966), 55—97. [Balanced identities in quasigroups. Mat. Sb. (N. S.) 70(112)(1966), 55—97.] MR 34 (1967) # 2757. *69*

[6] Неассоциативные бинарные системы. В Сб. "Алгебра. Топология. Геометрия. 1965." (Итоги науки) Москва (1967), 63—81. [Non-associative binary systems. Algebra, Topology, Geometry 1965, 63—81. Akad. Nauk SSSR Inst. Naučn. Tehn. Informacii, Moscow, 1967.] MR 35(1968) # 5537. *80*

[7] Основы теории квазигрупп и луп. Издательство Наука, Москва 1967. [Foundations of the theory of quasigroups and loops. Izdat. "Nauka", Moscow 1967.] MR 36(1968) # 1569. *41, 80, 124, 149, 251, 266*

[8] Продолжения квазигрупп. Изв. А. Н. Молд. ССР 8 (1967) 3—24. [Extensions of quasigroups. Bull. Akad. Ştiince RSS Moldoven (1967), No. 8., 3—24.] MR 38 (1969) # 4592. *32, 39, 41*

[9] Преобразования в сетях. Мат. Зап. Уральский УН — Т 6, Тетрадь 1(1967) 3—20. [Transformations in nets. Ural. Gos. Univ. Mat. Zap. 6, Tetrad' 1,(1967), 3—20.] MR 36 (1968) # 3906. *269*

[10] Системы ортогональных операций. Мат. Сборник (Н. С.) 77 (119) (1968), 38—58. [Systems of orthogonal operations. Mat. Sb. (N. S.) 77(119)(1968), 38—58.] Translated as Mathematics of the USSR. — Sbornik 6(1968), 33—52. MR 38(1969) # 1200. *175*

[11] Алгебраические сети и квазигруппы. Издательство Щтиинца, Кишинев, 1971, [Algebraic nets and quasigroups. Izdat. "Ştiince" Kishinev, 1971.] *251, 266*

[12] n-арные квазигруппы. Издательство Штиинца, Кишинев, 1972. [n-ary quasigroups. Izdat. "Ştiince", Kishinev 1972.] *251*

[Belousov V. D. and Belyavskaya G. B.] Белоусов В. Д. и Белявская Г. Б.

[1] О продолжении квазигрупп. Тезисы 1. Всесоюзного *41*

симпоз. по теории квазигрупп и её приближениям, Сухуми 1968. [On prolongations of quasigroups. Thesis I. All-Union Symposium on the theory of quasigroups and their applications, Suhumi, 1968.] Not reviewed in MR.

[Belousov V. D. and Gvaramiya A. A.] Белоусов В. Д. и Гварамия А. А.

[1] О квазигруппах Стейна. Сообщ. А. Н. Груз ССР, 44 (1966), 537—544. [Stein quasigroups. Sakharth SSR. Mecn. Akad. Moombe 44(1966), 537—544.] MR 34(1967) # 2758. *177*

[Belousov V. D. and Ryzkov V. V.] Белоусов В. Д. и Рыжков В. В.

[1] Об одном способе получения фигур замыкания. Мат. Исслед 1(1966), вып. 2, 140—150. [On a method of obtaining closure figures. Mat. Issled 1(1966), No. 2., 140—150.] MR 36(1968) # 6526. *269*

[Belyavskaya G. B.] Белявская Г. Б.

[1] О сужении квазигрупп. Десятый всесоюзный алгебраический коллоквиум. Новосибирск (1969). Резюме докладов. ТОМ 2, 106—107. [On contraction of quasigroups. Tenth All-Union Algebra Colloquium. Novosibirsk (1969). Summaries of Reports. Vol. 2, 106—107.] Not reviewed in MR *41*

[2] Цепно-изотопные квазигруппы. Мат. Исслед. 5(1970), №. 2., 13—27. [Chain-isotopic quasigroups. Mat. Issled. 5 (1970), No. 2, 13—27.] MR 44(1972) # 1757. *41*

[3] Обобщенное продолжение квазигрупп. Мат. Исслед. 5(1970), №. 2, 28—48. [On generalized prolongation of quasigroups. Mat. Issled. 5(1970), No. 2., 28—48.] MR 44 (1972) # 1758. *41*

[4] Сжатие квазигрупп I. Изв. А. Н. МССР. (1970). №. 1, 6—12. [Contractions of quasigroups I. Izv. Akad. Nauk Moldav. SSR (1970), No. 1, 6—12.] MR 43(1972) # 4942. *41*

[5] Сжатие квазигрупп II. Изв. А. Н. МССР (1970), №. 3, 3—17. [Contractions of quasigroups II. Izv. Akad. Nauk Moldav. SSR. (1970), No. 3, 3—17.] MR 44(1972) # 1756, *41*

*[6] Алгоритмы решения некоторых задач теории квазигрупп. «Вопросы теории квазигрупп и луп.» Ред. В. Д. Белусов. Акад. Наук Молдавской ССР, Кишинев, 1971, 20—30. [Algorithms for solving some problems in the theory of quasigroups. From "Questions in the theory of quasigroups and loops". Edited by V. D. Belousov. Acad. Nauk Moldav. SSR. Kishinev, 1971, 20—30.] *149, 471*

Berlekamp E. R. and Hwang F. K.

[1] Constructions of balanced Howell rotations for Bridge tournaments. J. Combinatorial Theory, Ser. A, 12(1972), 159—166. *229*

BEYNON G. W.

[1] Bridge Director's Manual. Coffin Publishing Co., Massachusetts, 1943. Not reviewed in MR. *219*

[2] Duplicate Bridge Direction. Stuyvesant Press, New York, 1944. Not reviewed in MR. *219, 221*

BLASCHKE W.

[1] Topologische Fragen der Differentialgeometrie I. Thomsens Sechseckgewebe. Zueinander diagonale Netze. Math. Z. 28(1928), 150—157. *251*

BLASCHKE W. and BOL G.

[1] Geometrie der Gewebe. Grundlehrung der Math. Wiss. 49, Berlin, 1938. *251*

BLUM J. R., SCHATZ J. A. and SEIDEN E.

$[1]_S$ On 2-level orthogonal arrays of odd index. J. Combinatorial Theory 9(1970), 239—243. MR 41(1971) # 8258. *191, 192*

BOL G.

[1] Topologische Fragen der Differentialgeometrie LXV. Gewebe und Gruppen. Math. Ann. 114(1937), 414—431. *251, 260, 265, 267*

BONDESEN A.

*[1] Er det en gruppetavle? Nordisk Mat. Tidskr. 17(1969), 132—136. Danish. [Is it a group table?] MR 41(1971) # 7016. *19,*

BOSE R. C.

*[1] On the application of the properties of Galois fields to the construction of hyper-Graeco-Latin squares. Sankhyā 3(1938), 323—338. *160, 167*

*[2] Strongly regular graphs, partial geometries, and partially balanced designs. Pacific J. Math. 13(1963), 389—419. MR 28(1964) # 1137. *312, 347*

BOSE R. C. and BUSH K. A.

$[1]_S$ Orthogonal arrays of strength 2 and 3. Ann. Math. Statist. 23(1952), 508—524. MR 14(1953), page 442. *191, 192*

BOSE R. C., CHAKRAVARTI I. M. and KNUTH D. E.

*[1] On methods of constructing sets of mutually orthogonal latin squares using a computer I. Technometrics 2(1960), 507—516. MR 23(1962) # A3099. *233, 481*

*[2] On methods of constructing sets of mutually orthogonal latin squares using a computer II. Technometrics 3(1961), 111—117. MR 23(1962) # A3100. *481*

BOSE R. C. and NAIR K. R.

[1] Partially balanced incomplete block designs. Sankhyā 4(1939), 337—372. Not reviewed in MR. *377*

*[2] On complete sets of latin squares. Sankhyā 5(1941), 361—382. MR 4(1943), page 33. *159, 161, 169, 281*

BOSE R. C. and SHRIKHANDE S. S.

*[1] On the falsity of Euler's conjecture about the non-existence of two orthogonal latin squares of order 4t + 2. Proc. Nat. Acad. Sci. USA. 45(1959), 734—737. MR 21(1960) #3343. *158, 397*

*[2] On the construction of sets of mutually orthogonal latin squares and the falsity of a conjecture of Euler. Trans. Amer. Math. Soc. 95(1960), 191—209. MR 22(1961) #2557. *404*

BOSE R. C., SHRIKHANDE S. S. and BHATTACHARYA K. N. *402*

[1] On the construction of group divisible incomplete block designs. Ann. Math. Statist. 24 (1953), 167—195. MR 15(1954), page 3.

BOSE R. C., SHRIKHANDE S. S. and PARKER E. T.

*[1] Further results on the construction of mutually orthogonal latin squares and the falsity of Euler's conjecture. Canad. J. Math. 12(1960), 189—203. MR 23(1962) #A69. *158, 230, 237, 404, 425, 426, 466*

BOSSEN D. C., HSIAO M. Y. and CHIEN R. T.

*[1] Orthogonal latin square codes. IBM. J. Res. D evelop. 14 (1970), 390—394. MR 43(1972) #4548. *359*

BRADLEY J. V.

*[1]$_S$ Complete counterbalancing of immediate sequential effects in a latin square design. J. Amer. Statist. Assoc. 53 (1958), 525—528. Not reviewed in MR. *83*

BRANDT H.

[1] Verallgemeinierung des Gruppenbegriffs. Math. Ann. 96(1927), 360—366. *19, 118*

BRAYTON R. K., COPPERSMITH D. and HOFFMAN A. J.

*[1] Self-orthogonal latin squares. To appear. *461*

BROCARD H.

*[1] Problème des 36 officiers. (Deuxième réponse à la question numéro 453.) Interméd. Math. 3(1896), 90. *158*

BROWN J. W.

*[1] Enumeration of latin squares with application to order 8. J. Combinatorial Theory 5(1968), 177—184. MR 37(1969) #5111. *144, 471*

*[2] An extension of Mann's theorem to a triple of mutually orthogonal latin squares of order 10. J. Combinatorial Theory, Ser. A, 12(1972), 316—318. MR 45(1973) #8544. *451*

BROWNLEE K. A. and LORAINE P. K.

*[1]$_S$ The relationship between finite groups and completely orthogonal squares, cubes and hypercubes. Biometrika 35 (1948), 277—282. MR 10(1949), page 313. *188*

BRUCK R. H.
[1] Some results in the theory of quasigroups. Trans. Amer. Math. Soc. 55(1944), 19—52. MR 5(1944), page 229. *31, 40, 68*
[2] Contributions to the theory of loops. Trans. Amer. Math. Soc. 60(1946), 245—354. MR 8(1947), page 134. *26, 45*
*[3] Finite nets I. Numerical invariants. Canad. J. Math. 3(1951), 94—107. MR 12(1951), page 580. *251, 270, 318*
[4] Difference sets in a finite group. Trans. Amer. Math. Soc. 78(1955), 464—481. MR 16(1955), page 1081. *454*
[5] A survey of binary systems. Springer Verlag, Berlin, 1958. MR 20(1959) # 76. *18, 26, 27, 32, 75, 80, 118, 120 126, 149, 251*
*[6] Finite nets. II. Uniqueness and embedding. Pacific J. Math. 13(1963), 421—457. MR 27(1964) # 4768. *173, 271, 331*
*[7] What is a loop? Studies in modern algebra, Chapter 4. Edited by A. A. Albert. Math. Assoc. of America and Prentice-Hall, 1963. MR 26(1963) # 3750. *32, 75, 223, 267*

BRUCK R. H. and RYSER H. J.
[1] The non-existence of certain finite projective planes. Canad. J. Math. 1(1949), 88—93. MR 10(1949), page 319. *160*

BRUCKHEIMER M., BRYAN A. C. and MUIR A.
[1] Groups which are the union of three subgroups. Amer. Math. Monthly 77(1970), 52—57. MR 41(1971) # 322. *47*

BRUEN A. and FISHER J. C.
*[1] Blocking sets, k-arcs and nets of order ten. Advances in Math. 10(1973), 317—320. *334, 484*

BRYANT B. F. and SCHNEIDER H.
[1] Principal loop isotopes of quasigroups. Canad. J. Math. 18(1966), 120—125. MR 32(1966) # 5772. *26, 145, 471*

BUGELSKI B. R.
*$[1]_S$ A note on Grant's discussion of the latin square principle in the design of experiments. Psychological Bulletin 46(1949), 49—50. Not reviewed in MR. *82*

BURMAN J. P. and PLACKETT R. L.
$[1]_S$ The design of optimum multifactorial experiments. Biometrika 33(1943—46), 305—325. MR 8(1947), page 44. *187, 191*

BUSH K. A.
[1] A generalization of a theorem due to MacNeish. Ann. Math. Statist. 23(1952), 293—295. MR 14(1953), page 125. *191*
[2] Orthogonal arrays of index unity. Ann. Math. Statist. 23(1952), 426—434. MR 14(1953), page 125. *191*

BUSSEMAKER F. C. and SEIDEL J. J.
*[1] Symmetric Hadamard matrices of order 36. Ann. N. Y. Acad. Sci. 175(1970), 66—79. MR 43(1972) # 1863. *312*

BYLEEN K.
[1] On Stanton and Mullin's construction of Room squares. Ann. Math. Statist. 41(1970), 1122—1125. *225*

BYLEEN K. and CROWE D. W.
[1] An infinite family of cyclic Room designs. Submitted to Rocky Mountain J. Math. (unpublished). *226, 459*

CAIRNS S.
[1] Computational attacks on discrete problems. Proc. Symp. on Special Topics in Appl. Math. Suppl. to Amer. Math. Monthly 61(1954), 29—31. MR 16(1955), page 77. *476*

CARLITZ L.
*[1] Congruences connected with three-line latin rectangles. Proc. Amer. Math. Soc. 4(1953), 9—11. MR 14(1953), page 726. *152*
[2] A note on abelian groups. Proc. Amer. Math. Soc. 4 (1953), 937—939. MR 15(1954), page 503. *34*

CARROL J. J., FISHER G. A., ODLYZKO A. M. and SLOANE N. J. A.
*[1] What are the latin square groups? Amer. Math. Monthly. To appear. *466*

CAYLEY A.
[1] On the theory of groups. Proc. Lond. Math. Soc. 9 (1877/8), 126—133. *15, 313*
[2] Desiderata and suggestions No. 1. The theory of groups. Amer. J. Math. 1(1878), 50—52. *15*
[3] Desiderata and suggestions No. 2. The theory of groups: graphical representation. Amer. J. Math. 1(1878), 174—176. *313*
*[4] On the theory of groups. Amer. J. Math. 11(1889), 139—157. *138*
*[5] On latin squares. Messenger of Math. 19(1890), 135 - 137. *139*

CHAKRABARTI M. C.
*$[1]_S$ Mathematics of design and analysis of experiments. Asia, London, 1962. MR 35(1968) # 7514. *385*
*$[2]_S$ Some recent developments in design of experiments. J. Annamalai Univ. Part B Sci. 25(1964), 54—72. MR 30 (1965) # 2628.

CHAKRAVARTI I. M.
*[1] On the construction of difference sets and their use in the search for orthogonal latin squares and error-correcting codes. Bull. Int. Statist. Inst. 41(1967), 957. Not reviewed in MR.

CHANG L. Q. and TAI S. S.
*[1] On the orthogonal relations among orthomorphisms of non-commutative groups of small orders. Acta Math. Sinica 14 (1964), 471—480. Chinese. Translated as Chinese. Math. Acta 5(1964), 506—515. MR 30(1965) # 4690. *480*

CHANG L. Q., HSIANG K. and TAI S.
*[1] Congruent mappings and congruence classes of orthomorphisms of groups. Acta Math. Sinica 14(1964), 747—756. Chinese. Translated as Chinese Math. Acta 6(1965), 141—152. MR 31(1966) # 1220. *480*

CHOUDHURY A. C.
[1] Quasigroups and non-associative systems I. Bull. Calcutta Math. Soc. 40(1948), 183—194. MR 10(1949), page 591. *317*
[2] Quasigroups and non-associative systems II. Bull. Calcutta Math. Soc. 41(1949), 211—219. MR 11(1950), page 417. *317*
[3] Quasigroups and non-associative systems III. Bull. Calcutta Math. Soc. 49(1957), 9—24. MR 20(1959) # 6478. *317*

CHOWLA S., ERDŐS P. and STRAUSS E. G.
*[1] On the maximal number of pairwise orthogonal latin squares of a given order. Canad. J. Math. 12(1960), 204—208. MR 23(1962) # A70. *174*

CLARKE G. M.
*$[1]_S$ A design for testing several treatments under controlled environmental conditions. Applied Statist. 4(1955), 199—206. Not reviewed in MR.

CLATWORTHY W. H.
*$[1]_S$ Some new families of partially balanced designs of the latin square type and related designs. Technometrics 9 (1967), 229—244. MR 35(1968) # 7515.

COCHRAN W. G. and COX G. M.
*$[1]_S$ Experimental designs. Wiley, New York, 1950 and Chapman & Hall, London, 1950. MR 11(1950), page 607. 2nd Edition 1957. MR 19(1958), page 75. *385*

COLE F. N., WHITE A. S. and CUMMINGS L. D.
[1] Complete classification of triad systems on fifteen elements. Mem. Nat. Acad. Sci. 14(1925), Second memoir, 89. *79*

COLLENS R. J. and MULLIN R. C.
[1] Some properties of Room squares — A computer search. Proc. Louisiana Conf. on Combinatorics, Graph Theory & Computing (Louisiana State Univ., Baton Rouge, La., 1970), 87—111. MR 42(1971) # 2958. *229*

COLOMBO U.
*[1] Sulle disposizioni a scacchiera construite con k campi di Galois. Giorn. Ist. Ital. Attuari. 26(1963), 106—117. MR 28(1964) # 5010. *465*

COXETER H. S. M. and MOSER W. O. J.
[1] Generators and relations for discrete groups. Springer-Verlag, Berlin, 2nd Edition 1965. MR 30(1965) # 4818. *147*, *314*

CRAMPIN D. J. and HILTON A. J. W.

*[1] The spectrum of latin squares orthogonal to their transposes. To appear. *461*

*[2] Remarks on Sade's disproof of the Euler conjecture with an application to latin squares orthogonal to their transposes. To appear. *461*

CRUSE A. B.

*[1] On embedding incomplete symmetric latin squares. J. Combinatorial Theory, Ser. A 16(1974), 18–22. *117*

CSIMA J.

[1] Restricted patterns. J. Combinatorial Theory, Ser. A, 12(1972), 346–356. *116*

CSIMA J. and ROSA A.

[1] An extremal problem on finite quasigroups. To appear. *39*

DAVENPORT H.

[1] The Higher Arithmetic. Hutchinson, London, 3rd Edition 1968. Not reviewed in MR. *170*

DAVIES I. J.

[1] Enumeration of certain subgroups of abelian p-groups. Proc. Edinburgh Math. Soc. 13(1962), 1—4. MR 26(1963) #3780. *148*

DELLANOY H. and BARBETTE E. [1]

See BARBETTE E. and DELLANOY H.

DEMBOWSKI P.

[1] Finite Geometries. Springer Verlag, Berlin, 1968. MR 38 (1969) #1597. *170, 284, 295, 385*

DÉNES J.

*[1] Candidature of Mathematical Sciences Thesis. 1961. *194, 217*

*[2] On a problem of L. Fuchs. Acta Sci. Math. (Szeged) 23(1962), 237—241. MR 27(1964) #1493. *16, 106*

[3] On some properties of commutator subgroups. Ann. Univ. Sci. Budapest. Eötvös Sect. Math. 7(1964), 123—127. MR 30(1965) #3912. *35*

[4] О таблицах умножения конечных квазигрупп и полугрупп. Мат. Исслед 2(1967), № 2, 172—175. [Multiplication tables for finite quasigroups and semigroups. Mat. Issled. 2(1967), No. 2, 172—175.] Translated as a University of Surrey preprint, 1967. MR 37(1969) #6391. *113*

[5] On commutator subgroups of finite groups. Comment. Math. Univ. St. Paul. 15(1966—67), 61—65. MR 35(1968) #4293. *35*

*[6] Algebraic and combinatorial characterization of latin squares I. Mat. Časopis Sloven. Akad. Vied. 17(1967), 249—265. MR 38(1969) #3164. *42, 48*

DÉNES J. and KEEDWELL A. D.
[1] On P-quasigroups and decompositions of complete undirected graphs. J. Combinatorial Theory, Ser. B, 13(1972), 270–275. *308*

DÉNES J. and PÁSZTOR Ené (K.)
*[1] A kvázicsoportok néhány problémájáról. Magyar Tud. Akad. Mat. Fiz. Oszt. Közl. 13(1963), 109—118. Hungarian. [Some problems on quasigroups.] MR 29(1965) # 180. *16, 29, 39, 40, 42, 45, 46, 56, 97, 112*

DÉNES J. and TÖRÖK É.
*[1] Groups and graphs. Combinatorial Theory and its applications. 257—289. North Holland, Amsterdam, 1970. MR 56(1973) # 91. *35, 83, 89, 300, 317*

DEVIDÉ V.
[1] Über eine Klasse von Gruppoiden. Hrvatsko Prirod. Društvo. Glasnik Mat.-Fiz. Astr. Ser. II. 10(1955), 265—286. MR 18(1957), page 872. *78*

DOYEN J.
[1] On the number of non-isomorphic Steiner systems $S(2, m, n)$. Combinatorial Structures and their Applications. (Proc. Calgary Internat. Conf., Calgary, Alta., 1969), 63—64. Gordon and Breach, New York, 1970. *80*
[2] Sur la croissance du nombre de systèmes triples de Steiner non isomorphes. J. Combinatorial Theory 8(1970), 424—441. MR 41(1971) # 1555. *80*

DOYEN J. and ROSA A.
[1] A bibliography and survey of Steiner systems. Bol. Un. Mat. Ital. To appear. *80*

DOYEN J. and VALETTE G.
[1] On the number of non-isomorphic Steiner triple systems. Math. Z. 120(1971), 178–192. MR 43(1972) # 4696. *80*

VAN DRIEL M. J.
[1] Magic Squares of $(2n + 1)^2$ Cells. Rider & Co., London, 1936. MR 1(1940), page 290. *194*
[2] A supplement to Magic Squares of $(2n + 1)^2$ Cells. Rider & Co., London, 1939. MR 1(1940), page 290. *194*

DUGUÉ D.
*[1] Carrés Latins. Mém. Sci. ingrs. civils France 1117 (1964), 21—24. Not reviewed in MR.

DULMAGE A. L., JOHNSON D. M. and MENDELSOHN N. S.
*[1] Orthogonal Latin squares. Canad. Math. Bull. 2(1959), 211—216. Not reviewed in MR.
*[2] Orthomorphisms of groups and orthogonal latin squares I. Canad. J. Math. 13(1961), 356—372. MR 23(1962) # A1544. *155, 232, 233, 479*

ELSPAS B. and KAUTZ W. H.
*[1] Single error-correcting codes for constant weight data words. IEEE. Trans. on Information Theory. IT-11 (1965), 132—141. Not reviewed in MR. *360*

ELSPAS B., MINNICK R. C. and SHORT R. A.
*[1] Symmetric latin squares. IEEE. Trans. on Elec. Comput. EC-12 (1963), 130—131. Not reviewed in MR. ***41, 304, 359, 361***

[EINGORINA T. N.] Эйнгорина Т. Н.
*[1] К вопросу о существовании множеств ортогональных латинских гиперпрямоугольников — Изв. Высших Учебных Заведений, Радиофизика 12(1969), 1732—1739. [On the question of the existence of sets of orthogonal latin hyperrectangles. J. of the Higher Education Institutions, Radiophysics 12(1969). 1732—1739. English translation: Radiophysics annd Quantum Electronics 12(1969), 1350—1355 (1972)] MR 46(1973) # 7057.

[EINGORINA T. N. and EINGORIN M. J.]
Эйнгорина Т. Н. и Эйнгорин М. Я.
*[1] Некоторые вопросы связи теории запоминающих устройств с многократным совпадением сигналов с теорией латинских гиперквадратов I, II. Изв. Высших Учебных Заведений. Радиофизика 10(1967), 1569—1575, 1576—1595. [Some questions connecting the theory of storage of multiple coincidences of signals with the theory of latin hyper-squares I, II. J. of the Higher Education Institutions, Radiophysics 10(1967), 1569—1575, 1576—1595]. Not reviewed in MR. *366*
*[2] К вопросу о существовании полных множеств ортогональных латинских гиперпрямоугольников. — Изв. Высших Учебных Заведений, Радиофизика 12 (1969), 1721—1731. [On the existence of complete sets of orthogonal latin hyper-rectangles. J. of the Higher Education Institutions, Radiophysics 12(1969), 1721—1731. English translation: Radiophysics and Quantum Electronics 12(1969), 1341—1349 (1972)] MR 46(1973) # 7056.

ERDŐS P. and GINZBURG A.
*[1] On a combinatorial problem in latin squares. Magyar Tud. Akad. Mat. Kutató Int. Közl. 8(1963), 407—411. MR 29(1965) # 2197. *22*

ERDŐS P. and KAPLANSKY I.
*[1] The asymptotic number of latin rectangles. Amer. J. Math. 68(1946), 230—236. MR 7(1946), page 407. ***95, 144, 153***

ERDŐS P., RÉNYI A. and SÓS V.
[1] Combinatorial Theory and its Applications. Colloquia Mathematica Societatis János Bolyai 4. North Holland, Amsterdam, 1970. ***115***

Erdős P. and Szekeres G.
[1] Über die Anzahl der Abelschen Gruppen gegebener Ordnung und über ein verwandtes zahlentheoretisches Problem. Acta Sci. Univ. Szeged 7(1934/5), 95—102. *147*

Etherington I. M. H.
[1] Note on quasigroups and trees. Proc. Edinburgh Math. Soc. (2) 13 (1962—63), 219—222. MR 28(1964) # 157 *61, 317, 318*
[2]Quasigroups and cubic curves. Proc. Edinburgh Math. Soc. (2) 14 (1964—65), 273—291. MR 33(1967) # 4170. *68, 73, 318*

Euler L.
*[1] De quadratis magicis. [Memoir presented to the Academy of Sciences of St. Petersburg on 17th October, 1776.] Published as (a) Mémoire de la Société de Flessingue, Commentationes arithmeticae collectae (elogé St. Petersburg 1783), 2(1849), 593—602; (b) Opera postuma 1(1862), 140—151; (c) Leonardi Euleri Opera Omnia, Série 1, 7(1923), 441—457. *194*
*[2] Recherches sur une nouvelle espèce de quarrés magiques. [Memoir presented to the Academy of Sciences of St. Petersburg on 8th March, 1779.] Published as (a) Verh. Zeeuwsch. Genootsch. Wetensch. Vlissengen 9(1782), 85—239; (b) Mémoire de la Société de Flessingue, Commentationes arithmetica collectae (elogé St. Petersburg 1783), 2(1849), 302—361; (c) Leonardi Euleri Opera Omnia, Série 1, 7(1923), 291—392. *29, 138, 155, 156, 194, 212, 445, 446*

Evans T.
[1] A note on the associative law. J. Lond. Math. Soc. 25 (1950), 196—201. MR 12(1961), page 75. *69, 78*
*[2] Embedding incomplete latin squares. Amer. Math. Monthly 67(1960), 958—961. MR 23(1962) # A68. *42, 97, 114, 116, 117, 118*
*[3] Latin cubes orthogonal to their transposes — A ternary analogue of Stein quasigroups. Aequationes Math. 9(1973), 296—297. *461*

Falconer E.
[1] Isotopes of some special quasigroup varieties. Acta Math. Acad. Sci. Hungar. 22(1971), 73—79. *69*

Faragó T.
[1] Contribution on the definition of a group. Publ. Math. Debrecen 3(1953), 133—137. MR 15(1954), page 851. *20*

Federer W. T.
*[1]$_S$ Experimental design: theory and application. Macmillan, New York, 1955. Not reviewed in MR. *103*

Federer W. T. and Balaan L. N. [1]
See Balaan L. N. and Federer W. T.

FEDERER W. T. and HEDAYAT A.

*[1] An application of group theory to the existence and non-existence of orthogonal latin squares. Biometrika 56 (1969), 547—551. MR 41(1971) # 6373. *446*

*[2]$_S$ An easy method for constructing partially replicated latin square designs of order n for all $n > 2$. Biometrics 26(1970), 327—330.

*[3] On embedding and enumeration of orthogonal latin squares. Ann. Math. Statist. 42(1971), 509–516. MR 45 (1973) # 2871. *465*

FEDERER W. T., HEDAYAT A. and PARKER E. T.

*[1]$_S$ Two families of designs for two successive experiments. Biometrika 57(1970), 351—355. MR 42(1971) # 3936. *173*

FENYVES F.

[1] Extra loops I. Publ. Math. Debrecen 15(1968), 235—238. MR 38(1969) # 5976. *64*

[2] Extra loops II. Publ. Math. Debrecen 16(1969), 187—192. MR 41(1971) # 7017. *64, 65*

FINNEY D. J.

*[1]$_S$ Some orthogonal properties of the 4×4 and 6×6 latin squares. Ann. Eugenics 12(1945), 213—219. MR 7(1946), page 107. *461*

*[2]$_S$ Orthogonal partitions of the 5×5 latin squares. Ann. Eugenics 13(1946), 1—3. MR 7(1946), page 407. *461*

*[3]$_S$ Latin squares of the sixth order. Experientia 2(1946), 404—405. MR 8(1947), page 190.

*[4]$_S$ Orthogonal partitions of the 6×6 latin squares. Ann. Eugenics. 13(1946), 184—196. MR 8(1947), page 247. *461*

*[5]$_S$ An introduction to the theory of experimental design. University of Chicago Press, Chicago, 1960. MR 22(1961) # 11493. *385*

FISHER R. A.

*[1] Some combinatorial theorems and enumerations connected with the numbers of diagonal types of a latin square. Ann. Eugenics 11(1942), 395—401. MR 4(1943), page 183. *142*

*[2] Completely orthogonal 9×9 squares. A correction. Ann. Eugenics 11(1942), 402—403. MR 4(1943), page 184. *169*

*[3]$_S$ A system of confounding for factors with more than two alternatives giving completely orthogonal cubes and higher powers. Ann. Eugenics 12(1945), 283—290. MR 7 (1946), page 107. *187*

*[4]$_S$ The design of experiments. Oliver & Boyd, Edinburgh, 8th edition 1966. Not reviewed in MR. *169, 187, 371, 385*

FISHER R. A. and YATES F.

*[1] The 6×6 latin squares. Proc. Camb. Phil. Soc. 30 (1934), 492—507. *141, 156, 471*

*[2]$_S$ Statistical tables for biological, agricultural and medical research. Oliver & Boyd, Edinburgh, 6th edition 1963. Not reviewed in MR.

FLEISHER E.
*[1] On Euler Squares. Bull. Amer. Math. Soc. 40(1934), 218–219. *158,*

FOG D.
[1] Gruppentafeln und abstrakte Gruppentheorie. Skand. Mat. Kongr. Stockholm (1934), 376—384. *16*

FORD G. G.
[1] Remarks on quasigroups obeying a generalized associative law. Canad. Math. Bull. 13(1970), 17—21. MR 41 (1971) # 5533. *78*

FREEMAN G. H.
*[1]$_S$ The addition of further treatments to latin square designs. Biometrics 20(1964), 713—729. Not reviewed in MR.

*[2]$_S$ Some non-orthogonal partitions of 4×4, 5×5 and 6×6 latin squares. Ann. Math. Statist. 37(1966), 666—681. MR 33(1967) # 839.

FRINK O.
[1] Symmetric and self-distributive systems. Amer. Math. Monthly 62(1955), 697—707. MR 17(1956), page 458. *68*

FROLOV M.
*[1] Recherches sur les permutations carrées. J. Math. Spéc. (3) 4(1890), 8—11. *19, 139*

*[2] Recherches sur les permutations carrées. J. Math. Spéc. (3) 4(1890), 25—30. *139*

*[3] Le Problème d'Euler et les carrés magiques. Gauthier-Villars, Paris, 1884. *194*

FUCHS L.
[1] Abelian groups. Akadémiai Kiadó, Budapest, 1958. MR 21(1960) # 5672. *106, 147*

GALLAGHER P. X.
[1] Counting finite groups of given order. Math. Z. 102 (1967), 236—237. MR 36(1968) # 5210. *148*

GEISSER S.
*[1]$_S$ The latin square as a repeated measurement design. Proc. Fourth Berkeley Sympos. June–July 1960. Vol. IV, Contributions to Biology and Problems of Medecine, 241–250. Univ. of California Press, 1961.

GERGELY E.

*[1] A simple method for constructing doubly diagonalized latin squares. J. Combinatorial Theory, Ser. A 16 (1974), 266–272. *201*

*[2] A remark on doubly diagonalized orthogonal latin squares. Discrete Math. To appear. *214*

GHURYE S. G.

*[1] A characteristic of species of 7×7 latin squares. Ann. Eugenics 14(1948), 133. *143*

GILBERT E. N.

*[1] Latin squares which contain no repeated diagrams. SIAM Rev. 7(1965), 189—198. MR 31(1966) # 3346. *85, 148*

GINZBURG A.

[1] Systèmes multiplicatifs de relations. Boucles quasi-associatives. C. R. Acad. Sci. Paris 250(1960), 1413—1416. MR 22(1961) # 4793. *22*

[2] A note on Cayley loops. Canad. J. Math. 16(1964), 77—81. MR 29(1965) # 2287. *22*

[3] Representation of groups by generalized normal multiplication tables. Canad. J. Math. 19(1967), 774—791. MR 35(1968) # 5503. *22*

GINZBURG A. and TAMARI D.

[1] Representation of binary systems by families of binary relations. Israel J. Math. 7(1969), 21—32. MR 40(1970) # 1315. *22*

[2] Representation of generalized groups by families of binary relations. Israel J. Math. 7(1969), 33—45. MR 40 (1970) # 1316. *22*

GLAUBERMANN G.

[1] On loops of odd order I. J. Algebra 1(1964), 374—396. MR 31(1966) # 267. *32*

[2] On loops of odd order II. J. Algebra 8(1968), 393—414. MR 36(1968) # 5250. *32, 63*

GLEASON A. M.

[1] Finite Fano planes. Amer. J. Math. 78 (1956), 797—807. MR 18(1957), page 593. *296*

GOLOMB S. W. and BAUMERT L. D. [1]

See BAUMERT L. D. and GOLOMB S. W.

GOLOMB S. W. and POSNER E. C.

*[1] Rook domains, latin squares, affine planes, and error-distributing codes. IEEE Trans. Information theory. IT-10 (1964), 196—208. MR 29(1965) # 5657. *353, 354*

GORDON B.

[1] Sequences in groups with distinct partial products. Pacific J. Math. 11(1961), 1309—1313. MR 24(1962) # A3193. *85, 89*

GRANT D. A.

*[1]$_S$ The latin square principle in the design and analysis of psychological experiments. Psychological Bull. 45(1948), 427—442. Not reviewed in MR.

GRECO D.

[1] I gruppi finiti che sono somma di quattro sottogruppi. Rend. Accad. Sci. Fis. Mat. Napoli (4)18(1951), 74—85. MR 14(1953), page 445. *47*

[2] Su alcuni gruppi finiti che sono somma di cinque sottogruppi. Rend. Sem. Mat. Univ. Padova 22(1953), 313—333. MR 15(1954), page 503. *47*

[3] Sui gruppi che sono somma di quatro o cinque sottogruppi. Rend. Accad. Sci. Fis. Mat. Napoli (4)23(1956), 49—59. MR 20(1959) # 75. *47*

GREENBERG L. and NEWMAN M.

[1] Some results on solvable groups. Arch. Math. (Basel) 21(1970), 349—352. MR 42(1971) # 6109. *148*

GRIDGEMAN N. T.

*[1] Latin square tiling. Mathematics Teacher 64(1971), 358—360.

*[2] Magic squares embedded in a latin square. J. Recreational Math. 5(1972), 250.

GROSS K. B., MULLIN R. C. and WALLIS W. D.

*[1] The number of pairwise orthogonal symmetric latin squares. To appear. *464*

GRUENTHER M.

[1] Duplicate Contract complete. Bridge World Inc., New York, 1933. *219*

GUÉRIN R.

*[1] Aspects algébraiques du problème de Yamamoto. C. R. Acad. Sci. Paris 256(1963), 583—586. MR 26(1963) # 3620. *442*

*[2] Sur une généralisation de la méthode de Yamamoto pour la construction des carrés latins orthogonaux. C. R. Acad. Sci. Paris 256(1963), 2097—2100. MR 26(1963) # 4934. *442*

*[3] Sur une généralisation de Bose pour la construction de c. l. m. o. I. M. S. Berne (1964). Not reviewed in MR. *442*

*[4] Existence et propriétés des carrés latins orthogonaux I. Publ. Inst. Statist. Univ. Paris 15(1966), 113—213. MR 35 (1968) # 73. *284*

*[5] Existence et propriétés des carrés latins orthogonaux II. Publ. Inst. Statist. Univ. Paris 15(1966), 215—293. MR 35 (1968) # 4118. *404, 405, 408, 425, 436, 442*

GUHA H. C. and HOO T. K.

[1] On a class of quasigroups. Indian J. Math. 7(1965), 1—7. MR 32(1966)#5773. *63*

GUNTHER S.

*[1] Mathematisch-historische Miscellen. II. Die magischen Quadrate bei Gauss. Z. Math. Phys. 21(1876), 61—64. *140*

HABER S. and ROSENFELD A.

[1] Groups as unions of proper subgroups. Amer. Math. Monthly 66(1959), 491—494. MR 21(1960)#2692. *39, 46, 47*

HAJÓS G.

[1] Über einfache und mehrfache Bedeckung des n-dimensionalen Raumes mit einer Würfelgitter. Math. Z. 47(1941), 427—467. MR 3(1942), page 302. *54*

HALE R. W. and YATES F.

*[1]$_S$ The analysis of latin squares when two or more rows or columns are missing. Suppl. J. Roy. Statist. Soc. 6(1939), 67-79 Not reviewed in MR.

HALL M.

[1] Projective planes. Trans. Amer. Math. Soc. 54 (1943), 229—277. MR 5(1944), page 72. *271, 273, 278, 481*

*[2] An existence theorem for latin squares. Bull. Amer. Math. Soc. 51(1945), 387—388. MR 7(1946), page 106. *96*

[3] Distinct representatives of subsets. Bull. Amer. Math. Soc. 54(1948), 922—926. MR 10(1949), page 238. *146*

[4] Correction to "Projective planes". Trans. Amer. Math. Soc. 65(1949), 473—474. MR 10(1949), page 618.

[5] A combinatorial problem on Abelian groups. Proc. Amer. Math. Soc. 3(1952), 584—587. MR 14(1953), page 350. *34*

[6] Uniqueness of the projective plane with 57 points. Proc. Amer. Math. Soc. 4(1953), 912—916. MR 15(1954), page 460. *169*

[7] Correction to "Uniqueness of the projective plane with 57 points". Proc. Amer. Math. Soc. 5(1954) 994—997. MR 16(1955), page 395. *169*

*[8] A Survey of Combinatorial Analysis, Chapter 4. Existence and construction of designs. From "Surveys in Applied Math. Vol. IV", pages 76—104. Wiley, New York 1958. MR 22(1961)#2556.

[9] The theory of groups. Macmillan, New York, 1959. MR 21(1960)#1996. *36*

[10] Numerical analysis of finite geometries. Proc. IBM Scientific Comput. Sympos. Combinatorial Problems (Yorktown Heights, New York, 1964), IBM Data Process Division, New York, 1966. MR 36(1968)#5812. *481*

*[11] Combinatorial Theory. Blaisdell, Massachusetts, 1967. MR 37(1969) # 80. *80 169, 385, 451*

HALL M. and KNUTH D. E.
*[1] Combinatorial analysis and Computers. Amer. Math. Monthly 72(1965), No. 2, Part 2. Computers and Computing, 21—28. MR 30(1965) # 3030. *468*

HALL M. and PAIGE L. J.
*[1] Complete mappings of finite groups. Pacific J. Math. 5(1955), 541—549. MR 18(1957), page 109. *35*

HALL M. and SENIOR J. K.
[1] The groups of order 2^n ($n \leq 6$). Macmillan, New York, 1964. MR 29(1965) # 5889. *147, 148*

HALL M. and SWIFT J. D.
[1] Determination of Steiner triple systems of order 15. Math. Tables Aids Comput. 9(1955), 146—152. MR 18 (1957), page 192. *79*

HALL M., SWIFT J. D. and KILLGROVE R.
[1] On projective planes of order 9. Math. Tables Aids Comput. 13(1959), 233—246. MR 21(1960) # 5933. *483*

HALL M., SWIFT J. D. and WALKER R. J.
[1] Uniqueness of the projective plane of order eight. Math. Tables Aids. Comput. 10(1956), 186—194. MR 18 (1957), page 816. *169, 483*

HALL P.
[1] A note on soluble groups. J. London Math. Soc. 3(1928), 98—105. *37*
[2] On representation of subsets. J. London Math. Soc. 10(1935), 26—30. *95*

HAMMEL A.
*[1] Verifying the associative property for finite groups. Math. Teacher 61(1968), No. 2, 136—139. Not reviewed in MR. *19*

HAMMING R. W.
[1] Error detecting and error correcting codes. Bell System Tech. J. 29(1950), 147—160. MR 12(1951), page 35. *349*

HANANI H.
*[1] On the number of orthogonal latin squares. J. Combinatorial Theory 8(1970), 247—271. MR 40(1970) # 5466. *404, 425*

[HARANEN L. J.] Харанен Л. Я.
*[1] Построение проективных плоскостей над латинскими квадратами, входящими в описание плоскости Хьюза по- *297*

пядка 9. — Пермский Государственный Педагогический Институт, Ученые Записки, Математика 1(1968), 47—62. [Construction of projective planes over the latin squares which are used in the description of the Hughes plane of order 9. Scientific memoirs of Perm State Pedagogical Inst., Mathematics 1(1968), 47—62].

HARARY F.

*[1] Unsolved problems in the enumeration of graphs. Magyar Tud. Akad. Mat. Kutató Int. Közl. 5(1960), 63—95. MR 26(1963) #4340. *311*

HAYNES J. L. and MINNICK R. C.

[1] Magnetic core access switches. IEEE Trans. Electronic Computers EC-11(1962), 352—368. Not reviewed in MR. *366*

HEDAYAT A.

*[1] A set of three mutually orthogonal latin squares of order 15. Technometrics, 13(1971), 696—698. *237*

*[2] An application of sum composition: a self-orthogonal latin square of order 10. J. Combinatorial Theory, Ser. A. 14(1973), 256—260. *460*

*[3] An algebraic property of the totally symmetric loops associated with Kirkman-Steiner triple systems. Pacific J. Math. 40(1972), 305—309. *149*

HEDAYAT A. and FEDERER W. T. [1], [2], [3]
See FEDERER W. T. and HEDAYAT A.

HEDAYAT A., FEDERER W. T. and PARKER E. T. [1]
See FEDERER W. T., HEDAYAT A. and PARKER E. T.

HEDAYAT A. and SEIDEN E.

*[1] On a method of sum composition of orthogonal latin squares II. Michigan State University Report, July 1970. *436, 442*

*[2] On a method of sum composition of orthogonal latin squares III. Michigan State University Report, August 1970. *436, 445,*

*[3] F-square and orthogonal F-square designs: a generalization of latin square and orthogonal latin square designs. Ann. Math. Statist. 41(1970), 2035—2044. MR 42(1971) #2604. See also MR 42(1971) #2956. *461*

*[4] On a method of sum composition of orthogonal latin squares. (Proc. Conf. on Combinatorial Geometry and its Applications, Perugia, 1970) Atti del Convegno di Geometria Combinatoria e sue Applicazioni, Gubbio, Perugia, 1971, 239—256. *436, 442*

HEDAYAT A. and SHRIKHANDE S. S.

[1] Experimental designs and combinatorial systems associated with latin squares. and sets of mutually orthogonal latin squares. Sankhyā A, 33(1971), 423—443. *385*

HEFFTER B.

*[1] Problème des 36 officiers. Interméd. Math. 5(1896), 176. *158, 463*

HEPPES A. and RÉVÉSZ P.

*[1] A latin négyzet és az ortogonális latin négyzet-pár fogalmának egy új általánosítása és ennek felhasználása kísérletek tervezésére. Magyar Tud. Akad. Mat. Int. Közl. 1 (1956), 379—390. Hungarian. [A new generalization of the method of latin squares and orthogonal latin squares and its application to the design of experiments.] Not reviewed in MR. *181, 183*

HESSENBURG G.

[1] Beweis des Desarguesschen Satzes aus dem Pascalschen. Math. Ann. 61(1905), 161—172. *265*

HIGMAN G.

[1] Enumerating p-groups I. Inequalities. Proc. London Math. Soc. (3)10(1960), 24—30. MR 22(1961) # 4779. *148*

HILTON A. J. W.

*[1] Embedding an incomplete diagonal latin square in a complete diagonal latin square. J. Combinatorial Theory, Ser. A 15(1973), 121–128. *117*

*[2] On double diagonal and cross latin squares. J. London Math. Soc., 6(1973), 679–689. *200*

*[3] On the number of mutually orthogonal double diagonal latin squares of order m. To appear. *214*

*[4] Some simple constructions for doubly diagonal orthogonal latin squares. To appear. *213*

*[5] A note on embedding latin rectangles. Proc. British Combinatorial Conf., Aberystwyth, 1973. *117*

HILTON A. J. W. and SCOTT S. H.

*[1] A further construction of double diagonal orthogonal latin squares. Discrete Math. 7(1974), 111–127. *213*

HINER F. P. and KILLGROVE R. B.

*[1] Subsquare complete latin squares. AMS Notices 17 (1970), 677-05, 758. *50*

*[2] Subsquares of latin squares. To appear. *51, 52*

HOBBY C., RUMSEY H. and WEICHSEL P. M.

[1] Finite groups having elements of every possible order. J. Washington Acad. Sci. 50(1960), No. 4, 11—12. MR 26 (1963) # 1356. *55*

HÖHLER P.

[1] Eine Verallgemeinerung von orthogonalen lateinischen Quadraten auf höhere Dimensionen. Diss. Dokt. Math. Eidgenöss. Techn. Hochschule Zürich, 1970, 595, 913, 350. MR 45(1973) # 8545. *181*

HORNER W. W.

*[1] Addition-multiplication magic squares. Scripta Math. 18(1952), 300—303. Not reviewed in MR. *215*

*[2] Addition-multiplication magic squares of order 8. Scripta Math. 21(1955), 23—27. MR 17(1956), page 227. *215*

HORTON J. D.

[1] Quintuplication of Room squares. Aequationes Math. 7(1971), 243—245. MR 46(1973) # 3344 *229*

[2] Room designs and one-factorizations. Aequationes Math. To appear. *228, 312*

*[3] Sub-latin squares and incomplete orthogonal arrays. J. Combinatorial Theory Ser. A 16(1974), 23—33.

HORTON J. D., MULLIN R. C. and STANTON R. G.

[1] A recursive construction for Room designs. Aequationes Math. 6(1971), 39—45. MR 45(1973) # 6649 *229*

HORTON J. D. and STANTON R. G.

[1] Composition of Room squares. Combinatorial Theory and its Applications, 1013—1021. North Holland, Amsterdam, 1970. MR 46(1973) # 73 *223*

[2] A multiplication theorem for Room squares. J. Combinatorial Theory, Ser. A 12(1972), 322—325. MR 45 (1973) # 6650. *223*

HOSSZÚ M.

[1] Belouszov egy tételéről és annak néhány alkalmazásáról. Magyar Tud. Akad. Mat. Fiz. Oszt. Közl. 9(1959), 51—56. Hungarian. [Concerning a theorem of Belousov and some of its applications.] MR 21(1960) # 4198. *78*

HOUSTON T. R.

*[1] Sequential counterbalancing in latin squares. Ann. Math. Statist. 37(1966), 741—743. MR 34(1967) # 905. *83, 97*

HSIAO M. Y., BOSSEN D. C. and CHIEN R. T. [1]

See BOSSEN D. C., CHIEN R. T. and HSIAO M. Y.

HUGHES D. R.

*[1] Planar division neo-rings. Trans. Amer. Math. Soc. 80 (1955), 502—527. MR 17(1956), page 451. *282*

[2] A class of non-desarguesian projective planes. Canad. J. Math. 9(1957), 378—388. MR 19(1958), page 444. *280*

[3] Combinatorial analysis. t-designs and permutation groups. Proc. Symp. Pure Math. 6(1962), 39—41. Amer. Math. Soc., Providence, R. I. MR 24(1962) # A 3206. *376*

[4] On t-designs and groups. Amer. J. Math. 87(1965), 761—778. MR 32(1966) # 5727 *376*

HUGHES N. J. S.

[1] A theorem on isotopic groupoids. J. London Math. Soc. 32(1957), 510—511. MR 19(1958), page 634. *26*

HUMBLOT L.

*[1] Sur une extension de la notion de carrés latins. C. R. Acad. Sci. Paris 273(1971), 795—798. MR 44(1972) #3892. *186*

HUNTER J. S. and YOUDEN W. J.

*[1]$_S$ Partially replicated latin squares. Biometrika 11(1955), 399—405. Not reviewed in MR.

[IBRAGIMOV S. G.] Ибрагимов С. Г.

[1] Из предистории теории квазигрупп. (О забытых работах Эрнста Щредера в XIV. в.) [On the history of quasigroup theory. (From the neglected work of Ernst Schroeder, 19th Century.) Abstracts of lectures: First All-Union Conference on quasigroups. Suhumi (1967), 15—16.] Not reviewed in MR. *16*

[ISTOMINA L. I.] Истомина Л. И.

*[1] Единственность конечной круговой плоскости порядка 4. Пермский Государственный Педагогический Институт, Ученые Записки, Математика. 1(1968) 40—46. [Uniqueness of the finite circular plane of order 4. Scientific Memoirs of Perm State Pedagogical Inst., Mathematics 1(1968), 40—46.] *297*

JACOB S. M.

*[1] The enumeration of the latin rectangle of depth three by means of a formula of reduction, with other theorems relating to non-clashing substitutions and latin squares. Proc. London Math. Soc. (2) 31(1930), 329—354. *141, 306*

JAIN N. C. and SHRIKHANDE S. S.

*[1]$_S$ The non-existence of some partially balanced incomplete block designs with latin square type association scheme. Sankhyā Ser. A 24(1962), 259—268. MR 33(1967) #7268.

JOHNSON D. M., DULMAGE A. L. and MENDELSOHN N. S. [2]
See DULMAGE A. L., JOHNSON D. M. and MENDELSOHN N. S.

JORDAN C.

[1] Traité des Substitutions. Gauthier-Villars, Paris, 1870 (reprinted 1957). *108*

JOSHI D. D.

[1] A note on upper bounds for minimum distance codes. Information and Control 1(1958), 289—295. MR 20(1959) #5705. *350*

Kalbfleisch J. G. and Weiland P. H.
[1] Some new results for the covering problem. Recent progress in combinatorics (Proc. Third Waterloo Conf. on Combinatorics, 1968), 37—45. Academic Press, New York, 1969. MR 41(1971) # 1547b. *362*

Kaplansky I.
[1] Solution of the "problème des ménages". Bull. Amer. Math. Soc. 49(1943), 784—785. MR 5(1944), page 86. *150, 152*

Kaplansky I. and Riordan J.
[1] Le problème des ménages. Scripta Math. 12(1946), 113—124. MR 8(1947), page 365. *150*

Kárteszi F.
*[1] Incidenciageometria. Mat. Lapok 14(1963), 246—263. Hungarian. [Incidence Geometry.] MR 31(1966) # 3905. *170*
*[2] Bevezetés a véges geometriákba. Disquisitiones mathematicae Hungaricae. Akadémiai Kiadó, Budapest, 1972. English translation: Introduction to finite geometries. Akadémiai Kiadó, Budapest, and North Holland, Amsterdam. To appear. *170, 251, 284*

Kautz W. H. and Elspas B. [1]
See Elspas B. and Kautz W. H.

Kautz W. H. and Singleton R. C.
[1] Non-random binary superimposed codes. IEEE Trans. Information theory IT-10(1964), 363—377. Not reviewed in MR. *366*

Keedwell A. D.
[1] On the order of projective planes with characteristic. Rend. Mat. e Appl. (5) 22(1963), 498—530. MR 30(1965) # 5213. *296*
[2] A geometrical proof of an analogue of Hessenberg's theorem. J. London Math. Soc. 39(1964), 424—426. MR 29 (1965) # 6359. *265*
*[3] A search for projective planes of a special type with the aid of a digital computer. Math. Comp. 19(1965), 317—322. MR 31(1966) # 3921. *49, 230, 297*
*[4] On orthogonal latin squares and a class of neofields. Rend. Mat. e Appl. (5) 25(1966), 519—561. MR 36(1968) # 3664; erratum MR 37(1969), page 1469. *160, 173, 230, 235, 237, 244, 246, 248, 274, 478, 489*
[5] On property D neofields. Rend. Mat. e Appl. (5) 26(1967), 383—402. MR 37(1969) # 5112. *248, 481*
*[6] On property *D* neofields and some problems concerning orthogonal latin squares. Computational Problems in Abstract Algebra (Proc. Conf., Oxford 1967), 315—319. Pergamon Press, Oxford, 1970. MR 41(1971) # 88. *173, 248*

[7] A note on planes with characteristic. (Proc. Conf. Combinatorial Geometry and its Applications, Perugia, 1970) Atti del Convegno di Geometria Combinatoria e sue Applicazioni, Gubbio, Perugia, 1971, 307–318. 296

*[8] Some problems concerning complete latin squares. Proc. Brit. Combinatorial Conf., Aberystwyth, 1973. To appear. 93

*[9] Some connections between latin squares and graphs. Atti del Colloquio Internazionale sulle Teorie Conbinatorie, Roma, 1973. To appear.

*[10] Row complete squares and a problem of A. Kotzig concerning P-quasigroups and Eulerian circuits. To appear. 83, 93

Kendall D. G. and Rankin R. A.

[1] On the number of Abelian groups of a given order. Quart. J. Math. Oxford Ser. 18(1947), 197—208. MR 9 (1948), page 226. 147

Kerewala S. M.

*[1] The enumeration of the latin rectangle of depth three by means of a difference equation. Bull. Calcutta Math. Soc. 33(1941), 119—127. MR 4(1943), page 69. 150

*[2] The asymptotic number of three-deep latin rectangles. Bull. Calcutta Math. Soc. 39(1947), 71—72. MR 9(1948), page 404. 153

*[3] Asymptotic solution of the "problème des ménages". Bull. Calcutta Math. Soc. 39(1947), 82—84. MR 9(1948), page 405. 153

*[4] A note on self-conjugate latin squares of prime degree. Math. Student 15(1947),16. MR 10(1949), page 347. 306

Kerr J. R., Pearce S. C. and Preece D. A.

*[1]$_S$ Orthogonal designs for three-dimensional experiments. Biometrika 60(1973), 349–358. 187

Kertész A.

[1] Kvazicsoportok. Mat. Lapok 15(1964), 87—113. Hungarian. [Quasigroups.] MR 30(1965) # 3935. 80

Kesava Menon P.

*[1] A method of constructing two mutually orthogonal latin squares of order $3n + 1$. Sankhyā Ser. A 23(1961), 281—282. Not reviewed in MR. 466

Killgrove R. B.

[1] A note on the non-existence of certain projective planes of order nine. Math. Comput. 14(1960), 70-71. MR 22(1961) # 4011. 483

[2] Completions of quadrangles in projective planes. Canad. J. Math. 16(1964), 63—76. MR 28(1964) # 513. 50, 53

[3] The K-configuration. AMS Notices 15(1968), 653—28, 15. Not reviewed in MR. *293, 294, 295*

[4] Private communication. *295*

KILLGROVE R. B. and HINER F. [1], [2]
See HINER F. and KILLGROVE R. B.

KILLGROVE R. B. and PARKER E. T.
[1] A note on projective planes of order nine. Math. Comput. 18(1964), 506—508. MR 29(1965) # 1573. *483*

KIRKMAN Rev. T. P.
[1] On a problem in combinations. Camb. & Dublin Math. J. 2(1847), 191—204. *79*

KISHEN K.
*[1] On latin and hyper-graeco cubes and hypercubes. Current Science 11(1942), 98—99. Not reviewed in MR. *187*

*[2] On the construction of latin and hyper-graeco-latin cubes and hypercubes. J. Indian Soc. Agric. Statistics 2 (1950), 20—48. MR 11(1950), page 637. *187, 188*

KLARNER D. A. and HAUTUS M. L. J.
*[1] Uniformly coloured stained glass windows. Proc. Lond. Math. Soc. (3) 23(1971), 613—628. MR 44(1972) # 6547 *466*

KLEINFELD E.
[1] Techniques for enumerating Veblen-Wedderburn systems. J. Assoc. Comput. Mach. 7(1960), 330—337. MR 23 (1962) # A2788. *485*

KLINGENBERG W.
[1] Beziehungen zwischen einigen affinen Schliessungssätzen. Abh. Math. Sem. Univ. Hamburg 18(1952), 120—143. MR 14(1953), page 786. *265*

[2] Beweis des Desarguesschen Satzes aus der Reidemeisterfigur und verwandte Sätze. Abh. Math. Sem. Univ. Hamburg 19(1955), 158—175. MR 16(1955), page 950. *265*

KNUTH D. E.
[1] Finite semifields and projective planes. J. Algebra 2 (1965), 182—217. MR 31(1966) # 218. *485*

[2] A class of projective planes. Trans. Amer. Math. Soc. 115(1965), 541—549. MR 34(1967) # 1916. *485*

KNUTH E.
*[1] Egy ortogonális latin négyzetekkel kapcsolatos problémáról. Közlemények No. 2 (1967), 26—40. Hungarian. [On a problem connected with orthogonal latin squares.] Not reviewed in MR. *366*

KOKSMA K. K.

*[1] A lower bound for the order of a partial transversal in a latin square. J. Combinatorial Theory 7(1969), 94—95. MR 39(1970)#1342. *99*

KOTZIG A.

[1] Groupoids and partitions of complete graphs. Combinatorial Structures and their Applications. (Proc. Calgary Internat. Conf., Calgary, Alta., 1969), 215—221. Gordon and Breach, New York, 1970. MR 42(1971)#4446. *303, 305*

KRAITCHIK M.

*[1] Mathematical recreations. Dover, New York, 2nd Edition 1953. MR 14(1953), page 620. *194, 212*

KRÄTZEL E.

[1] Die maximale Ordnung der Anzahl der wesentlich verschiedenen abelschen Gruppen n-ter Ordnung. Quart. J. Math. Oxford Ser. 21(1970), 273—275. MR 42(1941)#3171. *147*

LAUGEL L.

*[1] Sur le problème d'Euler, dit des 36 officiers. (Réponse à la question numéro 453.) Interméd. Math. 3(1896), 17. *158*

LAWLESS J. F.

*[1] Note on a family of BIBD's and sets of mutually orthogonal latin squares. J. Combinatorial Theory, Ser. A 11 (1971), 101–105. MR 44(1972)#3893

LEE C. Y.

[1] Some properties of non-binary error-correcting codes. IEEE Trans. Information Theory. IT-4 (1958), 77—82. MR 22(1961)#13353. *112*

LEHMER D. H.

[1] Teaching combinatorial tricks to a computer. Proc. Sympos. Appl. Math., American Mathematical Society, Providence, R. I. 10(1960), 179—193. MR 22(1961)#4127. *467, 468*

LEMOINE E.

*[1] Sur le problème des 36 officiers. Interméd. Math. 6(1889), 273. *158*

*[2] Sur le problème des 36 officiers. Interméd. Math. 7(1900), 311—312. *158*

LINDNER C. C.

*[1] On completing latin rectangles. Canad. Math. Bull. 13(1970), 65—68. MR 41(1971)#6702. *114, 115*

*[2] Comment on a note of J. Marica and J. Schönheim. Canad. Math. Bull. 13(1970), 539. MR 43(1972)#74. *115*

[3] Construction of quasigroups satisfying the identity $X(XY)=XY$. Canad. Math. Bull. 14(1971), 57–60. MR 45(1973#434. *433, 460*

*[4] The generalized singular direct product for quasigroups. Canad. Math. Bull. 14(1971), 61—64. MR 45(1973) #435. *434, 435, 460*

*[5] Quasigroups orthogonal to a given abelian group. Canad. Math. Bull. 14(1971), 117—118. MR 45(1973) #2070. *435*

*[6] Embedding partial idempotent latin squares. J. Combinatorial Theory, Ser. A 10(1971), 240—245. MR 43(1972) #1862. *117*

*[7] Finite embedding theorems for partial latin squares, quasigroups, and loops. J. Combinatorial Theory., Ser. A 13(1972), 339—345. *117, 118*

*[8] Small embeddings of partial commutative latin squares. *117*

*[9] Extending mutually orthogonal partial latin squares. Acta Sci. Math. (Szeged), 32(1971), 283—285. *116*

[10] Construction of quasigroups having a large number of orthogonal mates. Commentationes Math. Univ. Carolinae 12(1971), 611—618. MR 44(1972) #6890. *465*

[11] Construction of quasigroups using the singular direct product. Proc. Amer. Math. Soc. 29(1971), 263—266. MR 43 (1972) #6354. *74, 434, 435*

[12] Identities preserved by the singular direct product. Algebra Universalis 1(1971), 86—89. MR 44(1972) #5401. *433, 434*

[13] Identities preserved by the singular direct product II. Algebra Universalis 2(1972). 113—117 MR 46(1973) #5519. *434*

[14] An algebraic construction for Room squares. Siam J. Appl. Math. 22(1972), 574—579. MR 46(1973) #3671 *228*

*[15] On constructing doubly diagonalized latin squares. *201*

*[16] Construction of doubly diagonalized orthogonal latin squares. J. Discrete Math. 58(1973), 79—86. *213*

[17] On the construction of cyclic quasigroups. Discrete Math. 6(1973), 149—158. *434*

LINDNER C. C. and MENDELSOHN N. S.

[1] Construction of perpendicular Steiner quasigroups. Aequationes Math. 9(1973), 150—156. *192, 436, 465*

LINDNER C. C. and STRALEY T. H.

[1] Construction of quasigroups containing a specific number of subquasigroups of a given order. Algebra Universalis 1(1971), 238—247. MR 45(1973) # 3618. *436*

LOCKEMANN VON W.

*[1] Ein Rechenprogramm zur Erzeugung von lateinischen Quadraten. Elektronische Rechenanlagen 2(1960), 129—130. Not reviewed in MR. *469*

LONGYEAR J. A.

*[1] Certain M. O. L. S. as groups. Proc. Amer. Math. Soc. 36(1972), 379—384. MR 46(1973) # 7058 *466*

LORIGA J. D.

*[1] Question numéro 261. Intermed. Math. 1(1894), 146—147. *158*

[LUMPOV A. D.] Лумпов А. Д.

*[1] Построение проективных плоскостей порядка 9 над латинскими квадратами того же порядка, состоящими из подквадратов порядка 3 разного состава. Комбинаторный анализ 2(1972), 93—98. [Construction of a projective plane of order 9 over latin squares of the selfsame order which consist of subsquares of order 3 and are of various structures. Combinatorial Analysis, 2(1972), 93—98.] *297*

LUNN, A. C. and SENIOR J. K.

[1] A method of determining all the solvable groups of given order and its application to the orders 16p and 32p. Amer. J. Math. 56(1934), 319—327. *148*

[2] Determination of the groups of orders 101 to 161 omitting order 128. Amer. J. Math. 56(1934), 328—338. *147*

[LYAMZIN A. I.] Лямзин А. И.

*[1] Пример пары ортогональных латинских квадратов десятого порядка. Успехи мат. наук 18(1963), №.5(113), 173—174. [An example of a pair of orthogonal latin squares of order ten. Uspehi Mat. Nauk 18(1963), No.5(113), 173—174.] MR 28(1964) # 2979. *235*

MACINNES C. R.

[1] Finite planes with less than eight points on a line. Amer. Math. Monthly 14(1907), 171—174. *160, 169*

MACMAHON P. A.

*[1] A new method in combinatory analysis, with applications to latin squares and associated questions. Trans. Camb. Phil. Soc. 16(1898), 262—290. *140*

*[2] Combinatorial Analysis. The foundations of a new theory. Phil. Trans. Royal Soc. London, Ser. A 194(1900), 361—386.

*[3] Combinatory Analysis. Vols. I, II. Cambridge Univ. Press, 1915. *140, 141, 146*

MACNEISH H. F.

*[1] Das Problem der 36 Offiziere. Jber. Deutsch. Math. Verein 30(1921), 151—153. *158*

*[2] Euler squares. Ann. of Math. 23(1922), 221—227. *158, 390*

MACWILLIAMS F. J., SLOANE N. J. A. and THOMPSON W.

[1] On the existence of a projective plane of order 10. J. Combinatorial Theory. Ser. A 14(1973), 66—78. *371, 484*

MAILLET E.

*[1] Réponse à la question numéro 261. Interméd. Math. 1(1894), 262. *158, 212*

*[2] Sur les carrés latins d'Euler. C. R. Assoc. France Av. Sci. 23(1894), part 2, 244—252. *445, 446*

*[3] Sur une application de la théorie des groupes de substitutions à celle des carrés magiques. Mémoires de l'Académie des sciences de Toulouse (9) 6(1894), 258—281. *212*

*[4] Question numéro 453. Interméd. Math. 2(1895), 17. *158*

*[5] Application de la théorie des substitutions à celle des carrés magiques. Quart. J. Pure and Appl. Math. 27(1896), 132—144. *212*

*[6] Carrés d'Euler. L'Encyclopédie des sciences mathématiques pures et appliquées, édition française. Details unknown. (Paris & Leipzig, 1906).

*[7] Figures magiques. L'Encyclopédie des sciences mathématiques pures et appliquées, édition française t. I, vol. 3, fasc. 1, 62—75 (Paris & Leipzig, 1906). *194*

[MALIH A. E.] Малых А. Е.

*[1] Описание проективных плоскостей порядка n латинскими квадратами порядка $n-1$. Комбинаторный анализ, 2(1972), 86—92. [A description of projective planes of order n with the aid of latin squares of order $n - 1$. Combinatorial Analysis 2(1972), 86—92.] *297*

MANN H. B.

*[1] The construction of orthogonal Latin squares. Ann. Math. Statist. 13(1942), 418—423. MR 4(1943), pages 184, 340. *29, 156, 159, 231, 235, 391,*

*[2] On the construction of sets of orthogonal Latin squares. Ann. Math. Statist. 14(1943), 401—414. MR 5(1944), page 169.

*[3] On orthogonal latin squares. Bull. Amer. Math. Soc. 50(1944), 249—257. MR 6(1945), page 14. *157, 160, 169, 445*

*[4] Analysis and design of experiments. Dover, New York, 1949. MR 11(1950), page 262. *103, 385*

[5] On products of sets of group elements. Canad. J. Math. 4(1952), 64—66. MR 13(1952), page 720. *48*

MANO K.

*[1] On the reduced number of the latin squares of the nth order. Sci. Rep. Fac. Lit. Sci. Hirosaki Univ. 7(1960), 1—2. Not reviewed in MR. *147*

MARGOSSIAN A.

*[1] Carrés Latins et carrés d'Euler modules impairs. Enseignement Math. 30(1931), 41—49. *195*

MARICA J. and SCHÖNHEIM J.
*[1] Incomplete diagonals of latin squares. Canad. Math. Bull. 12(1969), 235. MR 40(1970) # 55. *115*

[MARKOVA E. V.] Маркова Е. В.
*[1]$_S$ Латинские квадраты в планировании эксперимента. Заводская Лаборатория 34(1968), 60—65. [Latin squares in the design of experiments. Zavodskaya Laboratoriya 34 (1968), 60—65.] Not reviewed in MR.
*[2]$_S$ Латинские прямоугольники и кубы в планировании эксперимента. Заводская Лаборатория 34 (1968), 832—837. [Latin rectangles and cubes in the design of experiments. Zavodskaya Laboratoriya 34(1968), 832—837.] Not reviewed in MR.
*[3]$_S$ О структурной связи между факторными экспериментами и латинскими квадратами. Заводская Лаборатория 37(1971), 60—66. [On structural connections between factorial experiments and latin squares. Zavodskaya Laboratoriya 37(1971), 60—66.] Not reviewed in MR.
*[4]$_S$ Руководство по применению латинских планов при планировании эксперимента с качественными факторами. Научный совет по кибернетике Академии Наук СССР. Уральский научно-исследовательский и проектный институт строительных материалов. Челябинск 1971. [Handbook of applications of latin designs for planning experiments with quality factors. Scientific Commitee for Cybernetics of the Academy of Sciences of the U.S.S.R. Uralian Scientific research and planning Institute for building materials. Chelyabinsk 1971.]

[MARKOVA E. V., SHTARKMAN B. P. and other authors] Маркова Е. В., Штаркман Б. П. и др.
*[1]$_S$ Примеры применения латинских квадратов. Заводская Лаборатория 37(1971), 316—319. [Examples of the applications of latin squares. Zavodskaya Laboratoriya 37(1971), 316—319.] Not reviewed in MR.

MARTIN G. E.
*[1] Planar ternary rings and latin squares. Matematiche (Catania) 23(1968), 305—318. MR 40(1970) # 3424. *160*

MCCLINTOCK E.
[1] On the most perfect forms of magic squares, with methods for their production. Amer. J. Math. 19(1897), 99—120. *214*

MCWORTER W. A.
[1] On a theorem of Mann. Amer. Math. Monthly 71(1964), 285—286. MR 28(1964) # 5135. *47*

MENDELSOHN N. S.
*[1] Hamiltonian decomposition of the complete directed n-graph. Theory of Graphs (Proc. Colloq., Tihany, 1966), 237—241. Academic Press, New York, 1968. MR 38(1969) #4361. *89, 91, 300, 475*
*[2] Combinatorial designs as models of universal algebras. Recent Progress in Combinatorics. (Proc. Third Waterloo Conf. on Combinatorics, 1968), 123—132. Academic Press, New York, 1969. MR 41(1971) #85. *456, 457*
*[3] Latin squares orthogonal to their transposes. J. Combinatorial Theory, Ser. A 11(1971), 187—189. MR 45 (1973) #88. *458, 459*
[4] Orthogonal Steiner Systems. Aequationes Math. 5(1970), 268—272. MR 44(1972) #1587. *192, 465*
*[5] On maximal sets of mutually orthogonal idempotent latin squares. Canad. Math. Bull. 14(1971), 449. *465*

MESNER D. M.
*[1]$_S$ An investigation of certain combinatorial properties of partially balanced incomplete block designs and association schemes, with a detailed study of latin square and related types. Ph. D. thesis, Michigan State University, 1956. *385*
*[2]$_S$ A new family of partially balanced incomplete block designs with some latin square design properties. Ann. Math. Statist. 38(1967), 571—581. MR 36(1968) #6072. *385*

MILIČ S.
*[1] A new proof of Belousov's theorem for a special law of quasigroup operations. Publ. Inst. Math. (Beograd) N. S. 11 (25)(1971), 89—91. MR 46(1973) #285. *78*

MILLER G. A.
[1] Determination of all the groups of order 64. Amer. J. Math. 52(1930), 617—634. *147*

MILLS W. H.
*[1] Three mutually orthogonal latin squares. J. Combinatorial Theory, Ser. A 13(1972), 79—82. MR 45(1973) #8546. *425*

MINNICK R. C. and HAYNES J. L. [1]
See HAYNES J. L. and MINNICK R. C.

MONJARDET B.
[1] Combinatoire et structures algébriques.I. Math. Sci. Humaines N°.18(1967), 33—40. MR 35(1968) #6573.
[2] Quasi-groupes finis, quasi-groupes orthogonaux, ensemble complet orthogonal. Math. Sci. Humaines N°.19 (1967), 13—20. MR 37(1969) #2610.

Mood A. M. and Parker E. T.

[1] Some balanced Howell rotations for duplicate bridge sessions. Amer. Math. Monthly 62(1955), 714—716. MR 17 (1956), page 449. *229*

Moser W. O. J.

*[1] The number of very reduced $4 \times n$ latin rectangles. Canad. J. Math. 19(1967), 1011—1017. MR 36(1968) # 61. *152*

Moufang R.

[1] Zur Struktur von Alternativkörpern. Math. Ann. 110 (1935), 416—430. *16*

Mullin R. C.

[1] On the existence of a Room design of side F_4. Utilitas Math. 1(1972) 111—120. MR 46(1973)—3336.

Mullin R. C. and Németh E.

[1] A counterexample to a direct product construction of Room squares. J. Combinatorial Theory 7(1969), 264—265. MR 40(1970) # 4391. *223*

[2] On furnishing Room squares. J. Combinatorial Theory 7(1969), 266—272. MR 41(1971) # 5228. *192, 225, 465*

[3] An existence theorem for Room squares. Canad. Math. Bull. 12(1969), 493—497. MR 40(1970) # 2560. *225*

[4] On the nonexistence of orthogonal Steiner systems of order 9. Canad. Math. Bull. 13(1970), 131—134. MR 41 (1971) # 3297. *192*

*[5] A construction for self-orthogonal latin squares from certain Room squares. Proc. Louisiana Conf. on Combinatorics, Graph Theory and Computing (Louisiana State Univ., Baton Rouge, La., 1970), 213—226, MR 42(1971) # 2957. *225, 229, 460*

Mullin R. C. and Stanton R. G.

[1] Construction of Room squares. Ann. Math. Statist. 39 (1968), 1540—1548. MR 38(1969) # 2904. *225, 229*

[2] Techniques for Room squares. Proc. Louisiana Conf. on Combinatorics, Graph Theory and Computing. (Louisiana State Univ., Baton Rouge, La., 1970), 445—464. MR 42 (1971) # 1679. *229*

*[3] Construction of Room squares from orthogonal latin squares. Proc. Second Louisiana Conf. on Combinatorics, Graph Theory and Computing (Louisiana State Univ., Baton Rouge, La., 1971), 375—386. *229*

[4] Room squares and Fermat primes. J. Algebra 20(1972), 83—89. MR 45(1973) # 8547 *229*

Mullin R. C. and Wallis W. D.

[1] On the existence of Room squares of order $4n$. Aequationes Math. 6(1971), 306—309. MR 44(1972) # 6516. *229, 464*

MURDOCH D. C.
[1] Quasigroups which satisfy certain generalized associative laws. Amer. J. Math. 61(1939), 509—522. *68*
[2] Structure of abelian quasigroups. Trans. Amer. Math. Soc. 49(1941), 392—409. MR 2(1941), page 218. *68*

NETTO E.
[1] Lehrbuch der Kombinatorik. Teubner, Leipzig, 1901. (2nd Ed., 1927). Chelsea reprint, New York, 1955. *74, 79*

NEUMANN M.
[1] Asupra unor teoreme de închidere Lucrăr Şti. Inst. Ped. Timişoara Mat.-Fiz. 1959 (1960), 85—93. Rumanian. [Some incidence theorems.] MR 24(1962) # A150. *269*
[2] Unele consecinte ale unei introduceri geometrice a quasigrupului. Lucrăr. Şti. Inst. Ped. Timişoara Mat.-Fiz. 1961(1962), 99—102. Rumanian. [Some consequences of a geometric approach to quasigroups.] MR 32(1966) # 4206. *269*

NORTON D. A.
*[1] Groups of orthogonal row-latin squares. Pacific J. Math. 2(1952), 335—341. MR 14(1953), page 235. *103, 105*
[2] Hamiltonian loops. Proc. Amer. Math. Soc. 3(1952), 56—65. MR 13(1952), page 720. *56*
[3] A note on associativity. Pacific J. Math. 10(1960), 591—595. MR 22(1961) # 6859. *20*

NORTON D. A. and STEIN S. S.
*[1] An integer associated with latin squares. Proc. Amer. Math. Soc. 7(1956), 331—334. MR 17(1956), page 1043. *465*

NORTON H. W.
*[1] The 7×7 squares. Ann. Eugenics 9(1939), 269—307. MR 1(1940), page 199. *29, 42, 140, 141, 158, 169*

[OLDEROGGE G. B.] Ольдерогге Г. Б.
[1] О некоторых специальных корректирующий кодах матричного типа. Радиотехника 18(1963), №. 7, 14—19. [On some special correcting codes of matrix type. Radiotechniques 18(1963), No. 7, 14—19.] Not reviewed in MR. *356*

OSBORN J. M.
[1] New loops from old geometries. Amer. Math. Monthly 68(1961), 103—107. MR 23(1962) # A1686. *40*

O'SHAUGHNESSY C. D.
[1] A Room design of order 14. Canad. Math. Bull. 11 (1968), 191—194. MR 37(1969) # 3940. *192, 226, 228, 465*
*[2] On Room squares of order $6m + 2$. J. Combinatorial Theory, Ser. A 13(1972), 306—314. MR 46(1973) # 1619. *229*

OSTROM T. G.

[1] Semi-translation planes. Trans. Amer. Math. Soc. 111 (1964), 1—18. MR 28(1964) #2472. *331*

[2] Nets with critical deficiency. Pacific J. Math. 14(1964), 1381—1387. MR 30(1965) #1446. *331, 347*

[3] Derivable nets. Canad. Math. Bull. 8(1965), 601—613. MR 33(1967) #3185. *347*

[4] Replaceable nets, net collineations, and net extensions. Canad. J. Math. 18(1966), 666—672. MR 35(1968) #4809. *347*

[5] Vector spaces and constructions of finite projective planes. Arch. Math. (Basel) 19(1968), 1—25. MR 37(1969) #2081. *251, 347*

OSTROWSKI R. T. and VAN DUREN K. D.

*[1] On a theorem of Mann on latin squares. Math. Comp. 15(1961), 293—295. MR 23(1962) #A1543. *451, 468*

PAIGE L. J.

[1] A note on finite abelian groups. Bull. Amer. Math. Soc. 53(1947), 590—593. MR 9(1948), page 6. *33, 34, 487*

[2] Neofields. Duke Math. J. 16(1949), 39—60. MR 10 (1949), page 430. *246*

[3] Complete mappings of finite groups. Pacific J. Math. 1(1951), 111—116. MR 13(1952), page 203. *33, 34, 38*

PAIGE L. J. and TOMPKINS C. B.

*[1] The size of the 10×10 orthogonal latin square problem. Proc. Sympos. Appl. Math. 10(1960), 71—83. American Mathematical Society, Providence, R. I. MR 22(1961) #6724. *173, 476*

PAIGE L. J. and WEXLER C.

*[1] A canonical form for incidence matrices of finite projective planes and their associated latin squares. Portugaliae Math. 12(1953), 105—112. MR 15(1954), page 671. *284, 286*

PANAYOTOPOULOS A.

*[1] Sur les carrés latins orthogonaux. Bull. Math. Soc. Sci. Math. Roumanie 14(62) (1970), 69—74.

[PANTELEVA L. I.] Пантелеева Л. И.

*[1] Построение проективной плоскости над латинским квадратом, входящим в описание плоскости Хьюза порядка 9. Пермский Государственный Педагогический Институт, Ученые Записки, Математика 1(1968), 63—67. [Construction of a projective plane over a latin square which occurs in the description of the Hughes plane of order 9. Scientific Memoirs of Perm State Pedagogical Inst., Mathematics 1(1968), 63—67.] *297*

PARKER E. T.

*[1] Construction o f some sets of mutually orthogonal latin squares. Proc. Amer. Math. Soc. 10(1959), 946—949. MR 22 (1961) # 674. *158, 393, 394*

*[2] Orthogonal latin squares. Proc. Nat. Acad. Sci. USA 45(1959), 859—862. MR 21(1960) # 3344. *158, 235, 397*

*[3] A computer search for latin squares orthogonal to latin squares of order 10. Abstract 564—571. Notices Amer. Math. Soc. 6(1959), 798. *173*

*[4] Computer searching for orthogonal latin squares of order 10. Abstract 61T-292. Notices Amer. Math. Soc. 8(1961), 617. *173*

*[5] Computer study of orthogonal latin squares of order 10. Computers and Automation 11(1962), 33—35. Not reviewed in MR. *173*

*[6] Nonextendibility conditions on mutually orthogonal latin squares. Proc. Amer. Math. Soc. 13(1962), 219—221. MR 25(1963) # 2968. *445, 452*

*[7] On orthogonal latin squares. Proc. Sympos. Pure Math. 6(1962), 43—46. American Mathematical Society, Providence, R. I. MR 24(1962) # A2541. *454*

*[8] Computer investigation of orthogonal latin squares of order ten. Proc. Sympos. Appl. Math. 15(1963), 73—81. American Mathematical Society, Providence, R. I. MR 31 (1966) # 5140. *155, 173, 478*

*[9] Pathological latin squares. Proc. Sympos. Pure Math. 19(1971), 177—181. American Mathematical Society, Providence, R. I. *445, 455*

PARKER E. T. and KILLGROVE R. B. [1]
See KILLGROVE R. B. and PARKER E. T.

PARKER E. T. and MOOD A. M. [1]
See MOOD A. M. and PARKER E. T.

PARKER F. D.

[1] When is a loop a group? Amer. Math. Monthly 72 (1965), 765—766. Not reviewed in MR. *19*

PETERSEN J.

*[1] Les 36 officiers. Annuaire des mathématiciens. Laisant et Buhl, Paris, 1902, 413—427. *158*

PHILLIPS J. P. N.

*[1] The use of magic squares for balancing and assessing order effects in some analysis of variance designs. Appl. Statist. 13(1964), 67—73. MR 31(1966) # 5299. *218*

PICKERT G.

[1] Sechseckgewebe und potenzassociative Loops. Proc. Internat. Congress Math. Amsterdam, 1954, Vol. 2, 245—246. Not reviewed in MR. *251*

[2] Projective Ebenen. Springer Verlag, Berlin, Göttingen, Heidelberg, 1955. MR 17(1956), page 399. *170, 251, 266, 278, 284, 295*

PIERCE W. A.

[1] The impossibility of Fano's configuration in a projective plane with eight points per line. Proc. Amer. Math. Soc. 4(1953), 908—912. MR 15(1954), page 460. *169*

PLACKETT R. L. and BURMAN J. P. [1]
See BURMAN J. P. and PLACKETT R. L.

[POSTNIKOV M. M.] Постников М. М.

[1] Магические квадраты. Наука, Москва, 1964. [Magic Squares. Nauka, Moscow, 1964.] Not reviewed in MR. *194*

PREECE D. A.

*[1]$_S$ Non-orthogonal Graeco-Latin designs. To appear.

PREECE D. A., PEARCE, S. C. and KERR J. R. [1]
See KERR J. R., PEARCE S. C. and PREECE D. A.

QUATTROCCHI P.

*[1] S-spazi e sistemi di rettangoli latini. Atti Sem. Mat. Fis. Univ. Modena 17(1968), 61—71. MR 38(1969) # 3759. *180*

RADÓ F.

On semi-symmetric quasigroups. Aequationes Math. To appear. *465*

RAGHAVARAO D.

*[1]$_S$ A note on the construction of GD designs from hyper-graeco-latin cubes of first order. Calcutta Statist. Assoc. Bull. 8(1959), 67—70. Not reviewed in MR.

*[2]$_S$ Constructions and combinatorial problems in design of experiments. Wiley, New York, 1971.

RAO C. R.

[1] Hypercubes of strength "d" leading to confounded designs in factorial experiments. Bull. Calcutta Math. Soc. 38(1946), 67—78. MR 8(1947), page 396. *187, 191*

[2] Factorial arrangements derivable from combinatorial arrangements of arrays. Suppl. J. Roy. Statist. Soc. 9(1947), 128—139. MR 9(1948), page 264. *191*

[3] On a class of arrangements. Proc. Edinburgh Math. Soc. (2)8(1949), 119—125. MR 11(1950), page 710. *190*

[4] A combinatorial assignment problem. Nature 191(1961), 100. Not reviewed in MR. *156*

RAY-CHAUDHURY D. K. and WILSON R. M.

[1] On the existence of resolvable balanced incomplete block designs. Proc. Colloq. Calgary (1969), Combinatorial Structures and their Applications, Gordon and Breach, New York, 1970, 331—341. MR 42(1971) # 1678. *79*

[2] Solution of Kirkman's school-girl problem. Proc. Sympos. Pure Math. 19(1971), 187—203. American Mathematical Society, Providence, R. I. *79*

RÉDEI L.

[1] Das "schiefe Produkt" in der Gruppentheorie mit Anwendung auf die endlichen nichtkommutativen Gruppen mit lauter kommutativen echten Untergruppen und die Ordnungszahlen, zu denen nur kommutative Gruppen gehören. Comment. Math. Helv. 20(1947), 225—264. MR 9 (1948), page 131. *147*

REIDEMEISTER K.

[1] Topologische Fragen der Differentialgeometrie V. Gewebe und Gruppen. Math. Z. 29(1929), 427—435. *251, 262*

REISS M.

[1] Über eine Steinersche combinatorische Aufgabe, welche in 45sten Bande dieses Journals, Seite 181, gestellte worden ist. J. reine angew. Math. 56(1858/9), 326—344. *79*

RÉNYI A.

*[1] A véges geometriák kombinatorikai alkalmazásai. Mat. Lapok 17(1966), 33—76. Hungarian. [Combinatorial applications of finite geometries I.] MR 36(1968) #2515. *89, 170*

RHEMTULLA A. R.

[1] On a problem of L. Fuchs. Studia Sci. Math. Hungar. 4(1969), 195—200. MR 40(1970) #1468. *35*

RIORDAN J.

*[1] Three-line latin rectangles. Amer. Math. Monthly 51 (1944), 450—452. MR 6(1945), page 113. *150, 151*

*[2] Three-line latin rectangles II. Amer. Math. Monthly 53 (1946), 18—20. MR 7(1946), page 233. *150, 151*

*[3] A recurrence relation for three-line latin rectangles. Amer. Math. Monthly 59(1952), 159—162. MR 13(1952), page 813. *151, 152*

*[4] Discordant permutations. Scripta Math. 20(1954), 14—23. MR 16(1955), page 104. *150*

*[5] An introduction to combinatorial analysis. Wiley, New York, 1958. MR 20(1959) #3077. *103, 153*

ROBINSON D. A.

[1] Bol loops. Trans. Amer. Math.Soc. 123(1966), 341—354. MR 33(1967) #2755. *32, 64*

ROGERS K.

*[1] A note on orthogonal latin squares. Pacific J. Math. 14 (1964), 1395—1397. MR 33(1967) #5501. *174*

ROJAS B. and WHITE R. F.
*[1]$_S$ The modified latin square. J. Roy. Statist. Soc. (B) 19 (1957), 305—317. Not reviewed in MR. *463*

ROKOVSKA B.
[1] Some remarks on the number of triple systems of Steiner. Colloquium Mathematicum 22(1971), 317–323. *80*
[2] On the number of different triple systems of Steiner. Prace Nauk. Inst. Mat. Fiz. Teoretycznej Politech. Wroclawskiej 6(1972), 41–57. *80*

ROOM T. G.
[1] A new type of magic square. Math. Gaz. 39 (1955), 307. Not reviewed in MR. *218, 219*

ROSA A.
[1] On the falsity of a conjecture on orthogonal Steiner triple systems. J. Combinatorial Theory, Ser. A 16(1974), 126–128. *192*

RUIZ F. and SEIDEN E.
*[1] Some results on construction of orthogonal latin squares by the method of sum composition. J. Combinatorial Theory, Ser. A 16(1974), 230–240. *445*

[RYBNIKOV A. K. and RYBNIKOVA N. M.] Рыбников А. К. and Рыбникова Н. М.
[1] Новое доказательство несуществования проективной плоскости порядка 6. Вестник Москов..Унив. сер. 1. мат. мех. 21(1966), №. 6, 20—24. [A new proof of the non-existence of a projective plane of order 6. Vestnik. Moscov. Univ. Ser. I Mat. Meh. 21(1966), No. 6, 20—24.] MR 34(1967) # 4982. *160*

RYSER H. J.
*[1] A combinatorial theorem with an application to latin rectangles. Proc. Amer. Math. Soc. 2(1951), 550—552. MR 13(1952), page 98. *96*
*[2] Combinatorial Mathematics. The Carus Mathematical Monographs. No. 14. Mathematical Association of America. Wiley, New York, 1963. MR 27(1964) # 51. *103*
*[3] Neuere Probleme der Kombinatorik. Vorträge über Kombinatorik Oberwolfach 24—29 Juli 1967. Matematischen Forschungsinstitute Oberwolfach. Not reviewed in MR. *32*

SADE A.
*[1] Énumération des carrés latins de côté 6. Marseille, 1948. MR 10(1949), page 278. *141*
*[2] Énumération des carrés latins. Application au 7e ordre. Conjecture pour les Ordres Supérieurs. Published by the author, Marseille, 1948. MR 10(1949), page 278. *142, 144*

[3] Quasigroupes. Published by the author, Marseille, 1950. MR 13(1952), page 203. *75, 178*

*[4] An omission in Norton's list of 7×7 squares. Ann. Math. Statist. 22(1951), 306—307. MR 12(1951), page 665. *142, 169*

*[5] Omission dans les listes de Norton pour les carrés 7×7. J. Reine Angew. Math. 189(1951), 190—191. MR 13(1952), page 813. *140, 142, 169*

[6] Contribution à la théorie des quasigroupes: diviseurs singuliers. C. R. Acad. Sci. Paris 237(1953), 372—374. MR 15(1954), page 98. *178*

[7] Contributions à la théorie des quasigroupes: quasigroupes obéissant à la "loi des keys" ou automorphes par certains groupes de permutations de leur support. C. R. Acad. Sci. Paris 237 (1953), 420—422. MR 15(1954), page 98. *61*

[8] Quasigroupes obéissant à certaines lois. Rev. Fac. Sci. Univ. Istanbul. Sér. A. 22(1957), 151—184. MR 21(1960) #4987. *57, 59, 75, 76, 80, 178*

[9] Quelques remarques sur l'isomorphisme et l'automorphisme des quasigroupes. Abh. Math. Sem. Univ. Hamburg 22(1958), 84—91. MR 20(1959) #77. *138*

[10] Quasigroupes automorphes par le groupe linéaire et géométrie finie. J. Reine Angew. Math. 199(1958), 100—120. MR 20(1959) #78. *269*

[11] Groupoïdes orthogonaux. Publ. Math. Debrecen 5(1958), 229—240. MR 20(1959) #5751. *174, 175, 177*

[12] Quasigroupes parastrophiques. Expressions et identités. Math. Nachr. 20(1959), 73—106. MR 22(1961) #5688. *65, 66*

[13] Système demosien associatif de multigroupoïdes avec un scalaire non-singulier. Ann. Soc. Sci. Bruxelles. Sér. I, 73(1959), 231—234. MR 21(1960) #3502. *78*

[14] Entropie demosienne de multigroupoïdes et de quasigroupes. Ann. Soc. Sci. Bruxelles. Sér. I, 73(1959), 302—309. MR 23(1962) #A1569. *68, 76*

*[15] Produit direct-singulier de quasigroupes orthogonaux et anti-abéliens. Ann. Soc. Sci. Bruxelles. Sér. I, 74(1960), 91—99. MR 25(1963) #4017. *31, 177, 428, 430, 431, 433, 456, 458*

[16] Théorie des systèmes demosiens de groupoïdes. Pacific J. Math. 10(1960), 625—660. MR 25(1963) #2019. *76*

[17] Demosian systems of quasigroups. Amer. Math. Monthly 68(1961), 329—337. MR 25(1963) #2020. *78*

[18] Paratopie et autoparatopie des quasigroupes. Ann. Soc. Sci. Bruxelles. Sér. 1, 76(1962), 88—96. MR 27(1964) #2576. *20, 138*

[19] Isotopies d'un groupoïde avec son conjoint. Rend. Circ. Mat. Palermo (2)12(1963), 357—381. MR 29(1965) #4831. *31*

[20] Le groupe d'anti-autotopie et l'équation $Q(X, X^{-1}, 1) = QP^{12} = Q$. J. Reine Angew. Math. 216(1964), 199—217. MR 29(1965) #4832. *432*

[21] Critères d'isotopie d'un quasigroupe avec un quasigroupe demi-symétrique. Univ. Lisboa Revista Fac. Ci A (2)11(1964/65), 121—136. MR 34(1967) # 1437. *61, 62*
[22] Quasigroupes demi-symétriques. Ann. Soc. Sci. Bruxelles Sér. I, 79(1965), 133—143. MR 34(1967) # 2760. *61, 62*
[23] Quasigroupes demi-symétriques. II Autotopies gauches. Ann. Soc. Sci. Bruxelles Sér. I, 79(1965), 225—232. MR 34 (1967) # 2761. *61*
[24] Autotopies d'un quasigroupe isotope à un quasigroupe demi-symétrique. Univ. Beograd. Publ. Elektrotehn. Fak. Ser. Mat. Fiz. No. 175—179(1967), 1—8. MR 35 (1968) # 4326. *61*
[25] Quasigroupes demi-symétriques. III Constructions linéaires, A-maps. Ann. Soc. Sci. Bruxelles Sér. I, 81(1967), 5—17. MR 35(1968) # 5539. *61*
[26] Quasigroupes demi-symétriques. Isotopies préservant la demi-symétrie. Math. Nachr. 33(1967), 177—188. MR 35 (1968) # 5540. *61*
[27] Quasigroupes isotopes. Autotopies d'un groupe. Ann. Soc. Sci. Bruxelles Sér. I, 81(1967), 231—239. MR 38 (1969) # 5978. *124*
[28] Quasigroupes parastrophiques. Groupe des automorphismes gauches. Ann. Soc. Sci. Bruxelles Sér. I, 82(1968), 73—78. MR 38(1969) # 260. *61*
[29] Autotopies des quasigroupes et des systèmes associatifs. Arch. Math. (Brno) 4 (1968), 1—23. MR 42(1971) # 1930. *124, 138*
[30] Morphismes de quasigroupes. Tables. Univ. Lisboa Revista Fac. Ci A(2), 13,Fasc. 2,(1970/71), 149—172. *138, 146*
[31] Une nouvelle construction des quasigroupes orthogonaux à leur conjoint. Notices Amer. Math. Soc. 19(1972), A434, 72T-A105. *461*

SAFFORD F. H.
*[1] Solution of a problem proposed by O. Veblen. Amer. Math. Monthly 14(1907), 84—86. *160*

[SAMOILENKO S. I.] Самойленко С. И.
[1] Применение магических квадратов для коррекции ошибок. Электросвязь 20(1965), 11, 24—32. [The application of magic squares in error correcting. Electrical Communication 20(1965), 11, 24—32.] Not reviewed in MR. *217, 369*
[2] Помехоустойчивое кодирование. Наука, Москва, 1966. [Encoding procedures providing protection against errors. Nauka, Moscow, 1966.] Not reviewed in MR. *371*

SAXENA P. N.
*[1] A simplified method of enumerating latin squares by MacMahon's differential operators. I. The 6×6 latin squares. J. Indian Soc. Agric. Statist. 2(1950), 161—188. MR 12 (1951), page 312. *141*

*[2] A simplified method of enumerating latin squares by MacMahon's differential operators II. The 7×7 latin squares. J. Indian Soc. Agric. Statist. 3(1951), 24—79. MR 13(1952), page 200. *141*

*[3] On the latin cubes of the second order and the fourth replication of the three-dimensional or cubic lattice design. J. Indian Soc. Agric. Statist. 12(1960), 100–140. MR 23 (1962) #A 3013. *189*

SCHAUFFLER R.

[1] Über die Bildung von Codewörter. Arch. Elektra. Übertragung 10(1956), 303–314. MR 18(1957), page 368 *76, 364, 365*

[2] Die Assoziativität im Ganzen, besonders bei Quasi. gruppen. Math. Z. 67(1957), 428—435. MR 20(1959) # 1648. *76, 78, 364, 365*

SCHÖNHARDT E.

*[1] Über lateinische Quadrate und Unionen. J. Reine Angew. Math. 163(1930), 183—229. *16, 23, 121, 125, 128, 141, 231*

SCHUBERT H.

[1] Mathematische Mussestunden. Göschen, Leipzig, First edition 1898. New edition: De Gruyter, Berlin, 1941. *194*

SCORZA G.

[1] I gruppi che possono pensarsi come somme di tre loro sottogruppi. Boll. un Mat. Ital. (1) 5(1926), 216—218. *47*

SEIDEL J. J.

*[1] Configurations. (Dutch) Colloquium on Discrete Mathematics, Mathematisch Centrum, Amsterdam, 1968, 3—50 and 104—106. MR 41(1971) # 3298.

SEIDEN E.

*[1] On the problem of construction of orthogonal arrays. Ann. Math. Statist. 25(1954), 151—156. MR 15(1954), page 495.

[2] On the maximum number of constraints of an orthogonal array. Ann. Math. Statist. 26(1955), 132—135: correction 27(1956), 204. MR 17(1956), page 227. *191*

[3] Further remark on the maximum number of constraints of an orthogonal array. Ann. Math. Statist. 26 (1955), 759—763. MR 17(1956), page 571. *191*

SEIDEN E. and ZEMACH R.

[1] On orthogonal arrays. Ann. Math. Statist. 37(1966), 1355—1370. MR 33(1967) # 5061. *192*

SHAH K. R.

[1] Analysis of Room's square design. Ann. Math. Statist. 41(1970), 743—745. MR 42(1971) # 2605. *221*

SHEE S. C.

[1] On quasigroup graphs. Nanyang Univ. J. Part I. 4(1970), 44—66. MR 45(1973) # 6695. *316*

[2] The Petersen graph represented as a point symmetric quasigroup graph. Nanta Math. 5(1971), No. 1., 108—110. MR 45(1973) # 5031. *316*

SHEE S. C. and TEH H. H.

[1] On homogeneous quasigroup graphs and right associative subsets of a quasigroup. Nanyang Univ. J. Part I. 4(1970), 67—75. MR 45(1973) # 6696. *316*

SHIN-CHI-CHI

*[1] A method of constructing orthogonal latin squares. (Chinese) Shuxue Jinzhan 8(1965), 98—104. Not reviewed in MR. *445*

SHOLANDER M.

[1] On the existence of the inverse operation in alternation groupoids. Bull. Amer. Math. Soc. 5(1949), 746—757. MR 11(1950), page 159. *68*

SHRIKHANDE S. S.

$[1]_S$ The uniqueness of the L_2-association scheme. Ann. Math. Statist. 30(1959), 781—798. MR 22(1961) # 1048. *385*

*[2] A note on mutually orthogonal latin squares. Sankhyā Ser. A. 23(1961), 115—116. MR 25(1963) # 703. *330, 347, 385*

*[3] Some recent developments on mutually orthogonal latin squares. Math. Student 31(1963), 167—177. MR 31 (1966) # 3348.

SHRIKHANDE S. S. and JAIN N. C. [1]

See JAIN N. C. and SHRIKHANDE S. S.

ŠIK F.

[1] Sur les décompositions créatrices sur les quasigroupes. Publ. Fac. Sci. Univ. Masaryk (1951), 169—186. MR 15 (1954), page 7. *45*

SIMS C. C.

[1] Enumerating p-groups. Proc. London Math. Soc. (3) 15(1965), 151—166. MR 30(1965) # 164. *148*

SINGER J.

[1] A theorem in finite projective geometry and some applications to number theory. Trans. Amer. Math. Soc. 43 (1938), 377—385. *220, 395*

*[2] A class of groups associated with latin squares. Amer. Math. Monthly 67(1960), 235—240. MR 23(1962) # A1542. *29, 149*

[3] Some remarks on the early history of combinatorial analysis. Ann. New York Acad. Sci. 175(1970), 354—362. MR 44(1972) # 98. *395*

DE SOUSA F.

*[1] Systèmes complets de représentants et carrés latins orthogonaux. Univ. Lisboa Revista Fac. Ci. A (2)12, Fasc. 1(1967/68), 47—54. MR 40(1970) # 1289. *426, 466*

SPECNICCIATI R.

[1] Tavole di multiplicatione ridotte di un gruppo Bolletino della Unione Mathematica Italiana (1966), 86—89.

SPEISER A.

[1] Theorie der Gruppen von endlicher Ordnung. Springer-Verlag, Berlin, 1927. (3rd ed., 1937). Dover reprint, New York, 1945. *19*

STANTON R. G.

[1] Covering theorems in groups. Recent Progress in Combinatorics (Proc. Third Waterloo Conf. on Combinatories, 1968), 21—36. Academic Press, New York, 1969. MR 41 (1971) # 1547a. *362*

STANTON R. G. and HORTON J. D. [1], [2]
See HORTON J. D. and STANTON R. G.

STANTON R. G. and MULLIN R. C. [1], [2]
See MULLIN R. C. and STANTON R. G.

STEIN S. K.

[1] Foundations of quasigroups. Proc. Nat. Acad. Sci. U.S.A. 42(1956), 545—546. MR 18(1957), page 111. *65*

[2] On the foundations of quasigroups. Trans. Amer. Math. Soc. 85(1957), 228—256. MR 20(1959) # 922. *65, 66, 68, 175, 456*

[3] Homogeneous quasigroups. Pacific J. Math. 14(1964), 1091—1102. MR 30(1965) # 1206. *177*

[4] Factoring by subsets. Pacific J. Math. 22(1967), 523—541. MR 36(1968) # 2517. *54*

STEINER J.

[1] Combinatorische Aufgabe. J. Reine Angew. Math. 45 (1852/3), 181—182. *79*

STEVENS W. L.

*[1] The completely orthogonalized latin square. Ann. Eugenics. 9(1939), 82—93. Not reviewed in MR. *167*

SUSCHKEWITSCH A.

[1] On a generalization of the associative law. Trans. Amer. Math. Soc. 31(1929), 204—214. *78, 104*

SZELE T.

[1] Über die endlichen Ordnungszahlen, zu denen nur eine Gruppe gehört. Comment. Math. Helv. 20(1947), 265—267. MR 9(1948), page 131. *147*

TAMARI D.

[1] Les images homomorphes des groupoïdes de Brandt et l'immersion des semi-groupes. C. R. Acad. Sci. Paris 229 (1949), 1291—1293. MR 11(1950), page 327. *21*

[2] Représentations isomorphes par des systèmes de relations. Systèmes associatifs. C. R. Acad. Sci. Paris 232 (1951), 1332—1334. MR 12(1951), page 583. *21*

[3] "Near-Groups" as generalized normal multiplication tables. Abstract 564—279. Notices Amer. Math. Soc. 7(1960), 77. Not reviewed in MR. *22*

TAMARI D. and GINZBURG A. [1]

See GINZBURG A. and TAMARI D.

TARRY G.

*[1] Sur le problème d'Euler des n^2 officiers. Interméd. Math. 6(1899), 251—252. *140, 158*

*[2] Le problème des 36 officiers. C. R. Assoc. France Av. Sci. 29(1900), part 2, 170—203. *140, 156, 173*

*[3] Sur le problème d'Euler des 36 officiers. Interméd. Math. 7(1900), 14—16. *140, 156*

*[4] Les permutations carrées de base 6. Mém. Soc. Sci. Liège Série 3, 2(1900), mémoire No. 7. *140, 156*

*[5] Les permutations carrées de base 6. [Extrait des "Mémoires de la Société royale des Sciences de Liège", Série 3, 2(1900)]. Mathésis, Série 2, 10(1900), Supplément 23—30. *140, 156*

*[6] Carrés cabalistiques Eulériens de base 8N. C. R. Assoc. France Av. Sci. 33(1904), part 2, 95—111. *195, 212, 213*

*[7] Carrés magiques. (Réponse à la question numéro 767). Interméd. Math. 12(1905), 174—176. *212, 213*

TAUSSKY O. and TODD J.

[1] Covering theorems for groups. Ann. Soc. Polon. Math. 21(1948), 303—305. MR 11(1950), page 7. *362*

TAYLOR W.

[1] On the coloration of cubes. Discrete Math. 2(1972), 187—190. MR 45(1973) # 6684. *205*

THOMSEN G.

[1] Topologische Fragen der Differentialgeometrie. XII Schnittpunktsätze in ebenen Geometrie. Abh. Math. Sem. Univ. Hamburg 7(1930), 99—106. *251, 262*

THOMSON G. H.

*$[1]_S$ The use of the latin square in designing educational experiments. Brit. J. Educ. Psychol. 11(1941), 135—137. Not reviewed in MR.

TOMPKINS C. B.
*[1] Methods of successive restrictions in computational problems involving discrete variables. Proc. Sympos. Appl. Math. 15(1962), 95—106. American Math. Soc., Providence R. I., 1963. MR 31(1966) # 2159. *468*

TOUCHARD J.
[1] Sur un problème de permutations. C. R. Acad. Sci. Paris 198(1934), 631—633. *150*
[2] Permutations discordant with two given permutations. Scripta Math. 19(1953), 109—119. MR 15(1954), page 387. *150*

TÖRÖK É.
[1] Unpublished manuscript. *472*

TREASH C.
[1] The completion of finite incomplete Steiner triple systems with applications to loop theory. J. Combinatorial Theory, Ser. A 10(1971), 259—265. MR 43(1972) # 397. *97*

VAJDA S.
*[1]$_S$ The Mathematics of Experimental Design: Incomplete block designs and latin squares. Griffin's Statistical Monographs and Courses. No. 23. Hafner Publishing Co., New York, 1967. MR 37(1969) # 3928. *385*

VEBLEN O. and WEDDERBURN J. H. M.
[1] Non-desarguesian and non-pascalian geometries. Trans. Amer. Math. Soc. 8(1907), 379—388. *169, 280, 483*

WAGNER A.
[1] On the associative law of groups. Rend. Mat. e Appl. (5) 21(1962), 60—76. MR 25(1963) # 5120; erratum MR 26(1963), page 1543. *19*

WALKER R. J.
[1] An enumerative technique for a class of combinatorial problems. Proc. Sympos. Appl. Math. 10(1959), 91—94, American Math. Soc., Providence R. I., 1960. MR 22(1961) # 12048. *468*
[2] Determination of division algebras with 32 elements. Proc. Sympos. Appl. Math. 15(1962), 83—85. American Math. Soc., Providence R. I., 1963. MR 28(1964) # 1219. *485*

WALL D. W.
[1] Sub-quasigroups of finite quasigroups. Pacific J. Math. 7(1957), 1711—1714. MR 19(1958), page 1159. *46*

WALLIS W. D.
*[1] A remark on latin squares and block designs. J. Austral. Math. Soc. 13(1972), 205—207. MR 45(1973) #4997
[2] Duplication of Room squares. J. Austral. Math. Soc. 14(1972), 75—81. *229*

[3] A doubling construction for Room squares. Discrete Math. 3(1972), 397—399. MR 46(1973) # 7053. *229*

[4] On Archbold's construction for Room squares. Utilitas Math. 2(1972), 47—53. *229*

[5] Some results on Room squares isomorphism. Proc. First Austral. Conf. on Combinatorial Math., Tunra, 1972, 85—102. *229*

[6] On the existence of Room squares. Aequationes Math. 9(1973), 260—266. *228*

[7] Room squares with subsquares. J. Combinatorial Theory Ser. A 15(1973), 329—332. *229*

[8] A family of Room subsquares. Utilitas Math. To appear. *229*

[9] Solution of the Room squares existence problem. J. Combinatorial Theory. To appear. *228*

WALLIS W. D. and MULLIN D. C. [1]
See MULLIN D. C. and WALLIS J. S.

WALLIS W. D., STREET A. F. and WALLIS J. S.
[1] Combinatorics: Room squares, sum-free sets, Hadamard matrices. (Lecture Notes in Mathematics, No. 292). Springer Verlag, Berlin, 1972. *228*

WANG Y.
*[1] A note on the maximal number of pairwise orthogonal latin squares of a given order. Sci. Sinica. 13(1964), 841—843. MR 30(1965) # 3866. *174*

*[2] On the maximal number of pairwise orthogonal latin squares of order *s*; an application of the sieve method. Acta Math. Sinica 16(1966), 400—410. Chinese. Translated as Chinese Math. Acta 8 (1966), 422—432. MR 34(1967) # 2483. *174*

WARRINGTON P. D.
*[1] Graeco-Latin cubes. J. Recreat. Math. 6(1973), 47—53. *181*

WEISNER L.
*[1] Special orthogonal latin squares of order 10. Canad. Math. Bull. 6(1963), 61—63. MR 26(1963) # 3621. *235, 243*

[2] A Room design of order 10. Canad. Math. Bull. 7(1964), 377—378. MR 29(1965) # 4707. *225*

WELLS M. B.
*[1] The number of latin squares of order eight. J. Combinatorial Theory 3(1967), 98—99. MR 35(1968) # 5343. *144*

[2] Elements of combinatorial computing. Pergamon, Oxford, 1971. MR 43(1972) # 2819. *468*

WERNICKE P.
*[1] Das Problem der 36 Offiziere. Jber. Deutsch Math. verein. 19(1910), 264—267. *158*

WIELANDT H.
[1] Arithmetical and normal structure of finite groups. Proc. Sympos. Pure Math. 6(1962), 17—38. American Math. Soc., Providence R. I., 1962. MR 26(1963) # 5045. *16*

WILLIAMS E. J.
*[1]$_S$ Experimental designs balanced for the estimation of residual effects of treatments. Australian J. Sci. Research. Ser. A, 2(1949), 149—168. MR 11(1950), page 449. *82, 83*
*[2]$_S$ Experimental designs balanced for pairs of residual effects. Australian J. Sci. Research Ser. A, 3(1950), 351—363. MR 12(1951), page 726. *83*

WILSON R. M.
*[1] Concerning the number of mutually orthogonal latin squares. To appear. *174, 405, 425*

WITT E.
*[1] Zum Problem der 36 Offiziere. Jber. Deutsch. Math. verein 48(1939), 66—67. *158*

YAMAMOTO K.
*[1] An asymptotic series for the number of three-line latin rectangles. Sûgaku 2(1949), 159—162; J. Math. Soc. Japan 1(1950), 226—241. MR 12(1951), page 494. *153*
*[2] On the asymptotic number of latin rectangles. Jap. J. Math. 21(1951), 113—119. MR 14(1953), page 442. *153*
*[3] Note on the enumeration of 7×7 latin squares. Bull. Math. Statist. 5(1952), 1—8. MR 14(1953), page 610. *144*
*[4] Symbolic methods in the problem of three-line latin rectangles. J. Math. Soc. Japan 5(1953), 13—23. MR 15 (1954), page 3. *153*
*[5] Euler squares and incomplete Euler squares of even degrees. Mem. Fac. Sci. Kyūsyū Univ. Ser. A, 8(1954), 161—180. MR 16(1955), page 325. *156*
*[6] Structure polynomials of latin rectangles and its application to a combinatorial problem. Mem. Fac. Sci. Kūysyū Univ. Ser. A, 10(1956), 1—13. MR 17(1956), page 1174. *153*
*[7] On latin squares. Sûgaku 12(1960/61), 67—79. Japanese. MR 25(1963) # 20. *51*
*[8] Generation principles of latin squares. Bull. Inst. Internat. Statist. 38(1961), 73—76. MR 26(1963) # 4933. *51, 436*

YAP H. P.
[1] A simple recurrence relation in finite abelian groups. Fibonacci Quart. 8(1970), No. 3, 255—263. MR 41(1971) # 3387. *311*

YATES F.
*[1]$_S$ The formation of latin squares for field experiments. Emp. J. Expl. Agr. 1(1933), 235—244.
*[2]$_S$ Complex experiments. J. Roy. Statist. Soc. Suppl. 2(1935), 181—247. *463*
*[3]$_S$ Incomplete latin squares. J. Agr. Sci. 26(1936), 301—315. *463*
[4]$_S$ Incomplete randomized blocks. Ann. Eugenics 7(1936), 121—140. *373*

YATES F. and HALE R. W. [1]
See HALE R. W. and YATES F.

YOUDEN W. J. and HUNTER J. S. [1]
See HUNTER J. S. and YOUDEN W. J.

ZASSENHAUS H. J.
[1] The theory of groups. Chelsea Publishing Company, New York, 2nd edition 1958. MR 19(1958), page 939. ***20, 21, 36, 45***

ZHANG L.
*[1] On the maximum number of orthogonal latin squares I. Shuxue Jinzhan 6(1963), 201—204. Chinese. MR 32(1966) # 4027. *347*

ZHANG L. Q. and DAI S. S. [1]
See CHANG L. Q. and TAI S. S.

Subject index

D

E

F

G

H

I

J

K

L

T

U

V